FOOD REGULATION

FOOD REGULATION
LAW, SCIENCE, POLICY, AND PRACTICE

Neal D. Fortin, J.D.
Professor and Director, Institute for Food Law & Regulations,
Michigan State University
Adjunct Professor of Law, Michigan State University College of Law

A JOHN WILEY & SONS, INC., PUBLICATION

Published by John Wiley & Sons, Inc., Hoboken, New Jersey
Published simultaneously in Canada

For general information on our other products and services or for technical support, please contact our Customer Care Department within the United States at (800) 762-2974, outside the United States at (317) 572-3993 or fax (317) 572-4002.

Wiley also publishes its books in a variety of electronic formats. Some content that appears in print may not be available in electronic formats. For more information about Wiley products, visit our web site at www.wiley.com.

Library of Congress Cataloging-in-Publication Data:

Fortin, Neal D.
  Food regulation: law, science, policy, and practice / Neal D. Fortin.
    p. cm.
  Includes index.
  ISBN 978-0-470-12709-4 (cloth)
  1. Food law and legislation–United States. 2. Food industry and trade–Safety regulations–United States. 3. Food adulteration and inspection–United States. I. Title.
  KF3875.F67 2009
  344.7304'232—dc22

                                                                    2008021433

Printed in the United States of America

10 9 8 7 6 5 4 3 2 1

Dedicated to
Katherine Fortin
and
Helen Fortin

▇▇▇ TABLE OF CONTENTS

Part IV: Specialized Food Regulation

8 Dietary Supplements The regulation of this special class of food that is regulated neither as conventional food nor as food additives. Dietary Supplement Health and Education Act of 1994 (DSHEA), ephedrine; "street drug alternatives," and dietary supplement ads on the Internet.

9 Biotechnology and Genetically Engineered Organisms The regulation biotechnology, genetically modified foods, and cloning.

10 Food Terrorism A look at food security and the federal government's new bioterrorism powers.

11 Importation and Exportation The regulation of U.S. food imports and exports, including harmonization of domestic and foreign trade restrictions, and agency enforcement authority.

Part V: Inspection and Enforcement

12 Federal Enforcement An overview of the enforcement tools of the FDA and USDA-FSIS with a focus on seizures, recalls, and penalties.

13 Inspections An overview of the FDA and state inspections, and Fourth Amendment, and practical issues.

14 State Laws and Their Relationship to Federal Laws A discussion of state enforcement, the state food and drug laws, and federal pre-emption.

15 Private Actions Private causes of actions and their regulatory role.

Part VI: General Chapters

16 International Food Law The implications of international trade, international treaties, and the Codex Alimentarius.

17 Ethics Discussion of ethical issues in food law and in government regulation.

Knowledge is the food of the soul.
                    —Plato

Food regulation is a complex and fascinating field. Study in this area is richly rewarding. From a human-interest perspective, the range of products regulated touches the lives of nearly every American every day. Food regulatory issues often warrant headline news because this is a subject that commands the public's attention, whether it be a news flash on a foodborne illness outbreak or information on diet that can help one live a longer and healthier life.

In addition the regulation of food provides a snapshot of the political, social, and economic currents in our society. Thus the study of food law provides a incisive look at important policy decisions on vital aspects of people's everyday lives.

ABOUT THE TEXT

This text is designed to provide an accessible guide the United States food regulation—to be enlightening, without being light. While the text contains in-depth discussion of the federal statutes, regulations, and the regulatory agencies, the material is not dense, and remains accessible to the average reader. For this reason the text is appropriate for a wide audience of students and professionals.

A modified casebook method is used. The black letter law is livened with discussion of emerging issues and trends plus case studies that explore important issues. These materials explore not only regulation, but the science, policy, and practice. The reader is challenged to move beyond theory into application of the theory.

The focus is on the Federal Food, Drug, and Cosmetic Act (FD&C Act), 21 U.S.C. section 321 *et seq.*, and the Food and Drug Administration (FDA). A good number of the cases and references in this text are to pharmaceuticals, medical devices, and cosmetics. The FD&C Act regulates all these products, and there are commonalities in the regulatory framework for all. In fact, some drug or medical device cases illustrate a point about food law better than any case directly about food.

This casebook presents diverse materials from pertinent sources. Commentary and context are provided as needed, but often the materials can be digested without these aids. The novice may feel challenged at first to understand the materials, but after jumping around the various writing styles and contexts, and the relative value and weight of each source should become discernable. Stay with it, and you will find that the materials become easier.

A ROAD MAP FOR READING THE MATERIALS IN THIS COURSE

In keeping with the way information is encountered in practice, not all readings in this text are equal. Some may be read quickly, while some require close scrutiny. Moreover, materials come in varying levels of formality. Some materials I have condensed to make them easier to read.

Readers new to this teaching style may be disconcerted at first. Do not let this throw you. Persist and trust your instincts, and you will find that your effort quickly pays dividends. In the end, you will learn much more than the mere rules, but develop and hone critical skills that are not only vital in legal analysis but are extremely useful in winnowing through the mountains of information available on the Internet. In addition law and regulation are not static subjects, so developing these dynamic skills will be beneficial in the end.

Here are a few tips to readers who are new to this teaching method:

1. Review your road map of each chapter. Review the chapter title, the other headings, and the table of contents before reading. These will provide you with an overview of how the chapter material relates to the overall text.
2. Put the material in context. Note the source of the material quoted. Who wrote the material will tell you what type of perspective is offered. Often regulations reconcile conflicting interests, and understanding both sides can be key to a complete picture. Note the date when the material was written, as the date may indicate that the material is provided for historical perspective, or that part of the information may be pertinent but part may be outdated.
3. The statutes and regulations are the primary source of our food law. That is, food regulatory law is largely bound by statutes and regulations. Therefore these materials should be the beginning of your research to answer a food law question. Often a problem is solved by examination of the statutory definitions (particularly key definitions, e.g., food, drug, misbranding, and adulterated).

4. In reading the cases, develop the ability to understand how the court reasoned through the conflict to a solution. Identify the particular factors used by the court to decide the case the way it did. Check to see if those factors are present in a problem with which you are dealing. If the factors are not present, then ask yourself if that justifies a different result. If there are any changes in the social or economic conditions that surrounded an earlier decision, ask how that affects the problem now at hand.

In short, learn to analyze the materials, rather than merely read and memorize rules.

EDITING

I have edited out the footnotes and citations from most of the cases. Remaining footnotes may be renumbered with my own footnotes. Unless otherwise indicated, any footnotes with cases are those of the court. In addition materials may be edited for typographic style without notation in the text.

STATUTORY RESEARCH USING THE FEDERAL REGISTER, CFRs, AND STATUTES

Food regulation in the United States is primarily based on statutory law. So it is generally best to read or review the statutory language before reading the cases and secondary materials, which serve mainly to explain statutory issues. When reading the statutes or regulations, be sure that you also review the definitions of defined terms used—particularly the key definitions in section 201 [321], such as "food" and "drug," and the definitions of "adulterated" and "misbranded." In addition, when reading the statutory language, obtain at least a general idea of what is covered by any statutory cross-references.

A NOTE ON STATUTE CITATIONS

All federal statutes in force in the United States are codified in the United States Code (U.S.C.). The U.S.C. is organized into subject matter titles with numbering that is unique from the section numbering in the statutes as they were enacted into the public acts. For example, section 1 of the Food, Drug, and Cosmetic Act is codified as 21 U.S.C. § 301. You may also find this section cited with one or the other or both reference numbers, such as "Sec. 1 [301]."

Statutory citations used in this material are to the FD&C Act statutory sections (which is the way practitioners refer to them). The citation within the brackets is the U.S.C. number. Nonetheless, occasionally you will see reference to a United States Code citation.

FD&C ACT REFERENCES

Four free online locations for reference to the FD&C Act follow:

Cornell's LII: www4.law.cornell.edu/uscode/21/ch9.html
FDA: www.fda.gov/opacom/laws/fdcact/fdctoc.htm
GPO Access: http://www.gpoaccess.gov/uscode/index.html
University of Virginia: www.uvm.edu/nusc/nusc237/ffdcatc.html

Of course, Westlaw and Lexis-Nexis provide access to the most up to date text of the FD&C Act.

DISCUSSION QUESTIONS

The Discussion Questions are designed to encourage thought on the material presented or for class discussion. Often there is not a right or wrong answer but multiple viewpoints on these issues or public policy questions. A great deal of insight can be gained by having candid discussions of these different perspectives.

PROBLEM EXERCISES

The Problem Exercises are designed to encourage critical thinking. They take on a variety of forms, but usually revolve around a public policy question in food law.

INTERNET CITATIONS

The fluid nature of Internet addresses creates difficulty for a textbook of this nature. The food regulation information available on the Internet is far too valuable not to include many Internet addresses. Inevitably, however, some of these addresses will have changed or the documents will have been removed within days of this book's printing.

However, learning what *types* of materials are available is more valuable than finding a specific document. When you find a broken Internet address, take the opportunity to use search engines to find the new location, or to find similar material on the Web.

In the types of materials reference in this text, most of the broken Internet addresses result from reorganization of large document repositories. If search engines cannot find a particular document—and you believe it contains vital information—you may be able to find the document using Internet archives.[1] Nonetheless, this text offers a complete and appetizing menu for understanding food regulation in the United States.

CITATION FORMAT

Citations in this text generally follow *The Bluebook: A Uniform System of Citation (18th Ed.)*. However, some conventions are modified to save space and repetition.

I hope you find this text offers a complete and appetizing menu for understanding food regulation in the United States.

---

[1] For example, the Wayback Machine, which contains 55 billion Web pages archived from 1996, *available at*: http://www.archive.org/web/web.php (last accessed Sept. 12, 2006).

ACKNOWLEDGMENTS

It is impossible to write a text of this nature without owing many people a debt of gratitude. I cannot begin to list you all, but extend a thank you to everyone who furthered my scholarship on food law. I also wish to acknowledge:

P. Vincent Hegarty, a great mentor and scholar, who first hired me to teach food law, and unwittingly launched this endeavor.

Peter Barton Hutt and **Richard A. Merrill**, co-authors of FOOD AND DRUG LAW: CASES AND MATERIALS (1st and 2nd editions), are deans of food and drug law scholarship, and I am indebted to their work.

Caren Wilcox, a guest instructor in my food regulation in the U.S. course, provided me with insight into the problems of bioterrorism.

Elijah Milne, a gifted writer and brilliant student, provided invaluable editorial support and legal research for this book.

My wife, Kathy, and daughter, Helen, supported me through the many months of writing, without which this book would never have been finished.

The following publishers, journals, and authors are thanked for their generosity in granting permission for me to publish excerpts from the following publications:

- Consumers Union of U.S., Inc.: *The Truth about Irradiated Meat*, CONSUMER REPORTS 34–37 (Aug. 2003).
- Food and Drug Law Institute: Neal Fortin, *The Hang-up with HACCP: The Resistance to Translating Science into Food Safety Law*, 58 FOOD AND DRUG LAW JOURNAL 565–594 (2003).
- International Food Information Council: FDA/IFIC, *Food Additives* (1992).
- Journal of Food Law and Policy: Neal Fortin, *Is a Picture Worth More Than 1,000 Words?* 1 JOURNAL OF FOOD LAW AND POLICY 239–268 (Fall 2005).
- Thompson-West: JAMES T. O'REILLY, FOOD AND DRUG ADMINISTRATION (2d ed. 2004).

PART I

INTRODUCTORY CHAPTERS

CHAPTER 1

Introduction to Food Regulation in the United States

1.1 INTRODUCTION

This chapter provides basic information for students with greatly varied backgrounds. While this information may be repetitive or elementary for some readers, the reader is nevertheless encouraged to treat this material as a review and refresher. This introduction also provides a historical background that gives insight into the public policy decisions in food regulation.

This chapter also provides a general explanation of the legal system, regulatory law in general, and the legal basis of food regulation in the United States. To enhance an understanding of the legal structure, and to simplify its otherwise mysteriousness, this chapter begins with an overview of the history of food regulation in the United States. This history accounts for and explains much of the current organization of federal and state regulatory agencies.

This chapter further presents an overview of the major food statutes, regulations, and the jurisdictions of various agencies. This knowledge will allow you to enhance your communication and functioning within this legal framework. In addition, a better understanding of the functions, authority, and interrelationship of various regulatory agencies promotes improved relations with those agencies. This understanding will also improve your ability to function within the regulatory system.

1.2 A SHORT HISTORY OF FOOD REGULATION IN THE UNITED STATES

1.2.1 Why Do We Have Food Laws?

From the beginnings of civilization, people have been concerned about food quality and safety. The focus of governmental protection originated to protect against economic fraud and to prevent against the sale of unsafe food. As early

Food Regulation: Law, Science, Policy, and Practice, by Neal D. Fortin
Copyright © 2009 Published by John Wiley & Sons, Inc.

as the fourth century BC, Theophrastus (372–287 BC) in his ten-volume treatise, ENQUIRY INTO PLANTS, reported on the use of food adulterants for economic reasons. Pliny the Elder's (AD 23–79) NATURAL HISTORY provides evidence of widespread adulteration, such as bread with chalk, pepper with juniper berries, and even adulteration with cattle fodder.[1] Ancient Roman law reflected this concern for adulteration of food with punishment that could result in condemnation to the mines or temporary exile.[2]

Starting in the thirteenth century, the trade guilds advanced higher food standards. The trade guilds, which included bakers, butchers, cooks, fruiters, among the many tradecrafts, held the power to search for and seize unwholesome products.

Indeed, as the guilds policed the marketplace, they were most interested to ensure continued and strong markets for their goods. Nevertheless, the guilds provide an early demonstration how stringent product quality and safety standards can bring a competitive economic advantage to industries and nations. Trust in food's safety and wholesomeness is necessary for the market to prosper. A number of commentators have noted the commonality of interest between business self-interest and stringent product safety standards.[3]

This early Massachusetts Food Act was passed on March 8, 1785.[4]

An Act Against Selling Unwholesome Provisions

Whereas some evilly disposed persons, from motives of avarice and filthy lucre, have been induced to sell diseased, corrupted, contagious, or unwholesome provisions, to the great nuisance of public health and peace:

Be it therefore enacted by the Senate and House of Representatives, in General Court assembled, and by the authority of the same, That if any person shall sell any such diseased, corrupted, contagious or unwholesome provisions, whether for meat or drink, knowing the same without making it known to the buyer, and being thereof convicted before the Justices of the General Sessions of the Peace, in the county where such offence shall be committed, or the Justices of the Supreme Judicial Court, he shall be punished by fine, imprisonment, standing in the pillory, and binding to the good behaviour, or one or more of these punishments, to be inflicted according to the degree and aggravation of the offence.

Nearly all of the regulation of food in the United States in the colonial era was by the state and local governments. Federal activity was limited to imported foods. The first federal food protection law was enacted by Congress in 1883

---

[1] Peter Barton Hutt, *Government Regulation of the Integrity of the Food Supply*, 4 ANNUAL REVIEW OF NUTRITION 1 (1984).

[2] *Id.*

[3] *See, e.g.,* MICHAEL E. PORTER, THE COMPETITIVE ADVANTAGE OF NATIONS, 648–649 (1990).

[4] John P. Swann, HISTORY OF THE FDA, FDA History Office, *available at:* http://www.fda.gov/oc/history/historyoffda/default.htm (last accessed Dec. 17, 2001).

to prevent the importation of adulterated tea. This was followed in 1896 by the oleo-margarine statute, which was passed because dairy farmers and the dairy industry objected to the sale of adulterated butter and fats colored to look like butter.

Although adulteration and mislabeling of food had been a centuries-old concern, the magnitude of the problems increased in the last half of the nineteenth century. This was an era of rapid development in chemistry, bringing advancements in food science, new food additives and colorings, and new means of adulteration. Fortunately, these scientific advances also provided the tools for detecting adulteration.

> We face a new situation in history. Ingenuity, striking hands with cunning trickery, compounds a substance to counterfeit an article of food. It is made to look like something it is not; to taste and smell like something it is not; to sell like something it is not, and so deceive the purchaser.
> —Congressional Record, 49 Congress I Session 1886

Indeed, as food production began shifting from the home to the factory, from consumers buying basic ingredients from neighbors in their community, to food processors and manufacturers more often at a distance, it became harder for consumers to determine the safety and quality of their food. Inevitably the responsibility for ensuring the safety of foods had to be shifted from local to national government. The demand for legislative oversight arose as national markets grew and legitimate manufacturers became concerned that their markets were being harmed by the dishonest and unsafe goods.

1.2.2 The 1906 Pure Food and Drug Act

In 1883 Dr. Harvey Wiley became the chief chemist of the U.S. Bureau of Chemistry (at that time, part of the Department of Agriculture). Dr. Wiley expanded research and testing of food and documented the widespread adulteration.[5] He helped spur public indignation by his publications and by campaigning for a national food and drug law. Wiley dramatically focused concern about chemical preservatives as adulterants through his highly publicized "Poison Squad." The Poison Squad consisted of live volunteers who consumed questionable food additives, such as boric acid and formaldehyde, to determine the impact on health. Observation and documentation of the ill effects and symptoms of the volunteers provided an appalling crude gauge of food additive safety.[6] However crude by today's standards, Wiley's leadership with the only tools of the day helped galvanize public awareness and advanced food safety.

---

[5] FDA, FDA Backgrounder: Milestones in U.S. Food and Drug Law History, *available at*: http://www.fda.gov/opacom/backgrounders/miles.html (last accessed Aug. 5, 2002).
[6] The data are collected in the USDA, Bureau of Chemistry, bulletin no. 84 (1902–1908).

Public support for passage of a federal food and drug law grew as muckraking journalists exposed in shocking detail the frauds and dangers of the food industry, such as the use of poisonous preservatives and dyes in food. A final catalyst for change was the 1905 publication of Upton Sinclair's THE JUNGLE. Sinclair's portrayal of nauseating practices and unsanitary conditions in the meat-packing industry captured the public's attention.

On June 30, 1906, President Theodore Roosevelt signed both the **Pure Food and Drug Act**[7] and the **Meat Inspection Act**[8] into law. Passage of these two statutes began the modern era of U.S. food regulation. While neither act could be considered comprehensive, they responded to the concerns of the day.

The Pure Food and Drug Act added regulatory functions to the U.S. Bureau of Chemistry. The Meat Inspection Act of 1906 required the U.S. Department of Agriculture to inspect all cattle, sheep, swine, goats, and horses as they are slaughtered and processed into products for human consumption. The primary goals of the Meat Inspection Act were to prevent adulterated livestock from being processed into food, and to ensure that meat was slaughtered and processed under sanitary conditions.

1.2.3 Evolution of the Food Statutes

Not long after passage of the Pure Food and Drug Act, legislative battles began to expand and strengthen the law. For instance, the act did not prohibit false therapeutic claims, but only false and misleading statements about the ingredients or identity of a drug. FDA wanted broader power and authority. Leaders in the food industry called for more stringent product quality standards to create a level playing field. Congress called for better safety standards and fair dealing.

However, major revision stalled until a precipitous event fell while a significant segment of the public was paying attention. Sulfanilamide, one of the new sulfa drugs, was being used effectively to treat strep throat and other bacterial diseases (Figure 1.1). To increase the palatability of the bad tasting drug, a drug company mixed the antibiotic with diethylene glycol, a sweet tasting liquid. The mixture was called elixir of sulfanilamide and shipped in the fall of 1937. Within weeks, deaths were reported to FDA. The manufacturer admitted they performed no safety tests. None were required. At least 107 died, often an agonizing death. Many of the dead were children who received the elixir for strep throat.[9]

The tragedy spurred legislative action, and in 1938, the **Food, Drug, and Cosmetic Act (FD&C Act)** was enacted. The FD&C Act required premarketing approval and proof of the safety of drugs. The act also

---

[7] 21 U.S.C. 1 *et seq.*
[8] 21 U.S.C. 601 *et seq.*
[9] PHILIP J. HILTS, PROTECTING AMERICA'S HEALTH: THE FDA, BUSINESS, AND ONE HUNDRED YEARS OF REGULATION, 89–92 (2003).

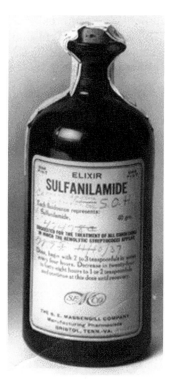

Figure 1.1 Elixir of sulfanilamide (Image courtesy FDA).

- extended government control to cosmetics and therapeutic devices;
- provided that safe tolerances be set for unavoidable poisonous substances in food;
- authorized standards of identity, quality, and fill-of-container for foods;
- authorized factory inspections; and
- added court injunctions to the previous penalties of seizures and prosecutions.

Food laws continued to evolve based on the concerns and issues of the times. In the 1950s, concerns over synthetic food additives, pesticides, and cancer were high. Consequently, in 1958, the **Food Additives Amendment** to the FD&C Act was enacted, requiring the evaluation of food additives to establish safety. The **Delaney Clause** forbade the use of any substance in food that was found to cause cancer in laboratory animals. In 1960, the **Color Additive Amendment** to the FD&C Act was enacted, which required manufacturers to establish the safety of color additives in foods, drugs, and cosmetics. A Delaney Clause also prohibited the approval of any color additive shown to induce cancer in humans or animals.

After a number of well-publicized outbreaks of botulism food poisoning from canned foods, the FDA issued **Low-Acid Food Processing Regulations** in 1973. After deaths from cyanide placed in Tylenol capsules, FDA issued the **Tamper-Resistant Packaging Regulations** in 1982. In 1983, Congress passed the **Federal Anti-tampering Act**, which makes it a federal crime to tamper with packaged consumer products.

Throughout the 1980s, there was a growing interest in the effect of nutrition on health along with increased marketing of foods to fulfill health concerns. At the same time, food processing continued a trend toward becoming nationally distributed rather than local. Various states implemented non-uniform laws to regulate health and nutrition claims, which the national industry found interfered with interstate commerce. In 1990, Congress enacted the **Nutritional Labeling and Education Act (NLEA)**, which requires nearly all packaged foods to bear nutritional labeling. The act also requires nutritional and health claims for foods to be consistent with terms defined by the FDA. NLEA preempts state requirements on food standards, nutrition labeling, and health claims.

With this background history, it is time to review some aspects of the U.S. legal system.

1.3 THE U.S. LEGAL SYSTEM

To understand the legal basis of food regulation in the United States, it is necessary to have an overall understanding of the U.S. legal system and some of the key concepts in American jurisprudence. First, let us look at the basic terminology.

Law: (1) a binding custom of a community; (2) a rule of conduct or action prescribed or enforced by a controlling authority; (3) the whole body of such rules; (4) the control brought about by the enforcement of such law; (5) the legal process; (6) the whole body of laws relating to one subject; (7) the legal profession; (8) legal knowledge and learning.

As you can quickly see, even defining the term "law" is not a simple proposition. To simplify the terminology, this text follows the predominant American meanings for the term "law" and its synonyms:

Law implies imposition by a sovereign authority. Law commonly refers to the entire body of law on the subject, but also as a synonym for "statute."

Statute means a law enacted by a legislative body.

Regulation implies prescription by administrative agency to carry out their statutory responsibilities. Federal regulations are first published in the Federal Register and later codified in the Code of Federal Regulations.

Rule applies to more restricted or more specific laws than statutes. "Rule" often is an abbreviated form of the term "administrative rule," which is a law promulgated by an administrative agency. Administrative rules are also called regulations. However, administrative rules are only one form of rules. Some administrative orders, resolutions, and formal opinions are also "rules."

Guideline suggests something advisory rather than binding.

Ordinance applies to an order enforced by a local unit of government, such as a city.

The system of U.S. laws can be divided into four parts:

- Constitution
- Statutes
- Regulations
- Common law and case law

These four types of laws are described below in reference to the federal law. However, a similar system of laws is observed by the various states.

1.3.1 The Constitution[10]

The U.S. Constitution provides the framework for the U.S. legal system. The Constitution both empowers and limits government. The Constitution provides the supreme law of the land, and it is, by design, difficult to alter as a way of protecting long-standing values.

The U.S. Constitution creates the federal government and divides the power among the three branches. The legislative power is vested in the U.S. Congress (Article I). (However, additional laws can be created by the executive and judicial branches.) The executive power is placed in a President (Article II). The judicial power is vested in the U.S. Supreme Court and lower courts (Article III). This division of power was designed to create checks and balances to protect against tyrannical rule.

This caution over the concentration of power is a theme that runs throughout U.S. law. The Constitution, in addition to granting powers to government, also limits government's powers and functions, particularly of the federal government. The first ten amendments of the Constitution are known as the Bill of Rights,[11] and they protect individual rights by setting restrictions on the activities of the federal government.

1.3.2 Statutes

Within their power granted by the U.S. and state constitutions, respectively, Congress and state legislatures enact public acts, also called statutes. (Cities

---

[10] Although the U.S. Constitution is at the root of all American law, the document is seldom read by nonlawyers. The U.S. Constitution can be read at http://findlaw.com/casecode/constitution/ (last accessed Sept. 1, 2007). Do not be intimidated by the document's importance. It is surprisingly simple language.

[11] See the appendix to this chapter.

and other municipalities generally call their enactments of law "ordinances.") All statutes must be consistent with the U.S. Constitution. State and local laws must also be consistent with the applicable state constitution.

1.3.3   Regulations

Although Congress and state legislatures have the primary authority to enact laws, they often delegate this authority to administrative agencies. This is particularly true for areas requiring technical expertise, such as health and science matters. The laws promulgated by administrative agencies are called regulations or administrative rules.

In theory, the administrative agencies merely execute the laws enacted by the legislature. However—because the legislatures often provide only a broad mandate—the agencies have considerable leeway in interpreting and applying their mandate. Typically an administrative agency promulgates the detailed regulations that are necessary to translate the legislative mandate into operating standards. The regulations must fall under the scope of authority delegated by the legislature in statute. Regulations must also be consistent with other relevant constitutional and statutory requirements. Generally, regulations have the full force of law found in the enabling statute.

The executive branch agencies have increased in number, size, and importance over the past half century. However, it is important to remember that the agencies can only carry out that which they are authorized to do by the legislature. In addition the legislature determines the amount of funding the executive branch receives. It is not unusual for legislatures to grant broad, noble sounding mandates by enacting popular statutes but fail to provide the necessary resources to carry out the legislative mandate.

1.3.4   Case Law and Common Law

Both case law and common law are based on judicial decisions. **Case law** is the law established by the precedents of judicial decisions in cases (as distinguished from laws created by legislatures). Case law is important because of the tradition of following precedents. When a court addresses a legal dispute, it is usually guided by what has been decided previously in similar cases. These precedents become the case law. The general concept is that judges should follow the principles of law set down in prior decisions, unless it would violate justice or fair play to do so. Reliance on precedent serves to promote uniformity, predictability, and foster trust in a rule by law, not by person.

Common law is the body of law based on legal tradition, custom, and general principles. Common law is embodied in case law and that serves as precedent or is applied to situations not covered by statute. U.S. common law was originally derived from English legal principles and traditions but now includes the precedents that have developed over time from the decisions of U.S. courts.

Common law generally applies only to areas of law where there is no statutory law. For example, if a firm discharges food-processing waste on a field, and a foul smell permeates nearby homes, this may violate the common law of nuisance. Private nuisance common law might allow individuals to sue the processing plant. Public nuisance common law might allow a government official to take action. However, if a statute regulates acceptable waste-handling methods for processing plants, then the legislative law can override the common law.

1.3.5 Federalism

To understand how the U.S. system of laws interrelates, one needs to understand federalism. The Constitution divides the power of government vertically between federal and state governments. **Federalism** is the term used to refer to this division of power. Federalism also limits the ability of a state to interfere or burden other states. An important example is that states cannot regulate or tax commerce in a way that places an undue burden on interstate commerce.

The Supremacy Clause of the Constitution provides that the Constitution and the federal laws are the supreme law of the land.[12] This provision, as a general matter, means that the federal laws preempt state and local laws if they conflict.[13] However, federal law can only preempt state law where there is authorization by the Constitution. The federal government only holds the powers delegated to it by the Constitution; other powers are reserved to the states or to the people.[14]

This division of power has been a great debate throughout U.S. history. However, the growth of national and international commerce and the problems of the modern age have led to a very expansive interpretation of the federal power. The Commerce Clause of the Constitution grants Congress plenary power to regulate commerce.[15] Commerce covers a wide range of activities, not only direct interstate commerce but also any activities that

---

[12] The U.S. Constitution Article VI provides that the Constitution, and the Laws of the United States which shall be made in pursuance thereof; and all Treaties made, or which shall be made, under the Authority of the United States, shall be the supreme Law of the Land; and the Judges in every State shall be bound thereby, anything in the Constitution or Laws of any State to the Contrary notwithstanding. U.S. CONST. Art. VI.

[13] Of course, state and federal laws may be different without direct conflict. Generally, states may pass more restrictive or stringent food safety laws (or weaker laws) than those promulgated at the federal level, so long as there is no direct conflict in the specifics of the laws.

[14] "The powers not delegated to the United States by the Constitution, nor prohibited by it to the States, are reserved to the States respectively or to the people." U.S. CONST. Amend. X.

[15] Article I of the Constitution authorizes Congress to make all laws that are necessary and proper for carrying into execution the government's constitutional powers. The "Commerce Clause," in Article I, section 8, clause 3 of the Constitution, authorizes Congress to regulate commerce with foreign nations, among the several States and with the Indian tribes.

indirectly affect interstate commerce. Today, given the nationally integrated economy of the United States, nearly all commerce is interstate or has an interstate impact; thus it is under federal purview.

However, states retain control over all matters not specifically delegated to the federal government.[16] The key area here is that *only* the states possess the power to regulate specifically for the health and welfare of the people.[17] **Police power** is the term used to refer to this exclusive state power, the broad powers traditionally possessed by governments and exercised to protect the health, safety, welfare, and general well-being of the citizenry.[18] Authority to make food inspection laws and health laws are part of the traditional police powers.

Nevertheless, often the federal government may regulate an activity that falls under the police power category because it also falls under federal authority via another power, such as the power to regulate interstate commerce. For example, the federal government could not regulate the minimum cold-holding temperatures of foods for health and safety reasons, but it may do so for the purpose of regulating interstate commerce.

The end result of federalism is the state's independent power creates more regional differences in the law and regulation than would occur if there were a single national legal standard. In addition states are free to legislate and regulate any arena that has not been preempted by federal law.[19] However, any additional restriction passed by a state must not place an unreasonable burden on interstate commerce.

Accordingly, firms shipping into various states must be careful that they meet both federal and state requirements. This patchwork of different laws has been criticized as being of burden to firms shipping to several states. This is one reason that cooperative and educational efforts at uniformity have been an important part of the legal landscape in food law. For example, the FDA issues a model Food Code for retail food establishment, and the Association of Food and Drug Officials issues a model Food, Drug, and Cosmetic Act. When the models or the federal laws are perceived as adequate by state governments, usually the states will adopt the model or federal regulations essentially word for word into state law.

This non-uniform approach can be troublesome from a commercial standpoint, but this decentralization of power was intentional to prevent against tyranny. There is also the benefit of different localities having the opportunity

---

[16] "The powers not delegated to the United States by the Constitution, nor prohibited by it to the States, are reserved to the States respectively or to the people." U.S. CONST. Amend. X.

[17] United States v. Lopez, 514 U.S. 549(1995).

[18] Gibbons v. Ogden, 22 U.S. 1 (1824). (Police powers "form a portion of that immense mass of legislation which embraces everything within the territory of the state, not surrendered to the general government; all of which can advantageously be exercised by the states themselves. Inspection laws, quarantine laws, health laws of every description . . . are component parts of this mass.")

[19] Alden v. Maine, 119 S.Ct. 2240 (1999).

to propose laws that best serve the needs of their community. For instance, coastal states often have closer scrutiny for seafood harvests than states without fisheries.[20]

The experience of trying out new ideas and conducting these experiments in democracy in local settings may yield useful information for future efforts to solve problems that face all communities.[21] For example, because sulfites can be dangerous to sensitive individuals, Michigan requires the labeling of sulfite use on salad bars.[22]

California, a major producer of canned food, adopted the first regulation for mandated thermal processing controls for canned food in 1920.[23] California's updated low-acid canning regulation eventually served as the model for the FDA low-acid canning regulation promulgated in 1973.

At beginning of the twentieth century, increased distribution of milk to growing population centers resulted in outbreaks of milk-borne diseases. The city of Chicago passed the first mandatory milk pasteurization law in 1908. In 1947 Michigan became the first state to require milk pasteurization.[24] Other states soon followed, but federal regulation did not prohibit unpasteurized milk until 1987.[25]

Consistent with the principles of federalism and of state's rights, courts have generally held that states may enact and enforce food laws that are different from the federal law so long as the state laws are not inconsistent with the federal law; and do not unreasonably burden interstate commerce. "Inconsistent" generally means direct or indirect conflict between state and federal law.

1.4   AGENCY PROCEDURAL REGULATION

The chief executive (the president or governor) bears the ultimate responsibility for executing the laws enacted by the legislative branch of government. This responsibility is carried out by the administrative agencies that are part of the executive branch of government.

---

[20] At least sixteen states have shellfish safety laws.

[21] New Ice Co. v. Liebman, 285 U.S. 262, 311 (1932); United States v. Lopez, 514 U.S. 549, 581 (1995).

[22] MCL §289.8103; for background on sulfites, *see* Ruth Papazian, *Sulfites: Safe for Most, Dangerous for Some*, FDA CONSUMER (Dec. 1996), *available at:* http://www.fda.gov/fdac/features/096_sulf.html.

[23] Food and Drug Branch, California Department of Public Health, *History of the California Cannery Inspection Program, available at*: http://www.dhs.ca.gov/fdb/HTML/food/indexcan.htm ("From 1899 to 1949, there were 483 outbreaks of botulism reported in North America (the United States and Canada) involving 1319 cases and 851 deaths.")

[24] Cornell University, *Heat Treatments and Pasteurization*, http://www.milkfacts.info/Milk%20Processing/Heat%20Treatments%20and%20Pasteurization.htm#PastHist (last accessed Apr. 2, 2008).

[25] 21 C.F.R. § 1240.61.

In addition to following the requirements of the Constitution and the enabling statutes, administrative agencies must comply with a number of procedural statutes. Three are the most important:

Administrative Procedure Act (APA), which specifies requirements for rulemaking (the process by which federal agencies make regulations) and agency adjudication.

Federal Advisory Committee Act (FACA), which requires that certain kinds of groups whose advice is relied upon by the government be chartered as advisory committees, that they be constituted to provide balance, to avoid a conflict of interest, and to hold committee meetings in public with an opportunity for comment from those outside the committee.

Freedom of Information Act (FOIA), which provides the public with a right to access agency information.

1.4.1   The Administrative Procedure Act

The federal Administrative Procedure Act (5 U.S.C. § 551 *et seq.*) provides for basic procedural safeguards in the federal regulatory system, and establishes and defines judicial review authority over the federal regulatory agencies. A major thrust of the APA is to ensure **due process** in the rulemaking and adjudication by administrative agencies.

In simplest terms, due process means fairness. The three most basic elements of due process are that those affected by the regulatory process are guaranteed **notice**, an **opportunity to be heard**, and a **record** for use in judicial appeals. The major statutory requirements of procedural fairness in the federal Administrative Procedure Act are paralleled in state administrative procedure acts.

1.4.2   Rulemaking

Rulemaking involves the development of administrative rules or regulations for future enforcement. Generally, regulations specify the technical details that are necessary to comply with a law's much broader requirements. For example, the FD&C Act, section 403, states in part "A food shall be deemed to be misbranded (a) If (1) its labeling is false or misleading in any particular. ..." Regulations are promulgated by the FDA to define specific information required on a label to avoid being false or misleading in any particular.

The APA specifies minimum procedural safeguards that agencies must follow when engaged in rulemaking. Notice of any proposed rule must be published by the proposing agency in the *Federal Register*. The agency must allow interested parties time to submit comments. In some instances, public hearings must be conducted with an official record and formal rules. Public comments must be reviewed and considered by the agency before final adoption of a regulation. The agency must explain why it did or did not incorporate suggestions in the final regulation. Final regulations must be published at least 30 days before they are to take effect, so as to allow an opportunity both for legal challenge and for adjustments necessary for compliance with the regulation. Note, however, that unless Congress specifies otherwise, federal agencies have some discretion under these procedural rules.

1.4.3 Adjudication

Judging noncompliance and imposing penalties for violation of regulations may also be a part of an agency's responsibility (if so authorized by statute). Agency adjudication is an agency hearing, somewhat similar to a judicial proceeding, but typically conducted before an agency official acting in the capacity of an administrative law judge (or hearing referee). Agency adjudication is less formal than most judicial proceedings. An adjudicatory hearing deals with specific parties and facts; it establishes what happened and prescribes what is to be done, including determining penalties. For example, a state agriculture department might conduct an adjudication proceeding in which it first establishes the facts as to whether a food establishment violated applicable sanitation standards and then whether revocation of the establishment's license is warranted.

Thus an administrative agency can serve as the lawmaker, the prosecutor, and the judge, all rolled into one. This does not necessarily violate the principle of separation of powers. The rationale is that administrative agencies have narrow areas of technical expertise, they are controlled by numerous procedural requirements, and these decisions always may be appealed to the court system. Due process and the APA specify that agencies, when engaged in adjudication, must provide a person notice of the case against him or her, and some sort of meaningful opportunity to present their case. In some cases the determination must be made by trial-type proceeding.[26]

While court challenges of agency adjudications are not uncommon, it should be noted that those challenges are usually based on procedural, rather than substantive grounds. The courts are enormously deferential to an agency's expertise, and are unlikely to interfere with the substantive decisions made by an agency.[27] Procedural challenges are much more likely to be successful, and also provide greater advantage for negotiated settlements or delays in the implementation of the agency's decision. For example, a grocery store may challenge an agency's decision to revoke their license due to insanitary conditions. However, the challenge is far less likely to be successful on the basis that the agency was incorrect in its professional judgment that the store was insanitary (a substantive challenge), as opposed to the challenge that the agency failed to consider all pertinent evidence in the record, because it failed to

---

[26] Mathews v. Eldridge, 424 U.S. 319; 96 S.Ct. 893 (1976).
[27] Chevron U.S.A., Inc. v. Natural Resources Defense Council, Inc., 467 U.S. 837 (1984).

properly notify the establishment (procedural challenges). A court is far less likely to overturn the agency's decision on the seriousness of the insanitation than to find there was a procedural deficiency.

1.4.4 Judicial Review

Administrative agency activity must also be consistent with the Constitution and relevant statutes. Judicial review of administrative agency activity oversees this consistency. Standards for judicial review of agency actions are outlined in the Administrative Procedure Act, which defines the basis and scope of judicial intervention and review. Generally, the courts will *not* consider whether an agency acted wisely, but only whether the agency has acted as follows:

- Stayed within its constitutional and statutory authority
- Properly interpreted the applicable law
- Conducted a fair proceeding
- Avoided arbitrary or capricious action
- Reached a decision supported by substantial evidence in the record

However, the Supreme Court has also ruled that the courts are to review agency decisions with a searching and careful inquiry to determine "whether the decision was based on consideration of the relevant factors and whether there has been a clear error of judgment."[28] This "Hard Look" doctrine leaves reviewing courts with considerable latitude for overseeing the actions of administrative agencies.

1.4.5 Federal Advisory Committee Act (FACA)

FACA requires that certain kinds of groups whose advice is relied upon by the government be chartered as advisory committees. Advisory committees must be constituted to provide balance and to avoid a conflict of interest. Committee meetings must also be held in public with an opportunity for comment from those outside the committee.

As science-based programs, the food-regulation agencies often rely on committees for scientific advice. Therefore effected parties may find it important to have a say in the deliberations and recommendations of these advisory committees. For example, USDA and HHS select members for the Dietary Guidelines Advisory Committee, which issues the nation's nutritional and dietary guidelines. These recommendations are the foundation for the nutritional standards in all federal food assistance programs, including school lunches and food stamps, and are used in developing the Food Guide Pyramid and nutritional classes. Various groups have contested the makeup of the

---

[28] Citizens to Preserve Overton Park, Inc. v. Volpe, 401 U.S. 402 (1971).

committee for lack of balance and for conflicts of interest. Because food companies are regular sponsors for educational activities of nutrition professional associations as well as nutrition research, finding nutrition academics without some connection to the food industry is difficult.[29]

DISCUSSION QUESTION

1.1. What type of conflicts of interest might arise in the composition of the Dietary Guidelines Advisory Committee?

1.4.6 Freedom of Information Act (FOIA)

A popular Government without popular information or the means of acquiring it, is but a Prologue to a Farce or a Tragedy or perhaps both. Knowledge will forever govern ignorance, and a people who mean to be their own Governors, must arm themselves with the power knowledge gives.

—James Madison

Federal executive branch agencies are required under the Freedom of Information Act (FOIA) to disclose records requested in writing by any person. FOIA applies only to federal agencies and does not create a right of access to records held by Congress, the courts, or by state or local government agencies. However, all states have passed their own public access laws that should be consulted concerning access to state and local records.

The Freedom of Information Act (FOIA) establishes a presumption that records in the possession of agencies are to be accessible to the people. However, agencies may withhold information pursuant to nine exemptions and three exclusions contained in the statute. Because agencies have the right in some circumstance to see sensitive materials held by food businesses, we will discuss FOIA disclosure and trade secrets further in a later chapter.

FOIA litigation is a complex area of law with thousands of court decisions interpreting the act. However, this should not intimidate you from understanding the fundamentals of the law or from making a request yourself.

1.4.7 Constitutional Limitations on Agency Power

Police power, specifically the power of state governments to regulate for the health and welfare of the people, has been upheld to be quite broad in reach and impact. Generally, these laws will be upheld if they are at all reasonable attempts to protect and promote the public's health, safety, or general welfare. The laws do not even need to be good laws, but merely avoid being arbitrary or capricious.

---

[29] Marion Nestle, Food Politics 112 (2002).

State authority to regulate health, safety, and general welfare has been sustained not only for laws aimed at protecting the public in general but also at protecting individuals. Such laws have been upheld even when restricting property rights and individual autonomy. The U.S. Supreme Court made it clear that "the police power is one of the least limitable of governmental powers . . . ," and that the states possess extensive authority to protect public health and safety.[30]

Although the courts have interpreted the state police power broadly, governmental authorities do have limits placed on their powers. Limitations on state and federal powers are found mainly in these three documents:

- The U.S. Constitution
- Constitutions of individual states
- Federal and state laws

In the case of a federal law, the federal government has limited, enumerated powers. If the subject matter of legislation does not fall within any of the enumerated areas of federal authority, then either the matter is one that is reserved to the states or it is a matter beyond the constitutional reach of government altogether. For example, Congress passed a law that required states to provide a disposal site for low-level radioactive waste by a specific date. Any state that failed to meet that deadline was required to take title to and be responsible for all low-level radioactive waste produced in the state. New York State contested the "take title" provision on the ground that it went beyond the enumerated powers of the federal government. The U.S. Supreme Court agreed that the act violated the Tenth Amendment of the U.S. Constitution.[31]

Food laws are sometimes challenged as infringing upon constitutionally protected individual rights. The first 10 amendments to the Constitution, the Bill of Rights, define those things that government cannot do to the individual. If Congress or a state legislature enacts a law inconsistent with any of these Constitutional provisions, the courts may be asked to invalidate the law as being "repugnant to the Constitution."

In the area of food safety, however, the courts historically have been hesitant to invalidate these laws, even for the sake of protecting individual rights. Nonetheless, foods laws have been challenged on this basis, and some important aspects highlighted below foreshadow issues that will rise in subsequent chapters. The cases illustrate how an individual's rights are balanced against society's need for protection from preventable harms.

The Bill of Rights is generally applicable to the states through the Fourteenth Amendment. Right by right, the Supreme Court has applied most, but

---

[30] Queenside Hills Realty Co., Inc. v. Saxl, Commissioner of Housing and Buildings of the City of New York, 328 U.S. 80 (1946).
[31] New York v. United States, 505 U.S. 144 (1992).

not all, of the Bill of Rights' restrictions to the state governments through the Fourteenth Amendment. For example, the states may not pass laws that abridge the freedom of speech, press, or assembly. Technically the state law would be in violation of the Fourteenth Amendment, but for ease of reference, this chapter will refer to the underlying Bill of Rights amendment (in this example the First Amendment's protections of the freedom of speech, press, and assembly).

Free Speech   Laws may be invalidated because they conflict with that part of the First Amendment, which protects the free communication of ideas: "Congress shall make no law . . . abridging the freedom of speech or of the press. . . ." As with all the Bill of Rights, the First Amendment rights are not absolute and may be abridged under certain circumstances. Justice Holmes noted that the First Amendment does not afford a right to cry "fire" in a crowded theater.

In *Cox v. New Hampshire*, 312 U.S. 569 (1941), the U.S. Supreme Court upheld an ordinance that required parade permits, although a group who challenged the law argued that it abridged their First Amendment rights of assembly and communication. The Court concluded:

> The authority of a municipality to impose regulations in order to assure the safety and convenience of the people in the use of public highways has never been regarded as inconsistent with civil liberties, but rather as one of the means of safe-guarding the good order upon which they ultimately depend. . . . The question in a particular case is whether that control is exerted so as not to deny or unwarrantedly abridge the right of assembly and the opportunities for the communication of thought and the discussion of public questions immemorially associated with resort to public places.

First Amendment issues will be discussed in later chapters regarding the right of free expression of commercial speech in conjunction with food advertising and claims.

Searches   The Fourth Amendment to the U.S. Constitution provides that:

> The right of the people to be secure in their persons, houses, papers, and effects, against unreasonable searches and seizures, shall not be violated, and no Warrants shall issue but upon probable cause supported by Oath or affirmation and particularly describing the place to be searched and the persons or things to be seized.

This is particularly relevant to how agencies conduct inspections. The courts have generally upheld the validity of laws granting government agencies the right to inspect food establishments; however, the scope of inspections is more controversial. The right to take photographs and the right to access records, such as complaint files, formulation files, and personnel files, will be discussed in later chapters.

The Fifth Amendment contains three provisions that are particularly pertinent to food regulation:

- **Self-incrimination.** No person shall be compelled to be a witness against himself in any criminal case.
- **Due process.** No person shall be deprived of life, liberty, or property without due process of law.
- **Just compensation.** No private property shall be taken for public use without just compensation.

Self-Incrimination   Under the Fifth Amendment's protection that no person shall be compelled to be a witness against himself in a criminal case, a person may refuse to answer official questions if the answers could be used as evidence against them in a criminal prosecution. This right applies not only to questioning by the federal government, but also through application of the Fourteenth Amendment, to questioning by state and local governmental agencies.

Compelled self-incrimination can become an issue when the records and reports required to be produced by food firms and supplied to food regulatory agencies could conceivably lead to criminal prosecution. For example, the Fifth Amendment might be implicated if a restaurant were compelled to produce a self-inspection report detailing food code violations and submit it to the regulatory agency. This potential conflict has been avoided by making it a criminal offense to fail to maintain and report such records, but forbidding their use for criminal prosecution. New York City took this approach in its self-inspection program for food establishments.[32]

On the other hand, the Fifth Amendment only prohibits being compelled to testify against oneself, not against providing access to records already produced. Therefore, if the same restaurant voluntarily produced self-inspection reports, the Fifth Amendment would not shield the records of those reports. In addition the Fifth Amendment does not provide protection to corporations, but only people.

Due Process   The Fifth Amendment due process provision provides that "no person shall be deprived of life, liberty, or property without due process of law." This clause, along with a similar provision in the Fourteenth Amendment applying due process to state governmental actions, establishes the principle that government must act fairly, according to clear procedures. In its most straightforward sense, due process means fairness in the procedural

---

[32] FRANK P. GRAD, THE PUBLIC HEALTH LAW MANUAL, 272–278, Washington, DC: American Public Health Association (2d ed. 1990) (N.Y.C. Health Code sections 81.39(a), 131.03(d), 131.05(b)).

application of the law. The most basic components of due process fairness are notice and an opportunity to be heard, which were also discussed above regarding the APA.

Additionally, notice means that the government must give adequate information about legal requirements to the persons affected so that they can avoid the consequences of noncompliance. Generally, fair notice means that a law must be published before being enforced. The law must also be written clearly enough so that those subject to the law can understand what the law requires. A law that is so vague that reasonable people may not understand its meaning lacks the basic fairness and violates due process. Such statutory or regulatory language could be invalidated by the courts as "void for vagueness" under the Due Process clause.

Due process also requires that when the government takes action affecting a person's rights or entitlements, the person must be given notice of the intended action and an opportunity to challenge the determination. For example, a government agency cannot revoke a food establishment license without giving the owner notice of the action and, under most circumstances, an opportunity to challenge the action before the license is revoked. In an emergency situation the agency may unilaterally revoke a license, but it must then give the owner an opportunity to challenge the action in a later hearing.

Just Compensation for the Taking of Private Property
The Fifth Amendment provides that no private property shall be taken for public use without just compensation. Agencies may seize or embargo food for being adulterated or misbranded. The purpose is protection of the public's health and welfare. However, seizures clearly interfere with people's use and enjoyment of their property.

Is a seizure a "taking" under the Fifth Amendment? If it is, then the government would be constitutionally required to compensate those persons whose private property rights were affected. However, in keeping with the broad authority the Constitution extends to government as the protector of public health and safety; the general rule is that government seizure of private property to prevent harm usually does not require compensation.

The Supreme Court balances the public interest involved against the reasonableness of the infringement on individual private interests. In *Mulger v. Kansas,* 123 U.S. 623 (1887), the U.S. Supreme Court noted:

> The power which the States have of prohibiting such use by individuals of their property as will be prejudicial to the health, the morals, or the safety of the public, is not—and, consistently with the existence and safety of organized society, cannot be—burdened with the condition that the State must compensate such individual owners for pecuniary losses they may sustain. The exercise of the police power by the destruction of property which is itself a public nuisance, or the prohibition of its use in a particular way, whereby its value becomes depreciated, is very different from taking property for public use, or from depriving a

person of his property without due process of law. In the one case, a nuisance only is abated; in the other, unoffending property is taken away from an innocent owner.

To illustrate this point, the state is not required to compensate the seller of adulterated meat for the salvage value of the protein. The courts have routinely upheld the exercise of the police power even when property will be confiscated or destroyed.

Equal Protection   The U.S. Supreme Court has also interpreted due process to mean that no person shall be denied equal protection of the laws. This guarantee is provided for explicitly in the Fourteenth Amendment, applicable to the states, and implicitly in the Fifth Amendment Due Process clause, applicable to the federal government. Equal protection of the law refers to an even-handed application of law. In its most basic sense this means that government and the legal system cannot arbitrarily discriminate. Equal protection may be violated in two ways: directly by the words of the law, or by the application of the law.

Equality before the law applies not only to the specifics of a law but also to how agencies implement the law. For example, under a local ordinance, which prohibited the construction of wooden laundries without a license, almost all Chinese applicants were denied licenses, while non-Chinese applicants routinely received them. Although the ordinance was a valid safety measure on its face, the implementation violated the equal protection clause of the Fourteenth Amendment.[33]

Nonetheless, equal protection does not require identical treatment. Government may classify people into groups and treat these groups differently. For example, regarding workers in food establishments, the law places special restrictions on persons suffering from certain communicable diseases. This distinction does not violate equal protection because the government may differentiate between individuals and groups if it has good reason to do so. The critical question is what is an acceptable reason for applying the law differently to persons in similar situations.

Privacy Rights   Although privacy right objections are frequently made against public health laws—such as immunization, fluoridation, and compulsory HIV testing—the argument is less common against food laws. The seminal case on privacy rights is the U.S. Supreme Court *Griswold v. Connecticut*, 381 U.S. 479 (1965), decision, where a Connecticut law prohibited the prescribing of contraceptives and their use by any person, including married couples. The Court declared the Connecticut statute unconstitutional. In the main opinion Justice William O. Douglas laid out the basis of a constitutional right to privacy.

---

[33] Yick Wo v. Hopkins, 118 U.S. 356 (1886).

The constitutional right to privacy has been applied by the Supreme Court only in situations involving the personal intimacies of the home, the family, marriage, motherhood, procreation, and child rearing. Efforts to expand the right of privacy to less intimate areas as a basis for invalidating public health and safety laws have not succeeded.

1.5 AGENCY JURISDICTION

Federal responsibility for the direct regulation of food in the United States has primarily been delegated to the Food and Drug Administration (FDA) and the U.S. Department of Agriculture (USDA). However, a number of other federal agencies become involved, depending on the type of food and the type of activity to be regulated. Although the involvement with food with some of these agencies is less direct than that of FDA and USDA, their roles are neither unimportant nor necessarily small.

THUMBNAIL COMPARISON OF AGENCY RESPONSIBILITIES FOR FOOD

Agency	*Responsibility*
Environmental Protection Agency (EPA)	• Drinking water • Pesticide residues
Food and Drug Administration (FDA)	• Food (but not meat) • Drug (OTC and prescriptions) • Dietary supplements • Cosmetics • Medical devices • Bottled water • Seafood • Wild game ("exotic" meat) • Eggs in the shell
Federal Trade Commission (FTC)	• Advertising
Alcohol and Tobacco Tax and Trade Bureau (TTB)	• Alcohol
U.S. Department of Agriculture (USDA)	• Raw vegetables grading • Raw fruit grading • Meats • Poultry • Eggs, processing and grading

The remainder of this chapter presents an overview of principal federal regulatory organizations responsible for food regulation[34] along with a summary of the major federal statutes.

1.5.1 Food and Drug Administration[35]

Oversees
- All domestic and imported food sold in interstate commerce, including shell eggs, but not meat and poultry.
- Bottled water.
- Wine beverages with less than 7 percent alcohol.

Food Safety Role   Food safety laws governing domestic and imported food, except meat and poultry, are enforced in a number of ways by:

- Inspecting food production establishments and food warehouses.
- Collecting and analyzing samples for physical, chemical, and microbial contamination.
- Reviewing safety of food and color additives before marketing.
- Reviewing animal drugs for safety to animals that receive them, and humans who eat food produced from the animals.
- Monitoring safety of animal feeds used in food-producing animals.
- Developing model codes and ordinances, guidelines and interpretations, and working with states to implement them.
- Establishing good food manufacturing practices and other production standards, such as plant sanitation, packaging requirements, and hazard analysis and critical control point programs.
- Working with foreign governments to ensure safety of certain imported food products.
- Requesting manufacturers to recall unsafe food products and monitoring those recalls.
- Taking appropriate enforcement actions.
- Educating industry and consumers on safe food-handling practices.

1.5.2 Centers for Disease Control and Prevention

Food Safety Role
- Investigates with local, state and other federal officials sources of foodborne disease outbreaks.
- Maintains a nationwide system of foodborne disease surveillance.

---

[34] *Derived from* FDA, FDA BACKGROUNDER: FOOD SAFETY: A TEAM APPROACH (Sept. 24, 1998).
[35] For a listing of the statutory responsibilities of the FDA, *see* 21 C.F.R. § 5.10.

- Develops and advocates public health policies to prevent foodborne diseases.
- Conducts research to help prevent foodborne illness.

For more information: www.cdc.gov

1.5.3   USDA Food Safety and Inspection Service (FSIS)

Oversees
- Domestic and imported meat and poultry and related products, such as meat- or poultry-containing stews, pizzas, and frozen foods.
- Processed egg products (generally liquid, frozen, and dried pasteurized egg products).

Food Safety Role   The Federal Meat Inspection Act, the Poultry Products Inspection Act, and the Egg Products Inspection Act, which regulate meat, poultry, and egg products are enforced by:

- Inspecting food animals for diseases before and after slaughter.
- Inspecting meat and poultry slaughter and processing plants.
- With USDA's Agricultural Marketing Service, monitoring and inspecting processed egg products.
- Collecting and analyzing samples of food products for microbial and chemical contaminants and infectious and toxic agents.
- Establishing production standards for use of food additives and other ingredients in preparing and packaging meat and poultry products, and for plant sanitation, thermal processing, and other processes.
- Ensuring all foreign meat and poultry processing plants exporting to the United States meet U.S. standards.
- Seeking voluntary recalls by meat and poultry processors of unsafe products.
- Educating industry and consumers on safe food-handling practices.

For more information: www.fsis.usda.gov

1.5.4   U.S. Environmental Protection Agency

Oversees
- Drinking water
- Pesticide safety

Food Safety Role
- Establishes safe drinking water standards.
- Regulates toxic substances and wastes to prevent their entry into the environment and food chain.

- Determines safety of new pesticides, sets tolerance levels for pesticide residues in foods, and publishes directions on safe use of pesticides.

For more information: www.epa.gov

1.5.5   National Marine Fisheries Service (NMFS)

Oversees
- Fish and seafood products (through a voluntary, fee-for-service system)

Food Safety Role
- The Seafood Inspection Program inspects and certifies fishing vessels, seafood processing plants, and retail facilities for federal sanitation standards.

For more information: www.seafood.nmfs.gov

1.5.6   Alcohol and Tobacco Tax and Trade Bureau (TTB)

The Alcohol and Tobacco Tax and Trade Bureau (TTB) of the U.S. Department of Treasury (formally the Bureau of Alcohol, Tobacco, and Firearms (BATF)) has jurisdiction over the labeling of alcoholic beverages under the Federal Alcohol Administration Act, 27 U.S.C. § 201 *et seq.*

Oversees
- Alcoholic beverages except wine beverages containing less than 7 percent alcohol.

Food Safety Role
- Enforces food safety laws governing alcoholic beverages.
- Investigates adulteration alcoholic products, sometimes with help from FDA.

For more information: www.ttb.gov/index.htm

1.5.7   U.S. Customs Service

Oversees
- Imported foods

Food Safety Role
- Works with federal regulatory agencies to ensure that all goods entering and exiting the United States do so according to U.S. laws and regulations.

For more information: www.customs.ustreas.gov

1.5.8 U.S. Department of Justice

Food Safety Role
- Prosecutes companies and individuals suspected of violating food safety laws.
- Through U.S. Marshals Service, seizes unsafe food products not yet in the marketplace, as ordered by courts.

For more information: www.usdoj.gov

1.5.9 Federal Trade Commission

Food Safety Role
- Enforces a variety of laws that protect consumers from unfair, deceptive, or fraudulent practices, including deceptive and unsubstantiated advertising.

For more information: www.ftc.gov

Other agencies and units become involved with food in some way as well. For example, the USDA has a number of programs that, though not regulatory by nature, can have an effect on food regulation. The USDA Agricultural Marketing Service (AMS) provides voluntary standardization, grading, and market news services for specific agricultural commodities. The Agricultural Research Service (ARS) is the main scientific research arm of USDA. The USDA Economic Research Service (ERS) provides economic analysis relating to agriculture, food, the environment, and rural development. The USDA Grain Inspection, Packers, and Stockyards Administration (GIPSA) provides grading and standardization programs for grains and related products, and regulates and maintains fair trade practices in the marketing of livestock.

The U.S. Codex Office is the point of contact in the United States for the Codex Alimentarius Commission and its activities. The Department of Commerce, National Marine Fisheries Service (NMFS), provides voluntary inspection and certification of fish operations, and administers grades and standards for fish and fish products (similar to the AMS grading and standards programs).

These food regulatory agencies also work with other government agencies when there are crossover responsibilities. For example, FDA works with the Consumer Product Safety Commission to enforce the Poison Prevention Packaging Act. FDA and USDA work with the FBI to enforce the Federal Anti-tampering Act, the Department of Transportation to enforce the Sanitary Food Transportation Act, and the U.S. Postal Service to enforce laws against mail fraud.

This federal delegation and organization of responsibilities is somewhat a haphazard patchwork. Just as the statutes were written to address specific problems at particular points in history, the delegation of food regulation was developed to address specific concerns. The delegation therefore represents an evolution rather than an organization by design.

A number of authors have called for an end to this patchwork system by creation of a unified food safety agency with paramount responsibility for the safety of the U.S. food supply.[36] Similarly, when large outbreaks of foodborne illnesses become public concerns, attention focuses on the organization of food safety regulation. For example, in August 1997, the largest recall of beef yet in the history of the United States occurred with the Hudson Foods Company, when a total of 25 million pounds of hamburger patties were recalled because of *E. coli* O157:H7 contamination. In May 1997, Vice President Al Gore announced the government's five-point plan to improve food safety and its commitment to the new food safety initiatives. The Vice President said, "We have built a solid foundation for the health of America's families. However, clearly we must do more. No parent should have to think twice about the juice they pour their children at breakfast, or a hamburger ordered during dinner out."

1.5.10   State and Local Governments

Allocation of resources is an additional reason state and local governments play a prominent role in food safety regulation in the United States. The combined food-related budget of the above-mentioned federal agencies amounts to only a small fraction of the total federal government budget. State and local officials far outnumber the federal food regulatory staff.

State and local government employ food inspectors, sanitarians, microbiologists, epidemiologists, food scientists, and more. Their precise duties are dictated by state and local laws. Some of these officials monitor only one kind of food, such as milk or seafood. Many work within a specified geographical area, such as a county or a city. Others regulate only one type of food establishment, such as restaurants or meat-packing plants.

State meat and poultry inspection programs must be assessed by the USDA-FSIS to determine whether the state inspection programs are at least equal to the federal program. However, meat and poultry products under state inspection may only be sold in that state.[37] FSIS assumes responsibility for inspection in a state that chooses to end its inspection program or cannot maintain the equivalent standard.

DISCUSSION QUESTION

1.2. The present U.S. food safety system is a patchwork of a dozen different federal agencies. In 1998 the National Academy of Sciences urged

---

[36] *See, e.g.,* U.S. GENERAL ACCOUNTING OFFICE (GAO), U.S. NEEDS A SINGLE AGENCY TO ADMINISTER A UNIFIED, RISK-BASED INSPECTION SYSTEM, GAO/T-RCED-99-256 (Aug. 4, 1999).
[37] Protecting the Public from Foodborne Illness, FSIS Backgrounder, the Food Safety and Inspection Service, April 2001.

Congress to establish a "unified, central framework for managing food safety programs" headed by a single individual. What are some of the pros and cons of creating a single federal food safety agency?

1.6 MAJOR FEDERAL LAWS

1.6.1 The Statutes

All statutes in force in the United States are codified in the United States Code (U.S.C.). The U.S.C. is organized into subject matter titles with numbering that is unique from the section numbering in the statutes as they were enacted into the public acts. For example, section 1 of the Food, Drug, and Cosmetic Act is codified as 21 U.S.C. § 301. Thus this section may be cited with one or the other or both reference numbers, such as "Sec. 1. [301]."

Food, Drug, and Cosmetic Act (FD&C Act)   The Federal Food, Drug, and Cosmetic Act of 1938 gives FDA authority over cosmetics and medical devices as well as food and drugs. The 1938 Act was adopted to correct imperfections of the 1906 Act and to respond to a change in technology and in societal demands from consumers who demanded ever-increasing information about food products. In particular, the 1938 act enacted a comprehensive set of standards by which food safety could be regulated.

Further amendments and revisions to the act after 1938 extended the coverage of the FD&C Act or enlarged FDA's authority over certain products. However, a few amendments have narrowed FDA's authority.

Many states have adopted the Uniform State Food, Drug, and Cosmetics Bill recommended by the Association of Food and Drug Officials, which bears many similarities to the federal FD&C Act. Adoption of this model law is voluntary, but most states have primary food laws that are largely the same as the federal law. AFDO has demonstrated that education and communication can achieve a large measure of cooperative uniformity.

Federal Meat Inspection Act (FMIA)[38]   Federal Meat Inspection Act of 1906 was substantially amended by the Wholesome Meat Act of 1967.[39] The FMIA requires USDA to inspect all cattle, sheep, swine, goats, and horses when slaughtered and processed into products for human consumption. The primary goals of the law are to prevent adulterated or misbranded livestock and products from being sold as food, and to ensure that meat and meat products are slaughtered and processed under sanitary conditions.

These requirements apply to animals and their products produced and sold within states as well as to imports, which must be inspected under

---

[38] A copy of the Federal Meat Inspection act is available at: http://www.fsis.usda.gov/ Regulations_&_Policies/FMIA/index.asp (last visited Jan. 14, 2006).
[39] Pub. L. 90-201 (1967).

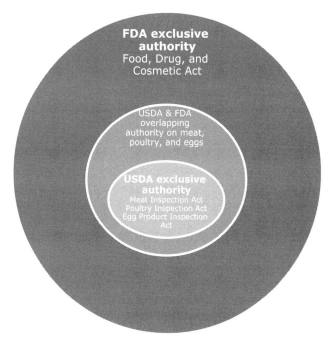

Figure 1.2 Overlapping statutory authority.

equivalent foreign standards. The Food and Drug Administration is responsible for all meats considered "exotic," including venison and buffalo (Figure 1.2).

1.6.2 Other Statutes

A number of other statutes form an important part of the food laws of the United States.

Poultry Products Inspection Act (PPIA)[40] The PPIA provides for the inspection of poultry and poultry products, and regulates the processing and distribution poultry to prevent the movement or sale of poultry products that are adulterated or misbranded.

Egg Products Inspection Act (EPIA)[41] EPIA provides for the inspection of certain egg products, restrictions on the certain qualities of eggs, and uniform standards for eggs, and EPIA otherwise regulates the processing and distribution of eggs and egg products.

[40] A copy of the Poultry Products Inspection Act is available at: http://www.fda.gov/opacom/laws/pltryact.htm (last visited Jan. 14, 2006).
[41] A copy of the Egg Products Inspection Act is available at: http://www.fsis.usda.gov/regulations_&_policies/Egg_Products_Inspection_Act/index.asp (last visited Jan. 14, 2006).

Food Quality Protection Act (FQPA)[42]   FQPA passed by Congress in 1996 amends prior pesticide legislation to establish a more consistent, protective regulatory scheme, based on sound science. It mandates a single, health-based standard for all pesticides in all foods; provides special protections for infants and children; expedites approval of safer pesticides; and creates incentives for the development and maintenance of effective crop protection tools for American farmers. It also requires periodic re-evaluation of pesticide registrations and tolerances to ensure that the scientific data supporting pesticide registrations will remain up to date in the future.

FDA Modernization Act of 1997[43]   The FDA Modernization Act reformed many aspects of the regulation of food, medical products, and cosmetics. The most important food regulation aspect is that the act eliminated the requirement for FDA's premarket approval for most packaging and other substances that come in contact with food and may migrate into it. Instead, the law establishes a process whereby the manufacturer can self-determine safety and notify the agency about its intent to use certain food contact substances. Unless FDA objects within 120 days, the manufacturer may proceed with the marketing of the new product. The act also expanded the procedures under which FDA can authorize health claims and nutrient content claims on foods.

1.6.3   The Regulations

Regulations are promulgated by federal agencies to implement and interpret the laws that are passed by Congress. Regulations are codified in the Code of Federal Regulations (C.F.R.). Regulation typically have the same or similar title number as their corresponding enabling statute in the U.S.C. For example, the regulations that have been promulgated to interpret and implement Title 21 of the United States Code are, for the most part, located in Title 21 of the C.F.R.

Regulations are first published in the Federal Register to order to comply with the requirement for notice and comment of the Administration Procedure Act. Titles 7, 9, and 21 contain most of the laws regulating foods. However, titles 5, 15, 16, 19, 27, 42, and 49 contain other matters that may relate to food in a less direct manner.

Title 5	Governmental organizations and employees
Title 7	Agriculture
Title 9	Animal and animal products
Title 15	Commerce and trade

[42] A copy of the Food Quality Protection Act is available at: http://www.fda.gov/opacom/laws/foodqual/fqpatoc.htm (last visited Jan. 14, 2006).
[43] A copy of the Food and Drug Administration Modernization Act (FDAMA) is available at: http://www.fda.gov/oc/fdama/default.htm (last visited Jan. 14, 2006).

Title 16   Conservation
Title 19   Customs
Title 21   Food and drugs
Title 27   Alcohol, tobacco products, and firearms
Title 42   Public health and welfare
Title 49   Transportation

1.7   INFORMATIONAL RESOURCES

1.7.1   Government Agencies

The government agencies provide a wealth of information on food regulations. Examples of gateway sites are as follows:

- The Food and Drug Administration welcome page: www.fda.gov
- Government food safety information: www.foodsafety.gov/
- USDA FSIS Web site: www.fsis.usda.gov

1.7.2   Associations and Trade Groups

Trade and professional associations can provide important sources of information, particularly on law and policy issues. Some examples are as follows:

- Biotechnology Industry Organization (BIO): www.bio.org
- The Association of Food, Beverage and Consumer Products companies (previously the Grocery Manufacturers Association (GMA)): www.gmabrands.com/index_flash.cfm
- Institute of Food Technologists (IFT): www.ift.org
- National Food Processors Association (NFPA): www.nfpa-food.org

1.7.3   Other Sources

As you have learned, the local food laws and regulations can vary from state to state and even city to city. Therefore you need develop skill at accessing this information. In particular, do not overlook your contacts and acquaintances. The Internet is a growing source of information, but some more traditional sources of information should not be forgotten:

- Colleagues
- Contacts and acquaintances
- Elected and non-elected officials
- Public interest groups
- Trade groups

- Public records
- State registers (similar to the Federal Register)

APPENDIX   CONSTITUTIONAL AMENDMENTS I THROUGH X (THE BILL OF RIGHTS)

Amendment I

Congress shall make no law respecting an establishment of religion, or prohibiting the free exercise thereof; or abridging the freedom of speech, or of the press; or the right of the people peaceably to assemble, and to petition the Government for a redress of grievances.

Amendment II

A well regulated Militia, being necessary to the security of a free State, the right of the people to keep and bear Arms, shall not be infringed.

Amendment III

No Soldier shall, in time of peace, be quartered in any house, without the consent of the Owner, nor in time of war, but in a manner to be prescribed by law.

Amendment IV

The right of the people to be secure in their persons, houses, papers, and effects, against unreasonable searches and seizures, shall not be violated, and no Warrants shall issue, but upon probable cause, supported by Oath or affirmation, and particularly describing the place to be searched, and the persons or things to be seized.

Amendment V

No person shall be held to answer for a capital, or otherwise infamous crime, unless on a presentment or indictment of a Grand Jury, except in cases arising in the land or naval forces, or in the Militia, when in actual service in time of War or public danger; nor shall any person be subject for the same offence to be twice put in jeopardy of life or limb; nor shall be compelled in any criminal case to be a witness against himself, nor be deprived of life, liberty, or property, without due process of law; nor shall private property be taken for public use, without just compensation.

Amendment VI

In all criminal prosecutions, the accused shall enjoy the right to a speedy and public trial, by an impartial jury of the State and district wherein the crime shall have been committed, which district shall have been previously ascertained by law, and to be informed of the nature and cause of the accusation; to be confronted with the witnesses against him; to have compulsory process for obtaining witnesses in his favor, and to have the Assistance of Counsel for his defense.

Amendment VII

In suits at common law, where the value in controversy shall exceed twenty dollars, the right of trial by jury shall be preserved, and no fact tried by a jury, shall be otherwise reexamined in any Court of the United States, than according to the rules of the common law.

Amendment VIII

Excessive bail shall not be required, nor excessive fines imposed, nor cruel and unusual punishments inflicted.

Amendment IX

The enumeration in the Constitution, of certain rights, shall not be construed to deny or disparage others retained by the people.

Amendment X

The powers not delegated to the United States by the Constitution, nor prohibited by it to the States, are reserved to the States respectively, or to the people.

CHAPTER 2

What Is a Food?

2.1 INTRODUCTION TO THE FOOD, DRUG, AND COSMETIC ACT

Most foods are regulated under the Federal Food, Drug, and Cosmetic Act (FD&C Act).[1] The FD&C Act regulates more products that Americans use in our daily activities than any other federal statute, including pharmaceuticals, medical devices, and cosmetics. The regulations of these products share many similarities, but the requirements for food differ significantly from those for drugs. Accordingly, the classification of a product as a food or drug (or both) can determine how rigorously the product is regulated, or whether the product is even legal. Thus the statute's definitions of products deserve close attention.

After you complete this chapter, you will have an understanding of:

- what makes an article subject to FD&C Act;
- what makes an article a food, a drug, or a product outside the scope of the FD&C Act; and
- the central role of intended use.

2.1.1 Definitions

SEC. 201. [321][2] For the purposes of this Act—

. . .

(f) The term "**food**" means (1) *articles used for food or drink for man or other animals, (2) chewing gum, and (3) articles used for components of any other such article.*

---

[1] 21 U.S.C. § 321 *et seq.*

[2] Statutory citations used in this material generally are to the FD&C Act statutory sections (which is the way practitioners refer to them). The citation within the brackets is the U.S.C. number. The United States Code (U.S.C.) the U.S.C. is organized into subject matter titles with numbering that is unique from the section numbering in the statutes as they were enacted into the public acts. For example, section 1 of the Food, Drug, and Cosmetic Act is codified as 21 U.S.C. § 301. Thus you may find this section cited with one or the other or both reference numbers, such as "Sec. 1. [301]."

(g) (1) The term "**drug**" means

   (A) articles recognized in the official United States Pharmacopoeia, official Homoeopathic Pharmacopoeia of the United States, or official National Formulary, or any supplement to any of them; and

   (B) articles intended for use in the diagnosis, cure, mitigation, treatment, or prevention of disease in man or other animals; and

   (C) articles (other than food) intended to affect the structure or any function of the body of man or other animals; and

   (D) articles intended for use as a component of any article specified in clause (A), (B), or (C). A food or dietary supplement for which a claim, subject to sections 343(r)(1)(B) and 343(r)(3) of this title or sections 343(r)(1)(B) and 343(r)(5)(D) of this title, is made in accordance with the requirements of section 343(r) of this title is not a drug solely because the label or the labeling contains such a claim. A food, dietary ingredient, or dietary supplement for which a truthful and not misleading statement is made in accordance with section 343(r)(6) of this title is not a drug under clause (C) solely because the label or the labeling contains such a statement.

(h) The term "**device**" . . . means an instrument, apparatus, implement, machine, contrivance, implant, in vitro reagent, or other similar or related article, including any component, part, or accessory, which is—

   (1) recognized in the official National Formulary, or the United States Pharmacopeia, or any supplement to them;

   (2) intended for use in the diagnosis of disease or other conditions, or in the cure, mitigation, treatment, or prevention of disease, in man or other animals; or

   (3) intended to affect the structure or any function of the body of man or other animals, and which does not achieve its primary intended purposes through chemical action within or on the body of man or other animals and which is not dependent upon being metabolized for the achievement of its primary intended purposes.

(i) The term "**cosmetic**" means (1) articles intended to be rubbed, poured, sprinkled, or sprayed on, introduced into, or otherwise applied to the human body or any part thereof for cleansing, beautifying, promoting attractiveness, or altering the appearance, and (2) articles intended for use as a component of any such articles; except that such term shall not include soap.

   . . . .

(s) The term "**food additive**" means any substance the intended use of which results or may reasonably be expected to result, directly or indirectly, in its becoming a component or otherwise affecting the characteristics of any food (including any substance intended for use in producing, manufacturing, packing, processing, preparing, treating, packaging, transporting, or holding food; and including any source of radiation intended for any such use), if such substance is not generally recognized, among experts qualified by scientific training and experience to evaluate its safety, as having been adequately shown through scientific procedures (or, in the case of a substance used in food prior to January 1, 1958, through either scientific procedures or experience based on common use in food) to be safe under the conditions of its intended use; except that such term does not include—

(1) a pesticide chemical in or on a raw agricultural commodity; or

(2) a pesticide chemical to the extent that it is intended for use or is used in the production, storage, or transportation of any raw agricultural commodity; or

(3) a color additive; or

(4) any substance used in accordance with a sanction or approval granted prior to the enactment of this paragraph pursuant to this . . . or

(5) a new animal drug.

. . . .

(ff) The term "**dietary supplement**"—

(1) means a product (other than tobacco) intended to supplement the diet that bears or contains one or more of the following dietary ingredients:

(A) a vitamin;

(B) a mineral;

(C) an herb or other botanical;

(D) an amino acid;

(E) a dietary substance for use by man to supplement the diet by increasing the total dietary intake; or

(F) a concentrate, metabolite, constituent, extract, or combination of any ingredient described in clause (A), (B), (C), (D), or (E);

(2) means a product that—

(A) (i) is intended for ingestion in a form described in section 350(c)(1)(B)(i) of this title; or

(ii) complies with section 350(c)(1)(B)(ii) of this title;

(B) is not represented for use as a conventional food or as a sole item of a meal or the diet; and

(C) is labeled as a dietary supplement; and

(3) does—

(A) include an article that is approved as a new drug under section 355 of this title or licensed as a biologic under section 262 of title 42 and was, prior to such approval, certification, or license, marketed as a dietary supplement or as a food unless the Secretary has issued a regulation, after notice and comment, finding that the article, when used as or in a dietary supplement under the conditions of use and dosages set forth in the labeling for such dietary supplement, is unlawful under section 342(f) of this title; and

(B) not include—

(i) an article that is approved as a new drug under section 355 of this title, certified as an antibiotic under section 357 of this title, or licensed as a biologic under section 262 of title 42; or

(ii) an article authorized for investigation as a new drug, antibiotic, or biological for which substantial clinical investigations have been instituted and for which the existence of such investigations has been made public, which was not before such approval, certification, licensing, or authorization marketed as a dietary supplement or as a food unless the Secretary, in the Secretary's discretion, has issued a regulation, after notice and comment, finding that the article would be lawful under this chapter.

Except for purposes of paragraph (g), a dietary supplement shall be deemed to be a food within the meaning of this chapter.

*   *   *   *   *

2.1.2 FDA's Jurisdiction and the Definition of Food

FDA's authority over food derives from the Federal Food, Drug, and Cosmetic Act (FD&C Act). Thus the definition of "food" in the act has importance in determining the reach and limits of FDA's jurisdiction and authority.

The statutory definition of "food" in FD&C Act section 321(f) is a term of art that is clearly intended to be broader than the commonsense definition of food. This creates numerous pitfalls for the unwary. For instance, the definition of "food" includes chewing gum and food additives. "Food additives" can be any substance, the intended use of which results, or may reasonably result, in its becoming a component or otherwise affecting the characteristics of any food.[3]

To a large extent the use to which a product is put will determine the category into which it will fall. The manufacturer's representations and the intended use play an important part of determining the classification. On occasion a manufacturer may find benefits in changing its representations so that their product falls into a different category. For example, a laxative gum can

---

[3] *See* 21 U.S.C. § 321(s).

escape the definition of food by being represented unequivocally as a drug product.

DISCUSSION QUESTION

2.1. Note the broad definition of "food" under the FD&C Act. This provides a very broad scope of authority to the FDA. Would this broad scope conflict with the USDA FSIS's authority?

2.1.3 Specific Food Classifications

Meat, Poultry, and Eggs The U.S. Department of Agriculture (USDA) is responsible for meat, poultry, and processed eggs; however, which agency has jurisdiction over these foods is complex and sometimes uncertain. All foods are subject to the FD&C Act—meats are exempt from the FD&C Act provisions, but only to the extent that the Federal Meat Inspection Act (FMIA) applies.[4]

FDA has exclusive regulatory jurisdiction over the live meat animals intended for food. USDA has exclusive jurisdiction over the slaughter and processing of meat animals. Except that food additives are under the jurisdiction of the FD&C Act, so the USDA and FDA have joint jurisdiction for food additives in meat and poultry.

FDA and USDA also have joint jurisdiction over the transport of meat products after processing, but FDA has exclusive jurisdiction over retail establishments (when in federal jurisdiction). USDA may still regulate USDA-labeled packages that are found in retail establishments, but USDA lacks authority over the retail establishments directly.

For products that contain meat, the percentage of meat determines whether a product is subject to USDA jurisdiction. For example, a product containing 3% or less raw meat falls under FDA jurisdiction. This division of responsibility is based on the internal decisions of the two agencies.[5] The Memoranda of Understandings (MOUs) are available to the public. These are also summarized in both agencies compliance policy guides. These references can be found on the agencies' Web sites.

Water The Safe Drinking Water Act[6] places the responsibility for the safety and purity of drinking water on EPA. However, FDA retains the authority over bottled drinking water. Differences between these two standards

---

[4] 21 U.S.C. 392(b).
[5] FDA provides an FDA/USDA jurisdiction chart summarizing jurisdiction overlap and the percentages of meat, FDA INVESTIGATIONS OPERATIONS MANUAL 2006, Chapter 3, Exhibit 3-1, *available at*: http://www.fda.gov/ora/inspect_ref/iom/exhibits/3-1.pdf (last accessed Jan. 10, 2007).
[6] 88 Stat. 1660 (1974).

sometimes create consternation for the agencies, the bottled water industry, and municipal water agencies.

In addition water, when used as a food ingredient, is a food, and thus is subject to all the same requirements of the FD&C Act as any other food ingredient. Similarly, for ice added as an ingredient, the FDA has jurisdiction over packaged ice as a food.

2.2 WHAT MAKES AN ARTICLE A FOOD OR A DRUG?

The *Nutrilab* starch blockers case below highlights the importance of the definitions in determining how a product will be regulated. Nutrilab claimed their starch blockers were a food because the product was derived from beans. The court, however, found that starch blockers was a drug under the FD&C Act because the "tablets and pills at issue were not consumed primarily for taste, aroma, or nutritive value" . . . but "they are taken for their ability to block the digestion of food and aid in weight loss." Foods are normally digested, but starch blockers blocked the digestion, which shows intent to affect the structure or function of the body. Starch blockers were therefore deemed to be drugs under section 321(g)(1)(C) of the FD&C Act.

* * * * *

Nutrilab, Inc. v. Schweiker

713 F.2d 335 (1983)

Judges: CUMMINGS, Chief Judge; POSNER, Circuit Judge; and FAIRCHILD, Senior Circuit Judge
Opinion: CUMMINGS

Plaintiffs manufacture and market a product known as "starch blockers" which "block" the human body's digestion of starch as an aid in controlling weight. . . . The only issue on appeal is whether starch blockers are foods or drugs under the Federal Food, Drug, and Cosmetic Act. Starch blocker tablets and capsules consist of a protein which is extracted from a certain type of raw kidney bean. That particular protein functions as an alpha-amylase inhibitor; alpha-amylase is an enzyme produced by the body which is utilized in digesting starch. When starch blockers are ingested during a meal, the protein acts to prevent the alpha-amylase enzyme from acting, thus allowing the undigested starch to pass through the body and avoiding the calories that would be realized from its digestion.

Kidney beans, from which alpha-amylase inhibitor is derived, are dangerous if eaten raw. By August 1982, FDA had received seventy-five reports of adverse effects on people who had taken starch blockers, including complaints of gastro-intestinal distress such as bloating, nausea, abdominal pain, constipation and vomiting. Because plaintiffs consider starch blockers to be food, no testing as required to obtain FDA approval as a new drug has taken place. If starch blockers were drugs, the manufacturers would be required to file a new drug application pursuant to 21 U.S.C. § 355 and remove the product from the marketplace until approved as a drug by the FDA.

The statutory scheme under the Food, Drug, and Cosmetic Act is a complicated one. Section 321(g)(1) provides that the term "drug" means ...

... (B) articles intended for use in the diagnosis, cure, mitigation, treatment, or prevention of disease in man or other animals; and (C) articles (other than food) intended to affect the structure or any function of the body of man or other animals; and (D) articles intended for use as a component of any article specified in clauses (A), (B), or (C) of this paragraph; but does not include devices or their components, parts, or accessories.

The term "food" as defined in section 321(f) means

(1) articles used for food or drink for man or other animals, (2) chewing gum, and (3) articles used for components of any such article.

Section 321(g)(1)(C) was added to the statute in 1938 to expand the definition of "drug." The amendment was necessary because certain articles intended by manufacturers to be used as drugs did not fit within the "disease" requirement of section 321(g)(1)(B). Obesity in particular was not considered a disease. Thus "anti-fat remedies" marketed with claims of "slenderizing effects" had escaped regulation under the prior definition. The purpose of part C in section 321(g)(1) supra was "to make possible the regulation of a great many products that have been found on the market that cannot be alleged to be treatments for diseased conditions."

It is well established that the definitions of food and drug are normally not mutually exclusive; an article that happens to be a food but is intended for use in the treatment of disease fits squarely within the drug definition in part B of section 321(g)(1) and may be regulated as such. Under part C of the statutory drug definition, however, "articles (other than food)" are expressly excluded from the drug definition (as are devices) in section 321(g)(1). In order to decide if starch blockers are drugs under section 321(g)(1)(C), therefore, we must decide if they are foods within the meaning of the part C "other than food" parenthetical exception to section 321(g)(1)(C). And in order to decide the meaning of "food" in that parenthetical exception, we must first decide the meaning of "food" in section 321(f).

Congress defined "food" in section 321(f) as "articles used as food." This definition is not too helpful, but it does emphasize that "food" is to be defined in terms of its function as food, rather than in terms of its source, biochemical composition, or ingestibility. Plaintiffs' argument that starch blockers are food because they are derived from food—kidney beans—is not convincing; if Congress intended food to mean articles derived from food it would have so specified. Indeed some articles that are derived from food are indisputably not food, such as caffeine and penicillin. In addition all articles that are classed biochemically as proteins cannot be food either, because for example insulin, botulism toxin, human hair, and influenza virus are proteins that are clearly not food.

Plaintiffs argue that 21 U.S.C. § 343(j) specifying labeling requirements for food for special dietary uses indicates that Congress intended products offered for weight conditions to come within the statutory definition of "food." Plaintiffs misinterpret that statutory Section. It does not define food but merely requires that if a product is a food and purports to be for special dietary uses, its label must contain certain information to avoid being misbranded. If all products intended to affect underweight or overweight conditions were per se foods, no diet product could be regulated as a drug under section 321(g)(1)(C), a result clearly contrary to the intent of Congress that "anti-fat remedies" and "slenderizers" qualify as drugs under that Section.

If defining food in terms of its source or defining it in terms of its biochemical composition is clearly wrong, defining food as articles intended by the manufacturer to be used as food is problematic. When Congress meant to define a drug in terms of its intended use, it explicitly incorporated that element into its statutory definition. For example, section 321(g)(1)(B) defines drugs as articles "intended for use" in, among other things, the treatment of disease; section 321(g)(1)(C) defines drugs as "articles (other than food) intended to affect the structure or any function of the body of man or other animals." ... Further a manufacturer cannot avoid the reach of the FDA by claiming that a product which looks like food and smells like food is not food because it was not intended for consumption. ... In *United States v. Technical Egg Prods., Inc.*, the defendant argued that the eggs at issue were not adulterated food under the Act because they were not intended to be eaten. The court held that there was a danger of their being diverted to food use and rejected defendant's argument.

Although it is easy to reject the proffered food definitions, it is difficult to arrive at a satisfactory one. In the absence of clear-cut congressional guidance, it is best to rely on statutory language and common sense. The statute evidently uses the word "food" in two different ways. The statutory definition of "food" in section 321(f) is a term of art, and is clearly intended to be broader than the commonsense definition of food, because the statutory definition of "food" also includes chewing gum and food additives. Food additives can be any substance the intended use of which results or may reasonably result in its becoming a component or otherwise affecting the characteristics of any food. Paper food-packaging when containing polychlorinated biphenyls (PCBs), for example, is an adulterated food because the PCBs may migrate from the package to the food and thereby become a component of it. Yet the statutory definition of "food" also includes in section 321(f)(1) the common-sense definition of food. When the statute defines "food" as "articles used for food," it means that the statutory definition of "food" includes articles used by people in the ordinary way most people use food—primarily for taste, aroma, or nutritive value. To hold as did the district court that articles used as food are articles used solely for taste, aroma, or nutritive value is unduly restrictive since some products such as coffee or prune juice are undoubtedly food but may be consumed on occasion for reasons other than taste, aroma, or nutritive value. . . .

This double use of the word "food" in section 321(f) makes it difficult to interpret the parenthetical "other than food" exclusion in the section 321(g)(1)(C) drug definition. As shown by that exclusion, Congress obviously meant a drug to be something "other than food," but was it referring to "food" as a term of art in the statutory sense or to foods in their ordinary meaning? Because all such foods are "intended to affect the structure or any function of the body of man or other animals" and would thus come within the part C drug definition, presumably Congress meant to exclude commonsense foods. Fortunately, it is not necessary to decide this question here because starch blockers are not food in either sense. The tablets and pills at issue are not consumed primarily for taste, aroma, or nutritive value under section 321(f)(1); in fact, as noted earlier, they are taken for their ability to block the digestion of food and aid in weight loss. In addition, starch blockers are not chewing gum under section 321(f)(2) and are not components of food under section 321(f)(3). To qualify as a drug under section 321(g)(1)(C), the articles must not only be articles "other than food" but must also be "intended to affect the structure or any function of the body of man or other animals." Starch blockers indisputably satisfy this requirement for they are intended to affect digestion in the people who take them. Therefore starch blockers are drugs under section 321(g)(1)(C) of the Food, Drug, and Cosmetic Act.

Affirmed.

* * * * *

2.3 THE CENTRAL ROLE OF INTENDED USE

In the *Nutrilab* starch blockers case, the manufacturer's intent was clear. As was the fact that the product was not consumed for its taste, aroma, or nutritive value. Thus starch blockers were deemed other than a conventional food.

Other products, however, might not present such clear distinctions. Vitamins and minerals have generally been classified as foods unless therapeutic claims have been made for them. However, in the 1970s, reports of human toxicity emerged from consumption of large doses of the vitamins A and D. These fat-soluble vitamins create special concern because they can accumulate in the fatty tissue.

To deal with this problem, in 1972 and 1973 FDA promulgated regulations classifying certain high dosages of vitamin A and D as drugs and requiring that they be sold by prescription.[7] However, in *National Nutritional*

[7] 37 Fed. Reg. 26618 (Dec. 14, 1972) and 38 Fed. Reg. 20723 (Aug. 2, 1973).

Foods Ass'n v. Mathews,[8] the court questioned FDA's approach and found FDA's administrative record incomplete. In particular, the court questioned whether FDA could classify vitamins as drugs when no intended therapeutic use was offered by the vendors, the labeling, or any promotional material. The court upheld the regulations, but FDA nonetheless later rescinded them.[9]

This subject is discussed in more depth in future chapters, but it is important to understand that the intended use of a product may determine whether it is a conventional food, a dietary supplement, or a drug. A generation ago, any health claim for a food or supplement moved the regulation of the product to "drug" status. Food-drug distinctions are somewhat less clear today because health claims no longer automatically move a food or dietary supplement over to regulation as a drug. FDA-approved health claims are permitted, for instance, without triggering drug status. In addition structure-function claims are a category of health-related claims that are not regulated as health claims (e.g., "calcium helps build strong bones").

This statutory organization is murky because, at times, it is difficult to draw distinctions between structure-function claims and drug claims. The *Nutrilab* case provides what remains one of the best rules of thumb for determining whether a product is a food or a drug. First ask, is the product a commonsense food? If not, is it consumed primarily for taste, aroma, or nutrition? If the answer is no to both these questions, then the product may not be a food. There can be other factors, but this commonsense rule still provides excellent guidance.

2.4 OTHER CONSIDERATIONS

2.4.1 Products Ordinarily Considered Foods

There have been a number of cases where products—ordinarily considered foods—were classified as drugs because of the product's therapeutic claims:

Honey[10]
Vinegar and honey[11]
Tea[12]
Water[13]

---

[8] 557 F.2d 325 (2nd Cir. 1977).
[9] 43 Fed. Reg. 10551 (Mar. 14, 1978).
[10] United States v. 250 Jars . . . Cal's Tupelo Blossum U.S. Fancy Pure Honey, 344 F.2d 288 (6th Cir. 1965).
[11] Sterling Vinegar and Honey, 338 F.2d 157 (2nd Cir. 1964).
[12] United States v. Hohensee, 243 F.2d 367 (3d. Cir. 1957).
[13] United States v. 500 Plastic Bottles . . . Wilfrey's Bio Water (D. Or. 1989).

Blue-green algae[14]
Mussels[15]

2.4.2 Products Intended to Be Processed into Food

A number of articles have been deemed to be "food" within the meaning of the FD&C Act definition because they are intended to be processed into a food or a component of food.

- **Green coffee beans.** It makes no difference if the beans require further roasting and processing before they would be ready for consumption.[16]
- **Live beef cattle.** The edible tissues of live calves constitute "food" as defined by the FD&C Act and are therefore subject to the adulteration provisions of the act.[17]

2.4.3 Products No Longer Fit for Food

A product that is generally regarded as a food is considered a food under the FD&C Act, when it is in food form, even if the product is decomposed or otherwise unfit for consumption. For example, a shipment of incubator reject shell eggs was still "food" although a large percentage of them were inedible eggs.[18] The product might not be intended to be eaten, but if there is a danger of the product being diverted to food use, the product is considered a food. Note that the intended use of the product is irrelevant to this determination, which is based on the product being in the form of a food.

2.4.4 Packaging Materials

The definition of "food" is significantly broadened by the inclusion of food additives within the definition of food. Food additives can be any substance whose intended use results or *may reasonably result in its becoming a component or otherwise affecting the characteristics of any food.*[19] Thus the definition of food includes any substances that migrate to the food from the packaging materials or containers.[20]

---

[14] United States v. Kollman, (DC Or. 1985, 1986).
[15] United States v. Articles of Drug. . . . Neptone (ND Cal. 1983).
[16] United States v. Green Coffee Beans, 188 F.2d 355 (1951).
[17] United States v. Tomahara Enterprises, Ltd., (DC ND N.Y. 1983).
[18] United States v. Technical Egg Prods., Inc., 171 F. Supp. 326 (N.D. Ga. 1959).
[19] *See* 21 U.S.C. § 321(s).
[20] *See* Natick Paperboard Corp. v. Weinburger, 525 F.2d 1103 (1st Cir. 1975).

2.4.5 Evidence of Intended Use

In determining whether a product is a "drug" because of intended therapeutic use, FDA is not bound by a manufacturer's subjective claims of intent.[21] Actual therapeutic intent may be found on the basis of any objective evidence. Such evidence may be inferred from "labeling, promotional material, advertising, and *any other relevant source.*"[22]

The FD&C Act definition of "food" lacks any reference to intent. Nonetheless, a court may consider the intended use of the product in considering whether it is a food. A manufacturer's subjective intent that a product is not intended for consumption will not allow it to avoid the reach of the FD&C Act if the product looks like food and smells like food.

DISCUSSION QUESTIONS

2.2. Could bottled water be characterized as a food, a drug, a dietary supplement, or all three? How?

2.3. Would your answer change if the product were cherry juice concentrate?

2.4. When would blackboard chalk be a drug?

2.5. "SkyHigh" brand glue is not only efficacious as glue, but is widely known to induce a high when sniffed. The manufacturer advertises the adhesive properties of the glue heavily in magazines that are popular in the drug ulture. Can the glue be regulated by FDA? Explain briefly.

2.6. Is the definition of "food" good statutory drafting?

2.7. Coffee is often consumed for its stimulant effect. Coffee is not consumed for its nutritional value. If a manufacturer promoted its coffee for the stimulant effect, would it be a drug?

2.8. If coffee was only promoted as a stimulant, would it still be regulated as a food?

---

[21] National Nutritional Foods Ass'n v. Mathews, 57 F.2d 325 (2nd Cir. 1977).
[22] *Id.*

PART II

REGULATION OF LABELING AND CLAIMS

■■■■■ CHAPTER 3

Food Labeling

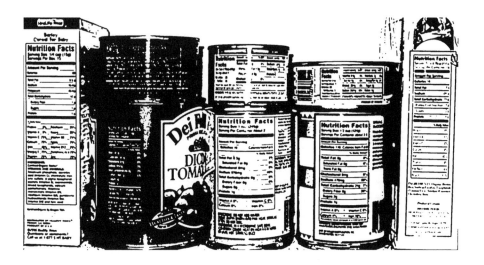

3.1   INTRODUCTION

Labeling, perhaps surprisingly, has been at the center of many aspects of food regulation. In addition incorrect labeling consistently ranks as the leading cause of food recalls and import denials. This chapter examines food labeling regulation that is designed to protect the economic expectations of both consumers and the food industry. In Chapter 4, we will cover the regulation of the nutritional content and labeling of food. In Chapter 5, we will cover the regulation of the identification of foods in more depth, including the standards of identity requirements and the requirement for common and usual names.

The food labeling requirements designed to protect economic expectations cover both **prohibitive** and **affirmative** regulation. The prohibitive requirements protect against fraud and deception. Prevention of false and misleading

Food Regulation: Law, Science, Policy, and Practice, by Neal D. Fortin
Copyright © 2009 Published by John Wiley & Sons, Inc.

statements is at the historical foundation of labeling regulation. On the other hand, the affirmative requirements mandate that food manufacturers provide information on their labels that they otherwise might not include.

These affirmative requirements are intended to provide consumers with information they need to make informed choices about the food. What information has been deemed material to informed choice is a surprisingly small set that has remained relatively stable.

This chapter provides:

1. An overview of the labeling laws,
2. Basic knowledge needed to review a label for compliance with applicable requirements,
3. Knowledge of where to look up answers, and
4. Identification of reference materials.

3.2   LEGAL AUTHORITIES

FDA's authority to compel the labeling of food products primarily derives from the Food, Drug, and Cosmetics Act (FD&C Act). Many of the statutory labeling requirements come from section 403 of the FD&C Act, which lists circumstances when a food will be considered "misbranded." The definition of "misbranded" contains the major misbranding requirements:

1. Mandatory labeling of the name of the food, ingredient statement, net quantity, and the name and address of the manufacturer or distributor.
2. Mandatory standards of identity.
3. Labeling of imitation foods.
4. Nutrition information for special dietary foods.
5. Prohibition of any false or misleading claims.

The Fair Packaging and Labeling Act (FPLA), 15 U.S.C. 1451 *et seq.*, was enacted in 1966 to prevent unfair and deceptive trade practices, and to provide consumers with accurate information regarding the quantity and value of products. The FPLA is administered by the FDA for labels on foods, drugs, and cosmetics, while the Federal Trade Commission (FTC) administers the FPLA for most other consumer commodities.

FDA's labeling regulations are located in 21 C.F.R. § 101 and cover both the requirements of the Food, Drug, and Cosmetic Act and the Fair Packaging and Labeling Act.

Labeling of meat, poultry, and egg products is regulated under separate laws by the USDA. The major principles and many of the specifics are the same in both sets of requirements. This chapter provides an overview of the differences.

3.3 LABELING TERMINOLOGY

* * * * *

Federal Food, Drug, and Cosmetic Act

Chapter II—Definitions
SEC. 201. [321] For the purposes of this Act—
. . .
(k) The term "**label**" means a display of written, printed, or graphic matter upon the immediate container of any article; and a requirement made by or under authority of this Act that any word, statement, or other information appearing on the label shall not be considered to be complied with unless such word, statement, or other information also appears on the outside container or wrapper, if any there be, of the retail package of such article, or is easily legible through the outside container or wrapper.

(l) The term "**immediate container**" does not include package liners.

(m) The term "**labeling**" means all labels and other written, printed, or graphic matter (1) upon any article or any of its containers or wrappers, or (2) accompanying such article.

(n) If an article is alleged to be misbranded because the labeling or advertising is misleading, then in determining whether the labeling or advertising is misleading there shall be taken into account (among other things) not only representations made or suggested by statement, word, design, device, or any combination thereof, but also the extent to which the labeling or advertising fails to reveal facts material in the light of such representations or material with respect to consequences which may result from the use of the article to which the labeling or advertising relates under the conditions of use prescribed in the labeling or advertising thereof or under such conditions of use as are customary or usual. . . .

(r) The term "raw agricultural commodity" means any food in its raw or natural state, including all fruits that are washed, colored, or otherwise treated in their unpeeled natural form prior to marketing.

* * * * *

Food Labeling

21 C.F.R. Part 101

§ 101.1 Principal display panel of package form food

The term *principal display panel* as it applies to food in package form and as used in this part, means the part of a label that is most likely to be displayed, presented, shown, or examined under customary conditions of display for retail sale. The principal display panel shall be large enough to accommodate all the mandatory label information required to be placed thereon by this part with clarity and conspicuousness and without obscuring design, vignettes, or crowding. Where packages bear alternate principal display panels, information required to be placed on the principal display panel shall be duplicated on each principal display panel. . . .

§ 101.2   Information panel of package form food

(a) The term *information panel* as it applies to packaged food means that part of the label immediately contiguous and to the right of the principal display panel as observed by an individual facing the principal display panel with the following exceptions:

   (1) If the part of the label immediately contiguous and to the right of the principal display panel is too small to accommodate the necessary information or is otherwise unusable label space, e.g., folded flaps or can ends, the panel immediately contiguous and to the right of this part of the label may be used.

   (2) If the package has one or more alternate principal display panels, the information panel is immediately contiguous and to the right of any principal display panel.

   (3) If the top of the container is the principal display panel and the package has no alternate principal display panel, the information panel is any panel adjacent to the principal display panel.

(b) All information required to appear on the label of any package of food under . . . this chapter shall appear either on the principal display panel or on the information panel, unless otherwise specified by regulations in this chapter.

(c) All information appearing on the principal display panel or the information panel pursuant to this section shall appear prominently and conspicuously, but in no case may the letters and/or numbers be less than one-sixteenth inch in height unless an exemption pursuant to paragraph (f) of this section is established. . . . [A number of exemptions from size and placement requirements are omitted.]

(e) All information appearing on the information panel pursuant to this section shall appear in one place without other intervening material.

* * * * *

3.3.1   Label versus Labeling

Mark Twain noted that the distinction between the right word and the almost right word is the difference between lightning and a lightning bug.[1] Notice the

---

[1] Letter from Mark Twain to George Bainton (Oct. 15, 1888), *in* THE ART OF AUTHORSHIP: LITERARY REMINISCENCES, METHODS OF WORK, AND ADVICE TO YOUNG BEGINNERS, 87–88 (1891) (George Bainton ed., New York: Appleton 1891), *available at*: http://www.bartleby.com/73/540.html (last visited Sept. 27, 2005).

distinction between the terms "label" and "labeling" as defined in section 201 of FD&C Act. The slight difference in the words creates an important distinction in meaning.

3.3.2 The Scope of Labeling

The term "labeling" is defined broadly in section 201 of FD&C Act to include "*all labels and other written, printed, or graphic matter* (1) *upon any article* or any of its containers or wrappers, or (2) *accompanying* such article."

   Kordel v. United States is a landmark case dealing with the jurisdictional reach of FDA's authority of "labeling." The *Kordel* case involved health foods—compounds of vitamins, minerals, and herbs—that were supplied with brochures and other literature. These health foods were deemed drugs, as defined by the FD&C Act, because of their intended use. Kordel contended that the literature was not "labeling" and, therefore, was not subject to the misbranding provisions of the FD&C Act.

* * * * *

Kordel v. United States

335 U.S. 345 (1948)

Opinion of the Court by Mr. Justice Douglas, announced by Mr. Justice Reed
. . .
Kordel was charged by informations containing twenty counts of introducing or delivering for introduction into interstate commerce misbranded drugs. . . . Kordel writes and lectures on health foods from information derived from studies in public and private libraries. Since 1941, he has been marketing his own health food products, which appear to be compounds of various vitamins, minerals, and herbs. The alleged misbranding consists of statements in circulars or pamphlets distributed to consumers by the vendors of the products, relating to their efficacy. The petitioner supplies these pamphlets as well as the products to the vendors. Some of the literature was displayed in stores in which the petitioner's products were on sale. Some of it was given away with the sale of products; some sold independently of the drugs; and some mailed to customers by the vendors.

   . . . The question of whether the separate shipment of the literature saved the drugs from being misbranded within the meaning of the Act presents the main issue in the case.

   Section 301(a) of the Act prohibits the introduction into interstate commerce of any drug that is adulterated or misbranded. It is misbranded according to 502(a) if its "labeling is false or misleading in any particular" and unless the labeling bears "adequate directions for use" per 502(f). The term labeling is defined in 201(m) to mean "all labels and other written, printed, or graphic matter (1) upon any article or any of its containers or wrappers, or (2) accompanying such article." Section 303 makes the violation of any of the provisions of 301 a crime.

In this case, the drugs and the literature had a common origin and a common destination. The literature was used in the sale of the drugs. It explained their uses. Nowhere else was the purchaser advised how to use them. It constituted an essential supplement to the label attached to the package. Thus the products and the literature were interdependent, as the Court of Appeals observed.

It would take an extremely narrow reading of the Act to hold that these drugs were not misbranded. A criminal law is not to be read expansively to include what is not plainly embraced within the language of the statute . . . since the purpose fairly to apprise men of the boundaries of the prohibited action would then be defeated. . . . But there is no canon against using common sense in reading a criminal law, so that strained and technical constructions do not defeat its purpose by creating exceptions from or loopholes in it. . . .

It would, indeed, create an obviously wide loophole to hold that these drugs would be misbranded if the literature had been shipped in the same container, but not misbranded if the literature left in the next or in the preceding mail. The high purpose of the Act to protect consumers who under present conditions are largely unable to protect themselves in this field would then be easily defeated. The administrative agency charged with its enforcement has not given the Act any such restricted construction. The textual structure of the Act is not agreeable to it. Accordingly, we conclude that the phrase "accompanying such article" is not restricted to labels that are on or in the article on package that is transported.

The first clause of 201(m)—all labels "upon any article or any of its containers or wrappers"—clearly embraces advertising or descriptive matter that goes with the package in which the articles are transported. The second clause— "accompanying such article"—has no specific reference to packages, containers or their contents as did a predecessor statute. . . . It plainly includes what is contained within the package whether or not it is "upon" the article or its wrapper or container. But the second clause does not say "accompanying such article in the package or container," and we see no reason for reading the additional words into the text.

One article or thing is accompanied by another when it supplements or explains it, in the manner that a committee report of the Congress accompanies a bill. No physical attachment of one to the other is necessary. It is the textual relationship that is significant. The analogy to the present case is obvious. We need not labor the point.

The false and misleading literature in the present case was designed for use in the distribution and sale of the drug, and it was so used. The fact that it went in a different mail was wholly irrelevant whether we judge the transaction by purpose or result. . . .

Petitioner points out that in the evolution of the Act, the ban on false advertising was eliminated, the control over it being transferred to the Federal Trade Commission. . . . We have searched the legislative history in vain, however, to find any indication that Congress had the purpose to eliminate from the Act advertising which performs the function of labeling. Every labeling is in a sense an advertisement. The advertising which we have here performs the same function as it would if it were on the article or on the containers or wrappers. As we have said, physical attachment or contiguity is unnecessary under 201(m)(2) . . .

We have considered the other objections tendered by petitioner and find them without merit.

Affirmed.

* * * * *

The *Kordel* court's interpretation of labeling is considered an expansive one because the items were shipped at different times. Any doubt about the Court's intent was eliminated by *United States v. Urbuteit*,[2] which found that pamphlets shipped separately from medical devices were interrelated enough to be considered labeling: "The problem is a practical one of consumer protection, not dialectics. The fact that the literature leaves in a separate mail does not save the article from being misbranded."[3]

Note, however, that the *Kordel* defendant held responsibility for shipping both the products and the literature, and the literature was clearly a tool for marketing the products. Thus there was a clear connection between the literature and the products even if not shipped together.

When the activities are not integrated, the courts are less likely to find that literature is labeling. For example, in *United States v. 24 Bottles "Sterling Vinegar & Honey,"* 338 F.2d 157 (2nd Circ. 1964), the court found that no inference that books touting the health benefits of vinegar and honey were sold for the purpose of increasing the sales of Sterling Vinegar & Honey. For example, the books had been sold for two years prior to production of the Sterling product.

> . . . The distinguishing characteristic of a label is that, in some manner or another, it is presented to the customer in immediate connection with his view and his purchase of the product. Such a connection existed at both wholesale and retail levels in *Kordel*: Although the pamphlets and drugs were mailed to retailers separately, they were mailed in "integrated transactions"; the vendors in turn gave the pamphlets away with the sale of the drugs in some cases. . . .
>
> "Folk Medicine" was a bestselling book which Balanced Foods and health food shops could be expected to carry without regard to Vinegar and Honey, as

[2] 335 U.S. 355 (1948).
[3] *Id.*

they did prior to introduction of the latter product. The book made broad claims for a vinegar and honey mixture, which led ultimately to Sterling's marketing Vinegar and Honey. It is not disputed that these claims were misleading, but the Federal Food, Drug, and Cosmetic Act was not intended to deal generally with misleading claims; much more general proscriptions may be found in §§ 12–15 of the Federal Trade Commission Act, 15 U.S.C. §§ 52–55 (1958). In our view, the Food and Drug Act was intended to deal with such claims only when made in immediate connection with sale of the product. . . .

U.S. v. 24 Bottles "Sterling Vinegar & Honey" at 159–160.

It should also be noted that since 1982 the FDA's policy has recommended against the seizure of labeling when it is in the form of books. The agency has instead recommended the collection of an official sample of the book as evidence that the product is violative.[4]

The FDA's policy recognizes that certain First Amendment free speech protections apply to commercial speech. In particular, the Supreme Court has established that free speech protections generally prohibit prior restraint of speech. A prior restraint exists where the dissemination of speech is restricted or prohibited before its violative nature has been judicially determined. Accordingly, FDA's policy is to seek a court injunction before seizing books.

3.3.3   Labeling versus Food Advertising

The definition of "labeling" is broad enough that it clearly includes some items that would normally be considered as advertising. Brochures, flyers, and booklets that accompany or are associated with a food may fall under the scope of "labeling."

Before 1938 no federal agency was directly charged with the regulation of food advertising. The federal regulation of advertising began with passage of the Federal Trade Commission Act in 1914, which created the Federal Trade Commission (FTC), but false advertising was only prohibited when there was evidence of injury to a *competitor*.

In the years leading to passage of the FD&C Act, Congress debated which agency should have jurisdiction over food advertising. The issue was decided with the passage of the **Wheeler-Lea Amendment of 1938**,[5] which designated the FTC as the agency to regulate the advertising of food. The Wheeler-Lea Amendment amended section 5 of the FTC Act and empowered the FTC to act against unfair or deceptive acts or practices if there was evidence of injury to the public. Proof of injury to competition was no longer necessary. At the

---

[4]FDA, Compliance Policy Guide No. 7153.13 (Dec. 1, 1982). The revised version of this compliance policy guide is available at: http://www.fda.gov/ora/compliance_ref/cpg/cpggenl/cpg140-100.html.

[5]52 Stat. 111, 114 (1938), later incorporated into 15 U.S.C. 52658.

time delegation of authority over advertising to FTC, rather than FDA, was considered a victory for the regulated industries.[6]

Nonetheless, FDA also has authority over advertising when it is also "labeling." This creates overlapping authority on some advertising. The labeling requirements tend to be more proscriptive than the advertising requirements. Some statement permitted in advertising may be prohibited on labeling. This situation has resulted in considerable attention to the meaning and limits of term "labeling."

3.3.4 The Internet and Labeling

* * * * *

FDA Letter on Labeling Food Products

Presented or Available on the Internet

U.S. Food and Drug Administration
Center for Food Safety and Applied Nutrition
November 1, 2001

[To: Washington Legal Foundation]
This letter responds to your citizen petition . . . Your petition asked FDA to "formally adopt a rule, policy, or guidance stating that information presented or available on a company's Internet website, including hyperlinks to other third party sites, does not constitute 'labeling,'" as defined by the Federal Food, Drug, and Cosmetic Act (FD&C Act) at 21 U.S.C. § 321(m). In your petition, you further requested that the rule, policy, or guidance specify that such information may, but does not necessarily, constitute advertising. Alternatively, you asked FDA to adopt a rule, policy, or guidance "exempting Internet information of food companies from labeling requirements."

. . . FDA, however, disagrees that information presented or available on a company's website could never constitute labeling. "Labeling" is defined in section 201(m) of the FD&C Act (21 U.S.C. § 321(m)) as "all labels and other written, printed or graphic matter upon any article . . . or accompanying such article." In *Kordel v. United States*, 335 U.S. 345 (1948), the Supreme Court concluded that the phrase "accompanying such article" included literature that was shipped separately and at different times from the drugs with which they were associated. "One article or thing is accompanied by another when it supplements or explains it, in the manner that a committee report of the Congress accompanies a bill. No physical attachment one to the other is necessary. It is the textual relationship that is significant." *Id*. at 350. The Court

---

[6] Peter Barton Hutt & Richard A. Merrill, Food And Drug Law 43 (2d ed. 1991).

also noted that the literature and drugs were parts of an integrated distribution program.

Based on this authority, FDA and the courts have interpreted "labeling" to include "[b] rochures, booklets, . . . motion picture films, film strips, . . . sound recordings, . . . and similar pieces of printed, audio, or visual matter descriptive of a drug . . . which are disseminated by or on behalf of its manufacturer, packer, or distributor." . . .

Lower court cases after *Kordel* reinforce a broad reading of the term "acc ompanying." . . . In addition, the courts have considered whether the information and the product are part of an integrated distribution program, where, for example, the information and the product originate from the same source or the information is designed to promote the distribution and sale of the product, even if such sale is not immediate. . . .

Accordingly, FDA believes that, in certain circumstances, information about FDA-regulated products that is disseminated over the Internet by, or on behalf of, a regulated company can meet the definition of labeling in section 201(m) of the FD&C Act. For example, if a company were to promote a regulated product on its website, and allow consumers to purchase the product directly from the website, the website is likely to be "labeling." The website, in that case, would be written, printed, or graphic matter that supplements or explains the product and is designed for use in the distribution and sale of the product.

To provide an example from the other end of the spectrum, some product-specific promotion presented on non-company websites that is very much similar, if not identical, to messages the agency has traditionally regulated as advertisements in print media (e.g., advertisements published in journals, magazines, periodicals, and newspapers) would be viewed as advertising. These are just examples at the extremes and, as discussed below, the agency will proceed on case-by-case basis in determining what is "labeling." . . .

FDA has explored developing a guidance on promotion of FDA-regulated products on the Internet, but has decided not to issue a document at this time. The agency believes that any rule or guidance on this issue would be quickly outdated due to the ongoing rapid changes in the Internet and its use. As a result, issuing a rule or guidance may stifle innovation and create greater confusion among industry and the public. For the time being, FDA will continue to use a case-by-case approach based on the specific facts of each case. . . .

FDA appreciates your interest in this area.

Sincerely yours,

Margaret M. Dotzel

Associate Commissioner for Policy

* * * * *

QUESTIONS AND NOTES

3.1. In the definition of "labeling," what does "accompanying" mean?

3.2. For a illustrated overview of the food labeling terms and basic require-ments, see FDA's A FOOD LABELING GUIDE available at: www.cfsan.fda.gov/~dms/2lg-toc.html (last accessed Aug. 21, 2008).

3.4 AFFIRMATIVE LABEL REQUIREMENTS

All required label information must appear on the food label in the English language. With a few exceptions, if the label of a food bears representations in a foreign language, the label must also bear all of the required statements in the foreign language, as well as in English.[7] In addition the required label information must be conspicuously displayed and in terms that the ordinary consumer is likely to read and understand under ordinary conditions of purchase and use.[8]

3.4.1 Principle Display Panel (PDP)

The PDP is the portion of the package that is most likely to be displayed, presented, shown, or examined under customary conditions of display and purchase. Some containers are designed with two or more different surfaces suitable for the principal display panel; these are known as alternate principal display panels.[9] The statement of identity (product name) and net quantity (metric and inch-pound units) are required to be on the PDP.[10] All other required information must be on the PDP or the information panel.

3.4.2 Information Panel

The information panel is generally the area contiguous to and immediately to the right of the PDP. The following information must be placed on the infor-mation panel, unless placed on the PDP[11]:

- Ingredients list
- Nutrition labeling
- Responsible party—name and address of the manufacturer, packer, or distributor

[7] 21 C.F.R. § 101.15(c)(2). The Tariff Act of 1930, 19 U.S.C. § 1304, also requires all imported arti-cles to be marked with the English name of the country of origin.
[8] FD&C Act § 403(f).
[9] 21 C.F.R. § 101.1.
[10] 21 C.F.R. §§ 101.3(a) and 101.105(a).
[11] 21 C.F.R. § 101.2(b) and (d).

3.4.3   Statement of Identity

Name of the Food   The statement of identity (name of the food) must appear on the principal display panel.[12]

Prominence   The underlying requirement regarding prominence is that all mandatory information must be printed and arranged with prominence and conspicuousness, rendering it likely to be read and understood by the average consumer.[13]

    The name of a food must appear on the PDP in bold print or type. The type size must be reasonably related to the most prominent printed matter on the front panel, and should be one of the most important features on the principal display panel (generally, this is interpreted to be at least one-half the size of the largest print on the label).[14] The name of the food also must be in lines generally parallel to the base of the package as it is displayed.[15]

Common or Usual Name   The common or usual name of the food, if the food has one, should be used as the statement of identity. If there is none should be used, then an appropriate descriptive name should be used that is not misleading.[16]

Standardized Foods   If there is a standard for the food, the complete name designated in the standard must be used (section 403(g) and 21 C.F.R. § 101.3). When a standard of identity exists, that food must bear the name prescribed by the standard. The name prescribed consists of the common or usual name of the food plus any additional terms required to be declared. For example, the common or usual name of sweet corn is "corn," "sweet corn," or "sugar corn." The standard also requires that the name declare the style (whole kernel or cream style), the color type (if white), and the words "vacuum pack" or "vacuum packed" (if they meet that criteria). Therefore "Sweet Corn" is not a complete identification, whereas "Whole Kernel Sweet Corn" or "Whole Kernel Corn" is adequate among the prescribed variations. If not declared, the color must be yellow (declaration as "yellow" or "golden" is optional).[17]

Undefined Foods   When no standard of identity exists for a food, the product must be identified by its common or usual name, or in the absence of a common or usual name, by an appropriately descriptive phrase. The descriptive phrase must accurately identify or describe, in as simple and direct terms as possible, the basic nature of the food or its characterizing ingredients

---

[12] 21 C.F.R. § 101.3.
[13] FD&C Act § 403(f).
[14] 21 C.F.R. § 101.3(d).
[15] 21 C.F.R. § 101.3(d).
[16] 21 C.F.R. § 101.3(b).
[17] 21 C.F.R. § 155.13.

or properties (e.g., "Chocolate-Flavored Caramel Corn" but not "Praline Cruncher").[18]

If the name of a food mentions ingredients, generally they must be listed in order of descending predominance. For example, "Apple-Strawberry Pie" would be correct if apples predominate over strawberries.

Forms of a Food   When a food is offered in various forms (whole, sliced, diced), the particular form is required as part of the statement of identity unless the form is visible through the container or is depicted by an appropriate vignette.[19]

Fanciful Names   If the nature of the food is obvious, a fanciful name commonly understood and used by the public for that food may also be used.[20] For example, "Submarine Sandwich" may be used (as the identification of a large sandwich made with a small loaf of bread and containing lettuce, condiments, and a variety of meats and cheeses). "B52 Belly Bomber" would be likely considered insufficient because the name is not commonly used or understood by the public. Fanciful names, of course, if they are not misleading, may be used *in addition* to the required statement of identity.

Similarly a brand name may serve as the statement of identity if the name is commonly used and understood by the public to refer to a specific food, for example, Coca Cola and Pepsi Cola.

In the following case the appellate court upheld the invalidation of a USDA regulation that permitted "all meat" sausage to contain up to 15 percent water and other ingredients. Although the court stated that a regulation authorizing a false or misleading label would have to be invalidated as not in accordance with the law, note that the court found the regulation invalid because USDA had not provided a reasonable basis for calling 85 percent meat sausage "all meat."

* * * * *

Federation of Homemakers v. Butz

466 F.2d 462 (D.C. Cir. 1972)

Mr. Justice Clark of the Supreme Court of the United States, and Leventhal and Robb, Circuit Judges
Opinion: Robb, Circuit Judge
The appellee, Federation of Homemakers, brought this action in the district court to challenge a regulation ... prescribing the labeling to be employed on certain sausage products, permits frankfurters to be labeled "All Meat,"

---

[18] 21 C.F.R. § 101.3(b).
[19] 21 C.F.R. § 101.3(c).
[20] 21 C.F.R. § 101.3(b)(3).

"All Beef," "All Pork," or "All [species]"as the case may be, although they contain, in addition to meat, 10 percent water and 5 percent other ingredients, including corn syrup, spice flavoring, and curing additives. At the same time the regulation prohibits the use of the "All Meat" label on frankfurters containing binders and extenders, such as dried milk, cereal, or meat by-products aggregating not more than $3\frac{1}{2}$ percent of the ingredients of the frankfurters. . . .

For purposes of this case the relevant parts of the regulation can be summarized as follows: Sausage products labeled "All Meat" may contain, in addition to meat, added water, corn syrup, salt, spices, and curing agents in designated quantities. The non-meat ingredients in "All Meat" sausages constitute approximately 15 percent of the finished product. Frankfurters which cannot be labeled "All Meat" differ from the "All Meat" variety in that they contain binders and extenders such as dried milk, cereal, or meat by-products. These added ingredients cannot constitute in the aggregate more than $3\frac{1}{2}$ percent of the total ingredients of the frankfurters. Thus, the only difference between "All Meat" frankfurters and other frankfurters is the existence of up to $3\frac{1}{2}$ percent binders and extenders in the latter; in all other respects the two products are subject to identical standards of composition under the applicable regulations.

The question presented here is whether the label "All Meat," applied to a product containing 85 percent meat, and employed to distinguish such products from those containing $3\frac{1}{2}$ percent binders and extenders and $81\frac{1}{2}$ percent meat, is false or misleading under 21 U.S.C. § 607(d), which provides that:

"No article subject to this subchapter shall be sold or offered for sale by any person, firm, or corporation, in commerce, under any name or other marking or labeling which is false or misleading, or in any container of a misleading form or size, but established trade names and other marking and labeling and containers which are not false or misleading and which are approved by the Secretary are permitted."

If the "All Meat" label is false or misleading, the challenged regulation must be invalidated, for the Secretary's action in promulgating such a regulation would be in excess of his authority and "arbitrary, capricious, an abuse of discretion, or otherwise not in accordance with law." . . .

It is indisputable that the label "All Meat" as employed in this case is inaccurate. The words used are clear and unequivocal, and they import a description which cannot be attached to a product which is "Part Meat" or "All Meat, Water, Condiments, and Curing Agents." The fact is that frankfurters labeled "All Meat" are simply not all meat. . . .

We are thus confronted with the question whether there is a rational basis for the distinction in the labels that may be applied to the two types of frankfurters. If a frankfurter containing 85 percent meat may be labeled "All Meat," then why must a frankfurter containing $81\frac{1}{2}$ percent meat be denied that label? . . . We think it plain from this that "All Meat" frankfurters are preferred by consumers. The "All Meat" label is therefore an indication that a frankfurter bearing it occupies a preferred status, or is at least considered to be in some way superior to a frankfurter not so labeled. . . .

Do the words "All Meat" mean to an ordinary consumer, as distinguished from an expert, that a frankfurter in a package on which these words appear contains 85 percent meat and other components, and not $81\frac{1}{2}$ percent meat and other components? We think the answer to the question is plain, that the words do not convey that meaning and distinction, and that the Secretary could not reasonably conclude that they do. As employed, therefore, the "All Meat" label is misleading and deceptive. . . . the common meaning of the words is clear and unequivocal. . . .

The district court ordered the Secretary to discontinue the use of the "All Meat" label within six months. . . . We agree with this result but we think that in the interim the Secretary should develop, prescribe, and submit to the district court revised labels that accurately and without deception distinguish the different types of frankfurters from each other and from competitive meats. . . . As so modified the judgment is Affirmed.

* * * * *

QUESTIONS AND NOTES

3.3. Did the *Federation of Homemakers'* Court determine whether the USDA's regulation was beneficial? What standard did the court apply in reviewing the regulation?

3.4. The *Federation of Homemakers'* Court found no rational basis for USDA's "All Meat" regulation. What actions can a federal agency take to prevent similar court rulings?

3.5. Note: The USDA subsequently differentiated between added water and nonadded water by regulation.[21]

Artificially Flavored When artificial flavorings are used that simulate, resemble, or reinforce the characterizing flavor of the food, the product name must be accompanied by the phrase "artificially flavored" or "artificial" in type not less than one-half the size of the name of the food; for example, "Artificial Orange Flavored Punch" or "Artificially Flavored Strawberry Cheesecake."[22]

Imitation A food that is an imitation of another food must be labeled, in type of uniform size and prominence, with the word "imitation" immediately followed by the name of the food imitated. Any product that resembles and substitutes for a traditional food and contains less nutritional value than the

---

[21] 55 Fed. Reg. 7294 (Mar. 1, 1990).
[22] For more detailed information, refer to 21 C.F.R. § 101.3(d) and 21 C.F.R. § 101.22.

traditional food is considered an imitation.[23] For example, a new food that resembles a traditional food and is a substitute for the traditional food must be labeled as an imitation if the new food contains less protein or a lesser amount of any essential vitamin or mineral.

Beverages Containing Juice   Beverages that claim to contain juice must declare the total percentage of juice on the information panel. In addition FDA regulations set detailed criteria for naming juice beverages. For example, when the label of a multi-juice beverage states one or more—but not all—of the juices present, and the predominantly named juice is present in minor amounts, the product's name must (1) state that the beverage is flavored with that juice, or (2) declare the amount of the juice in a 5 percent range—for example, "raspberry-flavored juice blend" or "juice blend, 2 to 7 percent raspberry juice."

* * * * *

When Is Peach Juice Apple Juice?

Marian Segal, FDA Consumer, Special Issue, Focus on Food Labeling, also available at: http://www.fda.gov/fdac/special/foodlabel/ingred.html (May 1993).

When it comes to juice labeling, there are those who would disagree with Shakespeare's sentiment that "a rose by any other name would smell as sweet." If the label implies that it's peach juice, they contend, it shouldn't consist mostly of apple and white grape juice—especially without saying so on the label.

The final rule on percentage juice declaration published in the Jan. 6, 1993, Federal Register will help remedy this problem. Beginning May 8, 1993, juice manufacturers will have to declare the total amount of juice in a beverage. . . .

The rule-making process on declaration of percentages of juice goes back many years, beginning with debates over standards of identity for diluted juice beverages. In 1974, FDA proposed a regulation to establish common or usual names for juice drinks instead of developing standards.

After many objections, tie-ups, and reworkings—including a final regulation in 1980 that never had an effective date, and two more proposals in 1984 and 1987—the Nutrition Labeling and Education Act came along in 1990 requiring that "a food that purports to be a beverage containing juice must declare the percent of total juice on the information panel."

But this alone would not solve the problem of misleading labels. Many manufacturers today use bland juices, like apple or white grape, as diluents instead of water, and call the product a 100 percent juice blend.

[23] 21 C.F.R. § 101.3(e)(1).

"Some of these labels are just not informative," Campbell says. "The label says 100 percent juice blend or 100 percent natural juices, but only the expensive juices—the raspberry or strawberry, which are in smaller amounts—appear prominently on the principal display panel. You have to look for the grape and the apple in the fine print."

To correct this, the FDA elaborated on the 1990 law, proposing that manufacturers be required to declare not only the total percent of juice, but the percent of each juice named or pictured on the label of a multi-juice beverage.

In responding to the proposal, however, manufacturers protested that this requirement would be impractical and difficult to comply with. They explained that juice, as an agricultural product, varies in strength, flavor, solids, and color. If they were required to state a percentage, they wouldn't have the flexibility necessary to adjust the amount of juice—using a little bit less or a little bit more or a little sweetening—to get the desired flavor. Nor would they be able to vary their formulas as driven by fluctuations in cost or availability of individual juices.

In addition, they said the amount of juice they use in their formulations is proprietary information, and requiring them to reveal this information in 1 percent increments would force them to divulge their secret formulas.

The final rule allowing a statement that the beverage is flavored, or declaring the amount of juice named in a 5 percent range, addresses manufacturers' concerns, while providing more accurate information for consumers.

* * * * *

3.4.4 Net Quantity

Net quantity is the requirement for an accurate statement of the net amount of the contents of food in a package. The net quantity statement helps customers in two ways: it allows them to know how much food is in a container, and it aids in price comparison. "Net" refers to the quantity of edible food in a package or container. Therefore net content excludes any liquid or juice in which the food may be packed, unless the liquid is usually consumed as part of the food. Net also excludes the weight of the container or wrappers.

The Federal Food, Drug, and Cosmetic Act and the Fair Packaging and Labeling Act require that a food, in package form, bear a label with an accurate statement of the quantity of the contents in terms of weight, measure, or numerical count. Regulations interpreting these statutory requirements require that the statement appear on the principal display panel in terms of the customary inch-pound system of measure.[24]

---

[24] 21 C.F.R. § 101.105.

The statement must appear in lines generally parallel to the base of the package when displayed for sale. If the area of the principal display panel of the package is larger than five square inches, the statement must appear within the lower 30 percent of the label panel. Also, with certain limited exceptions, the statement must appear in conspicuous and easily legible boldface print or type in distinct contrast to other matter on the package. Further the statement must meet the minimum type size set in 21 C.F.R. § 101.105.

Metric   The Fair Packaging and Labeling Act was amended by Public Law 102–329 to require that labels printed on or after February 14, 1994, bear a statement of the quantity of the contents in terms of the SI metric system as well as in terms of the customary inch-pound system of measure. Because the FPLA pertains only to consumer commodities, metric statements of quantity are not required where products are not marketed to consumers.[25]

The FPLA requires both metric and inch-pound units in the net contents statement on packages regulated by the act (with a few exceptions).[26] The most important exceptions apply mostly to retail establishments, specifically:

> *random weight packages* (i.e., packages of varying weights), where each
> package's label is different, need not include a metric weight;[27] and
> *items packaged at a retail store* need not include metric measurements.[28]

The FDA proposed metric labeling regulations in 1993, but the proposal has never been finalized.[29] Therefore the metric labeling requirements of the FPLA were never incorporated into FDA's regulations. The result is that although foods are required to include a metric statement of contents, there are no details specified on how to format or place the metric measurement. Firms looking for guidance may want to review the details of the proposed regulations.[30]

Moisture Loss   Although the section 403 net weight labeling requirement of the FD&C Act goes back to the 1906 Act,[31] two difficult practical problems made implementation difficult. Packages can lose weight from the loss of moisture when dry products packed in a humid climate are stored in a dry

---

[25] The Fair Packaging and Labeling Act, *available at*: www.ftc.gov/os/statutes/fplajump.html.
[26] 15 U.S.C. § 1453(a)(2).
[27] 15 U.S.C. § 1453(a)(3)(A)(ii).
[28] 15 U.S.C. 1453(a)(6).
[29] Metric Labeling; Quantity of Contents Labeling Requirement for Foods, Human and Animal Drugs, Animal Foods, Cosmetics, and Medical Devices, 58 Fed. Reg. 67444 (Dec. 21, 1993) (Docket nos. 92N–0406 and 93N–0226).
[30] FDA withdrew the proposed metric regulations, but the withdrawn proposal still provides guidance and offers a sound position. 68 Fed. Reg. 19766 (Apr. 22, 2003).
[31] The Gould Amendment of 1913 to the 1906 Act, 37 Stat. 732 (1913).

climate. Additionally wet foods, such as meats, may lose liquid during storage and transportation. Arriving at reasonable allowable variations has been difficult. Both FDA and USDA have largely adopted the approach recommended by the National Conference on Weights and Measures (NCWM) and the National Institute of Standards and Technology (NIST, previously the National Bureau of Standards).[32]

3.4.5 Ingredient Labeling

Ingredient declaration is required on all foods that have more than one ingredient, even standardized foods. The ingredient statement allows consumers to identify foods that have ingredients they are allergic to or want to avoid for other reasons. The listing also helps consumers select foods with ingredients they want.

The ingredients in a food must be listed by their common or usual names in decreasing order of their predominance by weight. The word "ingredients" does not refer to the chemical composition but rather the individual food components of a mixed food. If a certain ingredient is the characterizing one in a food (e.g., shrimp in shrimp cocktail) the percent of that ingredient may be required as part of the name of the food.

Foods with two or more discrete components, such as cherry pie—which has filling and pie crust—may have a separate ingredient list for each of the components. Food additives and colors are required to be listed as ingredients, but the law exempts butter, cheese, and ice cream from having to show the use of color, with the exception of FD&C Yellow No. 5 whose presence must be declared on all foods. Spices, flavors, and colors may be listed generically, without naming the specific source, except that any artificial colors or flavors must be identified as artificial, and all certified colors must be named specifically.[33] Because people may be allergic to certain additives, the ingredient list must include, when appropriate,

1. all FDA-certified colors, such as FD&C Blue No. 1, named specifically;
2. sources of protein hydrolysates, which are used in many foods as flavors and flavor enhancers;[34] and
3. declaration of casein and caseinate as a milk derivative in the ingredient list of foods that claim to be nondairy, such as coffee whiteners.

* * * * *

[32] *See* NIST Handbook 133, Checking the Net Contents of Packaged Goods, *available at*: http://ts.nist.gov/WeightsAndMeasures/h1334-05.cfm (last accessed Feb. 29, 2008) (This publication includes procedures for testing packages labeled by weight, volume, measure, and count.)
[33] FD&C Act §§ 403(I) and 403(k).
[34] 21 C.F.R. § 102.22 on protein hydrolysates applies to FDA regulated foods. "The common or usual name of a protein hydrolysate shall be specific to the ingredient and shall include the identity

Ingredient Labeling: What's in a Food?

Adapted from Marian Segal, FDA CONSUMER, SPECIAL ISSUE, FOCUS ON FOOD LABELING, also available at: http://www.fda.gov/fdac/special/foodlabel/ingred.html (May 1993).

Mr. Doodle can call his hat whatever he likes. Pasta makers, however, have long had to be very specific about what they call "macaroni." That's because since shortly after the Federal Food, Drug, and Cosmetic Act was passed in 1938, macaroni, along with some other foods people commonly prepared at home in those days, was exempted from the law's requirement that food manufacturers list their products' ingredients on the food label. Instead, the new act provided for "standards of identity"—prescribed recipes—for these foods, which the manufacturers had to follow.

"The law resulted in standardized recipes for such foods as dairy products, mayonnaise, ketchup, jelly, and orange juice," says Elizabeth Campbell, director of the programs and enforcement policy division in the Office of Food Labeling of FDA's Center for Food Safety and Applied Nutrition. "When a consumer bought a jar of jelly she knew it would have at least 45 percent fruit, as the standard provided, because that's what it takes to make jelly," she explained. "It's roughly half fruit and half sugar. People knew that because they used to make it themselves."

Well, maybe so, but we're in the '90s now, and with the fast pace of today's lifestyles, homemade breads and jellies mostly exist in Grandma's memories. It can hardly be taken for granted that people still know what's in those standardized foods. And yet, more and more, health-conscious consumers and people with dietary restrictions want and need to know what's in the foods they buy.

So, the law [has changed] to catch up with the times. The FDA now requires that ingredients for all standardized foods be listed on the label, the same as for all other foods. . . . (The U.S. Department of Agriculture requires full ingredient labeling on all meat and poultry products, including standardized products, such as chili or sausages.)

Before passage of the NLEA, the Food, Drug, and Cosmetic Act did not require flavorings, colorings, or spices to be identified by their common or usual names. Instead, they could be declared collectively under the general terms "flavorings," "spices," or "colorings." Under the NLEA, however, color

---

of the food source from which the protein was derived. (a) 'Hydrolyzed wheat gluten,' 'hydrolyzed soy protein,' and 'autolyzed yeast extract' are examples of acceptable names. 'Hydrolyzed casein' is also an example of an acceptable name, whereas 'hydrolyzed milk protein' is not an acceptable name for this ingredient because it is not specific to the ingredient (hydrolysates can be prepared from other milk proteins). The names 'hydrolyzed vegetable protein' and 'hydrolyzed protein' are not acceptable because they do not identify the food source of the protein. (b) [Reserved]." At this time USDA regulations still allow listing protein hydrolysate as "flavoring." For example, see 9 C.F.R. § 381.118 for the ingredients statement on poultry.

additives that FDA certifies for food use—FD&C colors Yellow No. 5, Red No. 40, Red No. 3, Yellow No. 6, Blue No. 1, Blue No. 2, and Green No. 3, and their lakes (specially formulated nonsoluble colors)—now must be declared on all foods except butter, cheese, and ice cream. Colors exempt from certification, such as caramel, paprika, and beet juice, do not have to be specifically identified; they can still be listed simply as "artificial colors."

People often look to the ingredient label for health reasons—perhaps to avoid substances they are allergic or sensitive to—or for religious or cultural reasons. . . .

Caseinate

If it says "nondairy," does it mean no milk? Many people are not aware that certain products claiming to be nondairy, such as some coffee whiteners, contain a milk derivative called caseinate, in this case used to whiten effectively.

"People expect there to be no milk ingredients in products marketed as dairy substitutes," Campbell says, "but some states require the label 'nondairy.' This issue is particularly important for people with milk allergies. The nondairy label may lead consumers to think that caseinates are not milk derived. Furthermore it guides people away from even checking the label for milk-derived ingredients."

Under the new rule, caseinate will have to be identified as a milk derivative in the ingredient statement when it's used in foods that claim to be nondairy. This requirement will help to flag it for casein-sensitive people.

Protein Hydrolysates

Hydrolyzed proteins (proteins broken down by acid or enzymes into amino acids) are added to foods to serve various functions. They can be used as leavening agents, stabilizers (to impart body or improve consistency, for example), thickeners, flavorings, flavor enhancers, and as a nutrient (protein source), to name a few uses.

Since the law does not require flavors to be identified by their common or usual names, some in industry have made a practice of declaring protein hydrolysates as "flavorings" or "natural flavors" even when they are used as flavor enhancers—a use not exempt from declaration. After reviewing the data, FDA concluded that protein hydrolysates added to foods as flavorings always function as flavor enhancers as well and, as such, must be declared by their common or usual name.

The source of protein in hydrolysates used for flavor-related purposes also must be identified. Previously the general terms "hydrolyzed vegetable protein," "hydrolyzed animal protein," or simply "hydrolyzed protein" were permitted, but the new regulation requires identification of the specific protein source, such as "hydrolyzed corn protein" or "hydrolyzed casein." There are two reasons for this.

First, the law requires that the common or usual name of a food should adequately describe its basic nature or characterizing properties or ingredients. FDA reasoned that the more general terms "animal" and "vegetable" don't meet this requirement because protein hydrolysates from different sources best serve different functions. Manufacturers select protein hydrolysates from specific sources depending on how they will be used in a product. Hydrolyzed casein is generally used in canned tuna, for example, whereas hydrolyzed wheat protein is used in meat flavors.

Second, the source of the additive is particularly important to consumers who have special dietary requirements, whether for religious, cultural, or health reasons. If hydrolyzed casein is added to canned tuna, for example, it must be identified as such, rather than simply as "hydrolyzed protein" or "hydrolyzed milk protein."

Furthermore, after reviewing comments on the June 1991 proposal, the agency concluded that to minimize confusion, the source of protein in hydrolysates used for non–flavor-related purposes should also be identified. Thus, the source of all protein hydrolysates—regardless of use—will now have to be identified.

Other final provisions of the new rule will:

Permit voluntary inclusion of the food source in the names of sweeteners. For example, "corn sugar monohydrate" would be permitted in addition to names previously permitted, such as "dextrose" or "dextrose monohydrate."

Provide a uniform format for voluntary declaration of percentage ingredient information. Manufacturers who choose to declare ingredients by percent of content would present them by weight rather than volume to avoid inconsistent calculations. Firms may use percentage declarations for as many or as few ingredients as they choose, as long as the information is not misleading. Manufacturers must still list ingredients in descending order, by weight, as required by law.

Require label declaration of sulfiting agents in standardized foods. This is required because some people are sensitive to these preservatives. FDA has required listing of sulfiting agents in nonstandardized foods since 1986.

* * * * *

3.4.6 Name and Address of the Responsible Party

The labeling of a responsible party is required mostly so that consumers have a point of contact if they find something wrong with the product. The name, street address, city, state, and zIP code of the manufacturer, packer, or distributor is required. The street address may be omitted by a firm listed in a current city or telephone directory. A firm whose address is outside the United States may omit the zIP code.

If the food is not manufactured by the person or company whose name appears on the label, the name must be qualified by "manufactured for," "distributed by," or a similar expression.

3.4.7 Product Dates and Codes

Consumers can use the dates that are given on food packaging if the manufacturer is using "open dating." On the other hand, consumers cannot use "code dating."

In open dating, dates are stated alphanumerically, such as "Oct. 15," or numerically, such as "10–15" or "1,015." In code dating, the information is coded in letters, numbers, and symbols so that usually only the manufacturer can translate it.

Some dates for which open dating is used are as follows:

Pull Date: This is the last day that the manufacturer recommends that the product remain for sale. This date takes into consideration additional time for storage and use at home, so if the food is bought on the pull date, it still can be eaten at a later date. How long the product should be offered for sale and how much home storage is allowed are determined by the manufacturer, based on knowledge of the product and the product's shelf life.

Quality Assurance or Freshness Date: This date shows how long the manufacturer thinks a food will be of optimal quality. On the label, it may appear like this: "Best if used by October 1996." This doesn't mean, however, that the product shouldn't be used after the suggested date.

Pack Date: This is the date the food was packaged or processed. It may enable consumers to determine how old a product is.

Expiration Date: This is the last day on which a product should be eaten. State governments regulate these dates for perishable items, such as milk and eggs. FDA regulates only the expiration dates of infant formula.

A common type of code dating is the product code. This code enables the manufacturer to convey a relatively large amount of information with a few small letters, numbers, and symbols. It tells when and where a product was packaged. In the case of a recall, this makes it easier to quickly identify and track down the product and take it off the market. FDA encourages manufacturers to put product codes on packaging, especially for products with a long shelf life.

3.5 MISBRANDED FOOD: PROHIBITED REPRESENTATIONS

3.5.1 Section 403 on Misbranded Food

Section 403(a) of the FD&C Act prohibits statements in labels or labeling that are "false or misleading in any particular." Failure to reveal "material facts" about a food product can be misleading or can also be a violation under section 201(n) of the Act. Under section 403(a), (343) a food will be deemed misbranded if its labeling is "false or misleading in any particular." Additionally a product will be considered misbranded if:

- offered for sale under the name of another food product;
- it is an imitation of another food (unless clearly labeled as an imitation); or
- if the container is misleading in any particular such as in size, fill or form.

* * * * *

Federal Food, Drug, and Cosmetic Act

Misbranded Food

SEC. 403. [343] A food shall be deemed to be **misbranded**—

(a) If (1) its labeling is *false or misleading in any particular*, or . . .

(b) If it is offered for sale under the name of another food.

(c) If it is an imitation of another food, unless its label bears, in type of uniform size and prominence, the word "imitation" and, immediately thereafter, the name of the food imitated.

(d) If its container is so made, formed, or filled as to be misleading.

(e) If in package form **unless it bears a label containing** (1) the name and place of business of the manufacturer, packer, or distributor; and (2) an accurate statement of the quantity of the contents in terms of weight, measure, or numerical count: Provided, That under clause (2) of this paragraph reasonable variations shall be permitted, and exemptions as to small packages shall be established, by regulations prescribed by the Secretary.

(f) If any word, statement, or other **information required** by or under authority of this Act to appear on the label or labeling is not **prominently placed** thereon with such conspicuousness (as compared with other words, statements, designs, or devices, in the labeling) and in such terms as to render it likely to be read and understood by the ordinary individual under customary conditions of purchase and use.

(g) If it purports to be or is represented as a food for which a definition and **standard of identity** has been prescribed by regulations as provided by section 401, unless (1) it conforms to such definition and standard, and (2) its label bears the name of the food specified in the definition and standard, and, insofar as may be required by such regulations, the common names of optional ingredients (other than spices, flavoring, and coloring) present in such food.

(h) If it purports to be or is represented as—

   (1) a food for which a standard of quality has been prescribed by regulations as provided by section 401, and its quality falls below such standard, unless its label bears, in such manner and form as such regulations specify, a statement that it falls below such standard; or

   (2) a food for which a standard or standards of fill of container have been prescribed by regulations as provided by section 401, and it falls below the standard of fill of container applicable thereto, unless its label bears, in such manner and form as such regulations specify, a statement that it falls below such standard; . . .

(i) Unless its label bears (1) the common or usual name of the food, if any there be, and (2) in case it is fabricated from two or more **ingredients**, the common or usual name of each such ingredient; except that spices, flavorings, and colorings, other than those sold as such, may be designated as spices, flavorings, and colorings without naming each: To the extent that compliance with the requirements of clause (2) of this paragraph is impracticable, or results in deception or unfair competition, exemptions shall be established by regulations promulgated by the Secretary.

(j) If it purports to be or is represented for special dietary uses, unless its label bears such information concerning its vitamin, mineral, and other dietary properties as the Secretary determines to be, and by regulations prescribes, as necessary in order fully to inform purchasers as to its value for such uses.

(k) If it bears or contains any **artificial flavoring, artificial coloring, or chemical preservative**, unless it bears labeling stating that fact: except that, to the extent that compliance with the requirements of this paragraph is impracticable, exemptions shall be established by regulations promulgated by the Secretary. The provisions of this paragraph and paragraphs (g) and (i) with respect to artificial coloring shall not apply in the case of butter, cheese, or ice cream. . . .

* * * * *

DISCUSSION QUESTION

3.6. Section 403(a)(1) of the Federal Food, Drug, and Cosmetic Act (FD&C Act) deems a food misbranded if the labeling is false or misleading "in any particular." What does "in any particular" mean?

As the case below demonstrates, the courts have upheld a strict standard for misleading labels. Note that statements may be technically accurate but still mislead.

* * * * *

United States v. 95 Barrels of Alleged Apple Cider

265 U.S. 438 (1924)[35]

Mr. JUSTICE BUTLER delivered the opinion of the court: This case arises under Food and Drugs Act June 30, 1906 . . . . The United States filed information . . . for the condemnation of 95 barrels of vinegar. Every barrel seized was labeled:

"Douglas Packing Company Excelsior Brand **Apple Cider Vinegar Made from Selected Apples** Reduced to 4 Percentum Rochester, N.Y."

The information alleged that the . . . vinegar was made from dried or evaporated apples, and was misbranded in violation of section 8, in that the statements on the label were false and misleading, and in that it was an imitation of and offered for sale under the distinctive name of another article, namely, apple cider vinegar. . . .

The question for decision is whether the vinegar was misbranded. The substance of the agreed statement of facts may be set forth briefly. Claimant is engaged in the manufacture of food products from evaporated and unevaporated apples. During the apple season, from about September 25 to December 15, it makes apple cider and apple cider vinegar from fresh or unevaporated apples. During the balance of the year, it makes products which it designates as "apple cider" and "apple cider vinegar" from evaporated apples. The most approved process for dehydrating apples is used, and, in applying it, small quantities of sulphur fumes are employed to prevent rot, fermentation, and consequent discoloration. The principal result of dehydration is the removal of about 80 percent of the water. Whether, and to what extent, any other constituents of the apple are removed is not beyond controversy; in the present state of chemical science, no accepted test or method of analysis is provided for the making of such determination. Only mature fruit, free from rot and ferment, can be used economically and advantageously.

In manufacturing, claimant places in a receptacle a quantity of evaporated apples to which an amount of pure water substantially equivalent to that removed in the evaporating process has been added. A heavy weight is placed on top of the apples and a stream of water is introduced at the top of the receptacle through a pipe and is applied until the liquid, released through a vent at the bottom, has carried off in solution such of the constituents of the evaporated apples as are soluble in cold water and useful in the manufacture of vinegar. Such liquid, which is substantially equivalent in quantity to that which would have been obtained had unevaporated apples been used, carries a small and entirely harmless quantity of sulphur dioxide, which is removed during the process of fining and filtration by the addition of barium carbonate

---

[35] This case predates the FD&C Act (1938), but the standard was the same under the Pure Food and Drug Act of 1906.

or some other proper chemical agent. The liquid is then subjected to alcoholic and subsequent acetic fermentation in the same manner as that followed by the manufacturer of apple cider vinegar made from the liquid content of unevaporated apples. Claimant employs the same receptacles, equipment, and process of manufacturing for evaporated as for unevaporated apples, except that, in the case of evaporated apples, pure water is added as above described, and in the process of fining and filtration an additional chemical is used to precipitate any sulphur compounds present and resulting from dehydration.

The resulting liquid, upon chemical analysis, gives results similar to those obtained from an analysis of apple cider made from unevaporated apples, except that it contains a trace of barium incident to the process of manufacture. Vinegar so made is similar in taste and in composition to the vinegar made from unevaporated apples, except that the vinegar made from evaporated apples contains a trace of barium incident to the process of manufacture. There is no claim by libellant that this trace of barium renders it deleterious or injurious to health. It was conceded that the vinegar involved in these proceedings was vinegar made from dried or evaporated apples by substantially the process above described. There is no claim by the libellant that the vinegar was inferior to that made from fresh or unevaporated apples.

Since 1906, claimant has sold throughout the United States its product manufactured from unevaporated as well as from evaporated apples as "apple cider" and "apple cider vinegar," selling its vinegar under the brand above quoted, or under the brand "Sun Bright brand apple cider vinegar made from selected apples." Its output of vinegar is about 100,000 barrels a year. Before and since the passage of the Food and Drugs Act, vinegar in large quantities, and to a certain extent a beverage, made from evaporated apples, were sold in various parts of the United States as "apple cider vinegar" and "apple cider," respectively, by many manufacturers. Claimant, in manufacturing and selling such products so labeled, acted in good faith. The Department of Agriculture has never sanctioned this labeling, and its attitude with reference thereto is evidenced by the definition of "apple cider vinegar" set forth in Circulars 13, 17, 19, and 136, and Food Inspection Decision 140.1. It is stipulated that the juice of unevaporated apples when subjected to alcoholic and subsequent acetous fermentation is entitled to the name "apple cider vinegar."

Section 6 of the act provides that:

"... The term 'food,' as used herein, shall include all articles used for food, drink, confectionery, or condiment by man or other animals, whether simple, mixed, or compound."

Section 8 provides:

That the term 'misbranded,' as used herein, shall apply to all ... articles of food, or articles which enter into the composition of food, the package or label of which shall bear any statement, design, or device regarding such article, or the ingredients or substances contained therein which shall be false or misleading in any particular.... That for the purposes of this act an article shall also be deemed to be misbranded: . . . In the case of food: First. If it be an imitation of or offered for sale under the distinctive name of another article. Second. If it be labeled or branded so as to deceive or mislead the purchaser. . . . Fourth. If the package containing it or its label shall bear any statement, design, or device regarding the ingredients or the substances contained therein, which . . . shall be false or misleading in any particular. . . .

The statute is plain and direct. Its comprehensive terms condemn every statement, design, and device which may mislead or deceive. Deception may result from the use of statements not technically false or which may be literally true. The aim of the statute is to prevent that resulting from indirection and ambiguity, as well as from statements which are false. It is not difficult to choose statements, designs, and devices which will not deceive. Those which are ambiguous and liable to mislead should be read favorably to the accomplishment of the purpose of the act. The statute applies to food, and the ingredients and substances contained therein. It was enacted to enable purchasers to buy food for what it really is. . . .

The vinegar made from dried apples was not the same as that which would have been produced from the apples without dehydration. The dehydration took from them about 80 percent of the water content—an amount in excess of two-thirds of the total of their constituent elements. The substance removed was a part of their juice from which cider and vinegar would have been made if the apples had been used in their natural state. That element was not replaced. The substance extracted from dried apples is different from the pressed out juice of apples. Samples of cider fermented and unfermented made from fresh and evaporated apples, and vinegar made from both kinds of cider, were submitted to and examined by the District Judge who tried the case. He found that there were slight differences in appearance and taste, but that all had the appearance and taste of cider and vinegar. While the vinegar in question made from dried apples was like or similar to that which would have been produced by the use of fresh apples, it was not the identical product. The added water, constituting an element amounting to more than one-half of the total of all ingredients of the vinegar, never was a constituent element or part of the apples. The use of dried apples necessarily results in a different product.

If an article is not the identical thing that the brand indicates it to be, it is misbranded. The vinegar in question was not the identical thing that the statement, "Excelsior Brand Apple Cider Vinegar made from selected apples," indicated it to be. These words are to be considered in view of the admitted facts and others of which the court may take judicial notice. The words "Excelsior Brand," calculated to give the impression of superiority, may be put to one side at not liable to mislead. But the words, "apple cider vinegar made from selected apples" are misleading. Apple cider vinegar is made from apple cider. Cider is the expressed juice of apples and is so popularly and generally known. . . . It was stipulated that the juice of unevaporated apples when subjected to alcoholic and subsequent acetous fermentation is entitled to the name "apple cider vinegar." The vinegar in question was not the same as if made from apples without dehydration. The name "apple cider vinegar" included in the brand did not represent the article to be what it really was, and, in effect, did represent it to be what it was not-vinegar made from fresh or unevaporated apples. The words "made from selected apples" indicate that the apples used were chosen with special regard to their fitness for the purpose of making apple cider vinegar. They give no hint that the vinegar was made from dried apples, or that the larger part of the moisture content of the apples was eliminated and water substituted therefore. As used on the label, they aid the misrepresentation made by the words "apple cider vinegar."

The misrepresentation was in respect of the vinegar itself, and did not relate to the method of production merely. When considered independently of the product, the method of manufacture is not material. The act requires no disclosure concerning it. And it makes no difference whether vinegar made from dried apples is or is not inferior to apple cider vinegar.

The label was misleading as to the vinegar, its substance, and ingredients. The facts admitted sustain the charge of misbranding.

* * * * *

DISCUSSION QUESTIONS

3.7. Puffery. In the context of false and misleading labels, what is the meaning of the word "puffery"? What is the difference between opinion, puffery, and misleading statements?

3.8. Misleading to whom. Whose viewpoint determines what is false and misleading? Is it a single consumer, most consumers, an average consumer, a diligently skeptical consumer, or an average gullible consumer?

3.9. Actual injury. In the *Alleged Apple Cider* case, would it matter that no purchasers were misled or injured? Should this matter?

3.5.2 False or Misleading as a Matter of Law

In the 1960s the A. Freed Novelty company sold a variety of novelty items and gag gifts. One Freed Novelty's item was labeled "Liquor Flavored Lollypops," but they contained no liquor. The FDA contended that the product was misbranded under the FD&C Act because the labeling was false or misleading, as the name implied that the lollipops were flavored with real liquor whereas they were not.

Freed Novelty argued that their product was not a food under the meaning of the FD&C Act but rather a "novelty." The company also argued that their product's labeling—as a whole—was not false or misleading because the ingredient statement informed consumers that the lollipops contained no liquor. Freed Novelty also contended that the word "candy" on the label indicated that the lollipops contained no liquor.

The procedural posture of the case is important in understanding the opinion of the court. The court did *not* decide whether the lollipop labeling was misleading. This case was decided as a summary judgment. Therefore the only ruling by the judge was whether the case could be decided solely on the pleadings submitted by the parties, or whether the case must be ordered to a full trial.

* * * * *

United States v. 432 Cartons Individually Wrapped Candy Lollipops

292 F. Supp. 839 (1968)

MANSFIELD, District Judge

This is a libel for condemnation instituted under the Federal Food, Drug, and Cosmetic Act, 21 U.S.C.A. § 334(a), on the ground that the article of food seized was misbranded when introduced into interstate commerce. The complaint for forfeiture alleges that the labeling of the article is false or misleading and that therefore the food is misbranded under 21 U.S.C.A. § 343(a). . . .

The article of food in question consists of about 432 cartons each containing six lollipops. On the outside the carton is labeled on top "Candy * * * for one with Sophisticated Taste," on one side, "A. Freed Novelty, Inc., NYC," and on the other side, "Ingredients: Sugar, corn syrup, citric acid, natural and artificial flavors." The inside of the box contains the legend, "Liquor Flavored Lollypops," and the slogan, "Take Your Pick of a Liquor Stick." In addition the lollipops themselves are labeled, both in the box and on the cellophane in which they are individually wrapped, as "Scotch," "Bourbon," and "Gin."

The government contends that the internal labeling is false or misleading in that it implies and represents that "the article is flavored with liquor, which it is not." In response claimant does not allege that the lollipops are flavored with liquor, but by way of affirmative defenses contends that they are not misbranded because the cartons are clearly labeled "candy" and the ingredients are distinctly set forth, and that the ordinary purchaser would not read or understand it to represent that the lollipops contain any alcohol or liquor.

In approaching the question of whether the labeling here was false and misleading within the meaning of the statute, we recognize that the statute does not provide for much flexibility in interpretation, since it requires only that the labeling be false or misleading "in any particular." This represents a stricter substantive standard than that applied with respect to false advertising, which in order to be prohibited must be "misleading in a material respect." Furthermore the statute says "false or misleading." For instance, the use of the term "fruit flavored" on a pudding product has been held after a trial on the merits to be false and misleading, even though the product was manufactured from grain which, while botanically a fruit, was not a fruit in common parlance.

The issue of whether a label is false or misleading may not be resolved by fragmentizing it, or isolating statements claimed to be false from the label in its entirety, since such statements may not be deemed misleading when read in the light of the label as a whole. However, even though the actual ingredients are stated on the outside of a carton, false or misleading statements inside the carton may lead to the conclusion that the labeling is misleading, since a true statement will not necessarily cure or neutralize a false one contained in the label. ... Furthermore, the fact that purchasers of a product have not been misled, while admissible on the issue of whether the label is false or misleading, would not constitute a defense. ...

Applying these principles here, it cannot be concluded as a matter of law that no material issue exists with respect to the alleged false and misleading character of the label here before us. Although the labeling on the inside of each box of "candy," when read alone, might be misleading, the detailed description of the contents of the box listed on the outside of the carton could convince a jury, when the labeling or literature is read as a whole, that it is not "misleading in any particular," as that term is used in 21 U.S.C. § 343(a). ...

It appears that the government, although it has not so indicated in its papers, may be concerned with some potential abuse in the distribution of this product that has not been drawn to the attention of this Court. If this is so, it would seem appropriate for this factual aspect of the case to be developed at trial rather than to grant a judgment on the pleadings in favor of the government on the basis of a completely rigid reading of the words of the statute and a fragmentization of the labeling under attack here. The government's motion for a judgment on the pleadings is therefore denied.

So ordered.

* * * * *

DISCUSSION QUESTIONS AND NOTE

3.11. Resolution on remand. Although the Lollipop case was ordered to go to trial, "an order for discontinuance of the action was entered pursuant to stipulation of the parties." PETER BARTON HUTT, RICHARD A. MERRILL, and LEWIS A. GROSSMAN, FOOD AND DRUG LAW 109 (3d ed. 2007) (citing 5 FDA Papers, No. 3, at 42 (Apr. 1971)). Often in such cases the company will decide to relabel the product to address FDA's concerns. This would have gained the release of any seized product and saved the company the expense of trial. From a practical standpoint, the financial advantages of resolving the issue likely exceeded substantially any potential future value of a favorable court decision.

3.12. Novelty defense. How is the "novelty" nature of the lollipops relevant?

3.13. Curing misleading statements. Can a false or misleading statement be "cured" by other information on the label?

3.14. In any particular. Note the FD&C Act's strict standard only requires the labeling be false or misleading "in any particular."

3.6 DECEPTIVE PACKAGING

FD&C Act section 403(d) states that a food is misbranded "if its container is so made, formed, or filled as to be misleading." FDA has rarely taken enforcement action against misleading packaging under this section. As the following case illustrates, courts have been reluctant to find violations of this provision. Some courts have been reluctant to find deceptive packaging when the net contents of packages are declared on the label. In addition a certain level of slack filling is required for machine filling. Because the packages clearly do not have to be packed tightly, courts have been reluctant to find that packages should have been packed tighter. In the following case the court additionally held that deceptive packaging may be allowable if necessary for protection of the product from the condition handling and shipping.

* * * * *

United States v. 174 Cases Delson Thin Mints

287 F.2d 246 (1961)

Before BIGGS, Chief Judge, and GOODRICH and FORMAN, Circuit Judges
Opinion: BIGGS

Under Section 403(d) of the Federal Food, Drug and Cosmetic Act, 21 U.S.C.A. § 343(d), food must be held to be misbranded "if its container is so made, formed, or filled as to be misleading." The standard set up by Judge Wyzanski is "whether the container would be likely to mislead the ordinary purchaser of this type of merchandise . . ." We think this standard is the correct one.

The opinion of the court below . . . sums up the evidence of the United States that the containers were so slack-filled as to be misleading and that their structure rendered them no more effective but perhaps less effective in safeguarding their contents than less misleading forms and also the claimant's evidence that its containers were a more efficacious safeguard for its product than other less deceptive containers would have been.

There are two ways in which a trial court may hold for the claimant in cases such as that at bar. First, the court can find as a fact that the accused package is not made, formed, or filled in such a way that it would deceive the ordinary purchaser as to the quantity of its contents. Alternatively, the court may find as a fact that even though the form or filling of the package deceives the ordinary purchaser into thinking that it contains more food than it actually does, the form and filling of the package is justified by considerations of safety and is reasonable in the light of available alternative safety features.

Did the district court in the present case make either of these findings? We conclude that it did not do so.

First, the court below did not find that the Delson package did not deceive the ordinary purchaser by making him think that it contained more than it actually did contain. The court stated in respect to this issue: "The case is, in my opinion, lacking in adequate proof that the average adult, of normal intelligence, would be induced by the exterior appearance of the accused containers to buy a box of Delson mints with the expectation that it would contain any particular number of individual candies." This statement is beside the point. The question was not whether the ordinary purchaser would expect to find a particular number of individual candies in the box but whether such a purchaser would expect to find more of the Delson box filled. For example, the purchaser of a crate of apples opens the crate and finds it half filled. To determine whether he was deceived, we do not ask whether he expected to find a particular number of individual apples in the crate. We do ask whether he expected to find more of the crate filled. This is the pertinent question. People do not think in terms of the number of individual mints when buying them in containers.

As to the second issue we point out that evidence introduced by the United States tended to show that only 44 percent of the total volume of the accused container and that only 75 percent of its practical volume was filled with mints; that the remainder of the usable space was taken up with hollow cardboard dividers and hollow end pieces. The United States introduced substantial uncontradicted evidence to show that purchasers of the mints, opening the boxes, expected to find far more mints in them than were there. In view of this it is obvious, if there were nothing more in the case, that the containers might well fall within the interdiction of the statute.

But, and this is a point which we must emphasize, a showing by the United States that the ordinary purchaser, on viewing a container, will believe that it contains significantly more food than in fact it does contain, and was deceived, cannot be dispositive of the issues of such a case as that at bar. A claimant may go forward and show, as the claimant has attempted to do here, that the circumstantial deception was forced upon it by other considerations such as packaging features necessary to safeguard its product. But safety considerations, before they can be held to justify a slack package must be shown to be reasonably necessary in the light of alternative methods of safeguarding the contents. For example, some padding is obviously necessary in egg crates to safeguard the eggs. But, a two-inch cotton cushion between each of the eggs would certainly not be justified even though such excessive padding would serve fully the ends of safety. The deception would outweigh the asserted justification of safety when viewed in the light of a more reasonable alternative such as cardboard dividers.

The trial court did not make any finding that the Delson slack package was justified by considerations of safety. The court stated only: "From the evidence I conclude that the type of container construction employed by the claimant(s), which the Government accuses in this case, is efficacious to a degree for the protective purposes contended for by the claimant(s) and was not adopted and is not being used for the purpose of deceiving prospective purchasers respecting the contents of the container." The court did find that the container is "efficacious to a degree." But this is not enough. The court has to find that the container's efficacy outweighs its deceptive quality. Further, it has to find that the available alternative efficacious means are not less deceptive than those actually employed.

Since the court below has not made the necessary findings of fact to support the legal conclusions which it has reached, we will vacate the judgment and remand with the direction to proceed as the facts and the law require.

* * * * *

Note that *174 Cases Delson Thin Mints* precedes enactment of the Fair Packaging and Labeling Act (FPLA) in 1966. The FPLA authorizes FDA to promulgate regulations to prevent nonfunctional slack filling of food, drug, and cosmetic packages. FDA has not proposed such regulations.

3.7 WARNING STATEMENTS: PRODUCTS REQUIRING WARNING LABELS

A number of food products require warning statements:

- Self-pressurized containers.
- Certain protein dietary supplements.
- Iron dietary supplements.
- Shell eggs.
- Aspartame—food that contains aspartame must bear the declaration "Phenylktonurics: Contains Phenylalanine."
- Food with 50 grams or more of sorbitol.
- Diet beverages containing a combination of nutritive and nonnutritive sweeteners.
- Foods containing dry or incompletely hydrated psyllium seed husk, and bearing a health claim on the association between soluble fiber from psyllium husk and reduced risk of coronary heart disease.

- Ozone-depleting substances must follow labeling requirements established by the EPA.
- Nonpasteurized fruit and vegetable juices.
- Saccharin—any food product that contains saccharin must be labeled to indicate that the product may to hazardous to health because it contains saccharin, which has been shown to cause cancer in laboratory animals.

* * * * *

Food Labeling

21 C.F.R. Part 101

SEC. 101.17 Food labeling warning and notice statements.

(a) *Self-pressurized containers.* (1) The label of a food packaged in a self-pressurized container and intended to be expelled from the package under pressure shall bear the following warning:

WARNING: Avoid spraying in eyes. Contents under pressure. Do not puncture or incinerate. Do not store at temperature above 120 deg. F. Keep out of reach of children. ... [Certain exceptions and variations omitted.]

(b) *Self-pressurized containers with halocarbon or hydrocarbon propellants.* (1) In addition to the warning required by paragraph (a) of this section, the label of a food packaged in a self-pressurized container in which the propellant consists in whole or in part of a halocarbon or a hydrocarbon shall bear the following warning:

WARNING: Use only as directed. Intentional misuse by deliberately concentrating and inhaling the contents can be harmful or fatal. ... [Certain exceptions omitted.]

(c) *Food containing or manufactured with a chlorofluorocarbon or other ozone-depleting substance.* Labeling requirements for foods that contain or are manufactured with a chlorofluorocarbon or other ozone-depleting substance designated by the Environmental Protection Agency (EPA) are set forth in 40 C.F.R. part 82.

(d) *Protein products.* (1) The label and labeling of any food product in liquid, powdered, tablet, capsule, or similar forms that derives more than 50 percent of its total caloric value from either whole protein, protein hydrolysates, amino acid mixtures, or a combination of these, and that is represented for use in reducing weight shall bear the following warning:

WARNING: Very low calorie protein diets (below 400 Calories per day) may cause serious illness or death. Do Not Use for Weight Reduction in Such Diets Without Medical Supervision. Not for use by infants, children, or pregnant or nursing women. ...

(3) The label and labeling of food products represented or intended for dietary (food) supplementation that derive more than 50 percent of their total caloric value from either whole protein, protein hydrolysates, amino acid mixtures, or a combination of these, that are represented specifically for purposes other than weight reduction; and that are not covered by the requirements of paragraph (d)(1) and (2) of this section; shall bear the following statement:

Notice: Use this product as a food supplement only. Do not use for weight reduction.

. . . .

(g) *Juices that have <u>not</u> been specifically processed to prevent, reduce, or eliminate the presence of pathogens.* (1) For purposes of this paragraph (g), "juice" means the aqueous liquid expressed or extracted from one or more fruits or vegetables, purees of the edible portions of one or more fruits or vegetables, or any concentrate of such liquid or puree.

(2) The label of:

(i) Any juice that has not been processed in the manner described in paragraph (g)(7) of this section; or

(ii) Any beverage containing juice where neither the juice ingredient nor the beverage has been processed in the manner described in paragraph (g)(7) of this section, shall bear the following warning statement:

WARNING: This product has not been pasteurized and, therefore, may contain harmful bacteria that can cause serious illness in children, the elderly, and persons with weakened immune systems.

(3) The warning statement required by this paragraph (g) shall not apply to juice that is not for distribution to retail consumers in the form shipped and that is for use solely in the manufacture of other foods or that is to be processed, labeled, or repacked at a site other than originally processed, provided that for juice that has not been processed in the manner described in paragraph (g)(7) of this section, the lack of such processing is disclosed in documents accompanying the juice, in accordance with the practice of the trade.

(4) The warning statement required by paragraph (g)(2) of this section shall appear prominently and conspicuously on the information panel or on the principal display panel of the label of the container, except that:

(i) For apple juice or apple cider, the warning statement may appear in labeling, including signs or placards, until September 8, 1999;

(ii) For all juices other than apple juice or apple cider, the warning statement may appear in labeling, including signs or placards, until November 5, 1999.

(5) The word "WARNING" shall be capitalized and shall appear in bold type.

(6) The warning statement required by paragraph (g)(2) of this section, when on a label, shall be set off in a box by use of hairlines.

(7) (i) The requirements in this paragraph (g) shall not apply to a juice that has been processed in a manner that will produce, at a minimum, a reduction in the pertinent microorganism for a period at least as long as the shelf life of the product when stored under normal and moderate abuse conditions, of the following magnitude:

(A) A 5-log (i.e., 100,000-fold) reduction; or

(B) A reduction that is equal to, or greater than, the criterion established for process controls by any final regulation requiring the application of Hazard Analysis and Critical Control Point (HACCP) principles to the processing of juice.

(ii) For the purposes of this paragraph (g), the "pertinent microorganism" is the most resistant microorganism of public health significance that is likely to occur in the juice.

* * * * *

NOTE

3.15. Culture can play an important role in how warnings are applied and how effective they are. For example, the warnings on cigarette packs in Japan illustrate how Japan takes a gentle tone when it warns against smoking: "There is a fear it can damage your health, so let's be careful not to smoke too much. Let's obey smoking manners." Other countries have taken a different tack. In Malaysia, for example, the government found that smokers shrugged off government warnings, so now packages warn that "women smokers have more facial wrinkling than nonsmokers" and warn male smokers that the habit may make them impotent. Lawrence Bartlett, *Tobacco: One million Chinese deaths make it wrong*, THE AUSTRALIAN (Aug. 23, 2004), *available at*: http://www.theaustralian.news.com.au/common/story_page/0,5744,10533868%255E23289,00.html.

3.8 ALLERGENS

The Food Allergen Labeling and Consumer Protection Act of 2004[36] amends the Federal Food, Drug, and Cosmetic Act to require a food that contains, or

---

[36] Public Law No: 108-282 of 2004. A copy of the act is available at: http://frwebgate.access.gpo.gov/cgi-bin/getdoc.cgi?dbname=108_cong_public_laws&docid=f:publ282.108 or at: http://www.cfsan.fda.gov/~dms/alrgact.html.

is derived from, a major food allergen to specifically indicate that information on its label.

The act defines "major food allergen" as any of the following:

- Milk
- Eggs
- Fish
- Crustacea
- Tree nuts
- Wheat
- Peanuts
- Soybeans

The declaration that a food contains a major food allergen must be phrased in one of two ways:

1. By stating the common or usual name of the food allergen in the list of ingredients followed in parentheses by the name of the food source from which the major food allergen is derived (unless the common or usual name of the ingredient uses the name of the food source or the name of the food source appears elsewhere in the ingredient list).
2. By stating "contains" followed by the name of food source from which the major food allergen is derived is printed immediately after or is adjacent to the list of ingredients.

The act requires allergens in flavoring, coloring, or incidental additives to also be labeled in accordance with these requirements. FDA may write rules allowing other methods of declaring the presence of a major food allergen. In addition FDA must define and permit use of the term "gluten-free" on food labels.

The Food Allergen Labeling and Consumer Protection Act of 2004 contained a number of other provisions regarding food allergens. These provisions relate to reports to Congress on food allergens and research on food allergens. The new labeling requirements apply to any food that is labeled on or after January 1, 2006.

* * * * *

Notice to Manufacturers

Label Declaration of Allergenic Substances in Foods

FRED R. SHANK, Director, Center for Food Safety and Applied Nutrition, FDA, June 10, 1996.

This letter is to make you aware of the Food and Drug Administration's (FDA's) concerns regarding the labeling of foods that contain allergenic substances. Recently, FDA has received a number of reports concerning consumers who experienced adverse reactions following exposure to an allergenic substance in foods. These exposures occurred because the presence of the allergenic substance in the food was not declared on the food label.

The Food, Drug, and Cosmetic Act (the act) requires, in virtually all cases, a complete listing of all the ingredients of a food. Two of the very narrow exemptions from ingredient labeling requirements appear to have been involved in a number of the recent incidents, however. First, section 403(i) of the act provides that spices, flavorings, and colorings may be declared collectively without naming each one. Second, FDA regulations (21 C.F.R. § 101.100(a)(3)) exempt from ingredient declaration incidental additives, such as processing aids, that are present in a food at insignificant levels and that do not have a technical or functional effect in the finished food.

In some of the instances of adverse reactions, failure to declare an ingredient appears to have been the result of a misinterpretation of the exemption from ingredient declaration provided for incidental additives in 101.100(a)(3). FDA reminds manufacturers that to qualify for the exemption from ingredient declaration provided for incidental additives and processing aids, a substance must meet both of the requirements of 101.100(a)(3), i.e., it must be present in the food at an insignificant level, and it must not have any technical or functional effect in the finished food. Thus, incidental additives may include substances that are present in a food by virtue of their incorporation as an ingredient in another food. However, when an ingredient added to another food continues to have an effect in the finished food (e.g., egg white as a binder in breading used on a breaded fish product), the ingredient is not an incidental additive, and its use must be declared on the label.

The recent adverse reaction reports indicate that some manufacturers have also incorrectly interpreted what constitutes an insignificant level of a substance. Clearly, an amount of a substance that may cause an adverse reaction is not insignificant. Because evidence suggests that some allergenic substances can cause serious allergic responses in some individuals upon ingestion of very small amounts of the substance, it is unlikely that such an allergen, when it is present in a food, can be present at an insignificant level. Thus it follows that the requirements of 101.100(a)(3) cannot be met under such circumstances. . . .

We have also received reports of adverse reactions to foods in which likely allergenic substances were used as flavors, and not declared by name. Therefore, in addition to the exemption in 101.100(a)(3), the agency is also considering whether an allergenic ingredient in a spice, flavor, or color should be required to be declared, 403(i) notwithstanding. On a substance-by-substance basis, the agency has required ingredients covered by the exemption in section 403(i) to be declared when necessary to protect individuals who experience adverse reactions to the substance (e.g., FD&C Yellow No. 5). The agency is open to suggestions on how to best address this problem. . . .

While the agency does so, FDA asks manufacturers to examine their product formulations for ingredients and processing aids that contain known allergens that they may have considered to be exempt from declaration as incidental additives under 101.100(a)(3), and to declare the presence of such ingredients in the ingredient statement. Where appropriate, the name of the ingredient may be accompanied by a parenthetical statement such as "(processing aid)" for clarity.

The voluntary declaration of an allergenic ingredient of a color, flavor, or spice could be accomplished by simply naming the allergenic ingredient in the ingredient list. Because such ingredients are normally present at very low levels, the name of the ingredient could generally be placed at the end of the ingredient list and be consistent with its descending order of predominance by weight. Other, non-allergenic ingredients that are exempt from declaration would remain unlisted.

Another area of concern is the potential, inadvertent introduction of an allergenic ingredient to a food (e.g., in a bakery that is manufacturing two food products on one production line, one product with peanuts and one without, where traces of peanuts, or peanut products, may end up in the product that does not normally contain peanuts). FDA is considering options for providing consumers with information about the possible presence of allergens in these foods.

The agency is aware that some manufacturers are voluntarily labeling their products with statements such as "may contain (insert name of allergenic ingredient)." FDA advises that, because adhering to good manufacturing practice (GMP) is essential for effective reduction of adverse reactions, such precautionary labeling should not be used in lieu of adherence to GMP. The agency urges manufacturers to take all steps necessary to eliminate cross contamination and to ensure the absence of the identified food. The agency is open to suggestions on how best to address this issue.

* * * * *

For more information about food allergens, visit www.cfsan.fda.gov/~dms/wh-alrgy.html.

3.9 ALCOHOLIC BEVERAGES

The Alcohol and Tobacco Tax and Trade Bureau (TTB) of the U.S. Department of Treasury (formally the Bureau of Alcohol, Tobacco, and Firearms, BATF) has jurisdiction over the labeling of alcoholic beverages under the Federal Alcohol Administration Act, 27 U.S.C. § 201 *et seq.* The Federal Alcohol Administration Act requires importers and bottlers of beverage alcohol to obtain certificates of label approval or certificates of exemption from label approval (COLAs) for most alcohol beverages prior to their introduction into interstate commerce.

The TTB also examines formulas for wine and distilled spirits, process statements, and pre-import applications filed by importers and proprietors of domestic distilled spirits plants, wineries, and breweries for proper tax classification and to ensure that the products are manufactured in accordance with federal laws and regulations. For more information, visit the TTB Web site at http://www.ttb.gov/index.htm

However, TTB only regulates those wine products that contain 7 percent or more alcohol. FDA regulates wine products containing less than 7 percent alcohol. Wine coolers, therefore, are regulated by FDA.

3.9.1 Wine Coolers versus Flavored Wine

Wine coolers and similar beverages containing less than 7 percent alcohol by volume are regulated by the FDA. Therefore wine coolers that purport to contain unfermented fruit or vegetable juice are covered by 21 C.F.R. § 101.30 and are required to bear a percentage juice declaration.

Wine coolers that do *not* contain unfermented juice are not covered by this requirement unless they purport to contain juice by means of advertising, labeling statements, vignettes, or physical characteristics. Thus, if a wine cooler does not contain any juice, has labeling that makes clear that it contains flavors rather than juice, and does not bear a vignette that implies fruit juice content, it is not subject to 21 C.F.R. § 101.30. Noncarbonated beverages that purport to contain juice—but in fact do not contain any juice—are required by 21 C.F.R. § 102.30 to state that they contain no juice.[37]

3.9.2 A Double Standard

The requirement for a percentage juice declaration on wine coolers has been called unfair because the same requirement does not apply to most other alcoholic beverages including spirits-based and malt-based coolers, which compete directly against wine coolers. FDA has commented:

> The agency advises that the labels of alcoholic beverages (those that contain 7 percent or more alcohol by volume and malt beverages) are regulated in accordance with the Federal Alcohol Administration Act (27 U.S.C. 205) administered by the Bureau of Alcohol, Tobacco and Firearms [now TTB] and are controlled differently from wine coolers. The labeling of wine coolers, like other beverages that contain less than 7 percent alcohol by volume, is regulated under the [Food, Drug, and Cosmetic] act. To the extent that these statutes differ, the products are regulated differently in other labeling aspects as well as in declaration of percentage juice content. It is not up to FDA, but to Congress, to decide that the same

[37] 58 Fed. Reg. 2899 (Jan. 6, 1993).

requirements must apply to wine coolers, other alcoholic beverages, and malt based beverages.[38]

3.10 USDA

The FDA labeling requirements apply to all foods except meat, poultry, and egg products. The USDA regulates the labeling of most meat, poultry, and egg products. USDA regulation of the labeling of meat, poultry, and egg products generally parallels those for FDA-regulated foods. One most important distinction is that most USDA-regulated products require FSIS label approval prior to marketing.

3.10.1 Labeling Approval

FSIS's labeling approval regulation, 9 C.F.R. § 317.4(a), states in part: "No final labeling shall be used on any product unless the sketch labeling of such final labeling has been submitted for approval to the Food Labeling Division, Regulatory Programs, Food Safety and Inspection Service, and approved by such division, accompanied by FSIS form, Application for Approval of Labels, Marking, and Devices, except for generically approved labeling authorized for use in Sec. 317.5(b)."[39]

3.10.2 "Generic" Approvals (Labels without Prior Approval)

Effective July 1, 1996, FSIS regulations allow food establishments more flexibility for producing labels without prior FSIS approval.[40] These labels fall into what is termed the **generic approval** category.

Once a generic label is approved, the regulations provide for use of **final** labeling without further authorization from FSIS. It is the establishment's responsibility to prepare final labeling in accordance with applicable regulations, and to create and maintain records of final labeling. Only limited changes in the product are permitted under a generic approval.

3.10.3 Safe Food-Handling Instructions

The USDA requires safe handling and cooking instructions on raw meat and poultry products. These instructions must state that "some food products may contain bacteria that could cause an illness if the product is mishandled or cooked improperly."

3.10.4 Additional Information Required

The official inspection legend.

[38] 58 Fed. Reg. 2899 (Jan. 6, 1993).
[39] 9 C.F.R. §§ 317.4 and 381.132.
[40] 9 C.F.R. §§ 317.5 and 381.133 (for meat and poultry, respectively).

The establishment's inspection number.

Other applicable warning statements, such as "Keep refrigerated" and "Keep frozen."

3.10.5 Record Keeping

FSIS also sets forth requirements for label recordkeeping in regulations 9 C.F.R. sections 320.1(b)(11) and 381.175(b)(6).

3.11 OPTIONAL LABEL INFORMATION

There is a variety of information that may be voluntarily included on the food label. Although the labeling is voluntary, this information often is closely regulated once it is applied to labels or labeling. So that you are familiar with the regulation of such information, we will touch on a number of these categories in this section.

3.11.1 Health Claims and Nutrient Level Claims

Health claims and nutrient content claims are voluntary. We discuss the regulations of these claims in the next chapter.

3.11.2 Grades

Some foods, such as milk, butter, eggs, orange juice, and meat, carry a grade on their label that denotes their quality. The grades generally show up as letters, such as AA, A, and B for eggs; words, such as "choice" and "select" for meat, or "substandard" for some canned vegetables; or as some kind of logo or mark, such as the Grade A shield on orange juice containers.

USDA establishes some of these grade standards for foods. Under the Agricultural Marketing Act of 1946,[41] the USDA is empowered to establish a voluntary system of food grading, inspection, and certification. Participating producers request and pay for the USDA inspection and grading service.

These quality standards relate to factors such as color, size, shape, flavor, texture, and so forth. This grading is most important for wholesale buyers because it provides an independent determination of quality that allows proper pricing. Grading may also provide useful information to consumers.

FDA has also standards for a number of foods, including canned vegetables. The National Marine Fisheries Service grades fish.

3.11.3 Trademarks and Copyrights

The ® symbol on a label indicates that a trademark used on the label is registered with the U.S. Patent and Trademark Office (USPTO). A™ symbol means

---

[41] 7 U.S.C. § 1621 *et seq.* (2000).

that a trademark right is claimed, although the mark is not registered with the USPTO. A © symbol means that the literary or artistic work of the label is protected under U.S. copyright laws.

3.11.4 Religious Symbols

A number of symbols may appear on foods to indicate that the food has been processed according to religious dietary laws. One of the more common is a letter "U" inside the letter "O." This means that the food has been authorized as "kosher" by the Union of Orthodox Jewish Congregations of America.

3.11.5 Universal Product Code

The UPC is a bar code with a 10-digit number. It is used with computerized grocery store checkout equipment to give an automated inventory system. The Uniform Code Council, Inc. of Dayton, Ohio, administers this system.

3.11.6 Organic

In 1990 Congress passed the "Organics Foods Production Act" (OFPA) (Title 21 of Public Law 101-624), which authorized the **National Organics Program**. USDA defines organic agriculture as "ecological production management system that promotes and enhances biodiversity, biological cycles, and soil biological activity based on minimal use of off-farm inputs and on management practices that restore, maintain, and enhance ecological harmony."

Under the new standards, foods labeled "organic" cannot include bioengineered ingredients or be irradiated to kill bacteria and lengthen shelf life. Meats sold as organic cannot be produced from animals that receive antibiotics.

Consumers can recognize organic products by a USDA mark they will carry, similar to the "USDA prime" identification on beef or the grade labels on egg cartons. Foods will be labeled "100 percent organic," "organic," or "made with organic ingredients," depending on ingredients.

The label "organic" had previously fallen under a variety of state, regional, and private certifier standards, giving rise to confusion about its meaning. Under the new standards, all agricultural products labeled organic must originate from farms or handling operations certified by a state or private agency accredited by USDA. Farms and handling operations that sell less than $5,000 worth per year of organic agricultural products are exempt from certification.

The OFPA also provided that an advisory board, the National Organic Standards Board, be assembled to help USDA write the regulation. The Board is comprised of 14 members, each representing different segments of the organic industry. They make recommendations to the Secretary, especially regarding the substances that can be used in organic production and handling.

CHAPTER 4

Nutritional Labeling, Nutrient Level Claims, and Health Claims

Nutrition Facts	Amount/serving	% DV*	Amount/serving	% DV*
	Total Fat 1.5g	**2%**	**Total Carbohydrate** 26mg	0%
	Saturated Fat 0.5g	**3%**	Dietary Fiber 2g	**3%**
Serving Size 2 slices (56 g)	*Trans* Fat 0.5g		Sugars 1g	
Servings per container 10	**Cholesterol** 0mg	0%	**Protein** 4mg	
Calories 140	**Sodium** 280mg	12%		
Calories from Fat 15				
*Percent Daily Values are based on a 2,000 calorie diet.	Vitamin A 0% • Vitamin C 0%		Calcium 6% • Iron 6%	
	Thiamin 15% • Riboflavin 8%		Niacin 10%	

4.1 INTRODUCTION

This chapter covers the regulation of nutritional content, nutritional labeling, nutritional claims, and health claims. These topics warrant treatment in a complete chapter for several reasons. Nutrition and health are matters of keen consumer interest, and with the graying of the baby boom generation, this interest gains added focus. Consequently nutritional claims and health claims can be potent marketing tools for the sale of food.

Nutrition and health is also an area of fast-paced change, which is still evolving. Historically the Food and Drug Administration (FDA) prohibited health claims in food labeling. Under the Food, Drug, and Cosmetic Act (FD&C Act) prior to 1990, all health claims were considered illegal drug claims. The FD&C Act and FDA's policy on health claims reflected our past limited understanding of the link between nutrition and disease.

As science advances, increasing evidence establishes additional links between diet and health. New evidence substantiating claims of nutrient links to diseases and other health-related conditions allows a growing number of permitted claims. The law has evolved to keep pace. In 1990, the Nutrition Labeling and Education Act (NLEA) amended the FD&C Act to allow health claims for foods and dietary supplements under limited conditions. The FDA

Food Regulation: Law, Science, Policy, and Practice, by Neal D. Fortin
Copyright © 2009 Published by John Wiley & Sons, Inc.

Modernization Act of 1997 (FDAMA) further amended the FD&C Act to permit health claims based on an "authoritative statement" linking a nutrient to a disease made by a scientific body. In December 2002, FDA announced the availability for companies to petition the FDA to authorize qualified health claims.

Finally, proper nutrition is a matter of great public health concern. It has been a priority objective of both political parties and numerous administrations. Accordingly, government programs relating to nutrition rate of high importance among the various agency functions. The total costs attributed to people being overweight or obese amounted to $117 billion in the year 2000— $400 for every man, woman, and child in the United States.[1] Health care for overweight and obese individuals costs on average 37 percent more than for individuals of normal weight.[2]

4.1.1 Nutrition and Public Health Expenditures

The cost of treatment for illnesses related to obesity rivals the financial toll of smoking-related disease at about 9 percent of all health care expenditures.[3] This economic burden falls heavily on Medicaid and Medicare, the government health programs for the poor, disabled, and elderly. Therefore it is not surprising that the federal government has a stake on the issue.

> There has been a debate about whether obesity is a personal or societal issue and whether the government has any business being involved. . . . The fact that the government, and ultimately the taxpayer, is financing half the economic burden of obesity, suggests that the government has a clear justification to try to reduce obesity rates.
>
> As lawmakers face rising federal deficits, the study shines a light on where more tax dollars are going. An obese Medicare recipient spends on average $1,500 more on medical care each year than non-obese seniors. Medicaid recipients, who are mostly poor, may have a higher prevalence of obesity because they engaged in "riskier behaviors" such as poor diet, lack of exercise or alcohol consumption.[4]

In 2003, Health and Human Services (HHS) announced an initiative through *Steps to a Healthier US*, an HHS campaign to help Americans live longer, healthier lives. The two central pillars of *Steps* is the promotion of a healthy diet rich in fruits and vegetables and encouraging regular physical activity.[5]

[1] The President's Council on Physical Fitness and Sports, *Steps to Preventing Overweight and Obesity, available at:* http://www.fitness.gov/news/obesity_america.html (July 7, 2003).
[2] Ceci Connolly, *Obesity Adds $93 Billion to U.S. Health Costs*, WASHINGTON POST, May 21, 2003.
[3] *Id.*
[4] *Id.* (quoting Finkelstein.)
[5] The President's Council on Physical Fitness and Sports, *Steps to Preventing Overweight and Obesity, available at:* http://www.fitness.gov/news/obesity_america.html (July 7, 2003).

At the same time, FDA announced that it will require labels be easier for people to count calories. As a result of recommendations made by an FDA task force on obesity, FDA plans to revise its requirements for packaged food labels to make the caloric content easier to read and understand. FDA also sent letters to food manufacturers warning them not to label packaged foods with unrealistically small servings because this falsely reduces the apparent calorie count. FDA may also change the criteria for foods that can claim to be "reduced" or "low" in calories.

Among the alternate approaches that have been suggested, others are as follows:

- Advertising campaigns
- A tax on fatty foods
- Subsidies for fruit and vegetable purchases

DISCUSSION QUESTIONS

4.1. How could food labels be revised to make them more effective in reducing obesity?

4.2. In what specific ways do you think label revisions could help consumers eat healthier diets?

4.1.2 McFat Litigation

In litigation that has been dubbed the "McFat" cases, customers sued the McDonald's fast food chain, claiming that the restaurant's unhealthy food caused their obesity and their related health problems. The cause of action was based on a claim of false and deceptive advertising. The complaint was dismissed with leave to amend. Following amendment, the restaurant chain moved to dismiss. The District Court held that the statute of limitations barred some claims, but also that causal connection between false advertising and health problems was not sufficiently alleged, and the advertising was not objectively deceptive.

* * * * *

Pelman et al. v. McDonald's Corp.

S.D.N.Y. (Sept. 3, 2003)[6]

Judge ROBERT W. SWEET:

. . . .

---

[6] WL 22052778 (not Reported in F.Supp.2d).

The infant plaintiffs are consumers who have purchased and consumed the defendant's products in New York State outlets and, as a result thereof, such consumption has been a significant or substantial factor in the development of their obesity, diabetes, coronary heart disease, high blood pressure, elevated cholesterol intake, and/or other detrimental and adverse health effects and/or diseases. . . .

McDonald's Advertising Campaigns

In one survey of the frequency of purchases by visitors to McDonald's restaurants, McDonald's found that 72 percent of its customers were "Heavy Users," meaning they visit McDonald's at least once a week, and that approximately 22 percent of its customers are "Super Heavy Users," or "SHUs," meaning that they eat "at McDonald's ten times or more a month." Super Heavy Users make up approximately 75 percent of McDonald's sales. Many of McDonald's advertisements, therefore, are designed to increase the consumption of Heavy Users or Super Heavy Users. The plaintiffs allege that to achieve that goal, McDonald's engaged in advertising campaigns which represented that McDonald's foods are nutritious and can easily be part of a healthy lifestyle.

Advertising campaigns run by McDonald's from 1987 onward claimed that it sold "Good basic nutritious food. Food that's been the foundation of well-balanced diets for generations. And will be for generations to come." McDonald's also represented that it would be "easy" to follow USDA and Health and Human Services guidelines for a healthful diet "and still enjoy your meal at McDonald's." McDonald's has described its beef as "nutritious" and "leaner than you think." And it has described its french fries as "well within the established guidelines for good nutrition."

While making these broad claims about its nutritious value, McDonald's has declined to make its nutrition information readily available at its restaurants. In 1987, McDonald's entered into a settlement agreement with the New York State Attorney General in which it agreed to provide [nutritional] information in easily understood pamphlets or brochures which will be free to all customers so they could take them with them for further study [and] to place signs, including in-store advertising to inform customers who walk in, and drive-through information and notices would be placed where drive-through customers could see them.

Despite this agreement, the plaintiffs have alleged that nutritional information was not adequately available to them for inspection upon request.

Claims

. . . The three remaining causes of action are based on deceptive acts in practices in violation of the Consumer Protection Act, New York General Business Law §§ 349 and 250. Count I alleges that McDonald's misled the plaintiffs, through advertising campaigns and other publicity, that its food products were nutritious, of a beneficial nutritional nature or effect, and/or were easily part of a healthy lifestyle if consumed on a daily basis. Count II alleges that McDonald's failed adequately to disclose the fact that certain of its foods were substantially less healthier, as a result of processing and ingredient additives, than represented by McDonald's in its advertising campaigns and other publicity. Count III alleges that McDonald's engaged in unfair and deceptive acts and practices by representing to the New York Attorney General and to New York consumers that it provides nutritional brochures and information at all of its stores when in fact such information was and is not adequately available to the plaintiffs at a significant number of McDonald's outlets.

The plaintiffs allege that as a result of the deceptive acts and practices enumerated in all three counts, they have suffered damages including, but not limited to, an increased likelihood of the development of obesity, diabetes, coronary heart disease, high blood pressure, elevated cholesterol intake, related cancers, and/or detrimental and adverse health effects and/or diseases.

. . . .

Plaintiffs Have Successfully Stated Reliance on a Single Allegedly Deceptive Advertising Campaign

. . . The plaintiffs counter that they have alleged that their misconceptions about the healthiness of McDonald's food resulted from "a long-term deceptive campaign by the Defendant of misrepresenting the nutritional benefits of their foods over last approximate [sic] fifteen (15) years." Plaintiffs further argue that reliance is not an element of New York GBL § 349. . . .

While plaintiffs have alleged that McDonald's has made it difficult to obtain nutritional information about its products, they have not alleged that McDonald's controlled all relevant information. Indeed, the complaint cites the complete ingredients of several McDonald's products. Plaintiffs are therefore required to allege reliance in order to survive a motion to dismiss. . . .

Plaintiffs argue that it would be impracticable to require each of the tens of thousands of potential class members to state exactly when and where they observed the deceptive advertisements. Before a class has been certified, however, the number of infant plaintiffs is only two, making the task much more manageable. It is true that it would be unduly burdensome for plaintiffs, at this stage, to allege the particular time and place that they saw the advertisements which allegedly caused their injuries. It will therefore be considered sufficient for plaintiffs to allege in general terms that plaintiffs were aware of the false advertisement, and that they relied to their detriment on the advertisement.

Nowhere in the amended complaint is it explicitly alleged that plaintiffs witnessed any of the allegedly false advertisements cited. . . .

Making all reasonable inferences in favor of the plaintiffs, the complaint implicitly alleges only one instance in which the infant plaintiffs were aware of allegedly false advertisements. The plaintiffs implicitly allege that they were aware of McDonald's national advertising campaign announcing that it was switching to "100 percent vegetable oil" in its French fries and hash browns, and that McDonald's fries contained zero milligrams of cholesterol, when they claim that they "would not have purchased or consumed said french fries or hash browns, or purchased and consumed in such quantities," had McDonald's disclosed the fact that these products "contain beef or extracts and trans fatty acids." . . .

Plaintiffs Have Failed to Allege that Consumption of McDonald's Food Caused Their Injuries

The most formidable hurdle for plaintiffs is to demonstrate that they "suffered injury as a result of the deceptive act." . . .

The absence of a reliance requirement does not, however, dispense with the need to allege some kind of connection between the allegedly deceptive practice and the plaintiffs' injuries. If a plaintiff had never seen a particular advertisement, she could obviously not allege that her injuries were suffered "as a result" of that advertisement. Excusing the reliance requirement only allows the plaintiff to forgo the heightened pleading burden that is necessary for common law fraud claims. It cannot, however, create a causal connection between a deceptive practice and a plaintiff's injury where none has been alleged. Accordingly, this Court required that to state a claim under § 349 in an amended complaint, plaintiffs would "have to set forth grounds to establish . . . that they suffered some injury as a result of that particular promotion." . . .

Plaintiffs have failed, however, to draw an adequate causal connection between their consumption of McDonald's food and their alleged injuries. This Court noted that the original complaint did not adequately allege the causation of plaintiffs' injuries because it did "not specify how often the plaintiffs ate at McDonald's." In terms of causation, "the more often a plaintiff had eaten at McDonald's, the stronger the likelihood that it was the McDonald's food (as opposed to other foods) that affected the plaintiffs' health."

Unlike the initial complaint, the amended complaint does specify how often the plaintiffs ate at McDonald's. For example, Jazlyn Bradley is alleged to have "consumed McDonald's foods her entire life . . . during school lunch breaks and before and after school, approximately five times per week, ordering two meals per day." Such frequency is sufficient to begin to raise a factual issue "as to whether McDonald's products played a significant role in the plaintiffs' health problems."

What plaintiffs have not done, however, is to address the role that "a number of other factors other than diet may come to play in obesity and the health problems of which the plaintiffs complain." This Court specifically apprised the plaintiffs that in order to allege that McDonald's products were a significant factor in the plaintiffs' obesity and health problems, the Complaint must address these other variables and, if possible, eliminate them or show that a McDiet is a substantial factor despite these other variables. Similarly, with regard to plaintiffs' health problems that they claim resulted from their obesity . . . , it would be necessary to allege that such diseases were not merely hereditary or caused by environmental or other factors.

Plaintiffs have not made any attempt to isolate the particular effect of McDonald's foods on their obesity and other injuries. The amended complaint simply states the frequency of consumption of McDonald's foods and that each infant plaintiff "exceeds the Body Mass Index (BMI) as established by the U.S. Surgeon General, National Institutes of Health, Centers for Disease Control, U.S. Food and Drug Administration and all acceptable scientific, medical guidelines for classification of clinical obesity."

In their opposition brief, plaintiffs argue that "surveys and sampling techniques" may be employed to establish causation. While that may be true, it is irrelevant in the present context, where a small number of plaintiffs are alleging measurable injuries. Following this Court's previous opinion, the plaintiffs should have included sufficient information about themselves to be able to draw a causal connection between the alleged deceptive practices and the plaintiffs' obesity and related diseases. Information about the frequency with which the plaintiffs ate at McDonald's is helpful, but only begins to address the issue of causation. Other pertinent, but unanswered questions include: What else did the plaintiffs eat? How much did they exercise? Is there a family history of the diseases which are alleged to have been caused by McDonald's products? Without this additional information, McDonald's does not have sufficient information to determine if its foods are the cause of plaintiffs' obesity, or if instead McDonald's foods are only a contributing factor. . . .

The Advertising Campaign upon Which Plaintiffs Have Stated Reliance Is Not Objectively Deceptive

Even if plaintiffs were able sufficiently to allege that their injuries were causally related to McDonald's representations about its french fries and hash browns, that claim must still be dismissed because the plaintiffs have not alleged that those advertisements were objectively misleading. . . .

The essence of the plaintiffs' claim of deception with regard to McDonald's french fries and hash browns is that McDonald's represented that its fries are cooked in "100 percent vegetable oil" and that they contain zero milligrams of cholesterol whereas in reality they "contain beef or extracts and trans fatty acids." However, the citations in the amended complaint to McDonald's advertisements, and the appended copies of the advertisements, do not bear out the plaintiffs' claims of deception. The first citation is to an advertisement titled "How we're getting a handle on cholesterol," alleged to have commenced in 1987 and to have continued for several years thereafter. The text cited by the plaintiffs states:

. . . a regular order of french fries is surprising low in cholesterol and 4.6 grams of saturated fat. Well within established guidelines for good nutrition.

The text cited in the complaint, however, inexplicably drops several significant words from the text of the advertisement included in the appendix to the amended complaint. The actual advertisement states:

... a regular order of french fries is surprising low in cholesterol and saturated fat: only 9mg of cholesterol and 4.6 grams of saturated fat. Well within established guidelines for good nutrition.

The advertisement also states that McDonald's uses "a specially blended beef and vegetable shortening to cook our world famous french fries and hash browns." *Id.*

The plaintiffs next allege that beginning on or around July 23, 1990, McDonald's announced that it would change its french fry recipe and cook its fries in "100 percent vegetable oil," a change that rendered its fries cholesterol-free. They allege that from the time of the change until May 21, 2001, McDonald's never acknowledged "that it has continued the use of beef tallow in the french fries and hash browns cooking process." On its website, however, McDonald's is alleged to have "admitted the truth about its french fries and hash browns":

A small amount of beef flavoring is added during potato processing—at the plant. After the potatoes are washed and steam peeled, they are cut, dried, par-fried, and frozen. It is during the par-frying process at the plant that the natural flavoring is used. These fries are then shipped to our U.S. restaurants. Our french fries are cooked in vegetable oil at our restaurants.

While the plaintiffs do allege that the beef flavoring that McDonald's acknowledges using is equivalent to beef tallow, the complaint does not allege that the beef flavoring contains cholesterol. McDonald's maintains that its "cholesterol disclosure is regulated by the FDA and is entirely accurate and appropriate under the FDA's regulations."

Plaintiffs further allege that McDonald's claims that its french fries and hash browns are cholesterol-free is also misleading because the oils in which those foods are cooked contain "trans fatty acids responsible for raising detrimental blood cholesterol levels (LDL) in individuals, leading to coronary heart disease." However, plaintiffs have made no allegations that McDonald's made any representations about the effect of its french fries on blood cholesterol levels. As McDonald's argues,

The contents of food and the effects of food are entirely different things. A person can become "fat" from eating "fat-free" foods, and a person's blood sugar level can increase from eating "sugar-free" foods.

Because the plaintiffs have failed to allege both that McDonald's caused the plaintiffs' injuries or that McDonald's representations to the public were deceptive, the motion to dismiss the complaint is granted.

. . . .

* * * * *

DISCUSSION QUESTION

4.3. The *Pelman* case appears to be an exception, as this is the single obesity cases to have advanced this far. Considering that *Pelman* was dismissed, and no other obesity suits advanced so far, why have obesity lawsuits captured the public imagination? Over 20 states have passed "hamburger shield" laws to ensure that restaurants cannot be sued for making someone fat. Why is there such a concern about these suits?

4.2   THE NUTRITION LABELING AND EDUCATION ACT (NLEA)

Congress passed the Nutrition Labeling and Education Act (NLEA) in 1990. The NLEA amended the Food, Drug, and Cosmetic Act (FD&C Act) and mandated nutritional labeling on most food products regulated by FDA. NLEA is codified in part into the FD&C Act.[7] The NLEA also mandated changes in label declarations for collective terms, sulfites, sweeteners, colors, spices, nondairy and allergenic substances, net contents, and metric labeling.

The NLEA was enacted in response to the consumer's demand for more information about the nutritional content of food products and the presence of food additives and allergens. FDA promulgated regulations for the use of health and nutrient content claims, such as "heart smart." Most of these regulations went into effect in 1994. Certain nutrient information is mandatory, while other nutrients may be listed at the discretion of the manufacturer, unless the manufacturer makes a claim about the optional nutrient or indicates that the food product is fortified with an optional nutrient.

Although not required to do so by law, the USDA also established nutritional labeling requirements for meat and poultry products, which parallel FDA's requirements for other foods.

* * * * *

Good Reading for Good Eating

Paula Kurtzweil, FDA CONSUMER, SPECIAL ISSUE, Focus on Food Labeling[8]

It may not have the power of a Pulitzer prize-winning novel or the luridness of a checkout counter tabloid, but the new food label still promises to make for good reading. . . .

[T]erms used to describe a food's nutrient content—light," "fat-free," and "low-calorie," for example—will meet government definitions so that they mean the same for any product on which they appear. Health claims about the relationship between a nutrient or food and a disease that are supported by scientific evidence will be allowed for the first time. Serving sizes:

- are more consistent across product lines to make comparison shopping easier
- are expressed in common household and metric measures
- better reflect the amounts people really eat.

---

[7] At 21 U.S.C. §§ 343(Q), (R) [§§ 403(Q), (R) FD&C Act].
[8] Also *available at*: http://www.fda.gov/fdac/special/foodlabel/ingred.html (May 1993).

There will be many more products with labels to read because the regulations, for the first time, make nutrition labeling mandatory for almost all processed foods. Also, uniform point-of-purchase nutrition information will accompany many fresh foods, such as fruits and vegetables and raw fish, meat and poultry.

The new food label is reading that can be put to good use, too, because it's designed to help clear up much of the confusion that has prevailed on supermarket shelves. It also can help consumers choose more healthful diets. And it can serve as an incentive to food companies to improve the nutritional qualities of their products.

"[This isn't] just another government program," said FDA Commissioner David Kessler, M.D. "The new food label is an unusual opportunity to help millions of Americans make more informed, healthier food choices."

"We expect the labels also will provide more food companies with an incentive to improve the nutritional quality of their products," said H. Russell Cross, Ph.D., FSIS administrator.

. . . .

Advertising is not covered by the Nutrition Labeling and Education Act, but the Federal Trade Commission has indicated it may apply the same criteria to advertising that FDA and FSIS do to labels.

A Look Back

The changes will mark the first extensive renovation of the food label since 1974, when FDA and USDA established voluntary nutrition labeling and began requiring nutrition information on labels of products that contain added nutrients or that carry nutrition claims. Other than adding sodium as a mandatory and potassium as a voluntary component to the list of nutrients allowed in voluntary nutrition labeling in 1984, the nutrition label has remained essentially the same all that time.

Nutrition labeling wasn't ignored during the interim, though, as Congress, regulators, and consumer and industry groups put forth ideas to overhaul it. Their efforts intensified as consumers became more interested in nutrition, and food marketing strategies began to focus on that interest.

That marketing trend represented a departure from usual practice, according to Ed Scarbrough, Ph.D., director of the Office of Food Labeling in FDA's Center for Food Safety and Applied Nutrition.

"The line from industry used to be: 'Nutrition won't sell food. It's price, taste, and convenience,'" he said. "By the time we got into the 1980s, nutrition clearly was selling products. Industry recognized this and started making claims about the food."

That was both good and bad, Scarbrough said. On the one hand, it gave consumers more information about nutrition. But on the other, claims got pushed to their outer limits as manufacturers scrambled to gain a competitive edge for their products.

"Consumers reacted to that," he said. "They couldn't believe many of the claims being made."

At about the same time, the Surgeon General of the U.S. Public Health Service and the National Academy of Sciences' National Research Council released two reports that lent strong support to development of a new food label. These reports—the 1988 Surgeon General's Report on Nutrition and Health, and the 1989 National Research Council's Diet and Health: Implications for Reducing Chronic Disease Risk—concluded that evidence substantiates an association between diet and risk of chronic disease and recommended similar dietary changes.

Those recommendations reflected what many public health experts had been saying for years: for example, that Americans should reduce their intake of fat (especially saturated fat), cholesterol, and sodium; maintain appropriate body weight; and consume adequate amounts of calcium and fiber. The National Research Council's report went so far as to recommend quantitative amounts for certain nutrients.

It soon became apparent, however, that the current food label did not offer enough information to help consumers follow those guidelines. That, coupled with often questionable marketing practices, led to the first serious effort to revamp the food label. . . .

According to John Vanderveen, Ph.D., director of FDA's Office of Plant and Dairy Foods and Beverages, the law makes the United States the first country in the world to have mandatory nutrition labeling and to allow health claims on food labels. "We've been pioneers," he said. . . .

Economic Impact

It is estimated that the new food label will cost FDA-regulated food processors between $1.4 billion and $2.3 billion over the next 20 years. However, the benefits to public health—measured in monetary terms—are estimated to well exceed the costs. Potential benefits include decreased rates of coronary heart disease, cancer, osteoporosis, obesity, high blood pressure, and allergic reactions to food.

. . . .

* * * * *

4.3 NUTRIENT LEVEL CLAIMS

FDA regulations set conditions for the use of terms that describe a food's nutrient level. Twelve basic terms have been defined that relate to nutrients:

- Free
- Low
- Reduced

- Fewer
- Lean
- High
- Less
- More
- Extra lean
- Good source
- Light
- Healthy

These 12 terms are the core nutrient level descriptors. These descriptors are defined as follows:[9]

Free: Product contains no amount of, or only trivial or "physiologically inconsequential" amounts of, one or more of these components: fat, saturated fat, cholesterol, sodium, sugars, and calories. For example, "calorie-free" means fewer than 5 calories per serving, and "sugar-free" and "fat-free" both mean less than 0.5 g per serving. Synonyms for "free" include "without," "no" and "zero." A synonym for fat-free milk is "skim."

Low: Foods that can be eaten frequently without exceeding dietary guidelines for one or more of these components: fat, saturated fat, cholesterol, sodium, and calories.

Low-fat: 3 g or less per serving.

Low-saturated fat: 1 g or less per serving.

Low-sodium: 140 mg or less per serving.

Very low sodium: 35 mg or less per serving.

Low-cholesterol: 20 mg or less and 2 g or less of saturated fat per serving.

Low-calorie: 40 calories or less per serving. Synonyms for low include "little," "few," "low source of," and "contains a small amount of."

Lean and extra lean: Describe the fat content of meat, poultry, seafood, and game meats.

Lean: Less than 10 g fat, 4.5 g or less saturated fat, and less than 95 mg cholesterol per serving and per 100 g.

Extra lean: Less than 5 g fat, less than 2 g saturated fat, and less than 95 mg cholesterol per serving and per 100 g.

High: Food contains 20 percent or more of the Daily Value[10] for a particular nutrient in a serving.

---

[9] *See*, Food and Drug Administration, *The New Food Label: Better Information for Special Diets*, FDA CONSUMER (Jan.–Feb. 1995, Revised Jan. 1998), *available at*: http://www.cfsan.fda.gov/~dms/fdspdiet.html.

[10] Daily Values (DVs) are label reference value. Daily Values encompass both the Reference Daily Intakes (RDIs) for vitamins and minerals and the Daily Reference Values (DRVs) for macronutrients, such as fat, protein, and sodium. FDA determined that a single DV would be less confusing on the label than two references values.

Good source: One serving of a food contains 10 to 19 percent of the Daily Value for a particular nutrient.

Reduced: Nutritionally altered product that contains at least 25 percent less of a nutrient or of calories than the regular, or reference, product. However, a reduced claim can't be made on a product if its reference food already meets the requirement for a "low" claim.

Less: Food that, whether altered or not, contains 25 percent less of a nutrient or of calories than the reference food. For example, pretzels that have 25 percent less fat than potato chips could carry a "less" claim. "Fewer" is an acceptable synonym.

Light: (1) A nutritionally altered product that contains one-third fewer calories or half the fat of the reference food (if the food derives 50 percent or more of its calories from fat, the reduction must be 50 percent of the fat), or (2) the sodium content of a low-calorie, low-fat food has been reduced by 50 percent. ("light in sodium" may also be used on food in which the sodium content has been reduced by at least 50 percent). The term "light" still can be used to describe such properties as texture and color, as long as the label explains the intent—for example, "light brown sugar" and "light and fluffy."

More: A serving of food that, whether altered or not, contains a nutrient that is at least 10 percent of the Daily Value more than the reference food. The 10 percent of Daily Value also applies to "fortified," "enriched," and "added" "extra and plus" claims, but in those cases the food must be altered. Alternative spelling of these descriptive terms and their synonyms is allowed—for example, "hi" and "lo"—as long as the alternatives are not misleading.

Percent fat free: A low-fat or a fat-free product. In addition the claim must accurately reflect the amount of fat present in 100 g of the food. Thus, if a food contains 2.5 g fat per 50 g, the claim must be "95 percent fat free."

Implied:[11] Prohibited when they wrongfully imply that a food contains or does not contain a meaningful level of a nutrient. For example, a product claiming to be made with an ingredient known to be a source of fiber (e.g., "made with oat bran") is not allowed unless the product contains enough of that ingredient (for example, oat bran) to meet the definition for "good source" of fiber. As another example, a claim that a product contains "no tropical oils" is allowed—but only on foods that are "low" in saturated fat because consumers have come to equate tropical oils with high saturated fat.

---

[11] "Express" claims directly characterize the nature of a food; for example, "low fat" and "fat free." "Implied" claims indirectly characterize the nature of the food by inference or association, rather than by direct statement; for example, "baked, not fried" implies the food is lower in fat than an equivalent fried version. The context and the entire label are often necessary to determine if there is an implied claims.

4.3.1 Meals and Main Dishes

Claims that a meal or main dish is "free" of a nutrient, such as sodium or cholesterol, must meet the same requirements as those for individual foods. Other claims can be used under special circumstances. For example, "low-calorie" means the meal or main dish contains 120 calories or less per 100 g. "Low-sodium" means the food has 140 mg or less per 100 g. "Low-cholesterol" means the food contains 20 mg cholesterol or less per 100 g and no more than 2 g saturated fat. "Light" means the meal or main dish is low-fat or low-calorie.

4.3.2 Standardized Foods

Any nutrient content claim, such as "reduced fat," "low calorie," and "light," may be used in conjunction with a standardized term if the new product has been specifically formulated to meet FDA's criteria for that claim, if the product is not nutritionally inferior to the traditional standardized food, and if the new product complies with certain compositional requirements set by FDA. A new product bearing a claim also must have performance characteristics similar to the referenced traditional standardized food. If the product doesn't, and the differences materially limit the product's use, its label must state the differences (e.g., not recommended for baking) to inform consumers.

4.3.3 Healthy

A "healthy" food must be low in fat and saturated fat and contain limited amounts of cholesterol and sodium. In addition, if it is a single-item food, it must provide at least 10 percent of one or more of vitamins A or C, iron, calcium, protein, or fiber. Exempt from this "10 percent" rule are certain raw, canned and frozen fruits and vegetables and certain cereal-grain products. These foods can be labeled "healthy" if they do not contain ingredients that change the nutritional profile and, in the case of enriched grain products, conform to standards of identity, which call for certain required ingredients. If it is a meal-type product, such as frozen entrees and multi-course frozen dinners, it must provide 10 percent of two or three of these vitamins or minerals or of protein or fiber, in addition to meeting the other criteria. The sodium content cannot exceed 360 mg per serving for individual foods and 480 mg per serving for meal-type products.[12]

4.4 NUTRITION PANEL FORMAT

The details of the nutrition facts panel requirements are quite specific and numerous. Therefore this chapter only covers some salient points, not all the detail.

---

[12] For more information, see: FDA, A Food Labeling Guide—Appendix A, Definitions of Nutrient Content Claims.

Nutrients are declared as percentages of the Daily Values, which are label reference values. The amount, in grams or milligrams, of macronutrients (e.g., fat, cholesterol, sodium, carbohydrates, and protein) is listed to the immediate right of these nutrients. A column headed "% Daily Value" appears on the far right side.

Declaring nutrients as a percentage of the Daily Values is intended to prevent misinterpretations that arise with quantitative values. For example, a food with 140 milligrams (mg) of sodium could be mistaken for a high-sodium food because 140 is a relatively large number. In actuality, however, that amount represents less than 6 percent of the Daily Value for sodium, which is 2400 mg.

On the other hand, a food with 5 g of saturated fat could be construed as being low in that nutrient. In fact that food would provide one-fourth the total Daily Value because 20 g is the Daily Value for saturated fat.

* * * * *

The Food Label

FDA Backgrounder, http://www.cfsan.fda.gov/~dms/fdnewlab.html (May 1999)

. . . .

Format Modifications

In some circumstances, variations in the format of the nutrition panel are allowed. Some are mandatory. For example, the labels of foods for children under 2 (except infant formula, which has special labeling rules under the Infant Formula Act of 1980) may not carry information about saturated fat, polyunsaturated fat, monounsaturated fat, cholesterol, calories from fat, or calories from saturated fat.

The reason is to prevent parents from wrongly assuming that infants and toddlers should restrict their fat intake, when, in fact, they should not. Fat is important during these years to ensure adequate growth and development.

The labels of foods for children under 4 may not include the % Daily Values for total fat, saturated fat, cholesterol, sodium, potassium, total carbohydrate, and dietary fiber. They may carry percent Daily Values for protein, vitamins, and minerals, however. These nutrients are the only ones for which FDA has set Daily Values for this age group.

Thus, the top portion of the "Nutrition Facts" panels of foods for children under 4 will consist of two columns. The nutrients' names will be listed on the left and their quantitative amounts will be on the right. The bottom portion will provide the % Daily Values for protein, vitamins, and minerals. Only the calorie conversion information may be given as a footnote.

Some foods qualify for a simplified label format. This format is allowed when the food contains insignificant amounts of seven or more of the mandatory nutrients and total calories. "Insignificant" means that a declaration of zero could be made in nutrition labeling, or, for total carbohydrate, dietary fiber, and protein, the declaration states "less than 1 g."

For foods for children under 2, the simplified format may be used if the product contains insignificant amounts of six or more of the following: calories, total fat, sodium, total carbohydrate, dietary fiber, sugars, protein, vitamins A and C, calcium, and iron.

If the simplified format is used, information on total calories, total fat, total carbohydrate, protein, and sodium—even if they are present in insignificant amounts—must be listed. Other nutrients, along with calories from fat, must be shown if they are present in more than insignificant amounts. Nutrients added to the food must be listed, too.

Some format exceptions exist for small and medium-size packages. Packages with less than 12 square inches of available labeling space (about the size of a package of chewing gum) do not have to carry nutrition information unless a nutrient content or health claim is made for the product. However, they must provide an address or telephone number for consumers to obtain the required nutrition information.

If manufacturers wish to provide nutrition information on these packages voluntarily, they have several options: (1) present the information in a smaller type size than that required for larger packages, or (2) present the information in a tabular or linear (string) format.

The tabular and linear formats also may be used on packages that have less than 40 square inches available for labeling and insufficient space for the full vertical format.

Other options for packages with less than 40 square inches of label space are:

- abbreviating names of dietary components
- omitting all footnotes, except for the statement that "Percent Daily Values are based on a 2,000-calorie diet"
- placing nutrition information on other panels readily seen by consumers.

A select group of packages with more than 40 square inches of labeling space is allowed a format exception, too. These are packages with insufficient vertical space (about 3 inches) to accommodate the required information. Some examples are bread bags, pie boxes, and bags of frozen vegetables. On these packages, the "Nutrition Facts" panel may appear, in tabular format, with the footnote information appearing to the far right.

For larger packages in which there is not sufficient space on the principal display panel or the information panel (the panel to the right of the principal display), FDA allows nutrition information to appear on any label panel that is readily seen by consumers. This lessens the chances of overcrowding of information and encourages manufacturers to provide the greatest amount of nutrition information possible.

For products that require additional preparation before eating, such as dry cake mixes and dry pasta dinners, or that are usually eaten with one or more additional foods, such as breakfast cereals with milk, FDA encourages manufacturers to provide voluntarily a second column of nutrition information. This is known as dual declaration.

With this variation, the first column, which is mandatory, contains nutrition information for the food as purchased. The second gives information about the food as prepared and eaten.

Still another variation is the aggregate display. This is allowed on labels of variety-pack food items, such as ready-to-eat cereals and assorted flavors of individual ice cream cups. With this display, the quantitative amount and % Daily Value for each nutrient are listed in separate columns under the name of each food.

Serving Sizes

The serving size remains the basis for reporting each food's nutrient content. However, unlike in the past, when the serving size was up to the discretion of the food manufacturer, serving sizes now are more uniform and reflect the amounts people actually eat. They also must be expressed in both common household and metric measures.

FDA allows as common household measures: the cup, tablespoon, teaspoon, piece, slice, fraction (such as "$\frac{1}{4}$ pizza"), and common household containers used to package food products (such as a jar or tray). Ounces may be used, but only if a common household unit is not applicable and an appropriate visual unit is given—for example, 1 oz (28 g/about $\frac{1}{2}$ pickle). Grams (g) and milliliters (mL) are the metric units that are used in serving size statements.

NLEA defines serving size as the amount of food customarily eaten at one time. The serving sizes that appear on food labels are based on FDA-established lists of "Reference Amounts Customarily Consumed per Eating Occasion."

These reference amounts, which are part of the regulations, are broken down into 139 FDA-regulated food product categories, including 11 groups of foods specially formulated or processed for infants or children under 4. They list the amounts of food customarily consumed per eating occasion for each category, based primarily on national food consumption surveys. FDA's list also gives the suggested label statement for serving size declaration. For example, the category "breads (excluding sweet quick type), rolls" has a reference amount of 50 g, and the appropriate label statement for sliced bread or roll is "___piece(s) (___g)" or, for unsliced bread, "2 oz (56 g/___ inch slice)."

The serving size of products that come in discrete units, such as cookies, candy bars, and sliced products, is the number of whole units that most closely approximates the reference amount. Cookies are an example. Under the "bakery products" category, cookies have a reference amount of 30 g. The household measure closest to that amount is the number of cookies that comes closest to weighing 30 g. Thus the serving size on the label of a package of cookies in which each cookie weighs 13 g would read "2 cookies (26 g)."

If one unit weighs more than 50 percent but less than 200 percent of the reference amount, the serving size is one unit. For example, the reference amount for bread is 50 g; therefore, the label of a loaf of bread in which each slice weighs more than 25 g would state a serving size of one slice.

Certain rules apply to food products that are packaged and sold individually. If such an individual package is less than 200 percent of the applicable reference amount, the item qualifies as one serving. Thus a 360-mL (12-fluid-ounce) can of soda is one serving, since the reference amount for carbonated beverages is 240 mL (8 ounces).

However, if the product has a reference amount of 100 g or 100 mL or more and the package contains more than 150 percent but less than 200 percent of the reference amount, manufacturers have the option of deciding whether the product can be one or two servings.

An example is a 15-ounce (420 g) can of soup. The serving size reference amount for soup is 245 g. Therefore the manufacturer has the option to declare the can of soup as one or two servings.

Daily Values—DRVs

The new label reference value, Daily Value, comprises two sets of dietary standards: Daily Reference Values (DRVs) and Reference Daily Intakes (RDIs). Only the Daily Value term appears on the label, though, to make label reading less confusing.

DRVs have been established for macronutrients that are sources of energy: fat, saturated fat, total carbohydrate (including fiber), and protein; and for cholesterol, sodium, and potassium, which do not contribute calories.

DRVs for the energy-producing nutrients are based on the number of calories consumed per day. A daily intake of 2,000 calories has been established as the reference. This level was chosen, in part, because it approximates the caloric requirements for postmenopausal women. This group has the highest risk for excessive intake of calories and fat.

DRVs for the energy-producing nutrients are calculated as:

- fat based on 30 percent of calories
- saturated fat based on 10 percent of calories
- carbohydrate based on 60 percent of calories
- protein based on 10 percent of calories. (The DRV for protein applies only to adults and children over 4. RDIs for protein for special groups have been established.)
- fiber based on 11.5 g of fiber per 1,000 calories.

Because of current public health recommendations, DRVs for some nutrients represent the uppermost limit that is considered desirable. The DRVs for total fat, saturated fat, cholesterol, and sodium are:

- total fat: less than 65 g
- saturated fat: less than 20 g
- cholesterol: less than 300 mg
- sodium: less than 2,400 mg

Daily Values—RDIs

"Reference Daily Intake" replaces the term "U.S. RDA," which was introduced in 1973 as a label reference value for vitamins, minerals, and protein in voluntary nutrition labeling. The name change was sought because of confusion that existed over "U.S. RDAs," the values determined by FDA and used on food labels, and "RDAs" (Recommended Dietary Allowances), the values determined by the National Academy of Sciences for various population groups and used by FDA to figure the U.S. RDAs. However, the values for the new RDIs remain the same as the old U.S. RDAs for the time being.

Baby Foods

FDA is not allowing broad use of nutrient claims on infant and toddler foods. However, the agency may propose claims specifically for these foods at a later date. The terms "unsweetened" and "unsalted" are allowed on these foods, however, because they relate to taste and not nutrient content.

* * * * *

DISCUSSION QUESTION

4.4. Why did FDA decide not to allow nutrient claims on infant and toddler foods?

4.5 *TRANS* FATS

Trans fats like partially hydrogenated vegetable oil are the call girls of the food supply: they're cheap, they're easy, they're everywhere, they'll do whatever you want, and they'll leave you feeling lousy afterwards.

—Gersh Kuntzman, NEWSWEEK[13]

[13] Gersh Kuntzman, *The Cookie Crumbles*, NEWSWEEK (May 2, 2005), *available at*: http://www.msnbc.msn.com/id/7711463/site/newsweek/ (last accessed Sept. 15, 2007).

4.5.1   Background on *Trans* Fat

Findings from human feeding studies and epidemiological studies show a positive association between the intake of *trans* fatty acids and the incidence of coronary heart disease. Walter Willett, professor of epidemiology at Harvard School of Public Health, in 1997 estimated that the use of hydrogenated oils was resulting in 30,000 heart-disease deaths a year, representing "the biggest food processing disaster in U.S. history."[14]

The Dietary Guidelines for Americans 2000 makes the following statements regarding *trans* fatty acids and food sources of *trans* fatty acids ("*trans* fat"):

> Foods high in *trans* fatty acids tend to raise blood cholesterol. These foods include those high in partially hydrogenated vegetable oils, such as many hard margarines and shortenings. Foods with a high amount of these ingredients include some commercially fried foods and some bakery goods. Aim for a total fat intake of not more than 30 percent of calories, as recommended in previous Guidelines. If you need to reduce your fat intake to achieve this level, do so primarily by cutting back on saturated and *trans* fats.

Most *trans* fatty acids are created in the hydrogenation of vegetable oil. Hydrogenation is the forcing of hydrogen atoms into the double bonds of unsaturated oil. This saturation of the oil is accomplished with high pressure, heat, and catalysts. Unfortunately, partially hydrogenated fats, along with *trans* fat, can be found in "everything you love to eat: margarine, commercial cakes and cookies, doughnuts, potato chips, crackers, popcorn, nondairy creamers, whipped toppings, gravy mixes, cake mixes, frozen French fries and pizzas, fish sticks and virtually all fried foods, unless you fry them yourself in unhydrogenated oils."[15]

4.5.2   Petition to Ban Hydrogenated Oil

"In 2003, the National Academies' Institute of Medicine concluded that the only safe level of *trans* fat in the diet is zero, and in 2004 an FDA advisory panel concluded [that] *trans* fat is even more harmful than saturated fat."[16] For this reason the Center for Science in the Public Interest (CSPI) in 2004 proposed revoking GRAS status for hydrogenated oil that contains *trans* fatty acids.[17] "Unlike fats that occur in nature, partially hydrogenated vegetable oil is totally artificial and absolutely unnecessary in the food supply,"

---

[14] Amanda Spake, *The Truth on Foods and Fats*, 124, 126, U.S. News & World Reports (2004).

[15] Robert L. Wolke, *Trans Fat Translation*, Washingtonpost.com, Page F01 (Aug. 20, 2003).

[16] *CSPI petitions FDA to ban hydrogenated vegetable oil*, Food Chemical News Daily, Vol. 6, No. 96 (May 19, 2004).

[17] CSPI's entire petition to FDA is available on their Web site: www.cspi.org.

said CSPI's Michael Jacobson. "Food-processing companies should worry less about the shelf life of their products and more about the shelf life of their customers. Getting rid of partially hydrogenated vegetable oil is probably the single easiest, fastest, cheapest way to save tens of thousands of lives each year."[18]

The National Food Processors Association called the petition the wrong way to address the issue because "Nutrition experts—including FDA—have called for consumers to choose diets low in *trans* fats, not to eliminate them. Nutrition experts also have cautioned consumers, in their efforts to reduce *trans* fat intake, against making dietary choices that lead to a nutritionally inadequate diet or that have other unintended effects, such as replacing *trans* fats in their diets with saturated fats."[19]

4.5.3 Highlights of the *Trans* Fat Rule

Rather than ban *trans* fat, FDA took a more moderate approach. FDA promulgated a rule to require the labeling *trans* fat in packaged foods.[20] The FDA final rule requires that the amount of *trans* fat in a serving be listed on a separate line under saturated fat on the Nutrition Facts panel. However, *trans* fat does not have to be listed if the total fat in a food is less than half a gram per serving and provided that no claims are made about fat, fatty acids, or cholesterol content.

4.6 HEALTH CLAIMS

A **health claim** is defined as any claim made on the label or labeling that expressly or by implication characterizes the relationship of any substance to a disease or health-related condition.[21] Note how broad this definition is. Particularly note that an *implied* association may trigger health claim regulation.

Manufacturers may make certain claims linking the effect of a nutrient or food to a disease or health-related condition, but only those claims supported by scientific evidence are allowed. In addition these claims can be used only under specific conditions, such as when the food is an adequate source of the appropriate nutrients.

The ability to make a health claim on a food product is a substantial marketing tool in today's health-conscious society. Therefore the claims are regulated tightly. However, ameliorating somewhat this strictness is the fact that

---

[18] *Id.*

[19] *Id.* (quoting Regina Hildwine).

[20] *See* Trans Fatty Acids in Nutrition Labeling, Nutrient Content Claims, and Health Claims, 68 Fed. Reg. 41433–41506 (July 11, 2003), *also available at*: http://www.cfsan.fda.gov/~lrd/fr03711a.html (last accessed Mar. 10, 2008).

[21] 21 C.F.R. § 101.14(a)(1).

there are three different types of health-related claims that are not regulated as health claims. These are called statements of nutritional support:[22]

1. **Descriptions of general well-being** from consumption of the food.
2. **Classical nutrient-deficiency disease** and nutrition.
3. **Structure-function claims.**

In addition there are three different regulatory categories of health claims that may be used on a label or in labeling for a food:

4. **Pre-approved claims.** These are authorized by the FDA under the Nutrition Labeling and Education Act (NLEA) of 1990 through the promulgation of a regulation authorizing the health claim.
5. **Authoritative statements claims.** An authoritative statement from a scientific body of the U.S. government or the National Academy of Sciences may form the basis of a health claim under provision of the 1997 Food and Drug Administration Modernization Act (FDAMA). FDAMA authorizes health claims based on these authoritative statements after submission of a health claim notification to FDA. An example of an authoritative-statement claim permitted is, "Diets high in plant foods—i.e., fruits, vegetables, legumes, and whole-grain cereals—are associated with a lower occurrence of coronary heart disease and cancers of the lung, colon, esophagus, and stomach."
6. **Qualified claims.** If the quality and strength of the scientific evidence falls below that required for FDA to issue an authorizing regulation, the health claims must be qualified to assure accuracy and nonmisleading presentation to consumers. An example of a qualified health claim is, "Supportive but not conclusive research shows that eating 1.5 ounces per day of walnuts, as part of a low saturated fat and low cholesterol diet and not resulting in increased caloric intake, may reduce the risk of coronary heart disease. See nutrition information for fat [and calorie] content."[23]

The differences between these methods of oversight for health claims are summarized below.

4.6.1 General Well-Being Claims

General well-being claims are statements that describe general well-being from consumption of a nutrient or dietary ingredient. A key to general

---

[22] Some writers group all of statements of nutritional support as structure/function claims, but the author finds this categorization unhelpful.

[23] FDA, Qualified Health Claims: Letter of Enforcement Discretion—Walnuts and Coronary Heart Disease (Mar. 9, 2004), *available at*: http://www.cfsan.fda.gov/~dms/qhcnuts3.html (last visited Mar. 14, 2006). (Docket No 02P-0292).

well-being claims is they do not mention a disease or disease-related condition. An example of a general well-being claim would be a claim that a multi-vitamin contributes to general good health.

4.6.2 Classical Nutrient-Deficiency Disease and Nutrition

These statements that describe a benefit related to a nutrient deficiency disease (e.g., vitamin C and scurvy) are permitted as long as the statement also tells how widespread such a disease is in the United States. These claims have little use in the United States because fortification has eliminated most of the classical nutrient-deficiency diseases.

4.6.3 Structure-Function Claims

Although structure-function claims are health-related claims, they are not "health claims" under the law. Health claims characterize the relationship between a substance and its ability to reduce the risk of a disease or health-related condition.

Structure-function claims describe the effect that a substance has on the *normal* structure or function of the body. The critical distinction here centers on normal versus diseases. For example, "Calcium builds strong bones" is a structure-function claim about normal bone development. Mention of osteoporosis or other disease (or even implying relationship to disease) would create a health claim.

Other examples of structure-function claims are, "Fiber maintains bowel regularity," and "antioxidants maintain cell integrity." These claims focus on maintaining or supporting normal body structures or functions, and do not focus on disease.

Structure-function claims may appear on the labels of foods and dietary supplements without any formal review or premarket approval by FDA.[24] However, the general FD&C Act requirements still apply and the claims must be truthful and nonmisleading.

Structure-function claims have historically appeared on the labels of conventional foods and dietary supplements as well as drugs. When used with conventional foods, structure-function claims must be based on the "nutritive" value of the food. However, FDA has not defined "nutritive value."[25]

The Dietary Supplement Health and Education Act of 1994 (DSHEA) established some special regulatory procedures for such claims for dietary

---

[24] 21 U.S.C. § 343(r)(6) and 21 C.F.R. § 101.93.

[25] This intersection of drug, dietary supplement, and conventional food has become extremely complicated. In a rare moment of regulatory candor, FDA recognized that its distinctions in this area sometimes fly in the face of common sense. *See* FDA, CFSAN/Office of Nutritional Products, Labeling, and Dietary Supplements, Discussion of a Conceptual Framework for Structure and Function Claims For Conventional Foods, Meeting Summary (Feb. 16–17, 2000), *available at*: http://www.cfsan.fda.gov/~dms/labstru2.html (last accessed Mar. 14, 2008).

supplement labels. These are discussed in more detail in a later chapter, but in summary: if a dietary supplement label makes such a claim, it must include a "disclaimer" that FDA has not evaluated the claim. The disclaimer must also state that the dietary supplement product is not intended to "diagnose, treat, cure or prevent any disease." Manufacturers of dietary supplements that make structure-function claims on labels or in labeling must submit a notification to FDA no later than 30 days after marketing the dietary supplement that includes the text of the structure-function claim.

4.6.4   Pre-approved Health Claims (NLEA)

Authorized health claims under the Significant Scientific Agreement standard are those claims expressly authorized by an FDA regulation under the authority provided by NLEA. Under authorized health claims provision of the FD&C Act (as amended by the NLEA), no food product may make such a claim unless:

1. expressly authorized by a specific regulation,
2. the claim complies with the terms of the regulation.

Claims can be made in several ways: through third-party references (e.g., the National Cancer Institute), statements, symbols (e.g., a heart), and vignettes or descriptions. Whatever the type, the claim must meet the requirements for authorized health claims. For example, the claim cannot state the degree of risk reduction and can only use "may" or "might" in discussing the nutrient or food–disease relationship. And the claim must state that other factors play a role in that disease. The claims also must be phrased so that consumers can understand the relationship between the nutrient and the disease, and the nutrient's importance in relationship to a daily diet. An example of an appropriate claim is: "While many factors affect heart disease, diets low in saturated fat and cholesterol may reduce the risk of this disease."

* * * * *

Staking a Claim to Good Health. FDA and Science Stand Behind Health Claims on Foods

Paula Kurtzweil, FDA CONSUMER, SPECIAL ISSUE, FOCUS ON FOOD LABELING[26]

Health claims authorized by the Food and Drug Administration are one of several ways food labels can win the attention of health-conscious consumers.

These claims alert shoppers to a product's health potential by stating that certain foods or food substances—as part of an overall healthy diet—may

---

[26] Also *available at*: http://www.fda.gov/fdac/features/1998/698_labl.html (last visited Mar. 15, 2008).

reduce the risk of certain diseases. Examples include folic acid in breakfast cereals, fiber in fruits and vegetables, calcium in dairy products, and calcium or folic acid in some dietary supplements. But food and food substances can qualify for health claims only if they meet FDA requirements.

"Health claims are not your fad-of-the-week," says Jim Hoadley, Ph.D., a senior regulatory scientist in FDA's Office of Food Labeling. Instead, he says, for health claims to be used, there needs to be sufficient scientific agreement among qualified experts that the claims are factual and truthful. . . .

Under NLEA, companies petition FDA to consider new health claims through rule-making. However, this process may require more than a year to complete because of the necessary scientific review and the need to issue a proposed rule to allow for public comment. And, in an effort to speed more of this kind of information to consumers, the Food and Drug Administration Modernization Act of 1997 includes a provision that is intended to expedite the process that establishes the scientific basis for health claims.

Although food manufacturers may use health claims to market their products, the intended purpose of health claims is to benefit consumers by providing information on healthful eating patterns that may help reduce the risk of heart disease, cancer, osteoporosis, high blood pressure, dental cavities, or certain birth defects.

What Is a Health Claim?

Health claims are among the various types of claims allowed in food labeling. They show a relationship between a nutrient or other substances in a food and a disease or health-related condition. They can be used on conventional foods or dietary supplements.

They differ from the more common claims that highlight a food's nutritional content, such as "low fat," "high fiber," and "low calorie." . . .

Health claims can include implied claims, which indirectly assert a diet–disease relationship. Implied claims may appear in brand names (such as "Heart Healthy"), symbols (such as a heart-shaped logo), and vignettes when used with specific nutrient information. However, all labels bearing implied claims must also bear the full health claim [that is, the complete language required by the regulation].

Public Confidence

Health claims became a hot issue in the 1980s, when food marketing strategies began reflecting increased recognition of the role of nutrition in promoting health. At that time, some of the claims used were considered misleading, and many consumers began to doubt their truthfulness. NLEA's intent, in part, was to rein in exaggerated claims by reinforcing FDA's authority to regulate health claims and to require that claims be supported by sufficient scientific evidence.

According to an FDA study, consumer confidence in health claims grew in the months following implementation of NLEA. Thirty-one percent of consumers contacted by phone in November 1995–17 months after implementation of NLEA—said they believed health claims were accurate, compared with 25 percent in March 1994, two months before NLEA went into effect. And fewer respondents—39 percent in 1995 compared with 47 percent in 1994—agreed with the statement "Claims are more like advertising than anything else."

FDA's phone survey also indicated more consumers were using health claims to make more informed food choices: 25 percent in 1995 said they were using health claims, compared with 20 percent in March 1994.

According to Brenda Derby, a statistician in the consumer studies branch of FDA's Center for Food Safety and Applied Nutrition, a 1996 FDA label-reading study of more than 1,400 grocery shoppers found that, in general, the effectiveness of health claims is similar to that of nutrient claims and had no greater effect than nutrient claims alone in influencing shoppers' purchasing decisions. Health claims are most effective when they provide consumers with new information, the study found.

. . . .

* * * * *

Discussed next are the FDA-authorized health claims and some specifics on their use.[27]

Calcium and Osteoporosis[28]   Low calcium intake is one risk factor for osteoporosis. Lifelong adequate calcium intake helps maintain bone health by increasing as much as genetically possible the amount of bone formed in the teens and early adult life and by helping slow the rate of bone loss that occurs later in life.

Claim Requirements   Food or supplement must be "high" in calcium and must not contain more phosphorus than calcium. Claims must cite other risk factors, state the need for regular exercise and a healthful diet, explain that adequate calcium early in life helps reduce fracture risk later by increasing as much as genetically possible a person's peak bone mass, and indicate that those at greatest risk of developing osteoporosis later in life are white and Asian teenage and young adult women presently in their bone-forming years. Claims for products with more than 400 mg of calcium per day must state that a daily intake over 2,000 mg offers no added known benefit to bone health.

---

[27] Adapted from Paula Kurtzweil, *Staking a Claim to Good Health*, FDA CONSUMER, SPECIAL ISSUE, FOCUS ON FOOD LABELING (Nov.–Dec. 1998), *available at*: http://www.fda.gov/fdac/features/1998/698_labl.html (last visited Mar. 15, 2008).

[28] 21 C.F.R. § 101.72.

Sample Claim "Regular exercise and a healthy diet with enough calcium helps teen and young adult white and Asian women maintain good bone health and may reduce their high risk of osteoporosis later in life."

Dietary Fat and Cancer[29] Diets high in fat increase the risk of some types of cancer, such as cancers of the breast, colon, and prostate. While scientists do not know how total fat intake affects cancer development, low-fat diets reduce the risk. Experts recommend that Americans consume 30 percent or less of daily calories as fat.

Claim Requirements Foods must meet criteria for "low fat." Fish and game meats must meet criteria for "extra lean." Claims may not mention specific types of fats and must use "total fat" or "fat" and "some types of cancer" or "some cancers" in discussing the nutrient–disease link.

Sample Claim "Development of cancer depends on many factors. A diet low in total fat may reduce the risk of some cancers."

Dietary Saturated Fat and Cholesterol and Risk of Coronary Heart Disease[30] Diets high in saturated fat and cholesterol increase total and low-density (bad) blood cholesterol levels, and thus the risk of coronary heart disease. Diets low in saturated fat and cholesterol decrease the risk. Guidelines recommend that American diets contain less than 10 percent of calories from saturated fat and less than 300 mg cholesterol daily. The average American adult diet has 13 percent saturated fat and 300 to 400 mg cholesterol a day.

Claim Requirements Foods must meet criteria for "low saturated fat," "low cholesterol," and "low fat." Fish and game meats must meet criteria for "extra lean." Claims must use "saturated fat and cholesterol" and "coronary heart disease" or "heart disease" in discussing the nutrient–disease link.

Sample Claim "While many factors affect heart disease, diets low in saturated fat and cholesterol may reduce the risk of this disease."

Sodium and Hypertension (High Blood Pressure)[31] Hypertension is a risk factor for coronary heart disease and stroke deaths. The most common source of sodium is table salt. Diets low in sodium may help lower blood pressure and related risks in many people. Guidelines recommend daily sodium intakes of not more than 2,400 mg.

[29] 21 C.F.R. § 101.73.
[30] 21 C.F.R. § 101.75.
[31] 21 C.F.R. § 101.74.

Claim Requirements   Foods must meet criteria for "low sodium." Claims must use "sodium" and "high blood pressure" in discussing the nutrient–disease link.

Sample Claim   "Diets low in sodium may reduce the risk of high blood pressure, a disease associated with many factors."

Fiber-Containing Grain Products, Fruits, and Vegetables and Cancer[32]   Diets low in fat and rich in fiber-containing grain products, fruits, and vegetables may reduce the risk of some types of cancer. The exact role of total dietary fiber, fiber components, and other nutrients and substances in these foods is not fully understood.

Claim Requirements   Foods must meet criteria for "low fat" and, without fortification, be a "good source" of dietary fiber. Claims must not specify types of fiber and must use "fiber," "dietary fiber," or "total dietary fiber" and "some types of cancer" or "some cancers" in discussing the nutrient–disease link.

Sample Claim   "Low-fat diets rich in fiber-containing grain products, fruits, and vegetables may reduce the risk of some types of cancer, a disease associated with many factors."

Fruits, Vegetables, and Grain Products That Contain Fiber, Particularly Soluble Fiber, and Risk of Coronary Heart Disease[33]   Diets low in saturated fat and cholesterol and rich in fruits, vegetables, and grain products that contain fiber, particularly soluble fiber, may reduce the risk of coronary heart disease. (It is impossible to adequately distinguish the effects of fiber, including soluble fiber, from those of other food components.)

Claim Requirements   Foods must meet criteria for "low saturated fat," "low fat," and "low cholesterol." They must contain, without fortification, at least 0.6 g of soluble fiber per reference amount, and the soluble fiber content must be listed. Claims must use "fiber," "dietary fiber," "some types of dietary fiber," "some dietary fibers," or "some fibers" and "coronary heart disease" or "heart disease" in discussing the nutrient–disease link. The term "soluble fiber" may be added.

Sample Claim   "Diets low in saturated fat and cholesterol and rich in fruits, vegetables, and grain products that contain some types of dietary fiber, particularly soluble fiber, may reduce the risk of heart disease, a disease associated with many factors."

---

[32] 21 C.F.R. § 101.76.
[33] 21 C.F.R. § 101.77.

Fruits and Vegetables and Cancer[34]   Diets low in fat and rich in fruits and vegetables may reduce the risk of some cancers. Fruits and vegetables are low-fat foods and may contain fiber or vitamin A (as beta-carotene) and vitamin C. (The effects of these vitamins cannot be adequately distinguished from those of other fruit or vegetable components.)

Claim Requirements   Foods must meet criteria for "low fat" and, without fortification, be a "good source" of fiber, vitamin A, or vitamin C. Claims must characterize fruits and vegetables as foods that are low in fat and may contain dietary fiber, vitamin A, or vitamin C; characterize the food itself as a "good source" of one or more of these nutrients, which must be listed; refrain from specifying types of fatty acids; and use "total fat" or "fat," "some types of cancer" or "some cancers," and "fiber," "dietary fiber," or "total dietary fiber" in discussing the nutrient–disease link.

Sample Claim   "Low-fat diets rich in fruits and vegetables (foods that are low in fat and may contain dietary fiber, vitamin A, or vitamin C) may reduce the risk of some types of cancer, a disease associated with many factors. Broccoli is high in vitamins A and C, and it is a good source of dietary fiber."

Folate and Neural Tube Birth Defects[35]   Defects of the neural tube (a structure that develops into the brain and spinal cord) occur within the first six weeks after conception, often before the pregnancy is known. The U.S. Public Health Service recommends that all women of childbearing age in the United States consume 0.4 mg (400 mcg) of folic acid daily to reduce their risk of having a baby affected with spina bifida or other neural tube defects.

Typical Foods   Enriched cereal grain products, some legumes (dried beans), peas, fresh leafy green vegetables, oranges, grapefruit, many berries, some dietary supplements, and fortified breakfast cereals.

Claim Requirements   Foods must meet or exceed criteria for "good source" of folate—that is, at least 40 mcg of folic acid per serving (at least 10 percent of the Daily Value). A serving of food cannot contain more than 100 percent of the Daily Value for vitamin A and vitamin D because of their potential risk to fetuses. Claims must use "folate," "folic acid," or "folacin" and "neural tube defects," "birth defects spina bifida or anencephaly," "birth defects of the brain or spinal cord anencephaly or spina bifida," "spina bifida and anencephaly, birth defects of the brain or spinal cord," "birth defects of the brain and spinal cord," or "brain or spinal cord birth defects" in discussing the nutrient–disease link. Folic acid content must be listed on the Nutrition Facts panel.

[34] 21 C.F.R. § 101.78.
[35] 21 C.F.R. § 101.79.

Sample Claim   "Healthful diets with adequate folate may reduce a woman's risk of having a child with a brain or spinal cord birth defect."

Dietary Noncariogenic Carbohydrate Sweeteners and Dental Caries (Cavities)[36]   Between-meal eating of foods high in sugar and starches may promote tooth decay. Sugarless candies made with certain sugar alcohols do not.

Claim Requirements   Foods must meet the criteria for "sugar free." The sugar alcohol must be xylitol, sorbitol, mannitol, maltitol, isomalt, lactitol, hydrogenated starch hydrolysates, hydrogenated glucose syrups, erythritol, or a combination of these. When the food contains a fermentable carbohydrate, such as sugar or flour, the food must not lower plaque pH in the mouth below 5.7 while it is being eaten or up to 30 minutes afterward. Claims must use "sugar alcohol," "sugar alcohols," or the name(s) of the sugar alcohol present and "dental caries" or "tooth decay" in discussing the nutrient–disease link. Claims must state that the sugar alcohol present "does not promote," "may reduce the risk of," "is useful in not promoting," or "is expressly for not promoting" dental caries.

Sample Claim   Full claim: "Frequent between-meal consumption of foods high in sugars and starches promotes tooth decay. The sugar alcohols in this food do not promote tooth decay." Shortened claim (on small packages only): "Does not promote tooth decay."

Dietary Soluble Fiber, Such as That Found in Whole Oats and Psyllium Seed Husk, and Coronary Heart Disease[37]   When included in a diet low in saturated fat and cholesterol, soluble fiber may affect blood lipid levels, such as cholesterol, and thus lower the risk of heart disease. However, because soluble dietary fibers constitute a family of very heterogeneous substances that vary greatly in their effect on the risk of heart disease, FDA has determined that sources of soluble fiber for this health claim need to be considered case by case. To date, FDA has reviewed and authorized two sources of soluble fiber eligible for this claim: whole oats and psyllium seed husk.

Claim Requirements   Foods must meet criteria for "low saturated fat," "low cholesterol," and "low fat." Foods that contain whole oats must contain at least 0.75 g of soluble fiber per serving. Foods that contain psyllium seed husk must contain at least 1.7 g of soluble fiber per serving. The claim must specify the daily dietary intake of the soluble fiber source necessary to reduce the risk of heart disease and the contribution one serving of the product makes toward that intake level. Soluble fiber content must be stated in the nutrition label. Claims must use "soluble fiber" qualified by the name of the eligible source

[36] 21 C.F.R. § 101.80.
[37] 21 C.F.R. § 101.81.

of soluble fiber and "heart disease" or "coronary heart disease" in discussing the nutrient–disease link. Because of the potential hazard of choking, foods containing dry or incompletely hydrated psyllium seed husk must carry a label statement telling consumers to drink adequate amounts of fluid, unless the manufacturer shows that a viscous adhesive mass is not formed when the food is exposed to fluid.

Sample Claim   "Diets low in saturated fat and cholesterol that include 3 g of soluble fiber from whole oats per day may reduce the risk of heart disease. One serving of this whole-oats product provides ____ grams of this soluble fiber."

Soy Protein and Coronary Heart Disease[38]   There is an association between soy protein and reduced risk of coronary heart disease (CHD) when included in a diet low in saturated fat and cholesterol through the lowering of blood cholesterol levels.

Claim Requirements   In order to qualify for this health claim, a food must contain at least 6.25 g of soy protein per serving, the amount that is one-fourth of the effective level of 25 g per day.

Sample Claim   "Diets low in saturated fat and cholesterol that include 25 g of soy protein a day may reduce the risk of heart disease. One serving of (name of food) provides ____ g of soy protein."

Coronary Heart Disease and Plant Sterols and Plant Stanols[39]   Evidence indicates that plant sterol or plant stanol esters help reduce the risk of coronary heart disease (CHD). Plant sterols are present in small quantities in many fruits, vegetables, nuts, seeds, cereals, legumes, and other plant sources. Plant stanols, which occur naturally in even smaller quantities, are obtained from refined plant sources, such as vegetable oils.

Claim Requirements   In order to qualify for this health claim, a food must contain at least 0.65 g of plant sterol esters per serving or at least 1.7 g of plant stanol esters per serving. The claim must specify that the daily dietary intake of plant sterol esters or plant stanol esters should be consumed in two servings eaten at different times of the day with other foods. To qualify, foods must also meet the requirements for low saturated fat and low cholesterol, and must also contain no more than 13 g of total fat per serving and per 50 g. However, spreads and salad dressings are not required to meet the limit for total fat per 50 g if the label of the food bears a disclosure statement referring

[38] 21 C.F.R. 101.82.
[39] 21 C.F.R. 101.83.

consumers to the Nutrition Facts section of the label for information about fat content. In addition, except for salad dressing and dietary supplements, the food must contain at least 10 percent of the Reference Daily Intake (RDI) or Daily Reference Value (DRV) for vitamin A, vitamin C, iron, calcium, protein, or fiber. FDA is also requiring, consistent with other health claims to reduce the risk of CHD, that the claim state that plant sterol and plant stanol esters should be consumed as part of a diet low in saturated fat and cholesterol.

Sample Claim "Foods containing at least 0.65 g per serving of plant sterol esters, eaten twice a day with meals for a daily total intake of at least 1.3 g, as part of a diet low in saturated fat and cholesterol, may reduce the risk of heart disease. A serving of (name of the food) supplies ____ g of plant sterol esters."

4.6.5 Authoritative Statements—FDA Modernization Act of 1997

Before the Food and Drug Administration Modernization Act of 1997 (FDAMA) companies could not use a health claim or nutrient content claim in food labeling unless the FDA published a regulation authorizing such a claim. Two new provisions of FDAMA[40] permit distributors and manufacturers to use claims if based on current, published, authoritative statements from certain federal scientific bodies. The National Academy of Sciences (NAS), the National Institutes of Health (NIH), and the Centers for Disease Control and Prevention (CDC) are federal government agencies specifically identified as scientific bodies by FDAMA.

FDAMA's provisions were intended to expedite the process by which health claims can be established and used. FDA interpreted "authoritative statements" so that they must reflect a consensus within the identified scientific body and be based on a deliberative review by the scientific body of the scientific evidence. In theory, the authoritative-statement standard is slightly less stringent than FDA's prior requirement for "significant scientific agreement." Particularly, FDAMA allows companies to notify FDA of their intent to use a new health claim based on an authoritative statement of only one federal scientific body, rather than show scientific agreement. However, in application, the standards show little difference.

From a process standpoint, FDAMA did provide procedural change to expedite review. FDAMA gives FDA 120 days to respond to new health claim proposals. If the agency does not act to prohibit or modify the claim within that time, the claim can be used.

Nevertheless, FDAMA's provisions to expedite approval did not meet the desires of everyone in the food industry. FDAMA sped up FDA's review, but

---

[40] Specifically, §§ 303 and 304, which amend, respectively, §§ 403(r)(3) and 403(r)(2) (21 U.S.C. §§ 343(r)(3) and (2)) of the Food, Drug, and Cosmetic Act.

the expedited review often resulted in denial. Two years after FDAMA, two food marketers sued over such a denial. The result was the landmark case, *Pearson v. Shalala*, which is discussed later in this chapter.

FDA has prepared a guide on how a firm can make use of authoritative statement-based health claims.[41] FDAMA does not include dietary supplements in the provisions for health claims based on authoritative statements. Consequently this method of oversight for health claims cannot be used for dietary supplements at this time.

As this book was being written, three health claims based on authoritative statements were approved:[42]

- Whole grain foods and risk of heart disease and certain cancers
- Potassium and the risk of high blood pressure and stroke
- Whole grain foods with moderate fat content and heart disease[43]

4.6.6 Qualified Claims—*After Pearson v. Shalala*[44]

Pearson began when the FDA rejected four proposed health claims by the *Pearson* plaintiffs. These four claims linked the consumption of a particular food (supplement) to the reduction in risk of a particular disease:

1. "Consumption of antioxidant vitamins may reduce the risk of certain kinds of cancers."
2. "Consumption of fiber may reduce the risk of colorectal cancer."
3. "Consumption of omega-3 fatty acids may reduce the risk of coronary heart disease."
4. "0.8 mg of folic acid in a dietary supplement is more effective in reducing the risk of neural tube defects than a lower amount in foods in common form."

Relying on arguments grounded in the First Amendment and the Administrative Procedures Act (APA), the U.S. Court of Appeals invalidated the

---

[41] FDA, GUIDANCE FOR INDUSTRY, NOTIFICATION OF A HEALTH CLAIM OR NUTRIENT CONTENT CLAIM BASED ON AN AUTHORITATIVE STATEMENT OF A SCIENTIFIC BODY (June 11, 1998), *available at*: http://www.cfsan.fda.gov/~dms/hclmguid.html (last accessed Mar. 15, 2008).

[42] FDA, FOOD LABELING GUIDE, APPENDIX C (Revised Nov. 2000), *available at*: http://www.cfsan.fda.gov/~dms/flg-6c.html (last accessed Mar. 15, 2008). There is also a nutrient-content claim for choline available based on authoritative statements. *See supra* the discussion of nutrient content claims.

[43] FDA, HEALTH CLAIM NOTIFICATION FOR WHOLE GRAIN FOODS WITH MODERATE FAT CONTENT (Dec. 9, 2003), *available at*: http://www.cfsan.fda.gov/~dms/flgrain2.html (last accessed Mar. 15, 2008).

[44] 164 F.3d 650 (D.C. Cir. 1999).

FDA regulations prohibiting those health claims on foods and required the FDA to reconsider its disapproval of the plaintiffs' claims. To briefly summarize, the court ruled that the FDA (1) violated the First Amendment by banning misleading health claims without considering the use of curative disclaimers, and (2) violated the arbitrary and capricious standard of the Administrative Procedure Act (APA) by failing to clarify the standard of "significant scientific agreement."

Health claims are a form of "commercial speech," and under First Amendment protections, the FDA cannot unnecessarily restrain such speech. FDA argued that health claims lacking "significant scientific agreement" are inherently misleading to consumers and, therefore, are incapable of being cured by disclaimers. However, the Court of Appeals ruled that the FDA had no basis to reject the health claims without first assessing whether the use of a disclaimer could communicate meaningful, nonmisleading information to the consumer. Where commercial speech is potentially misleading but can be "presented in a way that is not deceptive," the government cannot ban it. For example, a disclaimer might be able to communicate that available scientific evidence is inconclusive regarding the dietary substance and disease relationship because the studies performed have been on foods containing those components and not on the dietary substances themselves.

The court also found that FDA had not followed appropriate administrative procedures because it failed to fully explain why the four health claims did not meet the "significant scientific agreement" standard applicable to health claims. The FDA had not defined the criteria being applied to determine whether such agreement exists. The Court noted the legal and practical need to provide a governing rationale for approving or rejecting proposed health claims on the basis of a lack of "significant scientific agreement." The court concluded that FDA's denial of these health claims without defining "significant scientific agreement" constituted arbitrary and capricious action under the APA. Accordingly, the court ordered FDA to explain the meaning of "significant scientific agreement." At a minimum, the FDA must make it possible "for the regulated class to perceive the principles which are guiding agency action."

The decision created legal hurdles to FDA's efforts to reject petitions filed in support of health claims. However, the decision did not permit the plaintiffs in *Pearson* to make their health claims with disclaimers without any further pre-clearance by FDA. The decision directed FDA to reconsider the plaintiffs' four proposed claims in light of possible value of disclaimers. Basically the decision invalidated FDA's regulations but put the *Pearson* plaintiffs back at square one in the FDA pre-clearance process. In addition the Court did not rule out the possibility that "where evidence in support of a claim is outweighed by evidence against the claim," FDA could deem the claim "incurable" by a disclaimer and, therefore, reject the claim as unlawful.

* * * * *

Pearson v. Shalala
164 F.3d 650 (D.C. Cir. 1999)

Before: WALD, SILBERMAN, and GARLAND, Circuit Judges
Opinion: SILBERMAN

Marketers of dietary supplements must, before including on their labels a claim characterizing the relationship of the supplement to a disease or health-related condition, submit the claim to the Food and Drug Administration for preapproval. The FDA authorizes a claim only if it finds "significant scientific agreement" among experts that the claim is supported by the available evidence. Appellants failed to persuade the FDA to authorize four such claims and sought relief in the district court, where their various constitutional and statutory challenges were rejected. We reverse.

I

Dietary supplement marketers Durk Pearson and Sandy Shaw, presumably hoping to bolster sales by increasing the allure of their supplements' labels, asked the FDA to authorize four separate health claims. . . . A "health claim" is a "claim made on the label or in labeling of . . . a dietary supplement that expressly or by implication . . . characterizes the relationship of any substance to a disease or health-related condition." 21 C.F.R. § 101.14(a)(1) (1998). Each of appellants' four claims links the consumption of a particular supplement to the reduction in risk of a particular disease:

(1) "Consumption of antioxidant vitamins may reduce the risk of certain kinds of cancers."
(2) "Consumption of fiber may reduce the risk of colorectal cancer."
(3) "Consumption of omega-3 fatty acids may reduce the risk of coronary heart disease."
(4) "0.8 mg of folic acid in a dietary supplement is more effective in reducing the risk of neural tube defects than a lower amount in foods in common form."

. . . .

The NLEA addressed foods and dietary supplements separately. Health claims on foods may be made, without FDA approval as a new drug, or the risk of sanctions for issuing a "misbranded" product, if it has been certified by the FDA as supported by "significant scientific agreement." Congress created a similar safe harbor for health claims on dietary supplements, but delegated to the FDA the task of establishing a "procedure and standard respecting the validity of [the health] claim." *Id.* § 343(r)(5)(D).

The FDA has since promulgated 21 C.F.R. § 101.14—the "significant scientific agreement" "standard" (quoted above)—and 21 C.F.R. § 101.70—a "procedure" (not particularly relevant to this case)—for evaluating the validity of health claims on dietary supplements. In doing so, the agency rejected arguments asserted by commenters—including appellants—that the "significant scientific agreement" standard violates the First Amendment because it precludes the approval of less-well supported claims accompanied by a disclaimer and because it is impermissibly vague. The FDA explained that, in its view, the disclaimer approach would be ineffective because "there would be a question as to whether consumers would be able to ascertain which claims were preliminary [and accompanied by a disclaimer] and which were not," and concluded that its prophylactic approach is consistent with applicable commercial speech doctrine. The agency, responding to the comment that "significant scientific agreement" is impermissibly vague, asserted that the standard is "based on objective factors" and that its procedures for approving health claims, including the notice and comment procedure, sufficiently circumscribe its discretion.

Then the FDA rejected the four claims supported by appellants. . . . The problem with these claims, according to the FDA, was not a dearth of supporting evidence; rather, the agency concluded that the evidence was inconclusive for one reason or another and thus failed to give rise to "significant scientific agreement." But the FDA never explained just how it measured "significant" or otherwise defined the phrase. The agency refused to approve the dietary fiber–cancer claim because "a supplement would contain only fiber, and there is no evidence that any specific fiber itself caused the effects that were seen in studies involving fiber-rich [foods]." The FDA gave similar reasons for rejecting the antioxidant vitamins–cancer claim, and the omega-3 fatty acids-coronary heart disease claim. As for the claim that 0.8 mg of folic acid in a dietary supplement is more effective in reducing the risk of neural tube defects than a lower amount in foods in common form, the FDA merely stated that "the scientific literature does not support the superiority of any one source over others." The FDA declined to consider appellants' suggested alternative of permitting the claim while requiring a corrective disclaimer such as "The FDA has determined that the evidence supporting this claim is inconclusive."

A more general folate–neural tube defect claim supported by appellants—that consumption of folate reduces the risk of neural tube defects—was initially rejected but ultimately approved for both dietary supplement and food labels. The parties disagree on what caused the FDA's change of position on this claim. Appellants contend that political objections—Senator Hatch was one of the complainers—concentrated the agency's mind. The FDA insists that its initial denial of the claim was based on a concern that folate consumption might have harmful effects on persons suffering from anemia, and that its concern was alleviated by new scientific studies published after the initial denial of the claim.

Appellants sought relief in the district court, raising APA and other statutory claims as well as a constitutional challenge, but were rebuffed.

II

Appellants raise a host of challenges to the agency's action. But the most important are that their First Amendment rights have been impaired and that under the Administrative Procedure Act the FDA was obliged, at some point, to articulate a standard a good deal more concrete than the undefined "significant scientific agreement." Normally we would discuss the nonconstitutional argument first, particularly because we believe it has merit. We invert the normal order here to discuss first appellants' most powerful constitutional claim, that the government has violated the First Amendment by declining to employ a less draconian method—the use of disclaimers—to serve the government's interests, because the requested remedy stands apart from appellants' request under the APA that the FDA flesh out its standards. That is to say, even if "significant scientific agreement" were given a more concrete meaning, appellants might be entitled to make health claims that do not meet that standard—with proper disclaimers.

Appellants also claim that the agency's "non-definition" runs afoul of Fifth Amendment concerns for vagueness. This contention is, however, closely connected to appellants' APA challenge and may well not be implicated if appellants' APA challenge affords ultimate relief. Therefore we will defer it until our APA analysis.

Disclaimers

It is undisputed that FDA's restrictions on appellants' health claims are evaluated under the commercial speech doctrine. It seems also undisputed that the FDA has unequivocally rejected the notion of requiring disclaimers to cure "misleading" health claims for dietary supplements. (Although the general regulation does not in haec verba preclude authorization of qualified claims, the government implied in its statement of basis and purpose that disclaimers were not adequate, and did not consider their use in the four subregulations before us.) The government makes two alternative arguments in response to appellants' claim that it is unconstitutional for the government to refuse to entertain a disclaimer requirement for the proposed health claims: first, that health claims lacking "significant scientific agreement" are inherently misleading and thus entirely outside the protection of the First Amendment; and second, that even if the claims are only potentially misleading, under *Central Hudson Gas & Elec. Corp. v. Public Serv. Comm'n of New York*, the government is not obliged to consider requiring disclaimers in lieu of an outright ban on all claims that lack significant scientific agreement.

If such health claims could be thought inherently misleading, that would be the end of the inquiry.

Truthful advertising related to lawful activities is entitled to the protections of the First Amendment. But when the particular content or method of the advertising suggests that it is inherently misleading or when experience has proved that in fact such advertising is subject to abuse, the States may impose appropriate restrictions. Inherently misleading advertising may be prohibited entirely. But the States may not place an absolute prohibition on . . . potentially misleading information . . . if the information also may be presented in a way that is not deceptive.

As best we understand the government, its first argument runs along the following lines: that health claims lacking "significant scientific agreement" are inherently misleading because they have such an awesome impact on consumers as to make it virtually impossible for them to exercise any judgment at the point of sale. It would be as if the consumers were asked to buy something while hypnotized, and therefore they are bound to be misled. We think this contention is almost frivolous. We reject it. But the government's alternative argument is more substantial. It is asserted that health claims on dietary supplements should be thought at least potentially misleading because the consumer would have difficulty in independently verifying these claims. We are told, in addition, that consumers might actually assume that the government has approved such claims.

Under *Central Hudson*, we are obliged to evaluate a government scheme to regulate potentially misleading commercial speech by applying a three-part test. First, we ask whether the asserted government interest is substantial. The FDA advanced two general concerns: protection of public health and prevention of consumer fraud. The Supreme Court has said "there is no question that [the government's] interest in ensuring the accuracy of commercial information in the marketplace is substantial," *Edenfield v. Fane*, and that government has a substantial interest in "promoting the health, safety, and welfare of its citizens," *Rubin v. Coors Brewing Co.* At this level of generality, therefore, a substantial governmental interest is undeniable.

The more significant questions under *Central Hudson* are the next two factors: "whether the regulation directly advances the governmental interest asserted," and whether the fit between the government's ends and the means chosen to accomplish those ends "is not necessarily perfect, but reasonable." We think that the government's regulatory approach encounters difficulty with both factors.

It is important to recognize that the government does not assert that appellants' dietary supplements in any fashion threaten consumer's health and safety. The government simply asserts its "commonsense judgment" that the health of consumers is advanced directly by barring any health claims not approved by the FDA. Because it is not claimed that the product is harmful, the government's underlying—if unarticulated—premise must be that consumers have a limited amount of either attention or dollars that could be devoted to pursuing health through nutrition, and therefore products that are not indisputably health enhancing should be discouraged as threatening to crowd out more worthy expenditures. We are rather dubious that this simplistic view of human nature or market behavior is sound, but, in any event, it surely cannot be said that this notion—which the government does not even dare openly to set forth—is a direct pursuit of consumer health; it would seem a rather indirect route, to say the least.

On the other hand, the government would appear to advance directly its interest in protecting against consumer fraud through its regulatory scheme. If it can be assumed—and we think it can—that some health claims on dietary supplements will mislead consumers, it cannot be denied that requiring FDA preapproval and setting the standard extremely, perhaps even impossibly, high will surely prevent any confusion among consumers. We also recognize that the government's interest in preventing consumer fraud/confusion may well take on added importance in the context of a product, such as dietary supplements, that can affect the public's health.

The difficulty with the government's consumer fraud justification comes at the final *Central Hudson* factor: Is there a "reasonable" fit between the government's goals and the means chosen to advance those goals? The government insists that it is never obliged to utilize the disclaimer approach because the commercial speech doctrine does not embody a preference for disclosure over outright suppression. Our understanding of the doctrine is otherwise. In *Bates v. State Bar of Arizona*, the Supreme Court addressed an argument similar to the one the government advances. The State Bar had disciplined several attorneys who advertised their fees for certain legal services in violation of the Bar's rule, and sought to justify the rule on the ground that such advertising is inherently misleading "because advertising by attorneys will highlight irrelevant factors and fail to show the relevant factor of skill." The Court observed that the Bar's concern was "not without merit," but refused to credit the notion that "the public is not sophisticated enough to realize the limitations of advertising, and that the public is better kept in ignorance than trusted with correct but incomplete information." Accordingly, the Court held that the "incomplete" attorney advertising was not inherently misleading and that "the preferred remedy is more disclosure, rather than less." In more recent cases, the Court has reaffirmed this principle, repeatedly pointing to disclaimers as constitutionally preferable to outright suppression.

. . . .

Our rejection of the government's position that there is no general First Amendment preference for disclosure over suppression, of course, does not determine that any supposed weaknesses in the claims at issue can be remedied by disclaimers and thus does not answer whether the subregulations, 21 C.F.R. § 101.71(a), (c), (e); id. § 101.79(c)(2)(i)(G), are valid. The FDA deemed the first three claims—(1) "Consumption of antioxidant vitamins may reduce the risk of certain kinds of cancers," (2) "Consumption of fiber may reduce the risk of colorectal cancer," and (3) "Consumption of omega-3 fatty acids may reduce the risk of coronary heart disease"—to lack significant scientific agreement because existing research had examined only the relationship between consumption of foods containing these components and the risk of these diseases. The FDA logically determined that the specific effect of the component of the food constituting the dietary supplement could not be determined with certainty. (The FDA has approved similar health claims on foods containing these components. See, e.g., 21 C.F.R. § 101.79 (folate–neural tube defects).) But certainly this concern could be accommodated, in the first claim, for example, by adding a prominent disclaimer to the label along the following lines: "The evidence is inconclusive because existing studies have been performed with foods containing antioxidant vitamins, and the effect of those foods on reducing the risk of cancer may result from other components in those foods." A similar disclaimer would be equally effective for the latter two claims.

The FDA's concern regarding the fourth claim—"0.8 mg of folic acid in a dietary supplement is more effective in reducing the risk of neural tube defects than a lower amount in foods in common form"—is different from its reservations regarding the first three claims; the agency simply concluded that "the scientific literature does not support the superiority of (concluding that "losses [of folic acid] in cooking and canning [foods] can be very high due to heat destruction"), and we suspect that a clarifying disclaimer could be added to the effect that "The evidence in support of this claim is inconclusive."

The government's general concern that, given the extensiveness of government regulation of the sale of drugs, consumers might assume that a claim on a supplement's label is approved by the government, suggests an obvious answer: The agency could require the label to state that "The FDA does not approve this claim." Similarly, the government's interest in preventing the use of labels that are true but do not mention adverse effects would seem to be satisfied—at least ordinarily—by inclusion of a prominent disclaimer setting forth those adverse effects.

The government disputes that consumers would be able to comprehend appellants' proposed health claims in conjunction with the disclaimers we have suggested—this mix of information would, in the government's view, create confusion among consumers. But all the government offers in support is the FDA's pronouncement that "consumers would be considerably confused by a multitude of claims with differing degrees of reliability." Although the government may have more leeway in choosing suppression over disclosure as a response to the problem of consumer confusion where the product affects health, it must still meet its burden of justifying a restriction on speech—here the FDA's conclusory assertion falls far short.

We do not presume to draft precise disclaimers for each of appellants' four claims; we leave that task to the agency in the first instance. Nor do we rule out the possibility that where evidence in support of a claim is outweighed by evidence against the claim, the FDA could deem it incurable by a disclaimer and ban it outright.[45] For example, if the weight of the evidence were against the hypothetical claim that "Consumption of Vitamin E

---

[45] Similarly we see no problem with the FDA imposing an outright ban on a claim where evidence in support of the claim is qualitatively weaker than evidence against the claim—for example, where the claim rests on only one or two old studies.

reduces the risk of Alzheimer's disease," the agency might reasonably determine that adding a disclaimer such as "The FDA has determined that no evidence supports this claim" would not suffice to mitigate the claim's misleadingness. Finally, while we are skeptical that the government could demonstrate with empirical evidence that disclaimers similar to the ones we suggested above would bewilder consumers and fail to correct for deceptiveness, we do not rule out that possibility.

B. The Unarticulated Standard

Wholly apart from the question whether the FDA is obliged to consider appropriate disclaimers is appellants' claim that the agency is obliged to give some content to the phrase "significant scientific agreement." Appellants contend that the agency's failure to do so independently violates their constitutional rights under the First and Fifth Amendments. The First, because producers of dietary supplements are assertedly subject to a "prior restraint" on their protected speech—the labeling of products. The Fifth, because the agency's approach is so vague as to deprive the producers of liberty (and property?) without due process.

Appellants do not challenge the concept of a pre-screening system per se; their complaint is with the FDA's lack of guidance on which health claims will survive the prescreening process. But appellants never connected their vagueness concern with their oblique First Amendment prior restraint argument, and for that reason we need not decide whether prior restraint analysis applies to commercial speech. On the other hand, appellants' Fifth Amendment vagueness argument is squarely presented. Still, by prevailing on their APA claim appellants would seem to gain the same relief—invalidation of the FDA's interpretation of the general standard and a remand for more guidance—as they would through a successful Fifth Amendment claim (or indeed a First Amendment prior restraint claim, if it had been properly presented and assuming arguendo that prior restraint analysis applies in the commercial speech context).

Consideration of this constitutional claim seems unnecessary because we agree with appellants that the APA requires the agency to explain why it rejects their proposed health claims—to do so adequately necessarily implies giving some definitional content to the phrase "significant scientific agreement." We think this proposition is squarely rooted in the prohibition under the APA that an agency not engage in arbitrary and capricious action. See 5 U.S.C. § 706(2)(A) (1994). It simply will not do for a government agency to declare—without explanation—that a proposed course of private action is not approved. ("The agency must . . . articulate a satisfactory explanation for its action. . . .") To refuse to define the criteria it is applying is equivalent to simply saying no without explanation. Indeed, appellants' suspicions as to the agency's real reason for its volte-face on the general folate-neural tube defect claim highlight the importance of providing a governing rationale for approving or rejecting proposed health claims.

To be sure, Justice Stewart once said, in declining to define obscenity, "I know it when I see it," which is basically the approach the FDA takes to the term "significant scientific agreement." But the Supreme Court is not subject to the Administrative Procedure Act. Nor for that matter is the Congress. That is why we are quite unimpressed with the government's argument that the agency is justified in employing this standard without definition because Congress used the same standard in 21 U.S.C.A. § 343(r)(3)(B)(i). Presumably—we do not decide—the FDA in applying that statutory standard would similarly be obliged under the APA to give it content.

That is not to say that the agency was necessarily required to define the term in its initial general regulation—or indeed that it is obliged to issue a comprehensive definition all at once. But see n.12 supra. The agency is entitled to proceed case by case or, more accurately, subregulation by subregulation, but it must be possible for the regulated class to perceive the principles which are guiding agency action. Accordingly, on remand, the FDA must explain what it means by significant scientific agreement or, at minimum, what it does not mean.

. . ..

For the foregoing reasons, we hold invalid the four sub-regulations, 21 C.F.R. § 101.71(a), (c), (e); § 101.79(c)(2)(i)(G), and the FDA's interpretation of its general regulation, *id.* § 101.14. The decision of the district court is reversed, and the case is remanded to the district court with instructions to remand in turn to the FDA for reconsideration of appellants' health claims.

So ordered.

* * * * *

DISCUSSION QUESTIONS

4.5. In *Pearson*, the government argued that the First Amendment rights of food manufacturers were not infringed because the manufacturers could still make their claims in published articles and books. Is the government saying that some forms of communication are inherently more misleading than others? Is labeling inherently so different from advertising that a different standard for misleading should apply?

4.6. Did the *Pearson* court authorize the plaintiffs to make their claims? What gains did the plaintiffs make?

FDA's Changes after Pearson   Following the *Pearson* decision, the FDA announced a number of significant decisions and policy changes regarding its regulation of health claims. In general, these changes provide new flexibility for approval of claims. Food companies now have greater opportunity to communicate information about potential health benefits and specific conventional foods or dietary supplements.

Foremost, FDA now allows qualified health claims in the labeling of conventional foods and dietary supplements. The standard for approval for qualified claims shifted from the significant scientific agreement to the weight of scientific evidence. The FDA still requires premarket approval, but has stated that it will "consider" exercising enforcement discretion for a health claim when the following conditions are met:

1. The claim is the subject of an appropriately filed health claim petition.
2. The scientific evidence in support of the claim outweighs the scientific evidence against the claim, the claim is appropriately qualified, and all statements in the claim are consistent with the weight of the scientific evidence.
3. Consumer health and safety are not threatened.
4. The claim meets the general requirements for a health claim in 21 C.F.R. § 101.14.

Note: The first and fourth criteria are requirements found in the FDA regulations cited. The second and third come directly from the court of appeals opinion in *Pearson*.

The Shalala Claims Revisited   Regarding the particular health claims proposed by the *Pearson* plaintiffs, the FDA developed "qualified" claims that would be appropriate on food labeling, even in the absence of evidence meeting the "significant scientific agreement" standard.

One of the agency's qualified claims for folate is:

> Healthful diets with adequate folate may reduce a woman's risk of having a child with a brain or spinal cord birth defect. Women capable of becoming pregnant should take 400 mcg of folate per day from a supplement or fortified foods and consume food folate from a varied diet. It is not known whether the same level of protection can be achieved by using lower amounts.

The agency's qualified claim for omega-3 fatty acids and coronary heart disease is:

> The scientific evidence about whether omega-3 fatty acids may reduce the risk of coronary heart disease (CHD) is suggestive, but not conclusive. Studies in the general population have looked at diets containing fish, and it is not known whether diets or omega-3 fatty acids in fish may have a possible effect on a reduced risk of CHD. It is not known what effect omega-3 fatty acids may or may not have on risk of CHD in the general population.

Regarding dietary fiber, the FDA found no basis to conclude that the available evidence permitted a comparably nonmisleading use of qualifying information.

DISCUSSION QUESTIONS

4.7. Do you think FDA's approved "qualified" claims prevent consumers from being misled?

4.8. How likely are food distributors to use these health claims?

4.9. Do you think this result is what the *Pearson* court had in mind?

Not everyone agrees with the decision of the U.S. Court of Appeals in *Pearson v. Shalala*.[46] It has been argued that disclaimers thwart the purpose of Congress when it enacted the NLEA to ensure that consumers would no longer be subjected to unreliable and unverifiable health claims for dietary supplements—that disclaimers will relegate consumers to a marketplace rife with unproved and unreliable health claims. It has also been argued that the reasoning of *Pearson* misconceives basic First Amendment commercial speech principles because the Supreme Court has never directed a government agency to permit potentially misleading speech so long as it is accompanied by a disclaimer.[47]

Qualified Health Claims   In the Federal Register of October 6, 2000,[48] the FDA issued guidance on qualified health claims in the labeling of conventional foods and dietary supplements. FDA also republished and expanded this information as a guidance document for industry to include conventional foods along with dietary supplements.[49] The document sets forth criteria for when the agency allows a qualified health claim in labeling. In addition FDA states that the agency will use the "reasonable consumer" standard in evaluating food labeling claims. Use of this standard makes the FDA's regulation of food labeling consistent with the FTC's regulation of advertising for these products.

FDA noted that consumers are more likely to respond to health messages in food labeling if the messages are specific with respect to the health benefits associated with particular substances in the food. According to the Bureau of Economics Staff of the Federal Trade Commission (FTC),[50] "consumers are not as responsive to simple nutrient claims" as they are to health claims. FDA stated that in the aggregate, decisions by individual consumers to incorporate

[46] *See, e.g.,* David C. Vladeck, *Devaluing Truth: Unverified Health Claims in the Aftermath of Pearson v. Shalala*, 54 FOOD DRUG L. J. 535 (1999).

[47] *Id.*

[48] 65 Fed. Reg. 59855.

[49] CFSAN, FDA, GUIDANCE FOR INDUSTRY: QUALIFIED HEALTH CLAIMS IN THE LABELING OF CONVENTIONAL FOODS AND DIETARY SUPPLEMENTS (Dec. 18, 2002), *available at:* http://www.cfsan.fda. gov/~dms/hclmgui2.html (last accessed Sept. 24, 2007) (This document was superseded by later FDA guidance but still contains important background).

[50] *Id.* (citing BUREAU OF ECONOMICS STAFF, FTC, ADVERTISING NUTRITION & HEALTH: EVIDENCE FROM FOOD ADVERTISING 1977–1997 (Sep. 2002)).

beneficial foods into their diets improve public health, "By making clear the lawfulness of conventional foods labeled with truthful and non-misleading health claims, FDA believes that this guidance will precipitate greater communication in food labeling of the health benefits of consuming particular foods, thereby enhancing the public's health."[51]

In FDA's "Better Nutrition Information for Consumer Health Initiative," FDA has "acknowledged that consumers will benefit from more information on food labels concerning diet and health and this, in turn, has prompted the agency to establish interim procedures whereby 'qualified' health claims can be made not only for dietary supplements but for conventional foods as well. . . . FDA began considering qualified health claims under its interim procedures on September 1, 2003."[52]

To sum up key provisions for use of health claims:

- All health claims must undergo review by FDA.
- All unqualified health claims must meet the Significant Scientific Agreement standard.
- Qualified health claims must be accompanied by a disclaimer or otherwise "qualified" in a way as to not mislead consumers.
- The interim procedures for qualified health claims are available on the FDA Web site.

Accepted Qualified Health Claims   Qualified health claims have been accepted by FDA for the following:[53]

- Selenium and cancer
- Antioxidant vitamins and cancer
- Nuts and heart disease
- Walnuts and heart disease
- Omega-3 fatty acids and coronary heart disease
- B vitamins and vascular disease
- Monounsaturated fatty acids from olive oil and coronary heart disease
- Phosphatidylserine and cognitive dysfunction and dementia
- 0.8 mg folic acid and neural-tube birth defects
- Tomatoes and/or tomato sauce and prostate, ovarian, gastric, and pancreatic cancers

---

[51] *Id.*

[52] CFSAN, FDA, GUIDANCE FOR INDUSTRY: FDA'S IMPLEMENTATION OF "QUALIFIED HEALTH CLAIMS": QUESTIONS AND ANSWERS (Aug. 27, 2003; May 12, 2006) *available at*: http://www.cfsan. fda.gov/~dms/labqhcqa.html (last accessed Sept. 24, 2007).

[53] FDA, QUALIFIED HEALTH CLAIMS SUBJECT TO ENFORCEMENT DISCRETION (Revised Apr. 2007), *available at*: http://www.cfsan.fda.gov/~dms/qhc-sum.html (last accessed Mar. 15, 2008).

- Calcium and colon/rectal cancer and calcium and recurrent colon/rectal polyps
- Green tea and cancer
- Chromium picolinate and diabetes
- Calcium and hypertension, pregnancy-induced hypertension, and preeclampsia
- Corn oil and heart disease

Each accepted qualified health claim includes specific standards that a food must meet to in addition to the general requirements for the claim. Some of the qualifications are long and elaborate. For instance, take these sample qualified claims for green tea and cancer:

1. Two studies do not show that drinking green tea reduces the risk of breast cancer in women, but one weaker, more limited study suggests that drinking green tea may reduce this risk. Based on these studies, FDA concludes that it is highly unlikely that green tea reduces the risk of breast cancer. Or,

2. One weak and limited study does not show that drinking green tea reduces the risk of prostate cancer, but another weak and limited study suggests that drinking green tea may reduce this risk. Based on these studies, FDA concludes that it is highly unlikely that green tea reduces the risk of prostate cancer.[54]

4.6.7   Nutritional Claims Grading Proposed

Recently the FDA and the FTC announced a new grading system for health claims on labels of traditional food products and dietary supplements. Even though FDA announced that it will begin accepting "health claim" petitions under the proposed grading system on September 1, 2003, FDA did indicate that the agency plans to adopt the new approach through rulemaking. FTC indicated that FTC would look to this grading system in reviewing advertising substantiation.

Under the proposed grading system, companies may place health claims if (1) such claims have been pre-approved by FDA and (2) the claims include FDA-specified language qualifying the claim based on the pre-approval grade. The proposed grading system applies only to "health claims," such as: "Regular exercise and a healthy diet with enough calcium helps teen and young adult white and Asian women maintain good bone health and may reduce their high risk of osteoporosis later in life." The proposed grading system does not apply to "structure-function" claims, such as: "Calcium helps build strong bones."

---

[54] FDA, Enforcement Discretion Letter, Docket No. 2004Q-0083 (June 30, 2005), *available at*: http://www.cfsan.fda.gov/~dms/qhc-gtea.html (last accessed Mar. 15, 2008).

Companies wishing to include a health claim on their label may participate in the FDA's pre-approval process for that claim by submitting their proposed health claim to the FDA. The FDA then has nine months to review and grade the "health claim." The FDA will assign the claim a grade of A, B, C, or D, based on the availability of scientific evidence to support the claim. A grade of "A" means that there is "significant scientific agreement" that the claim is true. A "B" grade indicates that the evidence supporting the claim is promising but "not conclusive." A "C" grade indicates that the evidence supporting the claim is "limited" and "not conclusive." Finally, a "D" grade indicates that "little scientific evidence" supports the claim.

Once a claim has received a grade from FDA, the applicant company (and presumably competitors) may include that claim on its label, but must include the exact qualifying language for that grade in the text of the claim. FDA is currently conducting field investigations of consumers to determine the exact qualifying language based on consumer understanding and utility. For example, a product with a "B" grade may have to include the following language in the text of its claim: "Although there is some scientific evidence supporting this claim, the evidence is not conclusive." This language would have to appear in the text of the claim, as opposed to another place on the label. Any company receiving a grade, even those receiving "C" and "D" grades, will be able to include their health claim on the label as long as also included is the appropriate qualifying language.

FDA and FTC indicated that the purpose of this proposed grading system is to provide more and better information to consumers regarding health claims.

DISCUSSION QUESTION

4.10. FDA has taken significant steps to ensure it does not unnecessarily restrain commercial speech. However, history provides examples of excessive and unsubstantiated claims, which would indicate close regulation is required. On the other hand, there evidence that increased access to health information plays a useful role in helping consumers make informed choices for good health. These forces play against each other. Do you think FDA achieved the proper balance?

4.6.8   Substantiation of Claims

Fact Sheet on FDA's Draft Guidance for Industry: Substantiation for Dietary Supplement Claims

FDA, CFSAN/Office of Nutritional Products, Labeling and Dietary Supplements (November 4, 2004)[55]

---

[55] FDA, CFSAN/Office of Nutritional Products, Labeling and Dietary Supplements (Nov. 4, 2004), *available at*: http://www.cfsan.fda.gov/~dms/dsclmfs.html.

FDA's Draft Guidance for Industry is intended to describe the amount, type, and quality of evidence FDA recommends a manufacturer have to substantiate a claim under the Food, Drug, and Cosmetic Act (the Act). The Act requires dietary supplement manufacturers to have substantiation that structure/function, nutrient deficiency, and general well-being claims on the label of a dietary supplement product are truthful and not misleading.

Although there is no formula as to how many or what type of studies are needed to substantiate a claim, FDA intends to apply a standard of "competent and reliable scientific evidence."

In determining whether the substantiation standard has been met with competent and reliable scientific evidence, FDA recommends that firms consider the following issues in their assessment:

- the meaning of the claim(s) being made;
- the relationship of the evidence to the claim;
- the quality of the evidence; and
- the totality of the evidence.

Background

The act does not define what constitutes "substantiation" for a claim made for a dietary supplement. For this draft guidance, FDA reviewed regulations, case law, the Federal Trade Commission (FTC) experience with its policy on substantiating claims made for dietary supplements in advertising, as well as recommendations from the Commission on Dietary Supplement Labels.

FDA's approach provides flexibility to manufacturers in the precise amount and type of evidence that constitutes adequate substantiation. Thereby providing a standard for substantiation may also help preserve consumer confidence in these products.

FTC has typically applied a substantiation standard of "competent and reliable scientific evidence" to claims made for dietary supplements in advertising. FDA intends to apply a standard consistent with FTC's approach.

FDA considers the following factors important in determining whether information would constitute "competent and reliable scientific evidence":

- Does each study or piece of evidence bear a relationship to the specific claim(s)?
- What are the individual study's or evidence's strengths and weaknesses?
- If multiple studies exist, do the studies that have the most reliable methodologies suggest a particular outcome?
- If multiple studies exist, what do most studies suggest or find? Does the totality of the evidence agree with the claim(s)?

The Meaning of the Claim

The first step in determining what information is needed to substantiate a claim for a dietary supplement is to understand the meaning of the claim and clearly identify each implied and express claim. Understanding the claim's meaning will help identify the appropriate study hypotheses and measurable endpoints, which can be used to ensure that the firm has appropriate studies to substantiate the claim.

The Relationship of the Evidence to the Claim

Whether studies or evidence have a relationship to the specific claim being made or to the dietary supplement product itself is an important consideration in determining if a claim is substantiated. The following are some threshold questions in determining this relationship:

- Have the studies specified and measured the dietary supplement or dietary ingredient that is subject of the claim?
- Have the studies appropriately specified and measured the nutritional deficiency, structure/function, or general well-being that is the subject of the claim?
- Were the studies based on a population that is similar to that which will be consuming the dietary supplement product?

The Quality of the Evidence

In deciding whether studies substantiate a claim, an important consideration is the scientific quality of studies. Scientific quality is based on several criteria including study type, study population, study design and conduct (e.g., presence of a placebo control), data collection (e.g., dietary assessment method), statistical analysis, and outcome measures. If the scientific study adequately addressed all or most of the above criteria, it would be considered of high quality. Generally accepted scientific and statistical principles should be used to determine the quality of the studies used as evidence to substantiate a claim.

Totality of the Evidence

In determining whether there is adequate evidence to substantiate a claim, firms should consider the strength of the entire body of evidence, including criteria such as quality, quantity (number of various types of studies and sample sizes), consistency, relevance of exposure, and persuasiveness.

Ideally the evidence used to substantiate a claim agrees with the surrounding body of evidence. Conflicting or inconsistent results raise serious questions as to whether a particular claim is substantiated.

There is no general rule for how many studies, or what combination of types of evidence, is sufficient to support a claim. However, the replication of research results in independently conducted studies adds to the persuasiveness of the evidence.

Although the quality and persuasiveness of individual pieces of evidence are important, each piece should be considered in the context of all available information; that is, the strength of the total body of scientific evidence is the critical factor in assessing whether a claim is substantiated.

* * * * *

4.7 NLEA AND RESTAURANTS

Restaurants are exempt from labeling requirements, generally. Initially the FDA decided to exempt restaurant menus from all NLEA nutrition and health claim requirements. In part, the FDA invoked the doctrine of administrative necessity and argued that the agency lacked the resources to enforce NLEA in restaurants. In the following case the court rejected FDA's reasoning and found that FDA must abide by the unambiguous meaning of the statute.

* * * * *

Public Citizen and CPSI v. Shalala

932 F. Supp. 13 (1996)

PAUL L. FRIEDMAN, United States District Judge

... Plaintiffs challenge the decision of the United States Food and Drug Administration to exempt restaurant menus from the Nutrition Labeling and Education Act of 1990 ("NLEA"), alleging that the decision violates the NLEA and is arbitrary and capricious under the Administrative Procedure Act.

I. The NLEA

In 1990, Congress amended the Federal Food, Drug, and Cosmetic Act ("FD&C Act"), 21 U.S.C. §301, *et seq.* ... The NLEA added two sections—"q" and "r"—to Section 403 of the FD&C Act, thereby creating two new food labeling provisions. Section 403(q) created new general nutritional labeling standards and requirements. Restaurants are completely exempt from these standards and requirements. Section 403(r) imposed new restrictions on the ability of purveyors of food to make affirmative health and nutritional claims about food. Restaurants are exempt from some but not all of these restrictions. ...

The dispute in this case revolves around the FDA's decision to exempt restaurant menus from the labeling requirements governing both nutrient content claims and health claims. ... [T]he FDA concluded that Section 403(r) of the

NLEA generally governs claims made about restaurant food, but nevertheless decided to regulate only those claims made on signs, placards or posters but not claims made on menus. The FDA reasoned that menus are subject to frequent change and that the requirements might deter restaurants, especially small ones, from providing useful nutrition-related information on menus. The FDA regulations accordingly provide:

Nutrition labeling in accordance with §101.9 shall be provided upon request for any restaurant food or meal for which a nutrient content claim . . . or a health claim . . . is made (except on menus).

On June 15, 1993, the FDA proposed new rules that would effectively have overruled this restaurant menu exemption, but those rules have not been adopted.

Plaintiffs argue that the FDA lacked authority under the NLEA to exempt restaurant menus from the nutritional and health claim labeling requirements contained in Section 403(r). They assert that Congress intended restaurants to be covered by Section 403(r), that Congress provided for specific exceptions to that coverage and that additional exceptions cannot be implied or promulgated by regulation. Plaintiffs rely on the language and structure of the statute and on legislative history purporting to show that Congress specifically considered excluding restaurants from the NLEA's nutritional claim requirements and declined to do so. Plaintiffs further argue that Section 405 of the FD&C Act bars the menu exemption. They point to the FDA's rationale for its own proposed rule and suggest that in proposing such a rule, the FDA has acknowledged that restaurant menus are properly governed by the NLEA's nutrition and health claims labeling requirements. Finally, plaintiffs argue that because nearly half the American food dollar is spent on food consumed away from home, because as much as 30 percent of the American diet is composed of foods prepared in food service operations, and because restaurant menus often make misleading or false representations about the nutritional and health value of their foods, the restaurant menu exception is arbitrary and capricious.

Defendant responds that the NLEA nowhere bars the FDA from creating the restaurant menu exception, that the FDA has adequate authority under the NLEA to create such an exception and that even if the NLEA on its face does not permit such an exception, the FDA could create one as part of its assessment of its enforcement priorities.

II. Discussion

The validity of the FDA's interpretation of the NLEA statutory scheme is, in the first instance, to be measured under the yardstick provided by *Chevron U.S.A., Inc. v. Natural Resources Defense Council.* As the Supreme Court has explained: "If the intent of Congress is clear, that is the end of the matter; for the court, as well as the agency, must give effect to the unambiguously expressed intent of Congress. . . ."

Applying this standard, the Court finds that the language of the NLEA is clear and that Congress intended to include restaurant menus in the NLEA nutrition and health labeling provisions.

On its face, the NLEA specifically designates the various provisions that do and do not apply to restaurants. . . . The plain meaning of these express exclusions is that Congress intended those subsections not expressly excluded to apply to restaurant food. The general rule is that "when a statute lists several specific exemptions to the general purpose, others should not be implied." Defendant's comment that the NLEA nowhere prohibits the FDA from creating such an exception does not abrogate this general rule of statutory construction.

The FDA's interpretation, namely that the NLEA governs only health and nutritional claims made on signs, placards, or posters but not on menus, requires a tortured reading of the statute as a whole and creates an implausible result. Under the FDA's approach, theoretically a restaurant could claim on its menu that a particular meal is "low fat" or "lite" without any nutritional basis for making the claim or otherwise triggering the requirements of the NLEA, but it could not make that same representation on a sign, poster or placard unless the food complied with FDA definitions of those terms and the restaurant was prepared to substantiate the claim as required by FDA regulations. There is no language in the statute or the legislative history to suggest that Congress intended or even contemplated creating such a large loophole. . . .

[Discussion of legislative history showing no intent to exempt menus omitted.]

The Court also rejects defendant's invocation of the doctrine of "administrative necessity" . . . The FDA has not borne its "especially heavy" burden of establishing the administrative impossibility of applying the nutrition content and health claims provisions of the NLEA to restaurant menus. It is true that in promulgating the final rule, the FDA twice stated that it "does not have resources to adequately enforce its regulations in restaurants," but this explanation was proffered in support of the agency's decision to hold restaurants to a lower standard for substantiating claims of nutrition content and health, not of its decision to exempt menus altogether. Rather, in justifying the menu exemption, the FDA cited the need for flexibility for small restaurants and the fact that "menus are subject to frequent, even daily change." The final rule's two references to the FDA's lack of enforcement resources thus are irrelevant to the menu exemption because they were made in another context. The FDA therefore has not satisfied its "heavy burden" under the administrative necessity doctrine.

. . . .

DECLARED that the defendant's final regulations implementing the NLEA violate Section 3 of the NLEA, 21 U.S.C. §343(r), and the Administrative Procedure Act, 5 U.S.C. §706, because the regulations exempting restaurant menus from the nutrient content and health claim provisions of the NLEA are contrary to the meaning of the statute; and it is

FURTHER ORDERED that the defendant shall amend its regulations within thirty days from this Memorandum Opinion and Order to require that all restaurant menus be included under FDA regulations for the labeling of nutrient content and health claims.

SO ORDERED.

* * * * *

FDA subsequently promulgated regulations specifying the "reasonable basis" for assurance that restaurant nutritional claims comply with the nutrient requirements for the claim.

21 C.F.R. §101.10 Nutrition Labeling of Restaurant Foods

Nutrition labeling in accordance with Sec. 101.9 shall be provided upon request for any restaurant food or meal for which a nutrient content claim (as defined in Sec. 101.13 or in subpart D of this part) or a health claim (as defined in Sec. 101.14 and permitted by a regulation in subpart E of this part) is made, except that information on the nutrient amounts that are the basis for the claim (e. g., "low fat, this meal provides less than 10 grams of fat") may serve as the functional equivalent of complete nutrition information as described in Sec. 101.9. Nutrient levels may be determined by nutrient data bases, cookbooks, or analyses or by other reasonable bases that provide assurance that the food or meal meets the nutrient requirements for the claim. Presentation of nutrition labeling may be in various forms, including those provided in Sec. 101.45 and other reasonable means.[56]

4.8 ADVERTISING

The scope of this text only allows space to briefly overview the regulation of food advertising, which is predominantly the responsibility of the Federal Trade Commission (FTC).

4.8.1 Federal Trade Commission

Under section 5 of the Federal Trade Commission (FTC) Act, the Federal Trade Commission has the mandate to act against unfair and deceptive advertising practices. The FTC describes its mission:

> The Federal Trade Commission (FTC) works to ensure that the nation's markets are vigorous, efficient, and free of restrictions that harm consumers. Experience demonstrates that competition among firms yields products at the lowest prices, spurs innovation, and strengthens the economy. Markets also work best when consumers can make informed choices based on accurate information. To ensure

---

[56] 61 FR 40332, Aug. 2, 1996.

the smooth operation of our free market system, the FTC enforces federal consumer protection laws that prevent fraud, deception, and unfair business practices.[57]

FTC's Bureau of Consumer Protection, Division of Advertising Practices, enforces federal truth-in-advertising laws. The division's enforcement includes advertising claims for foods, drugs, dietary supplements, and other products promising health benefits. The FTC covers advertising claims made in newspaper, magazines; in radio and TV commercials; direct mail to consumers; and on the Internet.

4.8.2 Deceptive Advertising and Unfairness

Sections 5 and 12 of the FTC Act (15 U.S.C. 45 and 52), which broadly prohibit unfair or deceptive commercial acts or practices, specifically prohibit the dissemination of false advertisements for foods, drugs, medical devices, or cosmetics. The FTC has issued two policy statements, the Deception Policy Statement[58] and the Statement on Advertising Substantiation,[59] that articulate the basic elements of the deception analysis employed by the FTC in advertising cases. According to these policies, in identifying deception in an advertisement, the FTC considers the representation from the perspective of a consumer acting reasonably under the circumstances: "The test is whether the consumer's interpretation or reaction is reasonable."

According to the FTC policy, deceptive representation, omission, or unfairness in a trade practice must be a material one. Deceptive advertising can take a number of forms ranging from intentional false or misleading claims by an advertiser to ads that may be true in a literal sense, but leave consumers with a false or misleading impression. In the FTC Deception Policy Statement, the FTC commission finds deception "if there is a misrepresentation, omission, or practice that is likely to mislead the consumer acting reasonably in the circumstances to the consumer's detriment." This definition contains three elements: (1) misrepresentation, omission, or practice likely to mislead the consumer; (2) considered from the perspective of the reasonable consumer; (3) materiality, which means that the deception influenced the consumer's decision in a detrimental way.

4.8.3 Overview of Other Regulatory Aspects of Advertising

Federal Communications Commission The Federal Communications Commission (FCC) has jurisdiction over the radio, television, telephone, and

---

[57]FTC, Guide to the Federal Trade Commission, *available at*: www.ftc.gov/bcp/conline/pubs/general/guidetoftc.htm (last visited Aug. 27, 2003).

[58]Appended to Cliffdale Assocs., Inc., 103 F.T.C. 110, 174 (1984), *available at*: http://ftc.gov/bcp/policystmt/ad-decept.htm (last accessed Aug. 21, 2008).

[59]Appended to Thompson Med. Co., 104 F.T.C. 648, 839 (1984), *available at*: http://ftc.gov/bcp/guides/ad3subst.htm (last accessed Aug. 21, 2008).

telegraph industries. Its authority over the airways gives it the power to control advertising content and to restrict what products and services can be advertised on radio and television. The FCC generally works closely with the FTC in the regulation of advertising.

The U.S. Postal Service The U.S. Postal Service regulates advertising involving the use of mail.

The Alcohol and Tobacco Tax and Trade Bureau (TTB) The TTB of the U.S. Department of Treasury (formally the Bureau of Alcohol, Tobacco, and Firearms (BATF)) regulates the advertising of alcoholic beverages.

State Attorney Generals A number of states have mini-FTC laws modeled after the federal FTC Act. These acts are typically enforced by the various state Attorney Generals (AGs). Other states have laws on unfair trade practices or consumer protection that empower the state to act against certain types of advertising.

Therefore a state attorney general could bring a similar enforcement action on a matter where the FTC or FDA might act. On occasion, a number of state AGs may bring an action together. These actions may be concurrent with action by the FDA and the FTC.

TABLE 4.1 Comparison of False and Misleading

	Labels	Ads
Agency	FDA	FTC
Statute	FD& C Act (21 U.S.C. 343(a)(1))	FTC Act (15 U.S.C. 45 and 52)
Standard	False or misleading "*in any particular*"	*Likely* to mislead
Injury	*Not* required that a consumer be injured or even misled	"Materiality" required, which means that the deception influenced a consumer's decision in a detrimental way
Whose Perspective	Varies depending on the court from "the ignorant, the unthinking, and the credulous" consumer to the ordinary person or reasonable consumer	Ordinary person or reasonable consumer. "The test is whether the consumer's interpretation or reaction is reasonable."
Specifications	Regulations (e.g., NLEA) provide specific requirements and definitions	More subjective and context based

States may have laws different than the federal law. In addition state AGs are free to interpret and enforce the law individually. Often state AGs will communicate with the FDA and FTC and coordinate actions, but states may not always follow the federal lead.

Private Enforcement—The Lanham Act   Most businesses rely on the FTC to deal with the problem of deceptive or misleading advertising by their competitors. However, companies may also file lawsuits under the Lanham Act against competitors who they feel are making false claims. The Lanham Act encompasses false advertising and provides individuals with the opportunity to file a civil suit against a competitor. Many companies are using the Lanham Act to sue competitors for their advertising claims, particularly since comparative advertising has become so common.

For instance, the U.S. Court of Appeals in Philadelphia upheld an injunction received by Novartis to bar Johnson & Johnson from marketing its over-the-counter heartburn medicine as "Mylanta Night Time Strength."[60] Novartis had sued J&J under the Lanham Act on the ground that "night time strength" implied that the product had been specially formulated to work at night time, when in fact the product's formulation has no such unique characteristic.[61]

---

[60] *See* Novartis Consumer Health, Inc. v. Johnson & Johnson-Merck Consumer, 290 F.3d 578 (3rd Cir. 2002).

[61] *Id.*

CHAPTER 5

Economic and Aesthetic Adulteration

5.1 INTRODUCTION

This chapter studies the regulation of the economic adulteration of food and aesthetic adulteration of foods. **Economic adulteration** is the illegal substitution of inferior or cheaper ingredients for profit and to undercut the competition. The topic of economic adulteration of food overlaps with our earlier discussion of food labeling and with the definition of misbranding.

Aesthetic adulteration is the contamination of food with filthy, putrid, or decomposed substance. Food that is aesthetically adulterated may not be unsafe, but it is nonetheless considered unfit for food. Included in this category is food that has been held under insanitary conditions whereby it may have become contaminated with filth. Thus the topic of aesthetic adulteration encompasses the topics of sanitation and good manufacturing practices.

This intertwining and overlapping of categories contributes complexity into the Food, Drug, and Cosmetic Act (FD&C Act). This structure largely is the result of the 100 year evolution of the act. However, some of the overlap is by design to prevent anything from slipping through the cracks.

5.2 DEFINITIONS

5.2.1 Food

Food is defined as: "(1) articles used for food or drink for man or other animals, (2) chewing gum, and (3) articles used for components of any other such article." FD&C Act § 201(f) [21 U.S.C. § 321(f)].

Intended Use Although not included in the FD&C Act definition of food, the meaning of intended use is commonly imputed by the courts. This is

---

Food Regulation: Law, Science, Policy, and Practice, by Neal D. Fortin
Copyright © 2009 Published by John Wiley & Sons, Inc.

just common sense. A manufacturer or distributor of a product generally represents the product for an intended use, and that representation may determine whether the product is a food. For example, chewing gum is a food, but if a product is represented as a laxative in chewing gum form, it would be regulated as a drug and not a food.

Decomposed Food   Once a food is so decomposed that it is unfit for food, it is generally still regarded as "food," as defined by FD&C Act. To hold otherwise would provide a loophole in the law against selling decomposed food.[1]

This is a fine example of how the statutory term must differ from our everyday use of the words. We would certainly not, in everyday life, call a slimy, smelly, decomposed fruit "food." Yet, under FD&C Act, such putrid material still is "food."

Food Packaging Materials   Note that the FD&C Act definition of food includes any substances that migrate to the food from the packaging materials or containers.[2] This results in another counterintuitive definition where packaging material is "food." Again, this unusual understanding of terms closes what would otherwise be a gap in the protection of food from adulteration.

5.2.2   Adulterated

Food, Drug, and Cosmetic Act (21 U.S.C. § 301 *et seq.*)

SEC. 402. [342] A food shall be deemed to be **adulterated**—

(a)(1)   if it bears or contains any poisonous or deleterious substance <u>which **may** render it injurious to health</u>; but in case the substance is <u>**not** an added substance</u> such food shall not be considered adulterated under this clause if the quantity of such substance in such food does not ordinarily render it injurious to health; or

(2)(A)   if it bears or contains <u>**any added** poisonous or added deleterious substance</u> (other than one which is (i) a pesticide chemical in or on a raw agricultural commodity, (ii) a food additive, (iii) a color additive, or (iv) a new animal drug) which is unsafe within the meaning of section 406; or . . .

(3)   if it consists in whole or in part of **any** filthy, putrid, or decomposed substance, or if it is otherwise unfit for food; or

---

[1] *See, e.g.*, U.S. v. H.B. Gregory Co., 502 F.2d 700 (7th Cir. 1974) and U.S. v. Thirteen Crates of Frozen Eggs, 215 Fed. 584 (2d Cir. 1914).
[2] *See, e.g.*, Natick Paperboard Corp. v. Weinburger, 525 F.2d 1103 (1st Cir. 1975).

(4)   if it has been prepared, packed, or held under insanitary conditions whereby it may have become contaminated with filth, or whereby it may have been rendered injurious to health; or

(5)   if it is, in whole or in part, the product of a diseased animal or of an animal which has died otherwise than by slaughter; or

(6)   if its container is composed, in whole or in part, of any poisonous or deleterious substance which may render the contents injurious to health; . . .

(b)(1) <u>If any valuable constituent has been in whole or in part omitted or abstracted</u> therefrom; or (2) if any substance has been substituted wholly or in part therefore; or (3) if damage or inferiority has been concealed in any manner; or (4) if any substance has been added thereto or mixed or packed therewith so as to increase its bulk or weight, or reduce its quality or strength, or make it appear better or of greater value than it is.

(c)   If it is, or it bears or contains, a color additive which is unsafe within the meaning of section 706(a).

Note: The Federal Meat Inspection Act, Egg Products Inspection Act, and the Poultry Products Inspection Act contain separate language defining how the term "adulterated" will be applied to the foods each of these laws regulates.

5.2.3 Misbranded

SEC. 403. [343] A food shall be deemed to be **misbranded**—

. . .

(b) If it is offered for sale under the name of another food.

(c) If it is an imitation of another food, unless its label bears, in type of uniform size and prominence, the word "imitation" and, immediately thereafter, the name of the food imitated.

(d) If its container is so made, formed, or filled as to be misleading. . . .

(g) If it **purports** to be—or is represented as a food for which a definition and **standard of identity** has been prescribed by regulations as provided by section 401, unless (1) it conforms to such definition and standard, and (2) its label bears the name of the food specified in the definition and standard, and, insofar as may be required by such regulations, the common names of optional ingredients (other than spices, flavoring, and coloring) present in such food.

(h) If it **purports** to be or is represented as—

(1) a food for which a standard of quality has been prescribed by regulations as provided by section 401, and its quality falls below such

standard, unless its label bears, in such manner and form as such regulations specify, a statement that it falls below such standard; or

(2) a food for which a standard or standards of fill of container have been prescribed by regulations as provided by section 401, and it falls below the standard of fill of container applicable thereto, unless its label bears, in such manner and form as such regulations specify, a statement that it falls below such standard. . . .

DISCUSSION QUESTIONS

5.1. Any filthy, putrid, or decomposed substance. The FD&C Act definition of adulterated includes a food that consists "in whole or in part of *any* filthy, putrid, or decomposed substance." What is the literal meaning of "any"? What would be practical implication if this were enforced literally?

5.2. Why do you think this standard is written so strictly?

5.3   FOOD STANDARDS: REGULATION OF FOOD IDENTITY AND QUALITY

Standards of identity were discussed in the chapter on food labeling because they define specific labeling requirements for many foods. This topic returns under the topic of economic adulteration because standards of identity are an important regulatory tool for maintaining the general quality of foods and preventing economic fraud.

In addition defining the names of food, standards of identity define what a given food product is and the ingredients that must be used, or may be used, in the manufacture of the foods. Standards do not usually relate to such factors as deleterious impurities, filth, and decomposition, which we will discuss later in this chapter.[3]

Section 401 of the Federal Food, Drug, and Cosmetic Act requires that whenever such action will promote honesty and fair dealing in the interest of consumers, regulations shall be promulgated fixing and establishing for any food, under its common or usual name so far as practicable, a reasonable definition and standard of identity, a reasonable standard of quality, and/or reasonable standards of fill for containers. Some standards for foods set nutritional requirements, such as those for enriched bread, or nonfat dry milk with added vitamins A and D, and so forth. A food that is represented or <u>purports</u> to be a food for which a standard of identity has been promulgated must comply with the specifications of the standard in every respect.

---

[3] Exceptions, however, exist; for example, the standards for whole egg and yolk products and for egg white products require these products to be pasteurized or otherwise treated to destroy all viable *Salmonella* bacteria.

DISCUSSION QUESTION

5.3. Purports. What does "purports" mean?

5.3.1 Historical Overview

Looking back in history provides insight into the reasons standards of identity were put into the FD&C Act. The roaring 1920s brought embellishments in advertising, and the Great Depression of the 1930s created a market for cheap goods. This combination resulted in downward spiral of food standards. Consumers often could not depend on the labeling or appearance of a food to guarantee its contents or quality.

Consumers were not the only group hurt by the lack of standards. The food industry also clamored to Congress and the USDA for the establishment of standards. The canning industry was able to get the "Canner's Amendment" (McNary–Mapes) enacted in 1930, which allowed the establishment and enforcement of canned food standards. Substandard products could be sold but had to bear a black label declaration that while a product was good food, it was poor quality.

Passage in 1938 of the Food, Drug, and Cosmetic Act provided for the establishment of standards of identity, standards of quality, and standards of the fill of containers. These standards were to be established "whenever in the judgment of the Secretary such action will promote honesty and fair dealing in the interest of consumers." Congress thought that standards of identity would resemble a "recipe." FDA followed this approach, and foods were defined in terms of recipes or standards with which the consumer could readily identify.

One of the great achievements through use of food standards was the elimination of a number of nutritional deficiency diseases by promulgating standards for enriched food products. The standards were also an important mechanism for FDA to control food additives prior to the passage of the Food Additive Amendment of 1958.

However, the "recipe" concept of standards of identity started to become unwieldy as there was rapid increase in the variety of food products available in the marketplace, beginning in the 1950s and continuing. In addition this recipe approach did little to promote innovation in the food industry. Thus our current era of food standards has often been a tug of war between allowing industry innovation and at the same time protecting consumer expectations.[4]

* * * * *

[4]For a more detailed history and photographs, read: *The Rise and Fall of Federal Food Standards in the United States: The Case of the Peanut Butter and Jelly Sandwich*, by Suzanne White Junod, Ph.D., Historian, U.S. Food and Drug Administration. The full text and her slides are available at: http://www.fda.gov/oc/history/slideshow/default.htm (Sept. 5, 2003).

Fake Food Fight: Substitution of a Valuable Ingredient, Paula Kurtzweil, FDA CONSUMER (March–April 1999)

It is true that you may fool all the people some of the time; you can even fool some of the people all the time; but you can't fool all of the people all of the time.

—Abraham Lincoln

When it comes to fraudulent food in the marketplace, Lincoln's sage observation has certainly rung true. In the Food and Drug Administration's experience, when hucksters try to cheat Americans out of millions of dollars of genuine foods, their schemes are ultimately exposed—by a sharp-eyed consumer, a competitive industry, or FDA itself.

Known as economic adulteration of food, this practice involves using inferior, cheaper ingredients to cheat consumers and undercut the competition. And even though the 1938 Federal Food, Drug, and Cosmetic Act specifically bans it, economic adulteration persists, challenging FDA's resourcefulness to remain vigilant against it.

In recent years, FDA has sought and won convictions against companies and individuals engaged in making and selling bogus orange juice, apple juice, maple syrup, honey, cream, olive oil, and seafood [see Table 5.1].

According to Martin Stutsman, a consumer safety officer in FDA's Center for Food Safety and Applied Nutrition, FDA relies heavily on industry and consumers to help identify instances of economic fraud. In addition, he says, FDA is helping to develop sophisticated laboratory tests and compiling computerized pictorial databases to help industry and consumers determine whether the products they buy are authentic. Stutsman explains that while the agency has fewer resources to monitor the sale of fraudulent food products, it is working with states and local governments and industry and responding to consumer complaints to weed out such practices.

TABLE 5.1   Examples of Economic Food Adulterants from FDA's Files

Food	Adulterant
Orange juice	Beet sugar, corn syrup
Olive oil	Canola oil
Apple juice	Sugar, water, flavoring, hydrolyzed inulin syrup
Dairy cream	Corn oil
Maple and sorghum syrups	Corn syrup
Honey	Corn syrup
Scallops	Water, sodium tripolyphosphate (STP)
Horseradish	Potato starch
Milk	Salt, water
Ginseng (dietary supplement)	Sawdust

Source: Paula Kurtzweil, *Fake Food Fight: Substitution of a Valuable Ingredient*, FDA CONSUMER (Mar.–Apr. 1999).

"It's not a major problem," says Allen Matthys, Ph.D., vice president of regulatory affairs for the National Food Processors Association, "but it is a problem. It's one of those things that keeps bothering you."

An Economic Issue

Economic food fraud involves substituting something of lesser value for something of higher value and then passing off the product as one of higher value—for example, adding coloring to trout and falsely calling it salmon (a more expensive product) or substituting corn syrup for orange juice concentrate (a more expensive ingredient) to make what will be falsely labeled 100 percent pure orange juice.

One of the earliest adulterants was water. "That's one reason FDA exists," says Ben Canas, a food adulterant chemist in FDA's Center for Food Safety and Applied Nutrition, alluding to the 1938 federal law, which was enacted partly in response to public concerns about use of water to adulterate such foods as milk.

Rarely do the adulterants present a health hazard. "This is an economic issue," Stutsman says, explaining that the practice cheats consumers out of their money.

"No one wants to pay for something they're not getting," says Robert Reeves, president of the Institute of Shortening and Edible Oils.

Also, cheaper adulterated products labeled as authentic undercut legitimate industry's prices, making it difficult for honest companies to compete in the marketplace and recoup the expenses they've incurred.

The primary motive in selling a fraudulent food is "greed," says Sandra Williams, a compliance officer in FDA's Detroit district office. "If you sell a product of a lesser value at a higher price, you'll make money."

According to FDA investigations, some companies and individuals have made hundreds of thousands, even millions, of dollars off of their fraudulent foods. The agency estimated that in one fraud case, a Midwestern orange juice manufacturer defrauded consumers of more than $45 million during an estimated 20-year period. Another orange juice company and its president netted $2 million in two years by substituting invert beet sugar for frozen orange juice concentrate. Still another orange juice manufacturer saw its earnings rise from zero in the company's second year of operation to $57 million in its fifth year before being convicted and sentenced for adulterating orange juice concentrate with liquid beet sugar. A family-owned honey- and syrup-making business netted nearly $500,000 from its bogus products between 1993 and 1995.

FDA learns about most cases of economic food adulteration from industry members, who become suspicious of products being offered at prices below fair market value. Many companies also test incoming food ingredients in a laboratory to make sure they're getting what they ordered. When they're not, according to Reeves, word quickly gets around to other industry people and FDA.

In 1996, for example, manufacturers of apple juice products informed FDA of reports that an apple juice concentrate imported from Europe and widely used in the U.S. industry contained hydrolyzed inulin syrup in place of some of the apple juice concentrate. While the product, a high-fructose syrup, was not considered a health hazard, FDA, with industry, began random sampling of apple juice products nationwide to determine whether any products labeled as apple juice on the U.S. market contained hydrolyzed inulin syrup. FDA also tested hydrolyzed inulin syrup and pure apple juice to help verify the accuracy of laboratory tests developed for detecting this high-fructose syrup. According to FDA's Stutsman, these efforts facilitated the quick identification and voluntary removal of adulterated products from U.S. grocery shelves.

"It's the competitiveness of the industry," Reeves says. "Companies [that buy these foods from manufacturers] want to avoid [fraudulent] products. They don't want to lose their customers because once they do, they'll never get them back."

Savvy consumers occasionally alert FDA to possible food adulteration. A lengthy investigation of a Mississippi business selling phony pure honey and pure syrups stemmed in part from complaints FDA received from consumers about the products not tasting like the real thing.

Detective Work

Much of the work of identifying potential adulterants takes place in government and industry laboratories, where chemists use sophisticated tests like gas chromatography and mass spectrometry to identify unique markers that distinguish one substance from another—for example, to distinguish inulin syrup from natural apple sugars. Once a method for detecting an adulterant is verified by several other laboratories, it is published and made available to industry for in-house use. Occasionally, adulterants are identified microscopically.

Even with the tests, detecting an adulterant can be difficult because adulterators develop unique ways to concoct mixtures that closely resemble the real thing; for example, they might add chemicals that when tested, give the product the desired chemical profile of the natural product.

FDA's detective work also takes place in suspect companies' manufacturing facilities, where FDA investigators observe food production, storage, and distribution practices for incongruities. For example, in one bogus orange juice case, an FDA investigator observed a company employee adding pulp wash, the residual orange pulp left after squeezing oranges to get juice, to a product that was to be labeled orange juice from concentrate. Pulp wash isn't permitted to be added to make orange juice.

In another inspection of a syrup company's operations, an FDA investigator identified a supply of "pure maple syrup" labels on the premises, even though he could not spot any raw maple syrup ingredients.

One orange juice company went so far as to hide its supply of an adulterant—liquid beet sugar—in a secret room and used pipes hidden in the ceiling to transport the sugar to the production area. The setup was so well hidden that FDA investigators were able to find it only after receiving explicit directions from a former-employee-turned-informant.

In some cases, FDA investigators have had to go undercover to document evidence of adulteration—for example, secretly observing the nighttime delivery of suspect adulterants.

Taking Action

FDA's efforts to stop a documented case of economic adulteration of food can range from issuing warning letters to seeking full-scale criminal prosecutions. Evidence collected by FDA has enabled federal prosecutors to obtain hefty sentences for individuals and companies found guilty of food adulteration. For example:

- A $100,000 fine and five-year prison sentence for the former president and chief executive officer of an orange juice company that put more than 40 million gallons of adulterated orange juice on the U.S. market over 11 years.
- Fines and forfeitures totaling $120,000 for a seafood company and two of its principals for adding water to scallops to increase their net weight and thus net profit, since scallops are priced according to weight.
- Fines of $20,000 each and prison terms of 19 months and 30 months for two Mississippi brothers for adulterating pure honey and pure maple, cane and sorghum syrups that they sold in old-fashioned tins at farmers' markets and produce stands around the country.
- A $2.18 million fine for an established baby food manufacturer for selling a product labeled "100 percent" apple juice but which actually contained only sugar, water, and flavoring.

. . . .

"It's not always easy," FDA's Stutsman says about detecting economic adulteration. "But it's FDA's job. We want to promote honesty and fair dealing in the interest of consumers—and industry."

* * * * *

5.3.2   FD&C Act § 401, Power to Set Food Standards of Identity

What Are the Requirements Regarding Food Standards?

Excerpted from FDA, Requirements of Laws and Regulations Enforced by the U.S. Food and Drug Administration (1997).

Food Standards are a necessity to both consumers and the food industry. They maintain the general quality of a large part of the national food supply and prevent economic fraud. Without standards, different foods could have the same names or the same foods could have different names. Both situations would be confusing and misleading to consumers and create unfair competition.

Section 401 of the Federal Food, Drug, and Cosmetic Act requires that whenever such action will promote honesty and fair dealing in the interest of consumers, regulations shall be promulgated, fixing and establishing for any food, under its common or usual name so far as practicable, a reasonable definition and standard of identity, a reasonable standard of quality, and/or reasonable standards of fill-of-container. . . .

Standards of identity define what a given food product is, its name, and the ingredients that must be used, or may be used in the manufacture of the food. Standards of quality are minimum standards only and establish specifications for quality requirements. Fill-of-container standards define how full the container must be and how this is measured. FDA standards are based on the assumption that the food is properly prepared from clean, sound materials. Standards do not usually relate to such factors as deleterious impurities, filth, and decomposition. There are exceptions. For example, the standards for whole egg and yolk products and for egg white products require these products to be pasteurized or otherwise treated to destroy all viable *Salmonella* bacteria. Some standards for foods set nutritional requirements such as those for enriched bread, or nonfat dry milk with added vitamins A and D, etc. A food which is represented or purports to be a food for which a standard of identity has been promulgated must comply with the specifications of the standard in every respect.

Foods Named by Use of a Nutrient Content Claim and a Standardized Term

FDA regulations include a "general standard of identity" (21 C.F.R. § 130.10) for modified versions of traditional standardized foods (the standards for traditional foods are contained in 21 C.F.R. §§ 131 through 169). Such modified versions (e.g., "reduced fat" or "reduced calorie" versions of traditional standardized foods) must comply with the provisions of 21 C.F.R. § 130.10, that is, the modified food must:

- Comply with the provisions of the standard for the traditional standardized food except for the deviation described by the nutrient content claim.
- Not be nutritionally inferior to be traditional standardized food.
- Possess performance characteristics, such as physical properties, flavor characteristics, functional properties, and shelf life, that are similar to those of the traditional standardized food, unless the label bears a state-

ment informing the consumer of a significant difference in performance characteristics that materially limits the use of the modified food (e.g., "not recommended for baking").

- Contain a significant amount of any mandatory ingredient required to be present in the traditional standardized food.
- Contain the same ingredients as permitted in the standard for the traditional standardized food, except that ingredients may be used to improve texture, prevent syneresis, add flavor, extend shelf life, improve appearance, or add sweetness so that the modified food is not inferior in performance characteristics to the traditional standardized food.

DISCUSSION QUESTION

5.4. Additional information. See FDA, Information Materials for the Food and Cosmetics Industries at: www.cfsan.fda.gov/~dms/industry.html (last accessed Sept. 12, 2006).

* * * * *

Code of Federal Regulations

21 C.F.R. § 102.23

Title 21—Food and Drugs
Part 102—Common or Usual Name for Nonstandardized Foods—Table of Contents
Subpart B—Requirements for Specific Nonstandardized Foods
Sec. 102.23 **Peanut spreads**.

(a) The common or usual name of a spreadable peanut product that does not conform to Sec. 164.150 of this chapter, and more than 10 percent of which consists of nonpeanut ingredients, shall consist of the term "peanut spread" and a statement of the percentage by weight of peanuts in the product in the manner set forth in Sec. 102.5(b), except that peanut percentages shall be based on the amount of peanuts used to make the finished food and shall be declared in 5 percent increments expressed as a multiple of 5, not to exceed the actual percentage of peanuts in the products.

(b) A spreadable peanut product that is nutritionally inferior to peanut butter shall be labeled as an imitation of peanut butter under Sec. 101.3(e)(2) of this chapter; a spreadable peanut product shall be considered nutritionally equivalent to peanut butter if it meets all of the following conditions:

 (1) Protein.

   (i) The protein content of the product is at least 24 percent by weight of the finished product, and the overall biological quality of the protein contained in the product is at least 68 percent that of casein; or

(ii) The protein content of the product is at least 16.6 percent by weight of the finished product, and the overall biological quality of the protein contained in the product is equal to or greater than that of casein.

(2) Other nutrients. The product contains the following levels of nutrients per 100 grams of product:

NUTRIENT	AMOUNT (MILLIGRAMS)
Niacin	15.3
Vitamin B6	0.33
Folic acid	0.08
Iron	2.0
Zinc	2.9
Magnesium	73.0
Copper	0.6

(c) Compliance with the requirements of paragraph (b) of this section shall be determined by methods described in the following references except that in determining protein quantity in products with mixed protein sources a nitrogen conversion factor of 6.25 may be used.

(1) Protein quantity: "Official Methods of Analysis of the Association of Official Analytical Chemists" (AOAC), 13th Ed. (1980), using the method described in section 27.007, which is incorporated by reference. Copies may be obtained from the Association of Official Analytical Chemists International, 481 North Frederick Avenue, suite 500, Gaithersburg, MD 20877-2504, or may be examined at the Office of the Federal Register, 800 North Capitol Street, NW, suite 700, Washington, DC.

(2) Biological quality of protein: AOAC, 13th Ed. (1980), using the method described in sections 43.212–43.216, which is incorporated by reference. The availability of this incorporation by reference is given in paragraph (c)(1) of this section.

(3) Niacin: AOAC, 13th Ed. (1980), using the method described in sections 43.044–43.046, which is incorporated by reference. The availability of this incorporation by reference is given in paragraph (c)(1) of this section.

(4) Vitamin B6: AOAC, 13th Ed. (1980), using the method described in sections 43.188–43.193, which is incorporated by reference.

The availability of this incorporation by reference is given in paragraph (c)(1) of this section.

(5) Folic acid: Using the method described in U.S. Department of Agriculture Handbook No. 29, modified by use of ascorbate buffer as described by Ford and Scott, Journal of Dairy Research, 35:85–90 (1968), which is incorporated by reference. Copies are available from the Center for Food Safety and Applied Nutrition (HFS-800), Food and Drug Administration, 5100 Paint Branch Pkwy., College Park, MD 20740, or available for inspection at the Office of the Federal Register, 800 North Capitol Street, NW, suite 700, Washington, DC.

(6) Iron: AOAC, 13th Ed. (1980), using the method described in sections 43.217–43.219, which is incorporated by reference. The availability of this incorporation by reference is given in paragraph (c)(1) of this section.

(7) Zinc: AOAC, 13th Ed. (1980), using the method described in sections 25.150–25.153, which is incorporated by reference. The availability of this incorporation by reference is given in paragraph (c)(1) of this section.

(8) Copper: AOAC, 13th Ed. (1980), using the method described in sections 25.038–25.043, which is incorporated by reference. The availability of this incorporation by reference is given in paragraph (c)(1) of this section.

(9) Magnesium: AOAC, 13th Ed. (1980), using the method described in sections 2.109–2.113, which is incorporated by reference. The availability of this incorporation by reference is given in paragraph (c)(1) of this section.

* * * * *

Corn Products Co. v. Dept. of HEW

427 F.2d 511 (1970)

Before: STALEY, SEITZ, and STAHL, Circuit Judges
Opinion: STALEY

Corn Products Company and Derby Foods, Inc., petition for review of an order of the Food and Drug Administration, Department of Health, Education and Welfare, which establishes a definition and standard of identity for the food product known as peanut butter. They seek this review because their products, as they were formulated at the time of the order, fail to conform to the standard.

The order was promulgated under section 401 of the Federal Food, Drug, and Cosmetic Act, 21 U.S.C. § 341. Basically, it limits the percentage by weight of optional ingredients which may be added to the peanut ingredient to a maximum of 10 percent. It allows for the addition or removal of peanut oil and limits the fat content to 55 percent. The standard also identifies allowable additives and specifies certain labeling requirements.

As originally constituted, peanut butter was composed of ground peanuts, salt, and sometimes sugar. However, this product had the disadvantages of oil separation, stickiness, short shelf-life, etc. These deficiencies have been diminished, if not eliminated, by the addition of stabilizing ingredients, hydrogenated vegetable oils. Today, peanut butter consists of the peanut ingredient, which has a solid component and an oil component, the stabilizer, and seasonings.

Petitioners are the major producers of peanut butter. Each has enjoyed a high degree of success. In 1965 Corn Products, the industry leader, claimed 22 percent of the market for its brand, Skippy. Derby as the second leading producer had 14 percent of the market from its product, Peter Pan. Their product formulations fail to qualify under the standard, since each uses in excess of 10 percent of optional ingredients as these are defined by the standard, but each for a different reason.

Both petitioners were unsuccessful in urging the Food and Drug Administration to adopt a standard which would allow 13 percent of optional ingredients, i.e., consist of 87 percent peanuts. Corn Products urges here that the adoption of the 90 percent standard was unreasonable and arbitrary and that the standard will not promote honesty and fair dealing in the interest of consumers. It also argues that the findings upon which the order is based are not supported by substantial evidence. Both petitioners contend that they were entitled to specific findings as to why their products were eliminated. Since this is an appeal from an order of an administrative agency, our first concern must be the extent of our authority to review the order.

The scope of review of the appellate court in considering such orders is defined by the Federal Food, Drug, and Cosmetic Act and the Administrative Procedure Act. Section 701(f)(3) of the Federal Food, Drug, and Cosmetic Act, 21 U.S.C. § 371(f)(3), provides:

"The findings of the Secretary as to facts, if supported by substantial evidence, shall be conclusive."

Section 10(e) of the Administrative Procedure Act, 5 U.S.C. § 706, provides:

". . . The reviewing court shall . . . (2) hold unlawful and set aside agency action, findings, and conclusions found to be . . .
"(E) unsupported by substantial evidence . . ."

. . . The Supreme Court has indicated that substantiality must be determined in the light of all that the record relevantly presents; that findings must be set aside when the record clearly precludes the agency's decision from being justified by a fair estimate of the worth of the testimony of witnesses or its informed judgment on matters within its special competence or both; and, that "reviewing courts must be influenced by a feeling that they are not to abdicate the conventional judicial function."

The Commissioner has concluded that adoption of a standard will promote honesty and fair dealing in the interest of consumers. Support for this conclusion is found in the findings. There is a general lack of information among consumers about the actual composition of peanut butter. It was found that a trend toward a decrease in peanut content has not always been in the interest of consumers. Another finding demonstrates that other ingredients are cheaper and that in some cases the reduced peanut content has resulted from competitive pressure. It was further found that some consumers and state agencies recognize a need for regulation in this area. These findings are supported by sufficient rational probative evidence to afford a sound basis for the exercise of the Commissioner's judgment to promulgate a standard of identity.

In support of its argument that the adoption of the standard requiring 90 percent peanuts is arbitrary and unreasonable, Corn Products cites its market success, market history, established trade practices, and urges that the purpose of the Act, to prevent confusion and deception among consumers, would be served by a standard which would allow its product to be sold as it is presently formulated. It is at once apparent that this argument is not aimed at debasing the findings and conclusions upon which the order is based, but is rather an argument in support of a standard which would not require Corn Products to change the composition of Skippy.

The court's function, however, is to review the findings to determine if there is substantial evidence to support them. Because the court must consider the evidence in keeping with the normal judicial function, the issue of reasonableness would not appear to be completely beyond judicial reach. However, due regard must be given to the integrity of the administrative function. Given a range of reasonable alternatives, the administrator is given the task of selecting the one which, in his judgment, is most appropriate. In such circumstances, the court must defer to his judgment.

Using an affirmative approach to the order under consideration, the issue becomes whether the findings upon which the 90 percent standard is based are supported by substantial evidence. Corn Products' argument that the standard should have designated partially hydrogenated peanut oil as peanut ingredient must be directed at those findings which equate them.

Skippy fails to comply with the standard because it contains 8.5 percent of partially hydrogenated peanut oil and an amount of seasonings which together exceed the 10 percent limit on optional ingredients. No distinction is made in the standard between hydrogenated peanut oil and other hydrogenated vegetable oils.

Nine findings of fact deal directly with hydrogenated oils. These hydrogenated vegetable oils were found to resemble each other more than the oils from which they were derived, although many of the properties of the source oils are retained. Hydrogenation, a process by which unsaturated fats are changed to saturated fats through the addition of hydrogen, causes the physical properties, e.g., melting points, to differ from the source oils. The hydrogenated oils are said to be odorless. Four expert witnesses, all chemists, testified to the dissimilarity between vegetable oil and hydrogenated oil. There was testimony that there is no nutritional variation between these oils. The basic function of the hydrogenated oil, to prevent oil separation in the product, is said to be served regardless of the source oil. The use of hydrogenated peanut oil does not add flavor to the product. From the foregoing, it is quite clear that there is substantial evidence to support a conclusion which makes no distinction between hydrogenated vegetable oils. This conclusion rests upon expert testimony and it is well settled that such testimony is sufficient. . . .

The Peter Pan formulation, by using 9.6 percent dextrose, 1.7 percent stabilizer, and 1.7 percent seasoning, exceeds the 10 percent optional ingredient limitation. Finding of Fact No. 16 indicates that sweetening agents of various intensities are available. Amounts of sweeteners used range as high as 9 percent of the more potent sweeteners and up to 14 percent of the least potent. From surveys conducted in 1963 and 1965, it was found that some producers increased the amount of sweetener with the resultant reduction in the amount of peanuts. Testimony of witnesses and surveys support these statements.

It was also found that the use of optional ingredients, while to some extent required for product improvement, was in response to competitive pressure, since peanuts are the most expensive component. By limiting the amount of optional ingredients, the effect of the order is to require the use of a more potent sweetener in smaller amounts in combination with stabilizer and salt. Since the Commissioner may act to prevent economic adulteration of a product, *Federal Security Administrator v. Quaker Oats Co., supra*, the optional ingredients limitation may be seen as an attempt to prevent such an occurrence.

Other findings relate to the 90 percent requirement. The surveys conducted in 1963 and 1965 support the finding that a majority of manufacturers produced peanut butter containing 90 percent of peanuts. Further, it was found that other manufacturers who then would not comply with the standard had in the past produced a 90 percent peanut product. It is noted that compliance with the standard will not require a change of equipment. Expert testimony indicates that for those presently not in compliance only an alteration of formula is necessary.

Inferentially, petitioners contend that an 87 percent standard would satisfy the purposes of the Act, and there may be substantial evidence to support a standard which would allow their products to be marketed as formulated. Assuming without deciding that to be so, this does not militate against the conclusion that the findings are supported by substantial evidence. In addition, it does not compel the conclusion that the choice of the 90 percent level is arbitrary and unreasonable. It would simply indicate that a reasonable standard could have been established which would not require petitioners to change their formulations. Where equally reasonable alternatives are available, the court must defer to the exercise of administrative discretion.

Perhaps the most troubling of the points raised by petitioners is that there exists within the terms of the standard and definition of identity the means for subverting its intent. It is clear that the intent is to provide a practical maximum of peanut ingredient. The standard provides: "During processing, the oil content of the peanut ingredient may be adjusted by the addition or subtraction of peanut oil. The fat content of the finished foods shall not exceed 55 percent . . ." Petitioner Corn Products demonstrates that this allows the production of a peanut butter containing only 68 percent peanuts. This is not disputed, but a government witness stated that such a possibility is more hypothetical than real. Petitioners contend that it is unreasonable to exclude their products, which contain approximately 87 percent peanuts as defined under the standard, while at the same time including the possibility for making 68 percent peanut butter.

This provision is based upon findings that some adjustment of the oil content of the peanuts is necessary to account for crop variations. Testimony indicated that the oil content of peanuts is a variable. A government witness testified that some provision for the addition and removal of oil was necessary and that it would reflect a good commercial practice. Corn Products admits that addition and removal of peanut oil is an established practice. Should a manufacturer market a 68 percent product, it is apparent that this would violate the spirit if not the letter of the order. Of course, the order is capable of being modified to meet such an eventuality. Certainly, it could not be asserted that a standard is only reasonable if it provides for every possibility. Such an assertion must fall of its own weight, for language which circumscribes conduct is no match for human ingenuity.

Finally, petitioners contend that that they are entitled to specific findings containing reasons for the exclusion of their product formulations. They are unable, however, to cite any direct authority for such a contention. Respondent answers that such findings are not required since this is a rulemaking activity. . . .

Upon a review of the record, we conclude that the Commissioner acted within his authority in promulgating the standard and definition of identity. The findings are supported by substantial evidence and the conclusions rationally follow from the findings.

The standard reflects the practice of a number of manufacturers and to those not in compliance there will be no economic hardship in complying. The fact of exclusion of the leading producers does not make the regulation unreasonable. Products have been excluded before. Skippy and Peter Pan will not be banned; merely a change in product formula will be required. "It is an essence of legislation, functionally speaking, that in its immediate effect, it hurts some and benefits other members of society."

The order will be affirmed.

* * * * *

DISCUSSION QUESTION

5.5. In the *Corn Products* case, the court upheld FDA's standard requiring 90 percent peanuts in peanut butter and rejected the plaintiffs' claim that 87 percent should be sufficient. *Corn Products* demonstrates how a standard of identity may be used to raise the quality of existing foods. Indirectly, the case also demonstrates how standards of identity can provide government agencies with regulatory tools. For example, what do you think would have happened before FDA promulgated a standard of identity for peanut butter if the agency had prosecuted a manufacturer for adulteration under FD&C Act section 402(b) for selling a peanut butter product containing only 87 percent peanuts?

* * * * *

Libby, McNeil & Libby v. United States

148 F.2d 71 (1945)

Before: HUTCHESON, SIMONS, and CLARK, Circuit Judges
Opinion: SIMONS, Circuit Judge

The Federal Security Administrator charged with enforcement of the Federal Food, Drug, and Cosmetic Act, acting under authority of § 401, 21 U.S.C.A. §§ 343(g), (k), 341, promulgated regulations establishing a definition and standard of identity for tomato catsup. The appellant produced and shipped in interstate commerce the condemned food product which concededly does not conform to the standard in that it contains sodium benzoate, a substance not permitted as an ingredient. The government's libel charged that the food was misbranded in violation of § 403(g), and this the appellant, as claimant, denies on the ground that the product was not sold as tomato catsup but as "tomato catsup with preservative," the labels upon the containers specifically declaring that the product does not conform to government standard for catsup, and contains 1/10 of 1 percent benzoate of soda.

Sections 403(g), (k), of the Act declare when a food is deemed to be misbranded, and insofar as the provisions are pertinent, they are printed in the margin. The sole contention urged upon appeal is that the seized product being truthfully labeled, not deceptively packaged, and sold under a name accurately descriptive of its composition, is not misbranded within the meaning of § 403(g), because of the presence in the food of the sodium benzoate. It is urged that the branding of a product, as relating to its characteristics and composition, is the sole basis for determining whether it is misbranded, and that the section does not have the effect, nor was it intended by Congress to have the effect, of excluding any product from interstate commerce when it is sold for what it is. As a supplementary proposition, it is urged that misbranding of the specific product seized is not to be established by designations of identical products applied to them not by their producer but the retail dealers to their customers.

As produced and shipped by the appellant, the condemned food if packed in #10 cans with the described labels thereon. It is catsup as defined by the

Administrator, to which there has been added the minute quantity of sodium benzoate as a chemical preservative. This preservative is harmless, is commonly used in other foods, including oleomargarine, preserves, and jellies, and does not affect the viscosity, taste, smell, or appearance of the catsup. It is explained that there is a wide variation in the degree of concentration of catsup, and a well-established practice in the trade to call a catsup of the higher concentration "fancy," and that of the lower concentration "standard." The difference in specific gravity between the two products is due to the difference in the quantity of added sugar, and the amount of added sugar is determined by the quantity of vinegar added. Catsup is rendered virtually sterile by heat processing, but will spoil after opening unless it contains a preserving agent. Vinegar, sugar, and salt, in combination, are good preserving agents when added in sufficiently large quantities. The amounts required by the standard are relatively small because added only as seasoning ingredients, so it had been the practice in the industry, quite generally, up to 1940, to add sodium benzoate to a lower concentration so as to give it a keeping quality comparable to catsup preserved by added sugar and vinegar.

While fancy catsup is packed in bottles for table use, standard catsup is packed in #10 cans and sold primarily to hotels, restaurants, and similar establishments, although standard catsup, to some extent, is used as table catsup in low priced restaurants. Generally, however, standard catsup is used in cooking and in the preparation of sauces. It costs about 25 percent less that table catsup because it contains less sugar which is a costly ingredient, and is in response to a demand for a less expensive product.

The district court found the product under seizure to conform in all respects to the definition and standard promulgated by the Administrator, except for the addition of the small quantity of benzoate of soda, but held that it purported to be catsup, and so, since it did not conform to the standard, was misbranded. Decision therefore turns upon the meaning of the word "purport" as used in § 403(g). The appellant contends that the label is controlling, that its product does not thereby purport to be catsup, even though it conforms in all respects to the standard, except for the added ingredient. It is a specific article, namely tomato catsup with preservative, and since its label truthfully so indicates, there is no misbranding. The label may be disregarded only if it is assumed that § 403(g) expresses an intent on the part of the Congress to outlaw the manufacture of foods not conforming to applicable standards which, but for the standard, would be sold under the same common and usual name.

It is impossible for us, in the light of controlling authority, to accept the contention. The condemned food is tomato catsup, and purports to be tomato catsup. If producers of food products may, by adding to the common name of any such product mere words of qualification or description, escape the regulation of the Administrator, then the fixing of a standard for commonly known foods becomes utterly futile as an instrument for the protection of the consuming public. Here is no arbitrary or fanciful name, neither "representative or misrepresentative" of a common food product. Such designations invite inquiry as to what the food really is. The present product is intended to satisfy the demand and supply the market for catsup. Emphasis is laid on its conforming to the standard except for the preservatives. The argument defeats itself, for if it is an article of food, distinguished from the standard by the qualification, then other ingredients may be added or defined ingredients or processes omitted without conflicting with the regulation, if containers are truthfully labeled.

In *Federal Security Administrator v. Quaker Oats Co.*, it was said that the statutory purpose to fix a definition of identity of an article of food, sold under its common or usual name, would be defeated if producers were free to add ingredients, however wholesome, which are not within the definition, and so it was not an unreasonable choice of standards for the Administrator to adopt one which defined the familiar farina of commerce without permitting vitamin enrichment, and at the same time a standard for "enriched" farina which permitted a restoration of vitamins removed from whole wheat by milling. The respondent in that case had marketed "Quaker Farina Wheat Cereal, Enriched with Vitamin D." Since this did not conform either to the standard adopted for farina or to the standard adopted for enriched farina, it was held to be misbranded, although the label there as truthfully described the product as does the present label. The district judge was unable to distinguish the present case from the *Quaker Oats* case, and neither can we.

In reviewing the text and legislative history of the present statute, Mr. Justice Stone, in the *Quaker Oats* case, pointed out that its purpose was not confined to a requirement of truthful and informative labeling. False and misleading labeling had already been prohibited by the 1906 Act. The remedy chosen was not a requirement of informative labeling, rather, it was the purpose to authorize the Administrator to promulgate definitions and standards of identity under which the integrity of food products could be effectively maintained, and to require informative labeling only where no such standard had been promulgated; where the food did not purport to comply with the standard; or where the regulations permitted optional ingredients, or required their mention on the label, and that the provision for such standards of identity reflect a recognition by Congress of the inability of consumers to determine, solely on the basis of informative labeling, the relative merits of a variety of products superficially resembling each other. The court was unable to say that such standard of identity, designed to eliminate a source of confusion to purchasers, will not promote honesty and fair dealing within the meaning of the statute.

Neither the decision nor its rationalization in the *Quaker Oats* case can be escaped by a product that looks, tastes, and smells like catsup, which caters to the market for catsup, which dealers bought, sold, ordered, and invoiced as catsup, without reference to the preservative, and which substituted for catsup on the tables of low priced restaurants. The observation in the opinion that it was the purpose of the Congress to require informative labeling, "where the food did not purport to comply with a standard" is not to be lifted out of its context, given a meaning repugnant to the decision, so as to limit "purport" to what is disclosed by the label and to that alone.

The contention that Congress did not intend to, and may not prohibit shipment of non-deleterious substances, is fully answered both in the *Quaker Oats* case ... where the regulation is in the interest of consumers. While the recent case in the Sixth Circuit ... it was there held that the appropriate inquiry is whether the ultimate purchaser will be misled. ... The argument that an affirmance of the decision below will prevent the development of new foods and "lay a dead hand on progress" is one that may more appropriately be addressed to the Administrator or to Congress than to the courts.

The order of condemnation is affirmed.

* * * * *

5.3.3 Current Issue

In the last ten years there has been serious discussion whether standards of identity were needed anymore. This debate has lessened somewhat after FDA defined descriptors like "reduced" and "low fat," which allows a products to be called "low fat ice cream," for example. (In the past, "low fat ice cream could not be sold because "ice cream" has a standard of identity that defines the butterfat content. Low fat dairy desserts were named "ice milk.")

DISCUSSION QUESTIONS

5.6. Do you think standards of identities should be eliminated?

5.7. Can you think of a way around the *Quaker Oats* dilemma today?

5.3.4 Penalties

Under FD&C Act section 403(g),[5] a food that is subject to a definition and standard of identity prescribed by regulation is misbranded if it does not conform to an applicable standard of identity. The potential penalties for shipping foods that deviate from their applicable standards are seizure, injunction, and criminal actions such as fines and imprisonment.

5.3.5 Temporary Marketing Permits

Section 401 of FD&C Act[6] directs FDA to issue regulations establishing definitions and standards of identity for food. The food industry, consumer groups,

[5]21 U.S.C. § 343(g).
[6]21 U.S.C. § 41.

or other interested persons may petition FDA to promulgate or amend a definition or standard of identity.

To enable the food industry to obtain data in support of petitions to amend food standards, FDA may issue temporary marketing permits for interstate shipment of experimental packs of food varying from requirements of standards of identity, in accordance with 21 C.F.R. § 130.17. This allows a manufacturer to conduct an investigation of a potential advance in food technology and acceptance by consumers of a variation in a food from an applicable standard of identity.

A temporary marketing permit is contingent on the submission of labels that alert consumers that the food may vary from their expectations of the standardized food, and also protect consumers against false and misleading labeling.

* * * * *

5.4 SANITATION AND AESTHETIC ADULTERATION

Adulteration of food with contaminants can result in unsafe food. However, adulteration from contaminants or insanitary conditions has a second aspect—that of wholesomeness and aesthetic adulteration. Foods may be contaminated with filth, for example, yet processing may result in a sterile product with no safety risk. Nonetheless, most people do not wish to eat sterilized filth. Therefore food is also regulated for wholesomeness and esthetic adulteration. FD&C Act paragraphs 402(a)(3) [filth] and (4) [unsanitary conditions].

This immediately raises the question: When will a food be considered unwholesome or aesthetically adulterated? The broadest possible reading of FD&C Act paragraph 402(a)(3) might render nearly all food adulterated because—even with the best methods and technology—few foods are free of defects.

Recognizing that a food may contain natural or unavoidable defects that at low levels are not hazardous to health, the FDA establishes maximum defect levels for these defects in foods produced under good manufacturing practices and uses these levels in deciding whether to recommend regulatory action.

Some courts have also recognized the dilemma of unavoidable defects.

* * * * *

U.S. v. 1,500 Cases . . . Tomato Paste

236 F.2d 208 (7th Cir. 1956)

Before: Duffy, Chief Judge, and Finnegan and Swaim, Circuit Judges
Opinion: Swaim, Circuit Judge

Despite the plain language of the section [402(a)(3)] it has been generally held that the two "if" clauses in subsection (3) above are disjunctive, and that the words "otherwise unfit for food" do not limit the first part of the subsection which bans food in whole or in part filthy, etc., as adulterated. ...

We find it impossible to agree with the accepted interpretation of section 342(a)(3), 21 U.S.C.A., without ignoring completely the word "otherwise" therein. ... It has also been suggested that Congress wanted to protect "the aesthetic tastes and sensibilities of the consuming public," and therefore intended that food containing "any filthy, putrid, or decomposed substance" be deemed adulterated whether it was "unfit for food" or not. Congress may also have wanted to set a standard or purity well above what was required for the health of the consuming public, knowing that not every food product can be individually inspected. If the standard is set at the level of what is "fit for food" or not injurious to health, the occasional substandard item that slips by both industry and government scrutiny will be hazardous to the health of the consumer. A minimum standard of purity above what is actually the level of danger will, however, allow fewer products to drop below that level. A high standard will also have the same effect by encouraging more careful industry inspection. Therefore, we prefer to follow the general rule in interpreting section 342(a)(3), although admitting that we are unable to answer Judge Frank as to why Congress put the word "otherwise" in the section.

The interpretation we have chosen has one serious disadvantage which most courts have recognized. It sets a standard that if strictly enforced, would ban all processed food from interstate commerce. A scientist with a microscope could find filthy, putrid, and decomposed substances in almost any canned food we eat. (The substances which it is claimed render the respondent "adulterated" were visible only through a microscope.) The conclusion is inescapable that if we are to follow the majority of the decisions which have interpreted 21 U.S.C.A. § 342(a)(3), without imposing some limitation, the Pure Food and Drug Administration would be at liberty to seize this or any other food it chose to seize. And there could be no effective judicial review except perhaps for fraud, collusion, or some such dishonest procedure. Such a position is not indefensible. Congress has obviously found it difficult, if not impossible, to express a definite statutory standard of purity that will receive uniform interpretation. And this court is acutely aware of the fact that it is not the proper body to more narrowly define broad standards in this area so that they can be applied in a particular case. Courts know neither what is necessary for the health of the consuming public nor what can reasonably be expected from the canning industry. Furthermore, this is not a determination that should be made individually for each case on the basis of expert testimony. The Food and Drug Administration should set definite standards in each industry which, if reasonable, and in line with expressed congressional intent, would have the force of law.

Despite our limitations as a court and the fact that section 342(a)(3), 21 U.S.C.A., does not give us any power to limit the inescapable force of the words, "if it consists in whole or in part of any filthy, putrid, or decomposed substance," we do not think that Congress intended to let the acts of the agency under this subsection go completely without limitation. In section 346, 21 U.S.C.A., Congress directed that the administrator provide tolerances for amounts of poisonous or deleterious substances that cannot be avoided and are not injurious to health. It would not be reasonable to think that Congress would direct the administrator to set tolerances for the allowance of safe amounts of poisons in food and then declare that the presence of small amounts of filth, etc., which would admittedly have no effect upon health "adulterates" food and justifies its seizure. We believe that if the fact that almost all food contains some filthy, putrid, and decomposed substances had been called to the attention of Congress, that body would have directed the administrator to provide reasonable and acceptable tolerances for these substances just as it did in the case of poisons.

The spirit of 21 U.S.C.A. §§ 346 and 346a demands that we give effect to what reasonable standards have been set by the Food and Drug Administration in the area involved in this case, and determine them as best we can where they have not yet been established. The decomposed tomato material which the respondent is accused of containing is commonly referred to as rot. A tomato containing rot is simply a tomato parts of which have begun to decompose. This is not at all uncommon and such fruits are perfectly good if all of the decomposed portions can be cut out. Several different things cause tomatoes to decompose but by far the most common cause is mold. . . . The Food and Drug Administrator with industry cooperation has arrived at a tolerance for tomato paste which is expressed as 40 percent under the Howard Mold Count method of measurement. The Administration has announced that it will not seize tomato paste on the basis of mold count alone unless that count is over 40 percent. We, in our search for standards in this area, accept this administrative tolerance as a proper measure of what approximated amount of decomposition is allowable in tomato paste. A properly obtained mold count of over 40 percent will, therefore, be considered sufficient grounds for seizing tomato paste if the Food and Drug Administrator chooses to do so.

The record in this case does not disclose any established tolerances for what is termed "filth" in tomato paste: worm fragments, insects and insect fragments, fly eggs, etc. We can only judge on the basis of the testimony of experts as to what amounts are usual or unavoidable. . . .

This court holds that as a matter of law all tomato paste having a mold count (or an average mold count where several valid counts are taken) of over 40 percent of positive fields found, is adulterated under 21 U.S.C.A. § 342(a)(3). The record shows that all the codes involved in this proceeding which were canned in October 1955, and bear the code letter "J," have an average mold count above 40 percent. The government should be allowed to seize these codes. All of the codes canned in September, bearing the code letter "I," have an average mold count of less than 40 percent, and therefore cannot be seized on that ground.

The old maxim that the law cares not for small things which the government thinks was the principle the trial court used in releasing some of the codes with average mold counts over 40 percent is not here applicable. The tolerance is admittedly a somewhat arbitrary standard, but one that has been agreed upon by all the parties involved. The line must be drawn somewhere, and it has been validly drawn at 40 percent. Forty-one percent is not just a slight amount of mold, it is a slight amount over a standard that already has allowed for a large margin of error. A definite line must be drawn, and we will apply the one that has been approved by the industry and the government. . . .

* * * * *

5.4.1   GMPs

* * * * *

Good Manufacturing Practice Regulations for Food (GMPs) 21 C.F.R. Part 110

The FDA promulgates the Good Manufacturing Practice (GMP) regulations for foods, which are compiled in Title 21 Code of Federal Regulations, Part 110 (21 C.F.R. § 110).[7] Violation of the GMP regulation can be grounds for finding food is adulterated under the FD&C Act.

Title 21 C.F.R., Part 110.110   Maximum levels of natural/unavoidable defects in food for human use that present no inherent health hazard.

Sec. 110.110   Natural or unavoidable defects in food for human use that present no health hazard.

(a) Some foods, even when produced under current good manufacturing practice, contain natural or unavoidable defects that at low levels are not hazardous to health. The Food and Drug Administration establishes maximum levels for these defects in foods produced under current good manufacturing practice and uses these levels in deciding whether to recommend regulatory action.

(b) Defect action levels are established for foods whenever it is necessary and feasible to do so. These levels are subject to change upon the development of new technology or the availability of new information.

(c) Compliance with defect action levels does not excuse violation of the requirement in section 402(a)(4) of the act that food not be prepared, packed, or held under unsanitary conditions or the requirements in this part that food manufacturers, distributors, and holders shall observe current good manufacturing practice. Evidence indicating that such a violation exists causes the food to be adulterated within the meaning of the act, even though the amounts of natural or unavoidable defects are lower than the currently established defect action levels. The manufacturer, distributor, and holder of food shall at all times utilize quality control operations that reduce natural or unavoidable defects to the lowest level currently feasible.

(d) The mixing of a food containing defects above the current defect action level with another lot of food is not permitted and renders the final food adulterated within the meaning of the act, regardless of the defect level of the final food. . . .

* * * * *

The GMP regulations are general requirements that apply to all foods. FDA, in addition, promulgated specific regulations that apply to specific food categories. These regulations are printed in title 21, Code of Federal Regulations, parts 100–169 (21 C.F.R. §§ 100–169). Four sets of regulations are of particular note:

- Quality control procedures for assuring the nutrient content of infant formulas. 21 C.F.R. § 106.
- GMP regulations for thermally processed low-acid foods in hermetically sealed (airtight) containers (21 C.F.R. § 113), and for acidified foods (21 C.F.R. § 114).
- GMP regulations for bottled water. 21 C.F.R. § 129.
- Title 21 C.F.R. §, part 110.110 allows the FDA to set maximum levels of natural or unavoidable defects in food that present no inherent health hazard.

5.4.2 Waiter, There's a Fly in My Soup—FDA Defect Action Levels

Action levels and tolerances represent limits at or above which FDA will take legal action to remove products from the market. Where no established action level or tolerance exists, FDA may take legal action against the product at the minimal detectable level of the contaminant.

Although use of the terms varies widely, it is beneficial to draw a distinction between action levels and tolerances.

Action level is the term for limits that FDA sets informally. These are guidelines, which basically are a warning to the food industry. Action levels are found in FDA's policy statements.

Tolerance is the term for limits that FDA sets through the formal rulemaking process. Tolerances are found in the regulations.

---

[7]FDA publishes the GMPs at: http://vm.cfsan.fda.gov/~lrd/cfr110.html (last accessed Sept. 12, 2006).

5.4.3 The FDA Food Defect Level Handbook

* * * * *

The Food Defect Action Levels:[8] **Levels of Natural or Unavoidable Defects in Foods that Present No Health Hazards for Humans**

Title 21, Code of Federal Regulations, part 110.110 allows the Food and Drug Administration (FDA) to establish maximum levels of natural or unavoidable defects in foods for human use that present no health hazard. These "Food Defect Action Levels" listed in this booklet are set on this premise—that they pose no inherent hazard to health.

Poor manufacturing practices may result in enforcement action without regard to the action level. Likewise, the mixing of blending of food with a defect at or above the current defect action level with another lot of the same or another food is not permitted. That practice renders the final food unlawful regardless of the defect level of the finished food.

The FDA set these action levels because it is economically impractical to grow, harvest, or process raw products that are totally free of non-hazardous, naturally occurring, unavoidable defects. Products harmful to consumers are subject to regulatory action whether or not they exceed the action levels.

It is incorrect to assume that because the FDA has an established defect action level for a food commodity, the food manufacturer need only stay just below that level. The defect levels do no represent an average of the defects that occur in any of the products—the averages are actually much lower. The levels represent limits at which FDA will regard the food product "adulterated"; and subject to enforcement action under section 402(a)(3) of the Food, Drug, and Cosmetics Act.

As technology improves, the FDA may review and change defect action levels on this list. Also, products may be added to the list. The FDA publishes these revisions as *Notices* in the Federal Register. It is the responsibility of the user of this booklet to stay current with any changes to this list.

Products without Defect Levels

"If there is no defect action level for a product, or when findings show levels or types of defects that do not appear to fit the action level criteria, FDA evaluates the samples and decides on a case-by-case basis. In this procedure, FDA's technical and regulatory experts in filth and extraneous materials use a variety of criteria, often in combination, in determining the significance and regulatory impact of the findings."

The criteria considered is based on the reported findings (e.g., lengths of hairs, sizes of insect fragments, distribution of filth in the sample, and

---

[8]FDA, Center for Food Safety and Applied Nutrition, The Food Defect Action Levels, *available at:* http://www.cfsan.fda.gov/~dms/dalbook.html (last accessed Sept. 2006).

combinations of filth types found). Moreover, FDA interprets the findings considering available scientific information (e.g., ecology of animal species represented) and the knowledge of how a product is grown, harvested, and processed.

Use of Chemical Substances to Eliminate Defect Levels

It is FDA's position that pesticides are not the alternative to preventing food defects. The use of chemical substances to control insects, rodents, and other natural contaminants has little, if any, impact on natural and unavoidable defects in foods. The primary use of pesticides in the field is to protect food plants from being ravaged by destructive plant pests (leaf feeders, stem borers, etc.).

A secondary use of pesticides is for cosmetic purposes—to prevent some food products from becoming so severely damaged by pests that it becomes unfit to eat.

. . .

[The following defect action levels are illustrative.]

PRODUCT	DEFECT (METHOD)	ACTION LEVEL
OLIVES:		
Pitted olives	Pits (MPM-V67)	Average of 1.3 percent or more by count of olives with whole pits and/or pit fragments 2 mm or longer measured in the longest dimension
DEFECT SOURCE: *Processing* SIGNIFICANCE: *Mouth/tooth injury*		
Imported green olives	Insect damage (MPM-V67)	7 percent or more olives by count showing damage by olive fruit fly
DEFECT SOURCE: *Preharvest insect infestation* SIGNIFICANCE: *Aesthetic*		
Salad olives	Pits (MPM-V67)	Average of 1.3 or more olives by count of olives with whole pits and/or pit fragments 2 mm or longer measured in the longest dimension
	Insect damage (MPM-V67)	9 percent or more olives by weight showing damage by olive fruit fly

PRODUCT	DEFECT (METHOD)	ACTION LEVEL

DEFECT SOURCE: *Pits—processing; insect damage—preharvest insect infestation*
SIGNIFICANCE: *Pits—mouth/tooth injury, Insect damage—aesthetic*

Salt-cured olives	Insects (MPM-V67)	Average of 10 percent or more olives by count with 10 or more scale insects each
	Mold (MPM-V67)	Average of 25 percent or more olives by count are moldy

DEFECT SOURCE: *Scale insects—preharvest infestation; mold—postharvest and/or processing infection*
SIGNIFICANCE: *Aesthetic*

Imported black olives	Insect damage (MPM-V67)	10 percent or more olives by count showing damage by olive fruit fly

DEFECT SOURCE: *Preharvest insect infestation*
SIGNIFICANCE: *Aesthetic*

Oregano, ground	Insect filth (AOAC 975.49)	Average of 1,250 or more insect fragments per 10 grams
	Rodent filth (AOAC 975.49)	Average of 5 or more rodent hairs per 10 grams

DEFECT SOURCE: *Insect fragments—preharvest and/or postharvest and/or processing insect infestation; rodent hair—postharvest and/or processing contamination with animal hair or excreta*
SIGNIFICANCE: *Aesthetic*

Peaches, canned and frozen	Mold/insect damage (MPM-V51)	Average of 3 percent or more fruit by count are wormy or moldy
	Insects (MPM-V51)	In 12 one-pound cans or equivalent, one or more larvae and/or larval fragments whose aggregate length exceeds 5 mm

DEFECT SOURCE: *Mold—preharvest and/or postharvest infection; insect damage—preharvest insect infestation; larvae—preharvest insect infestation*
SIGNIFICANCE: *Aesthetic*

PRODUCT	DEFECT (METHOD)	ACTION LEVEL
Peanut butter	Insect filth (AOAC 968.35)	Average of 30 or more insect fragments per 100 grams
	Rodent filth (AOAC 968.35)	Average of 1 or more rodent hairs per 100 grams
	Grit (AOAC 968.35)	Gritty taste and water-insoluble inorganic residue is more than 25 mg per 100 grams

DEFECT SOURCE: *Insect fragments—preharvest and/or post harvest and/ or processing insect infestation; rodent hair—postharvest and/or processing contamination with animal hair or excreta; grit—harvest contamination*
SIGNIFICANCE: *Aesthetic*

Pepper, whole (black and white)	Insect filth and/or insect-mold (MPM-V39)	Average of 1 percent or more pieces by weight are infested and/or moldy
	Mammalian excreta (MPM-V39)	Average of 1 mg or more mammalian excreta per pound
	Foreign matter (MPM-V39)	Average of 1 percent or more pickings and siftings by weight

DEFECT SOURCE: *Insect infested—postharvest and/or processing infestation; moldy—postharvest and/or processing infection; mammalian excreta—postharvest and/or processing animal contamination; foreign material—postharvest contamination*
SIGNIFICANCE: *Aesthetic, potential health hazard—mammalian excreta may contain* Salmonella

Pepper, ground	Insect filth (AOAC 972.40)	Average of 475 or more insect fragments per 50 grams
	Rodent filth (AOAC 972.40)	Average of 2 or more rodent hairs per 50 grams

DEFECT SOURCE: *Insect fragments—postharvest and/or processing insect infestation; rodent hair—postharvest and/or processing contamination with animal hair or excreta*
SIGNIFICANCE: *Aesthetic*

* * * * *

5.4.4 Blending

FDA officially does not permit the blending of a food containing a substance in excess of an action level or tolerance with another food. 21 C.F.R. § 110.110(d) reads, "The mixing of a food containing defects above the current defect action level with another lot of food is not permitted and renders the final food adulterated within the meaning of the act, regardless of the defect level of the final food." Thus, if a defective food is blended with wholesome food, the final product resulting is unlawful, regardless of the level of the contaminant.

For example, wheat contaminated with rodent feces cannot be blended into pure wheat because the finished product would be adulterated. Although the final product may be below defect action levels, FDA will take action on the addition of a deleterious substance to food.

Nonetheless, reprocessing of batches that fail to meet specifications, including commingling with pure products, is a common industry practice. Generally, however, the FD&C Act requires that adulterated food be reconditioned without blending, destroyed, or sent back to country of origin. The FDA has on occasion granted permission to a firm with adulterated food to blend the food or to divert it to use as animal feed. For example, in 1978 the corn crop of seven Southern states was contaminated with high aflatoxin levels. The FDA allowed blending of the corn for animal feed use under tight controls and with prior approval.

NOTES

5.8. Reconditioning. For more information, see the FDA Regulatory Procedures Manual, "Reconditioning" available at: http://www.fda.gov/ora/compliance_ref/rpm/default.htm (last accessed Mar. 16, 2008).

5.9. Diversion. For more information, see the FDA Compliance Policy Guide, "Diversion of Adulterated Food to Acceptable Animal Feed Use" (CPG 7126.20) *available at*: http://www.fda.gov/ora/compliance_ref/cpg/cpgvet/cpg675-200.html (last accessed Mar. 16, 2008).

5.4.5 *De minimis* Filth

United States v. 484 Bags, More or Less

423 F2d 839 (1970)

Before: JONES, BELL, and GODBOLD, Circuit Judges
Opinion: GODBOLD, Circuit Judge

This case concerns whether molded green coffee is adulterated, within the meaning of the Food, Drug and Cosmetic Act, 21 U.S.C. § 342(a)(3). . . . The coffee was imported from Brazil, admitted to the United States, and stored in a warehouse in New Orleans. Three or four days after arrival in September 1965, it was damaged by water during Hurricane Betsy. In an effort to impede the growth of mold on the beans, the consignee had them run through a dryer and resacked. In October 1965, the government filed a libel against the coffee under 21 U.S.C. § 334, alleging that it was adulterated. Almost three years later the District Court granted summary judgment for the government on the issue of adulteration and ordered the coffee condemned. However, under 21 U.S.C. § 334(d) the court granted the petition of the consignee-claimant for release of the beans in order that they be brought into compliance with the Act.

The beans were burnished, or brushed, in an effort to remove the mold. The government was dissatisfied with the result and filed a motion that the coffee be destroyed. . . .

21 U.S.C. § 342(a)(3) provides that a food is deemed adulterated "if it consists in whole or in part of any filthy, putrid or decomposed substance, or if it is otherwise unfit for food." . . . This court, along with others, has long held that the two clauses are independent and complementary, so that a food substance may be condemned as decomposed, filthy, or putrid even though it is not unfit for food, or condemned as unfit for food even though not decomposed, filthy, or putrid. . . . Thus the District Court's finding that the beans were not unfit for food does not preclude condemnation of them as adulterated.

We turn to consideration of the standards to be used in determining if coffee beans are adulterated. The appellee contends that the statute lays down a rule of reason, allowing seizure and condemnation of only foods which deviate from the norm of purity to the extent of going beyond fair and safe standards. We recognize that "It [the first phrase of § 342(a)(3)] sets a standard that if strictly enforced, would ban all processed food from interstate commerce. A scientist with a microscope could find filthy, putrid, and decomposed substances in almost any canned food we eat." But the majority, in fact almost unanimous, rule is that the Act confers the power to exclude from commerce all food products which contain in any degree filthy, putrid, or decomposed substances. . . .

Unjustifiably harsh consequences of a completely literal enforcement are tempered by discretion given the Secretary (now the Secretary of Health, Education, and Welfare). He is allowed to adopt administrative working tolerances for violations of which he will prosecute. The courts may accept the administrative tolerance as a proper judicial measure of compliance with the Act. . . .

We remand the case to the District Court for it to determine under a correct reading of the statute whether the coffee is adulterated. It may accept as a judicial standard the allowable tolerances now permitted by the Secretary, whether published or not. A court may apply a stricter standard than the Secretary and hold a food substance adulterated though within the Secretary's tolerances. Considering the positive command of the statute, the power of the court to allow a greater departure from purity than the administrative tolerances is less certain. . . .

If the coffee is found to be adulterated it must be destroyed. Disposition of it is controlled by the first sentence of § 334(d). The exception to that subsection, adopted by amendment in 1957, authorizes under limited and prescribed conditions the export of articles condemned under § 334. Those conditions are not met in this instance, since the adulteration occurred after the coffee was imported. The language of the statute and the legislative history permit no other conclusion. This Circuit already has held that 21 U.S.C. § 381 does not apply to allow reexport of coffee that has been imported and condemned as adulterated.

Vacated and remanded for proceedings not inconsistent with this opinion.

* * * * *

United States v. Capital City Foods, Inc.

345 F Supp 277 (1972)

Opinion: VAN SICKLE

This is a criminal prosecution by information, based on a claimed violation of 21 U.S.C. § 301 et seq. (The Federal Food, Drug and Cosmetic Act).

Specifically, the defendants are charged with having introduced, or delivered for introduction, into interstate commerce, food that was adulterated (21 U.S.C. § 331(a)). The food is claimed to be adulterated because it consisted in part of a filthy substance, i.e., insect fragments. 21 U.S.C. § 342(a)(3).

Section 342(a)(3) provides that the food is adulterated if it consists in whole or in part of any filthy, . . . substance, or if it is otherwise unfit for food. Insect fragments in other than infinitesimal quantity are filth.

I apply § 342(a)(3) disjunctively . . . That is, I do not require that the food is, by virtue of filth, unfit for human consumption.

But, the presentation of this case has squarely raised these problems:

1. Since the Food and Drug Administration has not promulgated standards of allowable foreign matter in butter, is that not in itself a standard of zero allowance of foreign matter?
2. If the standard is zero allowance of foreign matter, is such a standard reasonable?
3. In any event, has the government proved sufficient foreign matter to raise its proof above the objection of the maxim de minimis lex?

The government was allowed great freedom to introduce testimony pointing to unclean operating conditions in the creamery. But, this testimony was rebutted by the United States Agricultural Department Inspector, and, since the criminal charge was not laid under § 342(a)(4), the defendants, properly, had not prepared a rebuttal; and, in argument, counsel for the United States admitted that United States made no claim of improper operating conditions.

The facts show that miniscule insect fragments were discovered in the butter, and these fragments were identifiable under a 470 power microscope. Some of the fragments were discernable, although none were identifiable with the naked eye. The manufacturing process, while not condemned by the government, did not assure a filter between the raw milk and the pasteurization or cooking process (although there were two in-line filters between the pasteurization and churning units). Thus we can assume any fatty substance reasonably related to the miniscule insect fragments was cooked and distributed into the finished butter. The defendant manager was shown to be responsible for the conduct of the dairy.

Although the defendant corporation, and manager, were charged under a criminal information, the butter involved was allowed to continue through the chain of commerce.

As shown by the analysis of evidence, which I present later, my concern in this case is the claim of the government that:

1. The failure of the government to establish under 21 U.S.C. § 346, a standard of permissible deleterious substance which may be tolerably added to butter when in the manufacturing process it cannot be avoided, establishes as reasonable a standard of zero allowance, and

2. therefore, in effect, the maxim of "*de minimis non curat lex*" has no application in butter cases.

But, in its "Notice of Proposed Rule Making on Natural or Unavoidable Defects in Food for Human Use That Present No Health Hazard," of the Food and Drug Administration, published in the Federal Register, Volume 37, No. 62, March 30, 1972, the introduction language includes this:

"Few foods contain no natural or unavoidable defects. Even with modern technology, all defects in foods cannot be eliminated. Foreign material cannot be wholly processed out of foods, and many contaminants introduced into foods through the environment can be reduced only by reducing their occurrence in the environment."

I accept as a rational, workable approach, the reasoning of the writer in 67 *Harv. L. Rev.*, 632 at 644:

"Indeed, if the section were interpreted literally, almost every food manufacturer in the country could be prosecuted since the statute bans products contaminated 'in whole or in part.' This undesirable result indicates that the section should not receive so expansive a reading. In fact, in several cases judicial common sense has led to recognition that the presence of a minimal amount of filth may be insufficient for condemnation."

The foreign matter found was mainly miniscule fragments of insect parts. They consisted of 12 particles of fly hair (seta), 11 unidentified insect fragments, 2 moth scales, 2 feather barbules, and 1 particle of rabbit hair. The evidence showed that some of these particles were visible to the naked eye, and some, the fly hair, would require a 30× microscope to see. They were identifiable with the aid of a 470× microscope. The only evidence as to size showed that there was one hair, $1\frac{1}{2}$ millimeters long, and one unidentified insect fragment 0.02 millimeters by 0.2 millimeters.

In all, 4,125 grams (9.1 lb) of butter were checked and 28 miniscule particles were found. This is an overall ratio of 3 miniscule particles of insect fragments per pound of butter.

Thus, there having been no standard established, and no showing that this number of miniscule fragments is excludable in the manufacturing process, I find that this contamination is a trifle, not a matter of concern to the law.

The defendants are found not guilty. Judgment will be entered accordingly.

* * * * *

5.4.6 Decomposition

Decomposition is another listed criteria for adulteration under FD&C Act 402(a)(3). Decomposition, like drunkenness, "is easy to detect, but hard to define." *United States v. 1,200 Cans, Etc., Pasteurized Whole Eggs*, 339 F. Supp. 131, 137 (1972). Note that the definition is generally read as disjunctive, and the FDA need not prove that a food is unfit for consumption, but only decomposed, to find adulteration under § 402(a)(3).

The courts have recognized that organoleptic analysis—smell and taste—by a trained examiner can be a valid scientific test for decomposition.

* * * * *

United States v. An Article of Food . . . 915 Cartons of Frog Legs

No. 79 Civ. 6036, U.S. District Court, S.D. N.Y. (1981 U.S. Dist. LEXIS 11628) March 26, 1981

Opinion: CANNELLA, District Judge

. . .

Charles Cardile, an FDA chemist, testified that on December 13, 1977, he and Albert Weber, another trained FDA organoleptic examiner, conducted a joint organoleptic analysis of the eighteen subsamples to determine whether the shrimp were decomposed.[9] Their analysis consisted of thawing the

---

[9] It has been said that decomposition, like drunkenness, "is easy to detect, but hard to define." *United States v. 1,200 Cans, Etc., Pasteurized Whole Eggs*, 339 F. Supp. 131, 137 (N.D. Ga. 1972). Decomposition is "a bacterial separation or breakdown in the elements of the food so as to produce an undesirable disintegration or rot." *Id*. It is well recognized that organoleptic analysis of food, whereby the examiner relies on his trained sense of smell to detect different types of offensive food, if honestly administered, is a valid scientific test for decomposition. See *id*. at 137–38.

eighteen subsamples, selecting 100 shrimp from each subsample, and then breaking the flesh of each shrimp and smelling it. On the basis of their training and pursuant to FDA Guidelines, the examiners then classified each shrimp as either class one, good commercial shrimp; class two, decomposed shrimp; or class three, shrimp in advanced stages of decomposition. Under the FDA Guidelines establishing tolerances for decomposition, a subsample is classified as decomposed if (1) 5 percent or more of the shrimp tested is class three, (2) 20 percent or more of the shrimp tested is class two, or (3) the percentage of class two shrimp plus four times the percentage of class three shrimp equals or exceeds 20 percent. The FDA will take legal action against the entire shipment when four or more of the eighteen subsamples are found to be decomposed. Based upon their examination of the eighteen subsamples at issue, Cardile and Weber found seven of the eighteen subsamples to be decomposed.

Although not directly challenging the validity of organoleptic testing generally, Biswa argues that the FDA has not complied with its own Guidelines in testing the shrimp at issue because it did not conduct a chemical analysis known as the indole test to confirm the results of the organoleptic examinations. The Guidelines specifically provide, however, that the indole test is optional for imported shrimp, when originally tested. There is no dispute that the shrimp at issue originated in India and were imported from the Netherlands. Moreover there is no evidence that Biswa requested the FDA to perform the indole test or that its own expert performed that test. Therefore, the Court concludes that the organoleptic analysis conducted by the experts in this action is a reliable indicator of decomposition.

Biswa's expert witness, Bernard Tzall, the president and director of Certified Laboratories, Inc., a private testing laboratory, organoleptically tested the shrimp for decomposition under the standard set forth in the FDA Guidelines. On May 23, 1980, Tzall, in the mistaken belief that he was supposed to test the shrimp for *Salmonella*, took only fifteen subsamples from the same lots of shrimp as those examined by the FDA.

Although fifteen subsamples is a sufficiently large sample to test for *Salmonella*, it does not meet the FDA Guidelines for testing shipments of shrimp containing over 100 cases for decomposition. Tzall nonetheless organoleptically tested the fifteen subsamples for decomposition under the same procedures followed by the FDA examiners, except that Tzall performed the analysis alone and examined only fifty shrimp from each subsample instead of one hundred. He found that none of the fifteen subsamples exceeded the 20 percent limit set in the FDA Guidelines. After discovering his sampling error, Tzall returned to the warehouse on June 9, 1980, and took an additional eighteen subsamples from the same three lots. Tzall selected six cartons from each of the three lots at random from the top layers of the pallets. Upon returning to his laboratory, he thawed the eighteen subsamples, organoleptically examined fifty shrimp from each subsample and classified them pursuant to the FDA Guidelines. Based on his examination, Tzall found that three of the eighteen subsamples exceeded the 20 percent limit set by the FDA. Since the number of decomposed subsamples did not exceed four, Tzall concluded that the 506 cartons of shrimp were not decomposed under the FDA Guidelines.

Discussion

Section 304(a) of the Act, as amended, 21 U.S.C. § 334(a)(1), provides that an article of food that is adulterated within the meaning of 21 U.S.C. § 342, when introduced into interstate commerce or while held for sale, "shall be liable to be proceeded against while in interstate commerce, or at any time thereafter, on libel of information and condemned in any district court of the United States" within the jurisdiction of which the article is found. Biswa does not dispute that the shrimp at issue has been shipped in interstate commerce as defined in 21 U.S.C. § 321(b), or that it is food within the meaning of 21 U.S.C. § 321(f).

Under the Act, a food shall be deemed "adulterated," and hence subject to condemnation, "if it consists in whole or in part of any filthy, putrid, or decomposed substance, or if it is otherwise unfit for food." 21 U.S.C. § 342(a)(3). The majority rule is that the Act "confers the power to exclude from commerce all food products which contain in any degree filthy, putrid or decomposed substances." Moreover, although the government has the burden of proving adulteration by a fair preponderance of the credible evidence, it need not demonstrate that the food is injurious or unfit for consumption. Because all processed foods transported in interstate commerce may to some extent be decomposed or contain filth, strict enforcement of section 342(a)(3) would result in the banning of all food products. Therefore the issue in a section 342(a)(3) action is to determine the degree of decomposition that renders an article of food adulterated.

To avoid the harsh results of strict enforcement, Congress empowered the FDA, in its discretion, to decline to prosecute minor violations. 21 U.S.C. § 336. In the exercise of that discretion, the FDA has announced the Guidelines employed by the experts in this action. Harsh results are also avoided by application of a judicially created de minimus doctrine, whereby small quantities of filth and decomposition can be overlooked by a court where there is evidence that that amount is unavoidable within the industry.

In determining the degree of decomposition necessary to render a shipment adulterated, the Court may accept as a judicial standard the tolerances now permitted by the FDA in its Guidelines. Alternatively, a court may apply a stricter standard than the FDA's to hold a food substance adulterated although within the FDA's tolerances. But "[c]onsidering the positive command of the statute, the power of the Court to allow a greater departure from purity than the administrative tolerances is less certain." In weighing the evidence before it, the Court has relied upon the expertise of the FDA and accepts its Guidelines as a proper and reasonable defect tolerance level for shrimp.

Since the results of the FDA's joint organoleptic analysis revealed that seven of the subsamples tested contained more than the 20 percent decomposition tolerated by the FDA, with at least two subsamples scoring as high as 100 percent and 170 percent, the Court concludes that more than a de minimus amount of decomposition was present in the frozen shrimp and that it is "adulterated" within the meaning of 21 U.S.C. § 342(a)(3). The Court finds that the government has sustained its burden of proof despite the conflicting evidence of Biswa's expert for two reasons. First, although the FDA Guidelines do not specify the number of shrimp that should tested from each subsample and an examination of fifty shrimp is sufficient under the Guidelines, the Court credits Cardile's testimony that more accurate test results are obtained when one hundred shrimp are tested. Second, because of the subjective nature of organoleptic analysis, the government test is more reliable than Tzall's because it was jointly conducted by two examiners. Moreover, the Court is not convinced that defendant's expert has greater expertise in organoleptic analysis than the government's expert. Although Mr. Tzall has been active in the food testing field for many years and has earned a number of science degrees, Mr. Cardile has attended several FDA seminars on organoleptic testing as well as received the same on-the-job training that Mr. Tzall testified he received during the course of his career. Accordingly, on the basis of the evidence adduced at trial, the Court finds that the 506 cartons of frozen shrimp are adulterated within the meaning of section 342(a)(3).

Conclusion

In accordance with the foregoing, having found that the government has sustained its burden of proof that the defendant in rem, with the exception of the 84 cartons of shrimp previously released, are adulterated within the meaning of the Act.

* * * * *

5.4.7 Insanitary Conditions

Another ground for finding adulteration under the FD&C Act is if a food was prepared, packed, or held under insanitary conditions whereby it may have become contaminated with filth.[10] Unlike the adulteration provisions of (a)(3), this provision only indirectly relates to the nature of the food. Instead, the conditions of the facility where the food has been handled, processed, or stored are the primary focus.

For a violation to be established under (a)(4), the FDA must prove both that the food was exposed to insanitary conditions and that by reason of this exposure the food "may have become contaminated with filth" or "may have

---

[10] FD&C Act § 402(a)(4).

been rendered injurious to health." FDA need not prove that the food is contaminated in fact.

The broad provision was put into the act when shocking conditions at some food plants were exposed to the public. This definition is so broad that some courts are sought means to moderate the literal meaning. In *Berger v. United States*[11] the court held that there must be a *"reasonable"* possibility of contamination, not just a "mere possibility." In *United States v. Certified Grocers Co-Op*,[12] the court recognized that the standard for conviction under (a)(4) is "whether the insanitary conditions made it *reasonably possible*" that contamination would occur. However, generally, the courts provide considerable deference to FDA in its determination of insanitary conditions, "whereby it may have become contaminated with filth, or whereby it may have been rendered injurious to health" and, thus, adulterated.

* * * * *

U.S. v. 1,200 Cans Pasteurized Whole Eggs by Frigid Food Products

339 F.Supp. 131 (1972)

Opinion: SIDNEY O. SMITH Jr., Chief Judge

These five actions were brought in different parts of the United States pursuant to Section 304 of the Federal Food, Drug and Cosmetic Act (21 U.S.C. § 334) to condemn and destroy as adulterated various lots of pasteurized frozen whole eggs and sugar yolks processed and introduced into interstate commerce . . . the government contends that the lots were "adulterated" in one or more of the definitions prescribed by Congress in 21 U.S.C. § 342, which provides in part:

"A food shall be deemed to be adulterated—

(a)(1) If it bears or contains any poisonous or deleterious substance which may render it injurious to health; . . .

(3) If it consists in whole or in part of any filthy, putrid, or decomposed substance, or . . .

(4) If it has been prepared, packed, or held under insanitary conditions whereby it may have become contaminated with filth, or whereby it may have been rendered injurious to health; . . ."

. . .

The Organoleptic Evidence

Indicative of this attribute is the almost universal acceptance of organoleptic tests for determining decomposition. All of the experts in this case agree that,

---

[11] Berger v. United States, 200 F.2d 818 (8th Cir. 1952).
[12] United States v. Certified Grocers Co-Op, 546 F.2d 1308 (1976).

honestly administered, they are valid. To some extent, all of us have God-given organoleptic expertise. We exercise the powers of sight, smell, taste, and feel to reject unpalatable food. As used in food and drug matters, the organoleptic test is a mere refinement in that people can be trained to detect why the food is offensive, i.e., due to rot, mold, sours, etc. The government periodically conducts schools for training its inspectors in such procedures, but in the final result, the organoleptic examination is not far removed from that daily performed by the housewife. If the food smells bad, she rejects it. And yet, it is generally approved by the most exacting of scientists as proper. More importantly, in civil cases, it has been recognized by the courts for at least 50 years. Organoleptic smell tests have worked extremely well on unpasteurized egg products for years. The product is either "passable" or "rejected." However, the pasteurization process, which basically arrests decomposition, has posed a new problem.

The pasteurization process universal since 1966, plus refinements in the freezing process, have masked decomposition odors and made the test much more difficult. . . .

III. Insanitary Conditions under 21 U.S.C. § 342(a)(4).

"A food shall be deemed to be adulterated—if it has been prepared, packed, or held under insanitary conditions whereby it may have become contaminated with filth, or whereby it may have been rendered injurious to health."

While there are many similarities between (a)(3) and (a)(4) proceedings, the legislative thrust of the latter is entirely different. In essence, the (a)(3) section permits the seizure of foods which have actually decomposed irrespective of processing conditions; even if they were completely sanitary. On the other hand, the (a)(4) section allows the condemnation of foods processed under insanitary conditions, whether they have actually decomposed or become dangerous to health or not. The objective of (a)(4) is to "require the observance of a reasonably decent standard of cleanliness in handling of food products" and to insure "the observance of those precautions which consciousness of the obligation imposed upon producers of perishable food products should require in the preparation of food for consumption by human beings." It almost reaches the aim of removing from commerce those products produced under circumstances which would offend a consumer's basic sense of sanitation and which would cause him to refuse them had he been aware of the conditions under which they were prepared.

To that end, although the ultimate product may not be filthy or injurious to health, if it was processed under insanitary conditions whereby it "may" have been contaminated with filth or whereby it "may" have been rendered injurious to health, it is adulterated within the meaning of section (a)(4).

Again, it would be helpful if there were specific plant standards or tolerances to guide the court. The need has been expressed before. Some argument has been made that the regulations promulgated in 1969 answer this purpose.

21 C.F.R. §§ 128.1–128.9. With a few exceptions, they are inadequate to do so in that they fail to specify just what is "necessary," "needed," "effective," "sufficient," or the like. In the context of actual conditions in a particular industry, the regulations simply require an absolute standard, which well might be impossible to achieve. It is true that impossibility technically furnishes no defense. Certainly it is no defense that a processor is "doing the best he can under the conditions and circumstances." However, even without a specific measuring stick, the law must always be construed to be real and meaningful to the every day life of the citizenry. It has been done in (a)(4) cases. Thus the ultimate test is whether the conditions are such that it is "reasonably possible" the food may become contaminated with filth or may be rendered injurious to health.

In the absence of particular standards, the question must be determined from the totality of the circumstances as revealed by the evidence. In this regard, it is not necessary that the evidence of insanitary conditions absolutely coincide with the dates of processing provided they are not too remote in time or space. The proof should, however, justify the inference that such conditions actually existed on the dates in question. . . .

Measured by the above, the test has been met in this case. Reviewing the evidence as a whole, the court must conclude that the conditions existing at the Golden Egg plant on the critical dates were exactly those the Congress sought to prevent by the passage of (a)(4). . . .

Accordingly, the court finds that all lots are subject to condemnation under section (a)(4). . . .

* * * * *

5.4.8 Good Manufacturing Practices (GMPs)

In 1967, FDA proposed good manufacturing practice (GMP) regulations for the food industry.[13] FDA justified the authority for these regulations on the "insanitary conditions" provisions of section 402(a)(4) of the FD&C Act. FDA also promulgated GMPs for specific commodities but soon found itself in court defending the legality of the specific regulations.

* * * * *

United States v. Nova Scotia Food Products Corp.
568 F.2d 240 (1977)

Before: WATERMAN and GURFEIN, Circuit Judges, and BLUMENFELD, District Judge

[13] 32 Fed. Reg. 17980 (Dec. 15, 1967) and later promulgated by 32 Fed. Reg. 6977 (Apr. 26, 1969) and codified at 21 C.F.R. § 110.

GURFEIN, Circuit Judge

This appeal involving a regulation of the Food and Drug Administration is not here upon a direct review of agency action. It is an appeal from a judgment of the District Court for the Eastern District of New York (Hon. John J. Dooling, Judge) enjoining the appellants, after a hearing, from processing hot smoked whitefish except in accordance with time-temperature-salinity (T-T-S) regulations contained in 21 C.F.R. part 122 (1977). . . .

The injunction was sought and granted on the ground that smoked whitefish which has been processed in violation of the T-T-S regulation is "adulterated."

Appellant Nova Scotia receives frozen or iced whitefish in interstate commerce which it processes by brining, smoking and cooking. The fish are then sold as smoked whitefish. . . .

Government inspection of appellants' plant established without question that the minimum T-T-S requirements were not being met. There is no substantial claim that the plant was processing whitefish under "insanitary conditions" in any other material respect. Appellants, on their part, do not defend on the ground that they were in compliance, but rather that the requirements could not be met if a marketable whitefish was to be produced. They defend upon the grounds that the regulation is invalid (1) because it is beyond the authority delegated by the statute; (2) because the FDA improperly relied upon undisclosed evidence in promulgating the regulation and because it is not supported by the administrative record; and (3) because there was no adequate statement setting forth the basis of the regulation. We reject the contention that the regulation is beyond the authority delegated by the statute, but we find serious inadequacies in the procedure followed in the promulgation of the regulation and hold it to be invalid as applied to the appellants herein.

The hazard which the FDA sought to minimize was the outgrowth and toxin formation of *Clostridium botulinum* Type E spores of the bacteria which sometimes inhabit fish. . . .

The Commissioner of Food and Drugs ("Commissioner"), employing informal "notice-and-comment" procedures under 21 U.S.C. § 371(a), issued a proposal for the control of *C. botulinum* bacteria Type E in fish. For his statutory authority to promulgate the regulations, the Commissioner specifically relied only upon § 342(a)(4) of the Act which provides:

"A food shall be deemed to be adulterated
"(4) if it has been prepared, packed, or held under insanitary conditions whereby it may have become contaminated with filth, or whereby it may have been rendered injurious to health;" . . .

The Commissioner thereafter issued the final regulations in which he adopted certain suggestions made in the comments, including a suggestion by the National Fisheries Institute, Inc. ("the Institute"), the intervenor herein.[14]
. . .

---

[14] The final regulations are codified at 21 C.F.R. part 122 (1977).

When, after several inspections and warnings, Nova Scotia failed to comply with the regulation, an action by the United States Attorney for injunctive relief was filed on April 7, 1976, six years later, and resulted in the judgment here on appeal. The District Court denied a stay pending appeal, and no application for a stay was made to this court.

I

The argument that the regulation is not supported by statutory authority cannot be dismissed out of hand. The sole statutory authority relied upon is § 342(a)(4) quoted above. . . . Nor is the Commissioner's expressed reliance solely on § 342(a)(4) a technicality which might be removed by a later and wiser reliance on another subsection. For in this case, as the agency recognized, there is no other section or subsection that can pass as statutory authority for the regulation. The categories of "adulteration" prohibited in section 342 all refer to food as an "adulterated" product rather than to the process of preparing food, except for subsection (a)(4) which alone deals with the processing of food.

Appellants contend that the prohibition against "insanitary conditions" embraces conditions only in the plant itself, but does not include conditions which merely inhibit the growth of organisms already in the food when it enters the plant in its raw state. They distinguish between conditions which are insanitary, which they concede to be within the ambit of § 342(a)(4), and conditions of sterilization required to destroy micro-organisms, which they contend are not.

It is true that on a first reading the language of the subsection appears to cover only "insanitary conditions" "*whereby* it (the food) may have been rendered injurious to health" (emphasis added). And a plausible argument can, indeed, be made that the references are to insanitary conditions in the plant itself, such as the presence of rodents or insects . . .

Yet, when we are dealing with the public health, the language of the Food, Drug and Cosmetic Act should not be read too restrictively, but rather as "consistent with the Act's overriding purpose to protect the public health." As Justice Frankfurter said in *United States v. Dotterweich*, 320 U.S. 277:

"The purposes of this legislation thus touch phases of the lives and health of people which, in the circumstances of modern industrialism, are largely beyond self-protection. Regard for these purposes should infuse construction of the legislation if it is to be treated as a working instrument of government and not merely as a collection of English words."

Thus a provision concerning "food additives" has been held to include even poisonous substances which have not been "added" by human hands.

Section 371(a), applicable to rulemaking under § 342(a)(4), provides: "The authority to promulgate regulations for the efficient enforcement of this chapter, except as otherwise provided in this section, is vested in the Secretary." We read this grant as analogous to the provision "make . . . such rules and regulations as may be necessary to carry out the provisions of this Act," in which case "the validity of a regulation promulgated thereunder will be sustained so long as it is 'reasonably related to the purposes of the enabling legislation.' (citations omitted)" ... When agency rulemaking serves the purposes of the statute, courts should refuse to adopt a narrow construction of the enabling legislation which would undercut the agency's authority to promulgate such rules. The court's role should be one of constructive cooperation with the agency in furtherance of the public interest. ...

Appellant's argument, it should be noted, is not that there has been an unlawful delegation of legislative power, or even a delegation of "unfettered discretion." The argument, fairly construed, is that Congress did not mean to go so far as to require sterilization sufficient to kill bacteria that may be in the food itself rather than bacteria which accreted in the factory through the use of insanitary equipment.

There are arguments which can indeed be mustered to support such a broad-based attack under 5 U.S.C. § 706.

First, the Act deals with standards of identity and various categories that can render food harmful to health. Yet, so far as the category of harmful micro-organisms is concerned, there is only a single provision, 21 U.S.C. § 344, which directly deals with "micro-organisms." That provision is limited to emergency permit controls dealing with any class of food which the Secretary finds, after investigation, "may, by reason of contamination with micro-organisms *during* the manufacture, processing or packing thereof in any locality, be injurious to health, and that such injurious nature cannot be adequately determined after such articles have entered interstate commerce, (in which event) he then, and in such case only, shall promulgate regulations providing for the issuance ... of permits. ..." (Emphasis added). It may be argued that the failure to mention "micro-organisms" in the "adulteration" section of the Act, which includes § 342(a)(4), means that Congress intended to delegate no further authority to control micro-organisms than is expressed in the "emergency" control of section 344.

On the other hand, as Judge Dooling held, the manner of processing can surely give rise to the survival, with attendant toxic effects on humans, of spores which would not have survived under stricter "sanitary" conditions. In that sense, treating "insanitary conditions" in relation to the hazard, the interpretation of the District Court which described the word "sanitary" as merely "inelegant" is a fair reading, emphasizing that the food does not have to be actually contaminated during processing and packing but simply that "it may have been rendered injurious to health," § 342(a)(4), by inadequate sanitary conditions of prevention. ...

We do not discount the logical arguments in support of a restrictive reading of § 342(a)(4), but we perceive a larger general purpose on the part of Congress in protecting the public health.

We come to this conclusion, aside from the general rules of construction noted above, for several reasons: First, until this enforcement proceeding was begun, no lawyer at the knowledgeable Food and Drug bar ever raised the question of lack of statutory delegation or even hinted at such a question. Second, the body of data gathered by the experts, including those of the Technical Laboratory of the Bureau of Fisheries manifested a concern about the hazards of botulism. Third, analogously, the Meat Inspection Act of 1907 (now codified as amended at 21 U.S.C. § 608), which hardly provided a clearer standard than does the "insanitary conditions" provision in the Food and Drug Act, has regulations under it concerning mandatory temperatures for processing pork muscle tissue to eliminate the hazard of trichonosis. The statute permits the Secretary "to prescribe the rules and regulations of sanitation under which such establishments shall be maintained." The current regulation, 9 C.F.R. § 318.10 (1977), provides: "All parts of the pork muscle tissue shall be heated to a temperature not lower than 137 °F., and the method used shall be one known to insure such a result" 9 C.F.R. § 318.10(c)(1) (1977). The same regulation was codified as early as 1949 as 9 C.F.R. § 18.10(c)(1) (1949). These regulations have been assumed for years to have been properly promulgated by the Secretary of Agriculture under the statutory authority given to him.

Lastly, a holding that the regulation of smoked fish against the hazards of botulism is invalid for lack of authority would probably invalidate, to the extent that our ruling would be followed, the regulations concerning the purity of raw materials before their entry into the manufacturing process in 21 C.F.R. part 113 (1977) (inspection of incoming raw materials for microbiological contamination before thermal processing of low-acid foods packed in hermetically sealed containers), in 21 C.F.R. part 118 (1977) (pasteurization of milk and egg products to destroy *Salmonella* microorganisms before use of the products in cacao products and confectionery), and 21 C.F.R. part 129 (1977) (product water supply for processing and bottling of bottled drinking water must be of a safe, sanitary quality when it enters the process).

The public interest will not permit invalidation simply on the basis of a lack of delegated statutory authority in this case. A gap in public health protection should not be created in the absence of a compelling reading based upon the utter absence of any statutory authority, even read expansively. Here we find no congressional history on the specific issue involved, and hence no impediment to the broader reading based on general purpose.[15] We believe, nevertheless, that it would be in the public interest for Congress to consider in the light of existing knowledge, a legislative scheme for administrative regulation of the processing of food where hazard from micro-organisms in food in its natural state may require affirmative procedures of sterilization. This would entail, as

---

[15] In December 1972, FDA Chief Counsel Hutt, speaking to the Annual Educational Conference of the Food and Drug Law Institute said, "(T)he Act must be regarded as a constitution." "(T)he fact that Congress simply has not considered or spoken on a particular issue certainly is no bar to the (FDA) exerting initiative and leadership in the public interest." 28 Food Drug Cosmetic Law Journal 177, 178–79 (Mar. 1973). For a reply, see H. Thomas Austern, *id*. at 189 (Mar. 1973). We do not take sides on the issue tendered, but we think Mr. Hutt's language to be conscious hyperbole. The test is not "initiative" but whether delegation may be fairly inferred from the general purpose.

well, a decision on the type of rulemaking procedure Congress thinks fit to impose. . . .

* * * * *

5.4.9 Otherwise Unfit for Food

A last condition that can result in adulteration under (a)(3) is that the product is "otherwise unfit for food." This provision has been used by FDA successfully against product that was "so tough and rubbery that the average, normal person, under ordinary conditions, would not chew and swallow it." *United States v. 24 Cases, More or Less*, 87 F. Supp. 826 (1949). But contrast with *United States v. 298 Cases . . . Ski Slide Brand Asparagus*, 88 F. Supp. 450 (1949), where the court held against FDA, noting that the government should not be keeping a low price, nutritious product from the market.

PART III

FOOD SAFETY REGULATION

CHAPTER 6

Food Safety Regulation

6.1 INTRODUCTION

In Chapter three, we examined food-labeling regulation that is designed to protect the economic expectations of both consumers and the food industry. We discussed how the regulation of food labeling and misbranding overlaps with the regulation of food adulteration. In Chapter five, we covered the protection of economic and aesthetic expectations through the regulation of economic adulteration (FD&C Act section 402(b)) and aesthetic adulteration (FD&C Act paragraphs 402(a)(3) and (4)). In subsequent chapters, we will cover adulteration in more detail as related to food additives (402(a)(2)(C)), food colorings (402(c)), and irradiation.

This chapter covers the concept of food safety and adulteration from poisonous and deleterious substances. On this topic it is especially important to understand that U.S. food safety law is not a single standard but an amalgamation of various standards. Each regulatory standard is directed at a distinct concern, but often with overlapping span. Therefore, when examining a potential adulterant, the first question is, "In which category does this component fall?"

This chapter covers the main FD&C Act subdivisions of adulteration with toxicants:

- Section 402(a)(1)'s **may render injurious** standard for added components of food;
- Section 402(a)(1)'s **ordinarily injurious** standard for nonadded components of food;
- Section 406's **tolerances for the protection of public health** for added components whose use is necessary or unavoidable; and
- Section 408's **tolerances for pesticide residues** on raw agricultural commodities.

In addition this chapter covers important regulation of carcinogens under the Delaney Clause and food safety with HACCP.

After you complete this chapter, you will have an understanding of the following:

1. The FDA statutes and regulations regarding poisonous and deleterious substances in foods;
2. The issues concerning the presence of environmental contaminants in our food and regulation of pesticide residues; and
3. The distinction between unintentional and intentional adulteration and tampering.

6.1.1 Background—The Nature and Cost of Foodborne Illness[1]

To fully appreciate the benefits of food safety regulation, it is necessary to understand the burden of foodborne illness. In excess of 200 known diseases are transmitted through food.[2] These diseases include infections, intoxications, and chronic sequelae.[3] The foodborne infectious agents include bacteria, viruses, and parasites. The intoxications (commonly called poisonings) include bacterial toxins, heavy metals, insecticides, and other chemical contaminants. Disease symptoms range from mild gastrointestinal distress to life-threatening neurological, hepatic, and renal syndromes, and death.[4]

Over the past ten years, science has begun to reveal the grim potential of foodborne pathogens to cause chronic sequelae, secondary complications that may develop months, even years, after the first unpleasant bout of symptoms.[5] Growing evidence exists for a multitude of chronic illnesses resulting from an attack of foodborne disease, such as "arthropathies, renal disease, cardiac and neurological disorders, and nutritional and other malabsorbtive disorders (incapacitating diarrhea)."[6] Sequelae include the immediate aftereffects of foodborne disease, toxins with long delay in onset, antigenic and autoimmune effects, and intracellular sequestration. It is estimated that chronic sequelae may occur in 2 to 3 percent of foodborne illness cases.[7]

---

[1] Adapted from Neal Fortin, *The Hang-up with HACCP: The Resistance to Translating Science into Food Safety Law*, 58 FOOD AND DRUG LAW JOURNAL 565–594 (2003).

[2] *See* Paul S. Mead et al., *Food-Related Illness and Death in the United States,* 5 EMERGING INFECTIOUS DISEASES 607 (1999) (citing F. L Brian, *Diseases Transmitted by Food, Centers for Disease Control* (1982)).

[3] A sequela is an aftereffect of disease or injury, or a secondary result of a disease.

[4] *See* James A. Lindsay, *Chronic Sequelae of Foodborne Disease*, 3(4) EMERGING INFECTIOUS DISEASES at 1 (1997) *available at*: http://www.cdc.gov/ncidod/eid/vol3no4/lindsay.htm (last visited Feb. 12, 2002).

[5] *Id.*

[6] *Id.*

[7] *Id.* and U.S. GENERAL ACCOUNTING OFFICE (GAO), FOOD SAFETY, INFORMATION ON FOODBORNE ILLNESSES, GAO/RCED-96-96, at 8 (May 1996).

The burden of foodborne illness is estimated as high as 300 million cases per year[8] and patient-related costs in the billions of dollars per year.[9] Each year in the United States, foodborne illness causes an estimated 76 million illnesses,[10] 320,000 hospitalizations, and 5,000 deaths.[11] Contaminated food results in one of every 100 hospitalizations, and one of every 500 deaths in the United States.[12]

One estimate places the cost for just direct, patient-related costs of foodborne illness at $164 billion per year.[13] Other estimates attempt to calculate the costs for a limited number of foodborne pathogens (typically five to seven major pathogens), and in these, the estimated annual cost of medical treatment and lost productivity varies from $5.6 billion to $37.1 billion.[14]

Many people casually reference the available aggregate estimates as the total cost of foodborne illness. However, these estimates—by design—are partial estimates of the burden of foodborne illness. For example, the USDA's Economic Research Service (ERS) estimated the medical costs and losses in productivity of five major foodborne pathogens at between $5.6 billion and $9.4 billion.[15] However, this estimate does not include hepatitis A virus and other significant pathogens. In addition the ERS estimate and other available estimates do not include difficult to quantify costs, such as the expenditures on foodborne illness by public health agencies. Further these aggregate estimates of cost do not include the loss of food (i.e., recall and destruction), lost production, lost sales, or pain and suffering. The aggregate estimates also do not encompass foodborne illness that is too mild to require medical treatment, and they do not include the amount consumers are willing to pay to avoid mild

[8] *See* Chryssa V. Deliganis, *Death By Apple Juice: The Problem of Foodborne Illness, the Regulatory Response, and Further Suggestions for Reform*, 53 FOOD & DRUG LAW JOURNAL 681, 695 (1998) (noting the statements of Michael Osterholm, epidemiologist, Minnesota Department of Health, at American Medical Association press conference on public health (Dec. 2, 1997) http://www.yahoo.com/headlines/971202/health/stories/food.htm).

[9] *See* Sanford A. Miller, *The Saga of Chicken Little and Rambo*, 51 JOURNAL OF THE ASSOCIATION OF FOOD & DRUG OFFICIALS 196 (1987) and Jean C. Buzby & Tanya Roberts, *Economic Costs and Trade Impacts of Microbial Foodborne Illness,* 50(1/2) WORLD HEALTH STATISTICS QUARTERLY 57 (1997).

[10] Illness as used here means the disease is serious enough to require medical treatment.

[11] CDC, FOODNET SURVEILLANCE REPORT FOR 1999 (FINAL REPORT), at 6 and 19, (Nov. 2000) (citing Mead P., et al., *Food-Related Illness and Death in the United States*, 5 EMERGING INFECTIOUS DISEASES 607 (1999)).

[12] JEAN C. BUZBY et al., *Product Liability and Microbial Foodborne Illness*/AER-799, at 3 (Economic Research Service/USDA 2001).

[13] Miller, *supra* note 9, at 196.

[14] JACK GUZEWICH and MARIANNE P. ROSS, FDA, EVALUATION OF RISKS RELATED TO MICROBIOLOGICAL CONTAMINATION OF READ-TO-EAT FOOD BY FOOD PREPARATION WORKERS AND THE EFFECTIVENESS OF INTERVENTIONS TO MINIMIZE THOSE RISKS at 3 (citing J. C. Buzby and T. Roberts, *Economic Costs and Trade Impacts of Microbial Foodborne Illness*, 50(1/2) WORLD HEALTH STAT. QUARTERLY 57 (1997)); and GAO, FOOD SAFETY, INFORMATION ON FOODBORNE ILLNESSES, GAO/RCED-96-96, at 9 (May 1996).

[15] GAO, FOOD SAFETY, INFORMATION ON FOODBORNE ILLNESSES, GAO/RCED-96-96, at 9 (May 1996).

diarrhea and nausea. Finally, none of the aggregate estimates includes the costs of the chronic sequelae of foodborne illness. Estimates of health consequences of chronic sequelae indicate that the economic costs may be higher than those of the acute diseases.[16] In this light, even the highest estimate, $164 billion per year for direct medical costs, may be far below the total burden of foodborne illness.

The significant burden of foodborne illness highlights the importance of an effective and efficient food safety system, and the potential gains from the application of HACCP. Safe food is a goal shared by all. Consumers obviously benefit by having fewer illnesses. Society benefits from lower health care costs and lost productivity. Food businesses profit from lower liability, fewer production losses (e.g., recalls), and improved marketability of their product.

6.1.2 Poisonous and Deleterious Substances

When examining a potential adulterant, the first question is: In which category does this component fall? The first such distinction is whether a toxicant is added or non-added. The 1906 Food and Drug Act defined as adulterated a food that contained "any *added* poisonous or other *added* deleterious ingredient which may render such article injurious to health." In 1938, Congress eliminated this limitation to "added" ingredients when it passed the present Food, Drug, and Cosmetic Act (FD&C Act). However, Congress retained a distinction between toxicants that were "added" and those that were not: "but in case the substance is not an added substance, such food shall not be considered adulterated under this clause if the quantity of such substance in such food does not ordinarily render it injurious to health."

Read the definition of adulteration due to poisonous and deleterious substances and note the difference between the regulation of added and non-added components of food.[17] Note that there are two standards identified in paragraph 402(a)(1).

The primary difference between these two standards is that the FDA must show a greater probability of harm to restrict a natural component of food than an added one. Under the "may render injurious" standard a food containing a toxicant is considered adulterated unless "it cannot by any possibility" injure the health of any consumer.[18]

A second important difference is that the "may render" standard allows FDA to take in account especially vulnerable segments of the population. In addition FDA has a greater burden of proof that a natural toxicant is sufficient to render the food "ordinarily injurious" to health.[19]

---

[16] *See* Lindsay, *supra* note 4, at 2.

[17] FD&C Act sect. 402(a) [342].

[18] *See* United States v. Lexington Mill & Elevator Co., 232 U.S. 399 (1914).

[19] *See* Millet, Pit and Seed Co., Inc. v. United States, 436 F.Supp. 84 (1977) (Where amygdalin in apricot kernels was found to be anonadded substance because it was naturally occurring and the amount of the poison amygdalin in the kernels would not make the kernels ordinarily injurious.)

6.1.3 Added Substances

The case of *United States v. Lexington Mill & Elevator Co.* demonstrates the FDA need not prove that a food containing added poisonous or other added deleterious ingredients *must* affect the public health. The FD&C Act's burden of proof placed on the government is only that the added poisonous or deleterious substances *may* render the food injurious to health.

Congress provided such broad language so that FDA might consider various uses of foods and various consumers. For example, a food may be consumed "by the strong and the weak, the old and the young, the well and the sick; and it is intended that if any flour, because of any added poisonous or other deleterious ingredient, may possibly injure the health of any of these, it shall come within the ban of the statute."[20]

* * * * *

United States v. Lexington Mill & Elevator Co. 232 U.S. 399 (1914)

Mr. Justice WILLIAM DAY delivered the opinion of the court:

The petitioner, the United States of America, proceeding under § 10 of the food and drugs act . . . sought to seize and condemn 625 sacks of flour in the possession of one Terry, which had been shipped from Lexington, Nebraska, to Castle, Missouri, and which remained in original, unbroken packages. . . . The amended libel charged that the flour had been treated by the "Alsop Process," so called, by which nitrogen peroxide gas, generated by electricity, was mixed with atmospheric air, and the mixture then brought in contact with the flour, and that it was thereby adulterated under the fourth and fifth subdivisions of 7 of the act; namely (1) in that the flour had been mixed, colored, and stained in a manner whereby damage and inferiority were concealed and the flour given the appearance of a better grade of flour than it really was, and (2) in that the flour had been caused to contain added poisonous or other added deleterious ingredients, to-wit, nitrites or nitrite reacting material, nitrogen peroxide, nitrous acid, nitric acid, and other poisonous and deleterious substances which might render the flour injurious to health. . . .

The Lexington Mill & Elevator Company, the respondent herein, appeared, claiming the flour, and answered the libel, admitting that the flour had been treated by the Alsop Process, but denying that it had been adulterated, and attacking the constitutionality of the act.

A special verdict to the effect that the flour was adulterated was returned and judgment of condemnation entered. The case was taken to the circuit court of appeals upon writ of error. The respondent contended that, among other errors, the instructions of the trial court as to adulteration were erroneous and that the act was unconstitutional. The circuit court of appeals held that the

---

[20] Lexington Mill & Elevator Co., 232 U.S. 399, 411 (1914).

testimony was insufficient to show that by the bleaching process the flour was so colored as to conceal inferiority, and was thereby adulterated, within the provisions of subdivision 4. That court also held—and this holding gives rise to the principal controversy here—that the trial court erred in instructing the jury that the addition of a poisonous substance, in any quantity, would adulterate the article, for the reason that "the possibility of injury to health due to the added ingredient, and in the quantity in which it is added, is plainly made an essential element of the prohibition." It did not pass upon the constitutionality of the act, in view of its rulings on the act's construction.

The case requires a construction of the food and drugs act. . . . Without reciting the testimony in detail, it is enough to say that for the government it tended to show that the added poisonous substances introduced into the flour by the Alsop Process, in the proportion of 1.8 parts per million, calculated as nitrogen, may be injurious to the health of those who use the flour in bread and other forms of food. On the other hand, the testimony for the respondent tended to show that the process does not add to the flour any poisonous or deleterious ingredients which can in any manner render it injurious to the health of a consumer. On these conflicting proofs the trial court was required to submit the case to the jury. . . .

It is evident from the charge given and refused that the trial court regarded the addition to the flour of any poisonous ingredient as an offense within this statute, no matter how small the quantity, and whether the flour might or might not injure the health of the consumer. At least, such is the purport of the part of the charge above given, and if not correct, it was clearly misleading, notwithstanding other parts of the charge seem to recognize that, in order to prove adulteration, it is necessary to show that the flour may be injurious to health. The testimony shows that the effect of the Alsop Process is to bleach or whiten the flour, and thus make it more marketable. If the testimony introduced on the part of the respondent was believed by the jury, they must necessarily have found that the added ingredient, nitrites of a poisonous character, did not have the effect to make the consumption of the flour by any possibility injurious to the health of the consumer.

The statute upon its face shows that the primary purpose of Congress was to prevent injury to the public health by the sale and transportation in interstate commerce of misbranded and adulterated foods. The legislation, as against misbranding, intended to make it possible that the consumer should know that an article purchased was what it purported to be; that it might be bought for what it really was, and not upon misrepresentations as to character and quality. As against adulteration, the statute was intended to protect the public health from possible injury by adding to articles of food consumption poisonous and deleterious substances which might render such articles injurious to the health of consumers. If this purpose has been effected by plain and unambiguous language, and the act is within the power of Congress, the only duty of the courts is to give it effect according to its terms. This principle has been frequently recognized in this court. . . .

Furthermore, all the words used in the statute should be given their proper signification and effect.

"We are not at liberty," said Mr. Justice Strong, "to construe any statute so as to deny effect to any part of its language. It is a cardinal rule of statutory construction that significance and effect shall, if possible, be accorded to every word. As early as in Bacon's Abridgment, 2, it was said that 'a statute ought, upon the whole, to be so construed that, if it can be prevented, no clause, sentence, or word, shall be superfluous, void, or insignificant.' This rule has been repeated innumerable times."

Applying these well-known principles in considering this statute, we find that the fifth subdivision of 7 provides that food shall be deemed to be adulterated "if it contain any added poisonous or other added deleterious ingredient which may render such article injurious to health." The instruction of the trial court permitted this statute to be read without the final and qualifying words, concerning the effect of the article upon health. If Congress had so intended, the provision would have stopped with the condemnation of food which contained any added poisonous or other added deleterious ingredient. In other words, the first and familiar consideration is that, if Congress had intended to enact the statute in that form, it would have done so by choice of apt words to express that intent. It did not do so, but only condemned food containing an added poisonous or other added deleterious ingredient when such addition might render the article of food injurious to the health. Congress has here, in this statute, with its penalties and forfeitures, definitely outlined its inhibition against a particular class of adulteration.

It is not required that the article of food containing added poisonous or other added deleterious ingredients must affect the public health, and it is not incumbent upon the government in order to make out a case to establish that fact. The act has placed upon the government the burden of establishing, in order to secure a verdict of condemnation under this statute, that the added poisonous or deleterious substances must be such as may render such article injurious to health. The word "may" is here used in its ordinary and usual signification, there being nothing to show the intention of Congress to affix to it any other meaning. It is, says Webster, "an auxiliary verb, qualifying the meaning of another verb, by expressing ability, ... contingency or liability, or possibility or probability." In thus describing the offense, Congress doubtless took into consideration that flour may be used in many ways, in bread, cake, gravy, broth, etc. It may be consumed, when prepared as a food, by the strong and the weak, the old and the young, the well and the sick; and it is intended that if any flour, because of any added poisonous or other deleterious ingredient, may possibly injure the health of any of these, it shall come within the ban of the statute. If it cannot by any possibility, when the facts are reasonably considered, injure the health of any consumer, such flour, though having a small addition of poisonous or deleterious ingredients, may not be condemned under the act. This is the plain meaning of the words, and in our view needs no additional support by reference to reports and debates, although it may be said in passing that the meaning which we have given to the statute was well expressed by Mr. Heyburn, chairman of the committee having it in charge upon the floor of the Senate, "As to the use of the term 'poisonous,' let me state that everything which contains poison is not poison. It depends on the quantity and the combination. A very large majority of the things consumed by the human family contain, under analysis, some kind of poison, but it depends upon the combination, the chemical relation which it bears to the body in which it exists, as to whether or not it is dangerous to take into the human system." ...

* * * * *

6.1.4 Nonadded Substances

The last chapter discussed the significance of naturally occurring components of a food when determining whether a food is considered aesthetically adulterated (filth or unwholesome components).[21] "Naturally occurring defects" in food is a term of art applied generally to defects that create no health hazard. When discussing potential health hazards—poisonous or deleterious substances—a similar distinction with different terms is made between added and nonadded substances.[22]

U.S. v. 1,231 Cases American Beauty Brand Oysters

43 F. Supp. 749 (1942)

Opinion: REEVES, District Judge

This is a proceeding by the process of libel to condemn an alleged adulterated food product. Such food consists of 1,232 cases of oysters, each case containing 24 cans, marked "American Beauty Brand Oysters."

As a basis for condemnation, it is alleged by the government that said article "contains shell fragments, many of them small enough to be swallowed and become lodged in the esophagus, and that said shell fragments are sharp and capable of inflicting injury in the mouth."

The provision of the law invoked by the government is section 342, Title 21 U.S.C.A., and sundry subdivisions thereof. Said section provides, among other things, that:

"A food shall be deemed to be adulterated—
(a) (1) If it bears or contains any poisonous or deleterious substance which may render it injurious to health; but in case the substance is not an added

---

[21] FD&C Act paragraphs 402(a)(3) and (4).
[22] FD&C Act § 402(a).

substance, such food shall not be considered adulterated under this clause if the quantity of such substance in such food does not ordinarily render it injurious to health." . . .

The evidence in the case showed that in the processing of oysters for food there is a constant effort to eliminate shells and fragments thereof from the product. For this purpose many means and devices are used to reduce as nearly to a minimum as possible such shells and fragments in the product. The evidence, however, on behalf of both the government and the defense was that with present known means and devices it was impossible to free the produce entirely from the presence of part shells and shell fragments. Moreover, it not only appeared, but it is a matter of common knowledge, that an oyster is a marine bivalve mollusk with a rough and an irregular shell wherein it develops and grows, and that, in the processing of the food produce, it is necessary to remove this irregular, rough shell so far as that may be accomplished. The shells, therefore, are not artificially added for the purpose of growth or to aid in the processing operations.

The evidence on the part of the government was that parts of shell and shell fragments upon inspection were found in many of the cans taken from the article seized. Such parts of shell and fragments were exhibited at the trial.

There was evidence on behalf of the claimant that its processing operations were in accord with the best manufacturing practice, and there was even some testimony that the means employed by it for the elimination of shell fragments were superior to the means employed by other processors engaged in similar operations. The testimony on the part of the claimant further tended to show that within the Kansas City area over a period of ten years it had sold approximately 5 million cans of its product and that no complaint had ever been made concerning the presence of shell fragments. Claimant also proved that over 50 million cans had been processed by it and distributed in its trade territory and that no complaints had ever been made of the presence of part shells or shell fragments.

It seems proper at this point to comment that in this case involving considerable testimony there was no substantial controversy as to the facts and practically no difference of opinion as to the law. There was a contention by the government that the shells as a deleterious substance were added to the product while being processed. There was no evidence to support this contention.

1. The excerpt from the statute heretofore quoted contemplates that there may be of necessity food products containing deleterious substances. No one who has had the experience of eating either fish or oysters is unfamiliar with the presence of bones in the fish (a deleterious substance) and fragments of shell in the oysters (also a deleterious substance).

   The Congress, however, withdrew such foods from the adulterated class "if the quantity of such substance in such food does not ordinarily render it injurious to health."

The evidence on both sides was that by the greatest effort, and in the use of the most modern means and devices, shell fragments could not be entirely separated from an oyster food product. The government, in its brief, quite aptly and concisely stated its point by using the following language: "It is the character, not the quantity of this substance that controls its ability to injure."

This concession on the part of the government, properly made, upon the evidence removes the case immediately from that portion of the statute which says: "... such food shall not be considered adulterated under this clause if the quantity of such substance in such food does not ordinarily render it injurious to health."

Since it is the "character, not the quantity of this substance that controls its ability to injure," as stated by the government, then in the view that it is impossible to eliminate shell fragments in toto from the product, the use of oysters as a food must be entirely prohibited or it must be found that the presence of shell fragments is not a deleterious substance within the meaning of the law and must be tolerated to reject oyster products as a food in unthinkable. It would be as reasonable to reject fish because of the presence of bones. Even if a greater percentage of shells and shell fragments were found in claimant's product than in that of other processors, yet this fact, under the theory of the government, would not add to the deleterious nature of claimant's product. It should be stated, however, that there was no evidence that there was an excess of shell fragments in claimant's product over that of other processors. On the contrary, a preponderance of the evidence showed that the claimant's processing methods were superior.

2. It does not seem necessary to discuss other portions of said section 342 invoked by the government. It is charged in the libel complaint that other provisions of the statute were violated by substituting shell fragments for oysters, and that shell fragments had been mixed or packed with the oyster product so as to reduce its quality. There was no testimony to support these averments and so as to make applicable those provisions of the law directed against such acts.

3. Counsel for both the government and the claimant, at the trial and in their briefs, discussed the question of the right to a tolerance regulation as provided by section 346, Title 21 U.S.C.A. This provision is for tolerance of both poisonous and deleterious substances where the presence of such substance cannot be avoided. However, that section says: "(a) Any poisonous or deleterious substance added to any food, except where such substance is required in the production thereof or cannot be avoided by good manufacturing practice, shall be deemed to be unsafe for purposes of the application of clause (2) of section 342 (a)."

Adverting to clause 2 of said section 342 (a), it reads as follows: "... or (2) if it [food] bears or contains any added poisonous or added deleterious substance which is unsafe within the meaning of section 346."

It will be seen at once that this provision does not apply where the deleterious substance inheres in the product and is not added. Further quoting from section 346, however, note this language: ". . . but when such substance is so required or cannot be so avoided, the Administrator shall promulgate regulations limiting the quantity therein or thereon to such extent as he finds necessary for the protection of public health."

Upon the concession made by the government in this case, even if the tolerance section could be construed to apply, it is not the quantity of the substance but its character "that controls its ability to injure."

4. Upon the evidence in the case it must be found that the presence of shell fragments in the article sought to be condemned does not ordinarily render it injurious to health.

Under the statute and upon the evidence the government is not authorized to condemn the article seized for the reason that the processed article does not offend against the food and drug law. The claimant, therefore, should have restored to it the articles seized and the libel should be dismissed. It will be so ordered.

* * * * *

DISCUSSION QUESTIONS

6.1. Ordinarily injurious. The "ordinarily injurious" standard is much less stringent than the "may render injurious" standard. Why do you think Congress made this distinction?

6.2. Naturally occurring. Why are naturally occurring substances allowed in foods, even though those substances may be poisonous or deleterious?

6.3. Problem exercise: FDA publishes a advanced notice of proposed rulemaking (ANPR) defining *Listeria monocytogenes* (*Lm*) as an added substance. Healthy adults, by and large, can consume small quantities of *Lm* without adverse effect. On the other hand, children, the elderly, immunocompromised, and pregnant woman are susceptible to *Lm* at lose doses. The effects of listeriosis include septicemia, encephalitis, and intrauterine or cervical infections in pregnant women, which may result in spontaneous abortion (second/third trimester) or stillbirth.[23] You are counsel for a soft cheese manufacturer that sells some cheeses made

---

[23] FDA, BAD BUG BOOK: FOODBORNE PATHOGENIC MICROORGANISMS AND NATURAL TOXINS HANDBOOK, *available at*: http://www.cfsan.fda.gov/~mow/chap6.html (last accessed Mar. 16, 2008).

from unpasteurized milk, but labeled with the warning: "May be dangerous to those with compromised immune systems and pregnant women." Advise the cheese company of the potential impact of FDA proposal.

6.1.5 Tolerances for Unavoidable or Necessary Poisonous and Deleterious Substances

When the FD&C Act was enacted in 1938, it included a new provision banning all unnecessary and avoidable poisonous or deleterious substances added to food. To ameliorate the stringency of this provision, Congress provided that FDA could provide safe tolerance levels for poisonous or deleterious substances that were required in the manufacture of food or were unavoidable, such as pesticide residues or lead in the solder of cans. Exceeding these tolerances results in adulteration under FD&C Act section 402(a).[24]

* * * * *

Tolerances for Poisonous or Deleterious Substances in Food
FD&C Act § 406 [21 U.S.C. § 346]

Any poisonous or deleterious substance added to any food, except where such substance is required in the production thereof or cannot be avoided by good manufacturing practice, shall be deemed to be unsafe for purposes of the application of clause (2)(A) of section 342(a) of this title; but when such substance is so required or cannot be so avoided, the Secretary shall promulgate regulations limiting the quantity therein or thereon to such extent as he finds necessary for the protection of public health, and any quantity exceeding the limits so fixed shall also be deemed to be unsafe for purposes of the application of clause (2)(A) of section 342(a) of this title. While such a regulation is in effect limiting the quantity of any such substance in the case of any food, such food shall not, by reason of bearing or containing any added amount of such substance, be considered to be adulterated within the meaning of clause (1) of section 342(a) of this title. In determining the quantity of such added substance to be tolerated in or on different articles of food the Secretary shall take into account the extent to which the use of such substance is required or cannot be avoided in the production of each such article, and the other ways in which the consumer may be affected by the same or other poisonous or deleterious substances.

* * * * *

Section 406 tolerances results in the regulated substances being under more comprehensive control than the general "may render injurious" standard of section 402(a)(1). This provision set the stage for consideration of cumulative

---

[24] FD&C Act § 406 [21 U.S.C. § 346].

dietary exposure and modern day risk analysis. These tolerances are established based on the unavoidability of the poisonous or deleterious substances. Thus the tolerances are raised or eliminated as technology or other changes make previous levels avoidable (e.g., the elimination of lead solder in cans). In addition these tolerances never permit contamination under circumstances where it is avoidable. The poisonous-or-deleterious tolerance levels are established and revised according to criteria specified in 21 C.F.R. parts 109 and 509.[25]

* * * * *

Young v. Community Nutrition Institute et al.

476 U.S. 974 (1986)

Justice O'CONNOR delivered the opinion of the Court.

. . .

The Food and Drug Administration (FDA) enforces the Federal Food, Drug, and Cosmetic Act (Act) as the designee of the Secretary of Health and Human Services. 21 U.S.C. § 371(a). The Act seeks to ensure the purity of the Nation's food supply, and accordingly bans "adulterated" food from interstate commerce. Title 21 U.S.C. § 342(a) deems food to be "adulterated."

> "(1) If it bears or contains any poisonous or deleterious substance which may render it injurious to health; but in case the substance is not an added substance such food shall not be considered adulterated under this clause if the quantity of such substance in such food does not ordinarily render it injurious to health; or (2)(A) if it bears or contains any added poisonous or added deleterious substance (other than [exceptions not relevant here]) which is unsafe within the meaning of section 346a(a) of this title. . . ."

As this provision makes clear, food containing a poisonous or deleterious substance in a quantity that ordinarily renders the food injurious to health is adulterated. If the harmful substance in the food is an added substance, then the food is deemed adulterated, even without direct proof that the food may be injurious to health, if the added substance is "unsafe" under 21 U.S.C. § 346.

Section 346 states:

> "Any poisonous or deleterious substance added to any food, except where such substance is required in the production thereof or cannot be avoided by good manufacturing practice shall be deemed to be unsafe for purposes of the application of clause (2)(A) of section 342(a) of this title; but when such substance is so required or cannot be so avoided, the Secretary shall promulgate regulations limiting the quantity therein or thereon to such extent as he finds necessary for

[25] *See* FDA's Industry Activities Staff Booklet, ACTION LEVELS FOR POISONOUS OR DELETERIOUS SUBSTANCES IN HUMAN FOOD AND ANIMAL FEED, *available at:* http://www.cfsan.fda.gov/~lrd/fdaact.html (last accessed Sept. 18, 2007).

the protection of public health, and any quantity exceeding the limits so fixed shall also be deemed to be unsafe for purposes of the application of clause (2)(A) of section 342(a) of this title. While such a regulation is in effect . . . food shall not, by reason of bearing or containing any added amount of such substance, be considered to be adulterated. . . ."

Any quantity of added poisonous or added deleterious substances is therefore "unsafe," unless the substance is required in food production or cannot be avoided by good manufacturing practice. For these latter substances, "the Secretary shall promulgate regulations limiting the quantity therein or thereon to such extent as he finds necessary for the protection of public health." It is this provision that is the heart of the dispute in this case.

The parties do not dispute that, since the enactment of the Act in 1938, the FDA has interpreted this provision to give it the discretion to decide whether to promulgate a § 346 regulation, which is known in the administrative vernacular as a "tolerance level." Tolerance levels are set through a fairly elaborate process, similar to formal rulemaking, with evidentiary hearings. On some occasions, the FDA has instead set "action levels" through a less formal process. In setting an action level, the FDA essentially assures food producers that it ordinarily will not enforce the general adulteration provisions of the Act against them if the quantity of the harmful added substance in their food is less than the quantity specified by the action level.

B

The substance at issue in this case is aflatoxin, which is produced by a fungal mold that grows in some foods. Aflatoxin, a potent carcinogen, is indisputedly "poisonous" or "deleterious" under §§ 342 and 346. The parties also agree that, although aflatoxin is naturally and unavoidably present in some foods, it is to be treated as "added" to food under § 346. As a "poisonous or deleterious substance added to any food," then, aflatoxin is a substance falling under the aegis of § 346, and therefore is at least potentially the subject of a tolerance level.

The FDA has not, however, set a § 346 tolerance level for aflatoxin. It has instead established an action level for aflatoxin of 20 parts per billion (ppb). In 1980, however, the FDA stated in a notice published in the Federal Register:

"The agency has determined that it will not recommend regulatory action for violation of the Federal Food, Drug, and Cosmetic Act with respect to the interstate shipment of corn from the 1980 crop harvested in North Carolina, South Carolina, and Virginia and which contains no more than 100 ppb aflatoxin. . . ." 46 Fed.Reg. 7448 (1981).

The notice further specified that such corn was to be used only as feed for mature, nonlactating livestock and mature poultry.

. . .

The FDA's longstanding interpretation of the statute that it administers is that the phrase "to such extent as he finds necessary for the protection of

public health" in § 346 modifies the word "shall." The FDA therefore interprets the statute to state that the FDA shall promulgate regulations to the
extent that it believes the regulations necessary to protect the public health.
Whether regulations are necessary to protect the public health is, under this
interpretation, a determination to be made by the FDA.

Respondents, in contrast, argue that the phrase "to such extent" modifies
the phrase "the quantity therein or thereon" in § 346, not the word "shall."
Since respondents therefore view the word "shall" as unqualified, they interpret § 346 to require the promulgation of tolerance levels for added, but
unavoidable, harmful substances. The FDA under this interpretation of § 346
has discretion in setting the particular tolerance level, but not in deciding
whether to set a tolerance level at all.

Our analysis must begin with *Chevron U.S.A. Inc. v. Natural Resources
Defense Council, Inc.*, 467 U.S. 837 . . .

While we agree with the Court of Appeals that Congress in § 346 was speaking directly to the precise question at issue in this case, we cannot agree with
the Court of Appeals that Congress unambiguously expressed its intent
through its choice of statutory language. The Court of Appeals' reading of the
statute may seem to some to be the more natural interpretation, but the phras-
ing of § 346 admits of either respondents' or petitioner's reading of the statute.
As enemies of the dangling participle well know, the English language does
not always force a writer to specify which of two possible objects is the one
to which a modifying phrase relates. A Congress more precise or more pre-
scient than the one that enacted § 346 might, if it wished petitioner's position
to prevail, have placed "to such extent as he finds necessary for the protection
of public health" as an appositive phrase immediately after "shall" rather than
as a free-floating phrase after "the quantity therein or thereon." A Congress
equally fastidious and foresighted, but intending respondents' position to
prevail, might have substituted the phrase "to the quantity" for the phrase "to
such extent as." But the Congress that actually enacted § 346 took neither tack.
In the absence of such improvements, the wording of § 346 must remain
ambiguous.

The FDA has therefore advanced an interpretation of an ambiguous statutory provision. . . .

We find the FDA's interpretation of § 346 to be sufficiently rational to preclude a court from substituting its judgment for that of the FDA.

To read § 346 as does the FDA is hardly to endorse an absurd result. Like
any other administrative agency, the FDA has been delegated broad discretion
by Congress in any number of areas. To interpret Congress' statutory language
to give the FDA discretion to decide whether tolerance levels are necessary
to protect the public health is therefore sensible. . . .

The premise of the Court of Appeals is of course correct: the Act does
provide that when a tolerance level has been set and a food contains an added
harmful substance in a quantity below the tolerance level, the food is legally
not adulterated. But one cannot logically draw from this premise, or from the

Act, the Court of Appeals' conclusion that food containing substances not subject to a tolerance level must be deemed adulterated. The presence of a certain premise (i.e., tolerance levels) may imply the absence of a particular conclusion (i.e., adulteration) without the absence of the premise implying the presence of the conclusion. . . . The Act is silent on what specifically to do about food containing an unavoidable, harmful, added substance for which there is no tolerance level; we must therefore assume that Congress intended the general provisions of § 342(a) to apply in such a case. Section 342(a) thus remains available to the FDA to prevent the shipment of any food "[i]f it bears or contains any poisonous or deleterious substance which may render it injurious to health." . . .

Finally, we note that our interpretation of § 346 does not render that provision superfluous, even in light of Congress' decision to authorize the FDA to "promulgate regulations for the efficient enforcement of [the] Act." Section 346 gives the FDA the authority to choose whatever tolerance level is deemed "necessary for the protection of public health," and food containing a quantity of a required or unavoidable substance less than the tolerance level "shall not, by reason of bearing or containing any added amount of such substance, be considered to be adulterated." Section 346 thereby creates a specific exception to § 342(a)'s general definition of adulterated food as that containing a quantity of a substance that renders the food "ordinarily . . . injurious to health." Simply because the FDA is given the choice between employing the standard of § 346 and the standard of § 342(a) does not render § 346 superfluous.

. . .

Justice STEVENS, dissenting.

* * * * *

6.2 PESTICIDE RESIDUES

Although pesticides fell under the adulteration provisions of FD&C Act 402 and the tolerance provision of 406, in 1954 Congress passed the Pesticide Residue Amendment to add FD&C Act section 408 to deal exclusively with pesticide residues. Section 408 shifted the burden of proof of safety to the pesticide manufacturer (where before FDA needed to prove lack of safety). Specifically, a pesticide is considered unsafe in or on food until a tolerance is in effect. A raw agricultural commodity is also deemed unsafe and adulterated if it contains a level exceeding a tolerance established under section 408.

Section 408 empowers the Environmental Protection Agency (EPA) to establish tolerances for pesticide residues on raw agricultural commodities. This makes EPA, not the FDA, the largest federal agency responsible for evaluating the safety of chemicals in food.[26]

---

[26] Richard A. Merrill & Jeffrey K. Francer, *Organizing Federal Food Safety Regulation*, 31 SETON HALL LAW REVIEW 61–170 (2000).

6.2.1 FIFRA and EPA

The regulation of pesticide residues on food falls under both the Food, Drug, and Cosmetic Act (FD&C Act) and the Federal Insecticide, Fungicide, and Rodenticide Act (FIFRA). FIFRA was enacted in 1947 and requires the registration (approval) of all pesticides sold or distributed in the United States. Initially USDA was solely responsible for the implementation of FIFRA. However, in 1970 the Environmental Protection Agency (EPA) was established and given primary responsibility for pesticide regulation.

A third and overlapping control on pesticides residues on food is placed by FD&C Act sections 402(a)(2)(B) and 409, which empower FDA to regulate pesticide residues as food additives when the residue is in raw commodities that are used in processed food. However, if the residues in the processed food do not exceed the EPA section 408 tolerance, a food additive regulation is not required.

These FD&C Act provisions overlap and are designed to complement the power to register pesticides under FIFRA. For example, a pesticide may be registered for food use under FIFRA and be subject to a tolerance under FD&C Act section 408.

FDA also establishes the actions levels for pesticide contamination under section 406 and enforces all the FD&C Act requirements for pesticide residues in food. Section 406 is also used for establishing action levels for pesticide residue that occurs on nontarget crops (e.g., wind drift or soil absorption).

USDA remains responsible for monitoring levels of pesticide residues in processed foods to ensure compliance. The FDA and EPA cooperate with the USDA on farm compliance programs.

6.2.2 FQPA and Risk Assessment

The Food Quality Protection Act (FQPA) of 1996 amended both FIFRA and the FD&C Act and changed how pesticides are regulated. The FQPA set safety standard of reasonable certainty of no harm from the use of pesticides on food product. Before registration of a pesticide for use on foods or raw agricultural commodities, the manufacturer must conduct a risk assessment based on scientific tests to show that the pesticide "will not generally cause unreasonable adverse effects on the environment."[27]

Under the Food Quality and Protection Act the EPA must establish upper limits or tolerances for all pesticide residues in food under a new health based standards for cumulative exposure and risk to susceptible populations such as children. This also required that the Agency reexamine the safety of roughly 10,000 established tolerances including analyses of individual dietary exposure, exposure through drinking water and other sources, and through exposure in the home.

Specific toxic effects of pesticide residues in children, such as adverse impact on their neurological development, must be considered when setting pesticide tolerances. Unfortunately, the EPA is behind in implementing the FQPA provisions and has been sued both for failure to pull pesticides considered by some to be particularly dangerous.

NOTES

6.4. Pesticide informational resources.

> EPA: http://www.cfsan.fda.gov/~lrd/foodteam.html#EPA
>
> Pesticides: http://www.epa.gov/ebtpages/pesticides.html
>
> Pesticide legislation: http://www.epa.gov/ebtpages/pestpesticlegislation.html
>
> FQPA: http://www.epa.gov/oppfead1/fqpa/fqpafifr.htm
>
> Antimicrobial pesticide: http://www.epa.gov/oppfead1/fqpa/fqpafifr.htm#sec.221
>
> Tolerances: http://www.epa.gov/pesticides/regulating/tolerances.htm

6.3 ENVIRONMENTAL CONTAMINANTS

The FD&C Act contains no provision that explicitly provides a regulatory mechanism for substances that become constituents of food through environmental contamination. Many of these substances, such as mercury, PCBs, aflatoxin, and PBBs, can pose serious risk to public health. In part because the FD&C Act did not authorize FDA to set tolerances for these contaminants, FDA began to set informal section 406 "action levels" in the 1960s. These action levels are the highest level of contamination that will not trigger FDA enforcement action.

A later chapter covers food additives in greater detail, but it is worth noting here that the FDA has invoked the food additive regulations a means of controlling pesticide residues that occur in food as a result of environmental contamination. FDA generally has relied on section 406 action levels for such situations, but has also successfully charged that the residue of DDT in fish was illegal because an amount of DDT in excess of the 5 ppm action level was an unapproved food additive, which made it automatically illegal.[28]

Another enforcement strategy used by FDA has been to regulate some environmental contaminants of food as "added" substances under the FD&C

---

[27] 7 U.S.C. § 136(a)(5)(D).

[28] United States v. Ewig Bros. Co., 502 F.2d 715, 722 (7th Cir. 1974) ("Although it may seem odd to place the label 'additive' on a chemical substance which was a component of the raw product and which is not changed by processing, Congress' choice of that label does not result in any 'transmogrification.' Before processing, DDT is a 'pesticide chemical' on a raw product; after processing, it is an 'additive.'" At 722. Also finding that FDA need not prove that the DDT residue was unsafe, but the food is adulterated as a matter of law if the substance is an unapproved food additive. Thus essentially FDA needed only to prove that the DDT residue was not GRAS.)

Act. FDA promulgated regulations to clarify its use of "added" and "nonadded" substances in this context.[29] Basically, only a substance that is an inherent natural constituent of the food, and not the result of environmental, agricultural, or other human-caused contamination is considered "nonadded."

At the same time FDA proposed an action level for mercury in fish. It was not long before FDA was challenged in its enforcement action against swordfish. In *United States v. Anderson Seafoods, Inc.*,[30] fish distributors brought a class action for declaratory and injunctive relief. The United States District Court entered a judgment from which distributor appealed. The Court of Appeals held that: (1) the term "added," as used in the FD&C Act definition of adulterated foods, means artificially introduced, or attributable in some degree to the acts of man; (2) where some portion of a toxin present in a food has been introduced by man, the entirety of that substance present in the food will be treated as an "added substance" and so considered under the "may render injurious to health" standard of the FD&C Act; and (3) there was sufficient evidence to show that some mercury in swordfish is attributable to the acts of man.

* * * * *

United States v. Anderson Seafoods, Inc.

622 F.2d 157 (1980)

Before: WISDOM, POLITZ, and SAM D. JOHNSON, Circuit Judges
Opinion: WISDOM, Circuit Judge

This appeal poses the question whether mercury in the tissues of swordfish is an "added substance" within the meaning of the Food, Drug, and Cosmetic Act, 21 U.S.C. § 342(a)(1) (1975) (FDA), and is, therefore, subject to regulation under the relaxed standard appropriate to added substances. Only part of that mercury has been added by man.

In April 1977, the United States sought an injunction against Anderson Seafoods, Inc., and its president, Charles F. Anderson, to prevent them from selling swordfish containing more than 0.5 parts per million (ppm) of mercury, which it considered adulterated under the meaning of § 342(a)(1) of the FDA. Anderson responded in May 1977 by seeking a declaratory judgment that fish containing 2.0 ppm of mercury or less are not adulterated. Anderson also sought an injunction against the Food and Drug Administration commensurate with the declaratory judgment. Anderson's suit was certified as a class action, and these suits were consolidated for trial.

The district court denied the injunction that the government sought. In Anderson's suit, the court also denied an injunction, but issued a declaratory

---

[29] 42 Fed. Reg. 52814 (September 30, 1977) codified at 21 C.F.R. § 109.3.
[30] 622 F.2d 157 (1980).

judgment that swordfish containing more than 1.0 ppm mercury is adulterated under § 342(a)(1). In doing so, the court determined that mercury is an "added substance" under the Act and rejected Anderson's contention that a level of 2.0 ppm is acceptable. Anderson appealed from the judgment in the class action. The government appealed from the judgment in its enforcement action and cross-appealed in the class action. The government then withdrew its appeal and cross-appeal. This appeal now consists of Anderson's challenge to the way the district court parsed the statute and to the sufficiency of the evidence. We affirm.

I

Section 342(a)(1) of the Act provides:

"A food shall be deemed to be adulterated (a)(1) if it bears or contains any poisonous or deleterious substance which may render it injurious to health; but in case the substance is not an added substance, such food shall not be considered adulterated under this clause if the quantity of such substance in such food does not ordinarily render it injurious to health." 21 U.S.C. § 342(a)(1).

The Act does not define "added substance." Whether a substance is added or not is important because of the evidentiary showing that the Food and Drug Administration must make to succeed in an enforcement action. If a substance is deemed "added," then the Agency need show only that it "may render (the food) injurious to health" in order to regulate consumption of the food containing the substance. The "may render" standard has been interpreted to mean that there is a reasonable possibility of injury to the consumer. If, however, a substance is considered "not-added", the Agency must go further, and show that the substance would "ordinarily render (the food) injurious to health", 21 U.S.C. § 342(a)(1), before it can regulate its consumption.

In the trial of this case three theories about the meaning of the term "added" emerged. The Food and Drug Administration sponsored the first theory. It argues that an "added substance" is one that is not "inherent". According to FDA regulations:

(c) A "naturally occurring poisonous or deleterious substance" is a poisonous or deleterious substance that is an inherent natural constituent of a food and is not the result of environmental, agricultural, industrial, or other contamination.

(d) An "added poisonous or deleterious substance" is a poisonous or deleterious substance that is not a naturally occurring poisonous or deleterious substance. When a naturally occurring poisonous or deleterious substance is increased to abnormal levels through mishandling or other intervening acts, it is an added poisonous or deleterious substance to the extent of such increase.

Under this theory, all the mercury in swordfish is an added substance because it results not from the creature's bodily processes but from mercury in the environment, whether natural or introduced by man.

Anderson put forward a second theory. A substance, under this theory, is not an added substance unless it is proved to be present as a result of the direct agency of man. Further, only that amount of a substance the lineage of which can be so traced is "added." If some mercury in swordfish occurs naturally, and some is the result of man-made pollution, only that percentage of the mercury in fish proved to result directly from pollution is an added substance.

The district court adopted a third theory. Under the court's theory, if a de minimis amount of the mercury in swordfish is shown to result from industrial pollution, then all of the metal in the fish is treated as an added substance and may be regulated under the statute's "may render injurious" standard. The legislative history and case law, though sparse, persuade us that this is the proper reading of the statute.

The distinction between added and not-added substances comes from the "adulterated food" provisions of the original Food, Drug, and Cosmetic Act of 1906. The legislative history shows that "added" meant attributable to acts of man, and "not-added" meant attributable to events of nature.

The Supreme Court drew the same distinction . . . Construing the "added . . . ingredient" provisions of the 1906 Act, the Court said:

"Congress, we think, referred to ingredients artificially introduced; these are described as 'added.' The addition might be made to a natural food product or to a compound . . . we think that it was the intention of Congress that the artificial introduction of ingredients of a poisonous or deleterious character which might render the article injurious to health should cause the prohibition of the statute to attach."

The Food and Drug Administration argues that there need not be any connection between man's acts and the presence of a contaminant for it to be considered an added substance. The Agency points to the rule it recently promulgated interpreting § 342(a)(1), quoted above, which defines an added substance as one which is not "an inherent natural constituent of the food," but is instead the "result of an environmental, agricultural, industrial, or other contamination." Under the rule, mercury in swordfish tissue deriving from the mercury naturally dissolved in seawater would be an added substance, as would any substance not produced by or essential for the life processes of the food organism. In light of the legislative history and the *Coca Cola* case, however, we agree with the district court that the term "added" as used in § 342(a)(1) means artificially introduced, or attributable in some degree to the acts of man.

The Food and Drug Administration finds further support for its view in several cases in which the courts refer to not-added substances under the Act as "inherent." . . . however, the courts were not defining the statutory term "added substance." That they referred to not-added substances as being inherent does not mean that all non-inherent substances are added. These cases are consistent with the proposition that some non-inherent substances, present in a food organism but unconnected to man's acts, are not-added substances under the Act. A final case, *United States v. An Article of Food Consisting of Cartons of Swordfish*, reads the Act to mean that any material obtained from the environment is an added substance. As the district court pointed out, "FDA has not urged this rather extreme position upon the court and the ruling, contrary to the legislative history of the Act and the language of the Supreme Court, is not persuasive authority."

Determining that man must appear on the stage before a substance is an added, one does not determine the size of the role he must play before it is. The dichotomy in § 342(a)(1) is between two clear cases that bracket the present case. The Act considers added things such as lead in coloring agents or caffeine in Coca Cola. It considers not-added things like oxalic acid in rhubarb or caffeine in coffee. The Act did not contemplate, however, the perhaps rare problem of a toxin, part of which occurs "naturally", and part of which results from human acts. The section is designed, of course, to insure the scrutiny of toxins introduced by man. As Senator Heyburn said of the 1906 Act:

"Suppose you would say if there is poison in (a food) already it cannot do much harm to put in more. Suppose commercial cupidity should tempt someone to add to the dormant poison that is in a hundred things that we consume everyday, are they to be permitted to do it? This bill says they shall not do it."

Anderson argues that when a toxin derives in part from man and in part from nature, only that part for which man is responsible may be considered added and so regulated under the "may render injurious" standard. In such a case, however, neither the statute nor FDA regulations suggest that the amount of an added toxic substance be quantified and shown to have a toxic effect of its own if the total amount of the substance in a food is sufficient to render the food potentially hazardous to health. It may be possible as in this case to prove that man introduced some percentage of a toxin into a food organism, but difficult or impossible to prove that percentage.

Since the purpose of the "may render injurious" standard was to facilitate regulation of food adulterated by acts of man, we think that it should apply to all of a toxic substance present in a food when any of that substance is shown to have been introduced by man. Anderson argues that this reading of the statute would result "in the anomalous situation where a substance in a food can be 90 percent natural and 10 percent added if the entire substance is considered as added." There is no anomaly, however, in such a situation. The Act's "may render it injurious to health" standard is to be applied to the food, not to the added substance. The food would not be considered adulterated under our view unless the 10 percent increment creates or increases a potentiality of injury to health. If the increment does create or increase such a potentiality, then, because the increment that triggered the potentiality was introduced by man, the Food and Drug Administration ought to be able to regulate it under the standard designed to apply to adulterations of food caused by man. Anderson's argument proves too much. Anderson would argue that if a swordfish contained 0.99 ppm of natural mercury, and 0.99 ppm of mercury from human sources, the fish could be sold although it contained nearly twice as much mercury as the district court found to be a safe level. Such a reading of the statute hardly accords with its "overriding purpose to protect the public health." The reading we have adopted does accord with this purpose. It may be severe in practice. It may permit the Food and Drug Administration to regulate in some cases where the amount of substance contributed by man which triggers the potentiality of harm is minute. But it is the only alternative that fits into the statutory scheme. Congress should amend the statute if our reading produces impracticable results.

In sum, we hold that where some portion of a toxin present in a food has been introduced by man, the entirety of that substance present in the food will be treated as an added substance and so considered under the "may render injurious to health" standard of the Act.

II

In addition to its attack on the way the district court parsed the statute, Anderson raises a subsidiary argument. There was insufficient evidence to support the conclusion that man's acts contributed "substantial amounts" of mercury to the tissues of swordfish. And, indeed, the court did not find that the amounts were substantial, but rather that they were unknown and perhaps unquantifiable. Under our reading of the statute, however, the amount of mercury that man contributes need not be "substantial." The FDA need show only that some portion of the mercury is attributable to acts of man, and that the total amount may be injurious to health.

There was sufficient evidence to show that some mercury is attributable to the acts of man. There was evidence that mercury is dumped into rivers and washes onto the continental shelf, where some of it is methylated by bacteria and taken up by plankton. It thereby enters the food chain of swordfish, for the plankton is consumed by small organisms and fish, such as copepods, herring, and hake, which are in turn eaten by larger organisms, and eventually by swordfish, a peak predator. This evidence was enough to trigger the Act's "may render injurious to health" standard.

III

The district court set 1.0 ppm as the health limit for mercury in swordfish. It noted that the decision was:

"based only on the scientific and empirical data accepted into evidence in these cases. It may be that further studies will reveal the decisions here made were based on erroneous or insufficient data."

We noted above that the government withdrew its appeal and cross-appeal. It is apparently considering new evidence to determine whether its present action level should be reaffirmed or changed. Our decision does not engrave the district court's 1.0 ppm level in administrative stone. While the government may not now prevent the sale of swordfish containing 1.0 ppm or less of mercury, the durability of our order is founded on the evidence the district court accepted.

The order of the district court is AFFIRMED.

* * * * *

6.4   TAMPERING: THE ANTI-TAMPERING ACT

In the late 1970s and 1980s a number of incidents of intentional food and drug contamination led to rapid advances in tamper evident packaging for consumer products and the passage of anti-tampering laws. Following poisonings deaths from cyanide placed in Tylenol capsules, in 1982 FDA issued Tamper-Resistant Packaging Regulations.

The Federal Anti-tampering Act passed in 1983 makes it a crime to tamper with packaged consumer products or their labeling or containers.[31] The Act also makes it a crime to make believable threats about a consumer product tampering incident.[32] It is also a crime for an individual to intend to cause serious injury to a business, taint a consumer product, or render materially false or misleading labeling or containers for a consumer product.[33] In addition it is also a crime to knowingly communicate false information that a consumer product has been tainted, if the tainting, had it occurred, would have created a risk of death or bodily injury to another person.[34]

Food and consumer product companies have been acutely aware of the problem with product tampering for some time. Prudent firms have designed product withdrawal and recall programs to ensure that affected product are removed from the market quickly. Now, however, firms must implement plans to reduce the likelihood that their products or companies would become targets of terrorist activity. Food companies are possible targets because brand credibility is often the most important asset of a food company. Highly visible food companies have already been the target for bad publicity campaigns by "activist" groups who capitalize on a famous brand to make a political statement.

Food producers are strongly encouraged to consider security issues that could affect the safety of foods they produce and sell. Developing a food security plan can be an obvious extension of HACCP. Commonly considered

[31] 18 U.S.C. § 1365.
[32] 18 U.S.C. § 1365(d).
[33] 18 U.S.C. § 1365(b).
[34] 18 U.S.C. § 1365(c)(1).

security issues to address include the following: infrastructure concerns (electricity, water, fuel, and electronic data transfer), security of the physical premises, safety of ingredients and materials that could come into contact with the food, security of transit for food and ingredient, traceability, and employee and contractor screening.

6.5 CARCINOGENS: THE DELANEY CLAUSE

> Poison is in everything, and nothing is without poison. The dosage makes it either a poison or a remedy.
>
> —Paracelsus, sixteenth-century alchemist

Paracelsus's insight had to wait more than 400 years before science could effectively document its accuracy. The science of systemic and controlled study of toxicity began in the 1930s and 1940s. At the same time life expectancy was rising, primarily from success of public health measures in combating infectious diseases. Where tuberculosis and diarrhea were once leading causes of death, by the 1940s these diseases occurred less frequently and were less likely to be fatal. Because people were living longer, heart disease and cancer became the two leading causes of death.

Therefore it is not surprising the public concern over cancer also increased dramatically in the 1930s and 1940s. At the same time new chemical additives and ingredients were being added to food, and the public concern about the possible health effects was growing. Congress acted on these concerns when it passed the Delaney Clause.

The basic provision of the Delaney Clause is to prevent the addition to food any substance that has been shown to cause cancer in humans or laboratory animals. The Delaney Clause appears in three provisions of the FD&C Act: the Food Additive Amendments of 1958, the Color Additive Amendments of 1960, and the Animal Drug Amendments of 1968. The current language of these provision appear in sections 409(c)(3)(A), 706(b)(5)(B), and 512(d)(1)(I) of the FD&C Act, respectively.

The Delaney Clause is now one of the most notorious provisions of the FD&C Act, but it was invoked rarely by the FDA in more than a decade. In the 1950s and 1960s few substances were tested for carcinogenicity, and analytical techniques were less sensitive than today. Consequently trace amounts of known carcinogens could easily go undetected in foods.

Debate over the Delaney Clause intensified as the tests for chemicals became a thousand-fold more sensitive. Amid the controversy Congress exempted saccharin from the Delaney Clause restriction.

By the 1980s FDA and EPA adopted relaxed interpretations of the Delaney Clauses. This approach was approved in *Scott v. FDA*, 728 F.2d 322, 325 (6th Cir. 1984), which upheld FDA's interpretation that the Delaney Clause did not apply to a component of a color additive if the additive as a whole did not cause cancer in test animals, although the ptoluidine present in minute quantities is carcinogenic when tested separately.

However, courts generally considered the language and intent of Congress in the Delaney Clauses to be clear and refused to allow relaxed administrative interpretations. In *Public Citizen v. Young*, 831 F.2d 1108, 1113 (D.C. Cir. 1987), the court rejected FDA's de minimis interpretation of the color additives Delaney Clause.

In 1996 Congress passed the Food Quality Protection Act, which eliminated the Delaney Clause zero tolerance for carcinogenic pesticide chemical residues to "a reasonable certainty that no harm will result."[35]

* * * * *

Scott v. FDA

728 F.2d 322 (6th Cir. 1984)

Judges: ENGEL, KRUPANSKY, and WELLFORD, Circuit Judges
Opinion: Per curiam

Petitioner, acting pro se, seeks judicial review of 21 C.F.R. § 74.1205, a regulation issued by the Food and Drug Administration (FDA) authorizing the permanent listing and therefore the continued use of a color additive, D&C Green No. 5, in drugs and cosmetics. This regulation was promulgated by the FDA after it determined through tests required under the Color Additive Amendments of 1960 that D&C Green No. 5, then provisionally listed, was safe for said use. Judicial review of this regulation is authorized by 21 U.S.C. § 371(f)(1), (3).

D&C Green No. 5 contains another color additive, D&C Green No. 6, manufactured through the use of p-toluidine, which has been proven to be a carcinogenic when tested separately, and which is present in minute quantities as a chemical impurity in D&C Green No. 5. After extensive tests, the FDA determined that D&C Green No. 5, as a whole, did not cause cancer in test animals. It also determined that p-toluidine was not itself a color additive. It concluded, therefore, that the Delaney Clause, contained in the Food, Drug and Cosmetic Act, 21 U.S.C. § 301 *et seq.*, regulating the use of color additives, did not bar the permanent listing of D&C Green No. 5. This clause provides:

> A color additive (i) shall be deemed unsafe, and shall not be listed, for any use which will or may result in ingestion of all or part of such additive, if the additive is found by the Secretary to induce cancer when ingested by man or animal, or if it is found by the Secretary, after tests which are appropriate for the evaluation of the safety or additives for use in food, to induce cancer in man or animal, and (ii) shall be deemed unsafe, and shall not be listed, for any use which will not result in ingestion of any part of such additive, if, after tests which are appropriate

---

[35] 21 U.S.C. § 346a(b)(2)(A)(ii).

for the evaluation of the safety of additives for such use, or after other relevant exposure of man or animal to such additive, it is found by the Secretary to induce cancer in man or animal. . . .

21 U.S.C. § 376(b)(5)(B).

Having found the Delaney Clause inapplicable, the FDA then evaluated the risk posed by the presence of p-toluidine in D&C Green No. 5 under the General Safety Clause of the Food, Drug, and Cosmetic Act, which provides:

The Secretary shall not list a color additive under this section for a proposed use unless the data before him establish that such use, under the conditions of use specified in the regulations, will be safe. . . . 21 U.S.C. § 376(b)(4).

The FDA regulations governing approval of color additives define "safe" as meaning "that there is convincing evidence that establishes with reasonable certainty that no harm will result from the intended use of the color additive."

The FDA first isolated the trace amounts of p-toluidine contained in D&C Green No. 5 and determined that the maximum life-term average individual exposure to p-toluidine from use of D&C Green No. 5 would be 50 nanograms per day. The FDA then extrapolated from the level of risk found in animal bioassays to the conditions of probable exposure for humans using two different risk assessment procedures. Under the first procedure, the upper limit individual's life time risk of contracting cancer from exposure to 50 nanograms per day of p-toluidine through the use of D&C Green No. 5 was 1 in 30 million; the second procedure resulted in a calculation of a 1 in 300 million risk. The agency concluded "that there is a reasonable certainty of no harm from the exposure to p-toluidine that results from the use of D&C Green No. 5."

Petitioner asserts on appeal that by permanently listing the color additive D&C Green No. 5, the FDA violated the Delaney Clause and the General Safety Clause of the Food, Drug, and Administration Act. Petitioner does not contest the validity of the tests employed by the FDA in determining that D&C Green No. 5 was safe for its intended uses but rather asserts that the Delaney Clause, as a matter of law, prohibits approval of a color additive when it contains a carcinogenic impurity in any amount and that the FDA has no discretion to find D&C Green No. 5 "safe" under the General Safety Clause because "[it is not] possible to establish a safe level of exposure to a carcinogen." The Agency found essentially that D&C Green No. 5, after studying the tests, did not cause cancer in test animals.

The decision of the FDA to approve permanent listing of color additive D&C Green No. 5 may be overturned by this court only if that decision was arbitrary and capricious, an abuse of discretion, or not in accordance with the law. 5 U.S.C. § 706(2)(A). The FDA's interpretation of the Food, Drug, and Cosmetic Act is entitled to considerable deference. Even when there "is more than one reasonable interpretation of this . . . [Act], the court should follow the interpretation urged by the FDA."

We affirm the judgment of the Food and Drug Administration. Petitioner's arguments are found to be without merit. The FDA's finding that the Delaney Clause is inapplicable to the instant case because D&C Green No. 5 does not cause cancer in humans is in accordance with the law. In its final order, the FDA stated its rationale for its conclusion, and it was fully mindful of the Delaney Clause in making its decision:

> The Agency does not believe that it is disregarding the Delaney Clause. In drafting the Delaney Clause, Congress implicitly recognized that known carcinogens might be present in color additives as intermediaries or impurities but at levels too low to trigger a response in conventional test systems. Congress apparently concluded that the presence of these intermediaries or impurities at these low levels was acceptable. This legislative judgment accounts for the absence of any requirement in the Delaney Clause that the impurities and intermediaries in a color additive, rather than the additive as a whole, be tested or otherwise evaluated for safety. Thus, Congress drew a rough, quantitative distinction between a color additive that is deemed unsafe under the Delaney Clause because it causes cancer, and an additive that is not subject to the Delaney Clause because it does not cause cancer even though one of its constituents does. FDA's decision on D&C Green No. 5 is consistent with this distinction.

This interpretation of the Delaney Clause case is a reasonable one, and it is consistent with its legislative history. Congress distinguished between "pure dye" and its "impurities" in its list of factors for the FDA to consider under the General Safety Clause, but omitted "impurities" as a factor under the Delaney Clause. Although the Agency's regulatory interpretation of the Delaney Clause contains the words, "color additive including its components," it is clear that this regulation was aimed only at those additives containing impurities that produced cancer when tested together:

> The Commissioner shall determine whether, based on the judgment of appropriately qualified scientists [from the results of appropriate tests], cancer has been induced and whether the color additive, including its components or impurities, was the causative substance. If it is his judgment that the data do not establish these facts, the cancer clause [Delaney Clause] is not applicable; and if the data as a whole establish that the color additive will be safe under the conditions that can be specified in the applicable regulation, it may be listed for such use.

21 C.F.R. § 70.50.

Since in the instant case it was determined by the FDA that D&C Green No. 5, after testing as a whole, did not cause cancer in test animals, under the plain language of the Delaney Clause and the FDA's interpretation of that Clause, the FDA was not prohibited from permanently listing D&C Green No. 5.

The FDA's conclusion that the risk levels ascertained after testing D&C Green No. 5 by isolating p-toluidine were so low as to preclude a reasonable harm from exposure to the additive, within the meaning of the General Safety

Clause, is also in accordance with the law. Petitioner's assertion that the FDA has no discretion to determine that D&C Green No. 5 is safe for its intended use because it contains p-toluidine is without merit. This finding is consistent with the holding in *Monsanto v. Kennedy*. That case involved an impurity, found to produce adverse results in test animals, present in the substance used to make beverage containers, which migrated from the container to the beverage. In discussing whether that impurity was a food additive, the court observed: "The Commissioner may determine based on the evidence before him that the level of migration into food of a particular chemical is so negligible as to present no public health or safety concerns, even to assure a wide margin of safety. This authority derives from the administration discretion, inherent in the statutory scheme, to deal appropriately with de minimis situations."

We find this determination by the *Monsanto* court persuasive and relevant to the particular facts of the instant case. We agree with the FDA's conclusion that since it "has discretion to find that low-level migration into food of substances in indirect additives is so insignificant as to present no public health or safety concern . . . it can make a similar finding about a carcinogenic constituent or impurity that is present in a color additive." Accordingly, we hold that the FDA did not abuse its discretion under the General Safety Clause in determining that the presence of p-toluidine in D&C Green No. 5 created no reasonable risk of harm to individuals exposed to the color additive.

The decision of the FDA to permanently list D&C Green No. 5 is hereby AFFIRMED.

* * * * *

Public Citizen v. Young

831 F.2d 1108 (D.C. Cir. 1987)

Before: RUTH B. GINSBURG and WILLIAMS, Circuit Judges, and HAROLD H. GREENE, District Judge
Opinion: WILLIAMS, Circuit Judge

The Color Additive Amendments of 1960 (codified at 21 U.S.C. § 376 (1982)), part of the Food, Drug and Cosmetic Act (the "Act"), establish an elaborate system for regulation of color additives in the interests of safety. A color additive may be used only after the Food and Drug Administration ("FDA") has published a regulation listing the additive for such uses as are safe. Such listing may occur only if the color additive in question satisfies (among other things) the requirements of the applicable "Delaney Clause," § 706(b)(5)(B) of the Act, one of three such clauses in the total system for regulation of color additives, food and animal food and drugs. The Clause prohibits the listing of any color additive "found . . . to induce cancer in man or animal."

In No. 86-1548, Public Citizen and certain individuals challenge the decision of the FDA to list two color additives, Orange No. 17 and Red No. 19, based on quantitative risk assessments indicating that the cancer risks presented by these dyes were trivial. This case thus requires us to determine whether the Delaney Clause for color additives is subject to an implicit "de minimis" exception. We conclude, with some reluctance, that the Clause lacks such an exception. . . .

I. The Delaney Clause and "De minimis" Exceptions

A. *Factual Background*

The FDA listed Orange No. 17 and Red No. 19 for use in externally applied cosmetics on August 7, 1986. In the listing notices, it carefully explained the testing processes for both dyes and praised the processes as "current state-of-the-art toxicological testing." In both notices it specifically rejected industry arguments that the Delaney Clause did not apply because the tests were inappropriate for evaluation of the dyes. It thus concluded that the studies established that the substances caused cancer in the test animals.

The notices then went on to describe two quantitative risk assessments of the dyes, one by the Cosmetic, Toiletry and Fragrance Association ("CTFA," an intervenor here and the industry proponent of both dyes) and one by a special scientific review panel made up of Public Health Service scientists. Such assessments seek to define the extent of health effects of exposures to particular hazards. As described by the National Research Council, they generally involve four steps: (1) hazard identification, or the determination of whether a substance is causally linked to a health effect; (2) dose–response assessment, or determination of the relation between exposure levels and health effects; (3) exposure assessment, or determination of human exposure; and (4) risk characterization, or description of the nature and magnitude of the risk. All agree that gaps exist in the available information and that the risk estimator must use assumptions to fill those gaps. The choice among possible assumptions is inevitably a matter of policy to some degree.

The assessments considered the risk to humans from the substances when used in various cosmetics—lipsticks, face powders and rouges, hair cosmetics, nail products, bathwater products, and wash-off products. The scientific review panel found the lifetime cancer risks of the substances extremely small: for Orange No. 17, it calculated them as one in 19 billion at worst, and for Red No. 19 one in nine million at worst. The FDA explained that the panel had used conservative assumptions in deriving these figures, and it characterized the risks as "so trivial as to be effectively no risk." It concluded that the two dyes were safe.

The FDA candidly acknowledged that its safety findings represented a departure from past agency practice: "In the past, because the data and information show that D & C Orange No. 17 is a carcinogen when ingested by laboratory animals, FDA in all likelihood would have terminated the provisional listing and denied CTFA's petition for the externally applied uses . . . without any further discussion." It also acknowledged that "[a] strictly literal application of the Delaney Clause would prohibit FDA from finding [both dyes] safe, and therefore, prohibit FDA from permanently listing [them]. . . ." Because the risks presented by these dyes were so small, however, the agency declared that it had "inherent authority" under the de minimis doctrine to list them for use in spite of this language. It indicated that as a general matter any risk lower than a one-in-one-million lifetime risk would meet the requirements for a de minimis exception to the Delaney Clause.

Assuming that the quantitative risk assessments are accurate, as we do for these purposes, it seems altogether correct to characterize these risks as trivial. For example, CTFA notes that a consumer would run a one-in-a-million lifetime risk of cancer if he or she ate one peanut with the FDA-permitted level of aflatoxins once every 250 days (liver cancer). Another activity posing a one-in-a-million lifetime risk is spending 1,000 minutes (less than 17 hours) every year in the city of Denver—with its high elevation and cosmic radiation levels—rather than in the District of Columbia. Most of us would not regard these as high-risk activities. Those who indulge in them can hardly be thought of as living dangerously. Indeed, they are risks taken without a second thought by persons whose economic position allows them a broad range of choice.

According to the risk assessments here, the riskier dye poses one ninth as much risk as the peanut or Colorado hypothetical; the less risky one poses only one 19,000th as much.

It may help put the one-in-a-million lifetime risk in perspective to compare it with a concededly dangerous activity, in which millions nonetheless engage, cigarette smoking. Each one-in-a-million risk amounts to less than one 200,000th the lifetime risk incurred by the average male smoker. Thus, a person would have to be exposed to more than 2,000 chemicals bearing the one-in-a-million lifetime risk, at the rates assumed in the risk assessment, in order to reach 100th the risk involved in smoking. To reach that level of risk with chemicals equivalent to the less risky dye (Orange No. 17), he would have to be exposed to more than 40 million such chemicals.

B. Plain Language and the De minimis Doctrine

The Delaney Clause of the Color Additive Amendments provides as follows:

> a color additive . . . (ii) shall be deemed unsafe, and shall not be listed, for any use which will not result in ingestion of any part of such additive, if, after tests which are appropriate for the evaluation of the safety of additives for such use, or after other relevant exposure of man or animal to such additive, it is found by the Secretary to induce cancer in man or animal. . . .

21 U.S.C. § 376(b)(5)(B).

The natural—almost inescapable—reading of this language is that if the Secretary finds the additive to "induce" cancer in animals, he must deny listing. Here, of course, the agency made precisely the finding that Orange No. 17 and Red No. 19 "induce[ ] cancer when tested in laboratory animals." (Below we address later agency pronouncements appearing to back away from these statements.)

The setting of the clause supports this strict reading. Adjacent to it is a section governing safety generally and directing the FDA to consider a variety of factors, including probable exposure, cumulative effects, and detection difficulties. 21 U.S.C. § 376(b)(5)(A). The contract in approach seems to us significant. For all safety hazards other than carcinogens, Congress made safety the issue, and authorized the agency to pursue a multifaceted inquiry in arriving at an evaluation. For carcinogens, however, it framed the issue in the simple form, "If A [finding that cancer is induced in man or animals], then B [no listing]." There is language inviting administrative discretion, but it relates only to the process leading to the finding of carcinogenicity: "appropriate" tests or "other relevant exposure," and the agency's "evaluation" of such data. Once the finding is made, the dye "shall be deemed unsafe, and shall not be listed." 21 U.S.C. § 367(b)(5)(B).

Courts (and agencies) are not, of course, helpless slaves to literalism. One escape hatch, invoked by the government and CTFA here, is the de minimis doctrine, shorthand for *de minimis non curat lex* ("the law does not concern itself with trifles"). The doctrine—articulated in recent times in a series of decisions by Judge Leventhal—serves a number of purposes. One is to spare agency resources for more important matters. But that is a goal of dubious relevance here. The finding of trivial risk necessarily followed not only the elaborate animal testing, but also the quantitative risk assessment process itself; indeed, application of the doctrine required additional expenditure of agency resources.

More relevant is the concept that "notwithstanding the 'plain meaning' of a statute, a court must look beyond the words to the purpose of the act where its literal terms lead to 'absurd or futile results.'" Imposition of pointless burdens on regulated entities is obviously to be avoided if possible, especially as burdens on them almost invariably entail losses for their customers: here, obviously, loss of access to the colors made possible by a broad range of dyes.

We have employed the concept in construing the Clean Air Act's mandate to the Environmental Protection Agency to set standards providing "an ample margin of safety to protect the public health," 42 U.S.C. § 7412(b)(1) (1982). That does not, we said, require limits assuring a "risk-free" environment. Rather, the agency must decide "what risks are acceptable in the world in which we live" and set limits accordingly. Assuming as always the validity of the risk assessments, we believe that the risks posed by the two dyes would have to be characterized as "acceptable." Accordingly, if the statute were to permit a de minimis exception, this would appear to be a case for its application.

Moreover, failure to employ a de minimis doctrine may lead to regulation that not only is "absurd or futile" in some general cost-benefit sense but also is directly contrary to the primary legislative goal. In a certain sense, precisely that may be the effect here. The primary goal of the Act is human safety, but literal application of the Delaney Clause may in some instances increase risk.

No one contends that the Color Additive Amendments impose a zero-risk standard for noncarcinogenic substances; if they did, the number of dyes passing muster might prove miniscule. As a result, makers of drugs and cosmetics who are barred from using a carcinogenic dye carrying a one-in-20-million lifetime risk may use instead a noncarcinogenic, but toxic, dye carrying, say, a one-in-10-million lifetime risk. The substitution appears to be a clear loss for safety.

Judge Leventhal articulated the standard for application of de minimis as virtually a presumption in its favor: "Unless Congress has been extraordinarily rigid, there is likely a basis for an implication of de minimis authority to provide [an] exemption when the burdens of regulation yield a gain of trivial or no value." But the doctrine obviously is not available to thwart a statutory command; it must be interpreted with a view to "implementing the legislative design." Nor is an agency to apply it on a finding merely that regulatory costs exceed regulatory benefits.

Here, we cannot find that exemption of exceedingly small (but measurable) risks tends to implement the legislative design of the color additive Delaney Clause. The language itself is rigid; the context—an alternative design admitting administrative discretion for all risks other than carcinogens—tends to confirm that rigidity. Below we consider first the legislative history; rather than offering any hint of softening, this only strengthens the inference. Second, we consider a number of factors that make Congress's apparent decision at least a comprehensible policy choice.

1. Legislative History

The Delaney Clause arose in the House bill and was, indeed, what principally distinguished the House from the Senate bill. The House included it in H.R. 7624, 106 Cong. Rec. 14,353–56, and the Senate accepted the language without debate, 106 Cong. Rec. 15,133 (1960). The House committee gave considerable attention to the degree of discretion permitted under the provision. The discussion points powerfully against any de minimis exception, and is not contradicted either by consideration on the House floor or by a post-enactment colloquy in the Senate.

House Committee. The House Report on the Color Additive Amendments is the most detailed evidence as to Congress's intentions on this issue. In discussing the Clause, the report first explains the source of concern: "[T]oday cancer is second only to heart disease as a cause of death among the American people. Every year, approximately 250,000 people die of cancer in this country. Approximately 450,000 new cases of cancer are discovered each year." The report reflects intense congressional concern over cancer risks from man-made substances.

The report acknowledged the "many unknowns about cancer," but highlighted certain areas of general agreement: "Laboratory experiments have shown that a number of substances when added to the diet of test animals have produced cancers of various kinds in the test animals. It is this fact—namely that small quantities of certain materials over a period of time will cause abnormal cell growth in animals—that gave rise to the Delaney anticancer clause. . . ." The report quoted at length from the hearing testimony of Arthur S. Flemming, Secretary of Health, Education, and Welfare (the parent agency of the FDA and the predecessor of Health and Human Services). The Secretary took a very strong line on the absence of a basis for finding "threshold" levels below which carcinogens would not be dangerous:

> We have no basis for asking Congress to give us discretion to establish a safe tolerance for a substance which definitely has been shown to produce cancer when added to the diet of test animals. We simply have no basis on which such discretion could be exercised because no one can tell us with any assurance at all how to establish a safe dose of any cancer-producing substance.

Secretary Flemming also developed the theme that, with many cancer risks inescapably present in the environment, it made sense to remove unnecessary ones:

> Unless and until there is a sound scientific basis for the establishment of toler- ances for carcinogens, I believe the government has a duty to make clear—in law as well as in administrative policy—that it will do everything possible to put persons in a position where they will not unnecessarily be adding residues of carcinogens to their diet.
>   The population is inadvertently exposed to certain carcinogens. . . . In view of these facts, it becomes all the more imperative to protect the public from deliberate introduction of additional carcinogenic materials into the human environment.
>   . . . .
>   It is clear that if we include in our diet substances that induce cancer when included in the diet of test animals, we are taking a risk. In light of the rising number of cases of cancer, why should we take that risk?

Before adopting Flemming's no-threshold premise the House committee heard many witnesses on the opposite side of the debate, and its Report acknowledges their contentions. It also notes that some took the position that the ban should "apply only to colors that induce cancer when ingested in an amount and under conditions reasonably related to their intended use." Similarly, it notes support for making carcinogenicity simply one of the factors for the Secretary to consider in determining safety. Finally, it mentions a position taken by some scientific witnesses strikingly similar to that taken by FDA here. These experts suggested that, in spite of the diffi- culties in designing and evaluating tests for carcinogenicity, the Secretary "should have the authority to decide that a minute amount of a cancer-producing chemical may be added to man's food after a group of scientists consider all the facts and conclude that the quantity to be tolerated is probably without hazard."

The committee rejected all these positions on the ground that they would "weaken the present anticancer clause." The report responded to them with another quote from Secretary Flemming's hearing testimony, reflecting the view that agency discretion should cease once "a substance has been shown to produce cancer when added to the diet of test animals":

> The rallying point against the anticancer provision is the catch phrase that it takes away the scientists's [sic] right to exercise judgment. The issue thus made is a false one because the clause allows the exercise of all the judgment that can safely be exercised on the basis of our present knowledge. ... It allows the Department and its scientific people full discretion and judgment in deciding whether a substance has been shown to produce cancer when added to the diet of test animals. But once this decision is made, the limits of judgment have been reached and there is no reliable basis on which discretion could be exercised in determining a safe threshold dose for the established carcinogen.

Beyond this delineation of the intended scope of discretion, the House Report also addressed the possibility that its scientific premise—the absence of a threshold—might prove false. Its evident solution was that Congress, not the FDA, should examine the evidence and find a solution. The House Report quotes Secretary Flemming to precisely this effect:

> Whenever a sound scientific basis is developed for the establishment of tolerances for carcinogens, we will request the Congress to give us that authority. We believe, however, that the issue is so important that the elected representatives of the people should have the opportunity of examining the evidence and determining whether or not the authority should be granted.

The government and CFTA note that exempting substances shown by quantitative risk assessment to carry only trivial risks rests on a quite different foundation from establishing threshold levels below which no cancer is thought to occur. We agree that the two are distinguishable, but do not find the distinctions between them to cut in favor of a de minimis exception. If it is correct to read the statute as barring tolerances based on an assumed threshold, it follows a fortiori that the agency must ban color additives with real but negligible cancer risks.

House floor. In the House debate, little of substance occurred. Congressman Delaney contended that the anticancer provision was essential "if the public health is to be adequately protected," and asserted in conclusory terms the inability to establish a safe dose or tolerance, *id*. Congressman Rogers, describing the anticancer clause (which he supported), observed that "[t]he 'safe for use' principle does not apply to situations where carcinogenicity is at issue." One participant, Congressman Allen, expressed the view that the anticancer clause was "unnecessary and restrictive," and that the "decision on safety [should] be determined by the Secretary of Health, Education and Welfare rather than … determined by law." Accordingly, he urged passage of the Senate bill instead. Although Congressman Allen's view of the bill was negative, his interpretation seems to accord with that of its proponents: a ban follows automatically from a finding of carcinogenicity in man or animal.

Post-enactment Senate colloquy. The inferences of rigidity supported by the remarks above are drawn slightly in question—but ultimately, we think, not much—by an exchange that occurred the day after the Senate took final action on the final version of the Act. Senator Javits politely complained about the Senate's acting on this legislation in his absence. He secured unanimous consent for including in the Record the conclusions of a then-recent Report of the Panel on Food Additives of the President's Advisory Committee (the "Kistiakowsky Report"). He characterized the Report as stating that "authority such as that conferred by the amendment [the Report was addressed to the food additive Delaney Clause] should be used and applied within the "rule of reason." After Senators Dirksen and Hill assented to this proposition, Javits agreed to lay on the table a motion to reconsider the vote of the previous day.

Appellees interpret the rule-of-reason colloquy as squarely supporting their de minimis approach, but in fact it is ambiguous. The Kistiakowsky Report defined "rule of reason" by a quotation from *Rathbun v. United States*: "Every statute must be interpreted in the light of reason and common understanding to reach the results intended by the legislature." The proposition accords exactly with the way in which Judge Leventhal formulated the test for application of the de minimis doctrine: would the doctrine "implement[ ] the legislative design"? But that is the question, not the answer. Thus the exchange invoking the rule of reason appears to do no more than exhort us to pursue the inquiry we've been pursuing.

Indeed, although the Kistiakowsky Report itself points out some possible consequences of "a literal interpretation" of the food additive Delaney Clause, and states that in its interpretation the FDA "must employ the 'rule of reason' "as defined in Rathbun, it also acknowledges that clause may prevent the agency from "exercis[ing] discretion consistent with the recommendations of this report." Thus a commitment to the "rule of reason" in this context hardly carries an inexorable implication that the color additive Delaney Clause grants the FDA the discretion it now claims.

Taken as a whole, the remarks do not seem strong enough to undermine the inference we have drawn that the clause was to operate automatically once the FDA squeezed the scientific trigger. This is so even without regard to the usual hazards of post-enactment legislative history, which ordinarily lead to its being disregarded altogether.

2. Possible Explanations for an Absolute Rule

Like all legislative history, this is hardly conclusive. But short of an explicit declaration in the statute barring use of a de minimis exception, this is perhaps as strong as it is likely to get. Facing the explicit claim that the Clause was "extraordinarily rigid," a claim well supported by the Clause's language in contrast with the bill's grants of discretion elsewhere, Congress persevered.

Moreover our reading of the legislative history suggests some possible explanations for Congress's apparent rigidity. One is that Congress, and the nation in general (at least as perceived by Congress), appear to have been truly alarmed about the risks of cancer. This concern resulted in a close focus on substances increasing cancer threats and a willingness to take extreme steps to lessen even small risks.[36] Congress hoped to reduce the incidence of cancer by banning carcinogenic dyes, and may also have hoped to lessen public fears by demonstrating strong resolve.

A second possible explanation for Congress's failure to authorize greater administrative discretion is that it perceived color additives as lacking any great value. For example, Congressman Delaney remarked, "Some food additives serve a useful purpose. . . . However, color additives provide no nutrient value. They have no value at all, except so-called eye appeal." Representative Sullivan said, "we like the bright and light [lipstick] shades but if they cannot safely be produced, then we prefer to do without these particular shades." And Representative King: "The colors which go into our foods and cosmetics are in no way essential to the public interest or the national security. . . . [C]onsumers will easily get along without [carcinogenic colors]."

It is true that the legislation as a whole implicitly recognizes that color additives are of value, since one of its purposes was to allow tolerances for certain dyes—harmful but not carcinogenic—that would have been banned under the former law. There was also testimony pointing out that in some uses color additives advance health: they can help identify medications and prevent misapplications where a patient must take several. Nevertheless, there is evidence that Congress thought the public could get along without carcinogenic colors, especially in view of the existence of safer substitutes. Thus the legislators may have estimated the costs of an overly protective rule as trivial.

So far as we can determine, no one drew the legislators' attention to the way in which the Delaney Clause, interacting with the flexible standard for

---

[36] See Color Additives Hearings at 341 (testimony of representative of Consumers Union) ("we are faced with an epidemic, an epidemic of cancer, a chronic disease, and . . . all measures that will protect the public health should be taken, even at the cost of discomfort or sacrifice, financial sacrifice, to some segments of industry.").

determining safety of noncarcinogens, might cause manufacturers to substitute more dangerous toxic chemicals for less dangerous carcinogens. But the obviously more stringent standard for carcinogens may rest on a view that cancer deaths are in some way more to be feared than others.

Finally, as we have already noted, the House committee (or its amanuenses) considered the possibility that its no-threshold assumption might prove false and contemplated a solution: renewed consideration by Congress.

Considering these circumstances—great concern over a specific health risk, the apparently low cost of protection, and the possibility of remedying any mistakes—Congress's enactment of an absolute rule seems less surprising.

C. Special Arguments for Application of De minimis

Apart from their contentions on legislative history, the FDA and CTFA assert two grounds for a de minimis exception: an analysis of two cases applying de minimis concepts in the food and drug regulation context, and contentions that, because of scientific advances since enactment, the disallowance of de minimis authority would have preposterous results in related areas of food and drug law. . . . We are, ultimately, not persuaded.

1. De Minimis Cases

Monsanto Co. v. Kennedy, considered whether acrylonitrile in beverage containers was a "food additive" within the meaning of the Food, Drug, and Cosmetic Act's definition of that term:

> any substance the intended use of which results or may reasonably be expected to result, directly or indirectly, in its becoming a component or otherwise affecting the characteristics of any food . . . if such substance is not generally recognized . . . to be safe under the conditions of its intended use. . . . Section 201(s), Food, Drug and Cosmetic Act, 21 U.S.C. § 321(s) (1982).
>
> By operation of the second law of thermodynamics, any substance, obviously including acrylonitrile, will migrate in minute amounts from a bottle into a beverage within the bottle. Questions had been raised about its safety. The court found the FDA's decision to ban its use insufficiently well considered. In remanding the case for reconsideration, the court emphasized the FDA Commissioner's discretion to exclude a chemical from the statutory definition of food additives if "the level of migration into food . . . is so negligible as to present no public health or safety concerns."

The opinion makes no suggestion that anyone supposed acrylonitrile to be carcinogenic, or that the Delaney Clause governing food additives, 21 U.S.C. § 348(c)(3)(A), was in any way implicated. Thus the case cannot support a view that the food additive Delaney Clause (or, obviously, the color additive one) admits of a de minimis exception.

Scott v. Food and Drug Administration involves the color additive Delaney Clause, but is nonetheless distinguishable. Petitioner challenged the FDA's listing of Green No. 5, on the ground that it contained a chemical impurity in minute quantities that had been found to cause cancer in test animals. The dye as a whole, however, had been found not to induce cancer in test animals. The Sixth Circuit upheld the FDA's decision that the Delaney Clause of the Color Additive Amendments did not apply. The court cited *Monsanto* in support of upholding the FDA's view that it had discretion "to find that low-level migration into food of substances in indirect additives is so insignificant as to present no public health or safety concerns."

We must evaluate *Scott* in light of the possibility that the carcinogenic impurity in question acted as an "initiating agent" or was a "complete carcinogen," and, accordingly, would be subject to no threshold. If so, it would seem that if the impurity itself were carcinogenic, so would be any substance to which it was added.

Application of a de minimis exception for constituents of a color additive, however, seems to us materially different from use of such a doctrine for the color additive itself. As the *Scott* court noted, the FDA's action was completely consistent with the plain language of the statute, as there was no finding that the dye caused cancer in animals. Here, as we have observed, application of a de minimis exception requires putting a gloss on the statute qualifying its literal terms.

Monsanto and *Scott* demonstrate that the de minimis doctrine is alive and well in the food and drug context, even on the periphery of the Delaney Clauses. But no case has applied it to limit the apparent meaning of any of those Clauses in their core operation.

2. Scientific Advance and the Implications for Food Additive Regulation

The CTFA also argues that in a number of respects scientific advance has rendered obsolete any inference of congressional insistence on rigidity. CTFA notes that while in 1958 (date of enactment of the food additive Delaney Clause) there were only four known human carcinogens, by 1978 there were 37 substances known to produce cancer in humans and over 500 in animals. They identify an impressive array of food ingredients now found to be animal carcinogens and that appear in a large number of food products. These include many items normally viewed as essential ingredients in a healthy diet, such as vitamins C and D, calcium, protein, and amino acids. If the color additive Delaney Clause has no de minimis exception, it follows (they suggest) that the food additive one must be equally rigid. The upshot would be to deny the American people access to a healthy food supply.

As a historical matter, the argument is overdrawn: the House committee was clearly on notice that certain common foods and nutrients were suspected carcinogens.

Beyond that, it is not clear that an interpretation of the food additive Delaney Clause identical with our interpretation of the color additive clause would entail the feared consequences. The food additive definition contains an exception for substances "generally recognized" as safe (known as the "GRAS" exception), an exception that has no parallel in the color additive definition, 21 U.S.C. § 321(t)(1). That definition may permit a de minimis exception at a stage that logically precedes the FDA's ever reaching the food additive Delaney Clause. Indeed, Monsanto so holds—though, as we have noted, in a case not trenching upon the food additive Delaney Clause. Moreover, the GRAS exception itself builds in special protection for substances used in food prior to January 1, 1958, which may be shown to be safe "through either scientific procedures or experience based on common use in food." Indeed, the Kistiakowsky Report, filed with the House committee, stated that the grandfathering provision of the food additives Delaney Clause "considerably narrows [its] effect . . . on industry and the public."

The relationship of the GRAS exception and the food additive Delaney Clause clearly poses a problem: if the food additive definition allows the FDA to classify as GRAS substances carrying trivial risks (as *Monsanto* and our recent decision in *Natural Resources Defense Council v. EPA* seem to suggest), but the food additive Delaney Clause is absolute, then Congress has adopted inconsistent provisions. On the other hand, if (1) the GRAS exception does not encompass substances with trivial carcinogenic effect (especially if its special provision for substances used before 1958 does not do so for long-established substances), and (2) the food additive Delaney Clause is as rigid as we find the color additive clause to be, conceivably the consequences identified by the CTFA, or some of them, may follow. All these are difficult questions, but they are neither before us nor is their answer foreordained by our decision here.

Moreover we deal here only with the color additive Delaney Clause, not the one for food additives. Although the clauses have almost identical wording, the context is clearly different. Without having canvassed the legislative history of the food additive Delaney Clause, we may safely say that its proponents could not have regarded as trivial the social cost of banning those parts of the American diet that CTFA argues are at risk.

Finally, even a court decision construing the food additive provisions to require a ban on dietary essentials would not, in fact, bring about such a ban. As Secretary Flemming noted, in words selected by the House Report for quotation, the FDA could bring critical new discoveries to Congress's attention. If the present law would lead to the consequences predicted, we suppose that the FDA would do so, and that Congress would respond.

D. The Meaning of "[I]nduce Cancer"

After Public Citizen initiated the litigation in No. 86-5150, the FDA published a notice embellishing the preamble to its initial safety determinations. These notices effectively apply quantitative risk assessment at the stage of determining whether a substance "induce[s] cancer in man or animal." They assert that even where a substance does cause cancer in animals in the conventional sense of the term, the FDA may find that it does not "induce cancer in man or animal" within the meaning of 21 U.S.C. § 376(b)(5)(B). It is not crystal clear whether such a negative finding would flow simply from a quantitative risk assessment finding the risk to be trivial for humans under conditions of intended use, or whether it would require a projection back to the laboratory animals: i.e., an assessment that the risk would be trivial for animals exposed to the substance in quantities proportional to the exposure hypothesized for human risk assessment purposes. (Perhaps the distinction is without a difference.) In any event, the notices argued:

> The words "induce cancer in man or animal" as used in the Delaney Clause are terms of art intended to convey a regulatory judgment that is something more than a scientific observation that an additive is carcinogenic in laboratory animals. To limit this judgment to such a simple observation would be to arbitrarily exclude from FDA's consideration developing sophisticated testing and analytical methodologies, leaving FDA with only the most primitive techniques for its use in this important endeavor to protect public health. Certainly the language of the Delaney Clause itself cannot be read to mandate such a counterproductive limit on FDA's discharge of its responsibilities.

The notices acknowledged that the words "to induce cancer" had not been "rigorously and unambiguously" so limited in the previous notices. This is a considerable understatement. The original determinations were quite unambiguous in concluding that the colors induced cancer in animals in valid tests; the explanations went to some trouble to rebut industry arguments to the contrary. Despite these arguments, FDA concluded that the tests demonstrated that the dyes were responsible for increases in animal tumors.

The plain language of the Delaney Clause covers all animals exposed to color additives, including laboratory animals exposed to high doses. It would be surprising if it did not. High-dose exposures are standard testing procedure, today just as in 1960; such high doses are justified to offset practical limitations on such tests: compared to expected exposure of millions of humans over long periods, the time periods are short and the animals few. Many references in the legislative history reflect awareness of reliance on animal testing, and at least the more sophisticated participants must have been aware that this meant high-dose testing. A few so specified.

All this indicates to us that Congress did not intend the FDA to be able to take a finding that a substance causes only trivial risk in humans and work back from that to a finding that the substance does not "induce cancer in . . . animals." This is simply the basic question—is the operation of the clause automatic once the FDA makes a finding of carcinogenicity in animals?—in a new guise. The only new argument offered in the notices is that, without the new interpretation, only "primitive techniques" could be used. In fact, of course, the agency is clearly free to incorporate the latest breakthroughs in animal testing; indeed, here it touted the most recent animal tests as "state of the art." The limitation on techniques is only that the agency may not, once a color additive is found to induce cancer in test animals in the conventional sense of the term, undercut the statutory consequence. As we find the FDA's construction "contrary to clear congressional intent," we need not defer to it.

. . . .

In sum, we hold that the Delaney Clause of the Color Additive Amendments does not contain an implicit de minimis exception for carcinogenic dyes with trivial risks to humans. We based this decision on our understanding that Congress adopted an "extraordinarily rigid" position, denying the FDA authority to list a dye once it found it to "induce cancer in . . . animals" in the conventional sense of the term. We believe that, in the color additive context, Congress intended that if this rule produced unexpected or undesirable consequences, the agency should come to it for relief. That moment may well have arrived, but we cannot provide the desired escape.

II. Provisional Listing

The regulatory scheme of the Color Additive Amendments included grandfathering provisions for commercially established color additives. These allowed provisional listing of established dyes pending testing for a two-and-a-half year period. They empowered the Secretary to extend the listing "for such period or periods as he finds necessary to carry out the purpose of this section, if in the Secretary's judgment such action is consistent with the objective of carrying to completion in good faith, as soon as reasonably practicable, the scientific investigations necessary for making a determination as to listing such additive. . . ."

The process of completing these scientific investigations is only now being completed. When the litigation in No. 86-5150 began, ten color additives were on the provisional list. Today, only three—Red No. 3, Red No. 33, and Red No. 36—remain.

Public Citizen petitioned for a ban on the provisionally listed colors; when the petition was denied, it sued in the court below. The court granted summary judgment for the FDA (and other appellees supporting provisional listing).

In *McIlwain v. Hayes*, this court set forth the guidelines governing challenges to the speediness of the Secretary's evaluations of provisionally listed dyes. The *McIlwain* court determined that agency discretion to postpone the expiration of provisional listings was limited only as follows: "Such postponements must be consistent with the public health, and the Commissioner must judge that the scientific investigations are going forward in good faith and will be completed as soon as reasonably practicable." The majority acknowledged that it was doubtful that Congress foresaw the advances in testing technology that occasioned the delays, but saw no reason to depart from the statute's plain language.

McIlwain controls here. The FDA has found that the postponements for further evaluation of Red No. 3, Red No. 33, and Red No. 36 are consistent with the public health, that evaluations are going forward in good faith, and that they will be completed as soon as reasonably practicable. The agency carefully explained in its Federal Register notices and response to the rulemaking petition that extra time was needed for review of completed tests and in some cases the conduct of additional tests; a special scientific review panel was involved in this, and on completion of its work the agency would have to review its report. Announcing its most recent extension of Red No. 3, the agency explained that more time was needed "[b]ecause of the complexity of the scientific issues being considered." The most recent extensions for Red No. 33 and Red No. 36 announced that these reviews were essentially complete and the agency intended to list these dyes permanently, but that further time was necessary for the agency to prepare adequate explanations of its decisions. Although *McIlwain* dealt specifically with delays caused by the need for further testing, its logic applies with equal force where further evaluation of completed tests is required. To the extent that Public Citizen's complaint rests on the length of time already taken and anticipated for review of these dyes, it is foreclosed by *McIlwain*. Public Citizen's allegations of bad faith were not properly raised below, and in any event amount to no more than speculation.

Public Citizen also argues that provisional listing is permissible only when permanent listing is a reasonable possibility—an outcome precluded under this opinion if the outcome from the animal studies is positive. But this has not yet happened and may never happen. Neither Red No. 33 nor Red No. 36 has been found to induce cancer in humans or animals.

The situation is slightly less clear with regard to Red No. 3. The Commissioner explained, in denying Public Citizen's petition, that further evaluation was necessary to determine whether a carcinogenic effect observed in animal testing was caused by a secondary mechanism. There was, to be sure, evidence linking a statistically significant increase in tumors to the dye, but the chain of causation has yet to be established. There was a possibility, the Commission explained, that the dye might have affected the rats' thyroid glands, with that effect in turn causing the tumors. If this were established, then a no-effect level in rats might be established. Until the agency arrives at a final decision as to this question, the question of the Delaney Clause's application is not ripe. We therefore express no opinion as to the applicability of the provision in this secondary-effect situation, and decline to disturb the judgment of the District Court.

Conclusion

In sum, we hold that the agency's de minimis interpretation of the Delaney Clause of the Color Additive Amendments is contrary to law. The listing decisions for Orange No. 17 and Red No. 19 based on that interpretation must therefore be corrected. As for the colors still on the provisional list, we affirm the judgment of the court below in No. 85-5150, in view of *McIlwain* and the lack of a finding of carcinogenicity in the dyes at issue.

So ordered.

* * * * *

DISCUSSION QUESTION

6.5. Contrast the Delaney Clause approach to carcinogens with the *de minimis* doctrine to filth in food. Review FD&C Act section 402(a)(3)'s language, particularly the prohibition of "any filth." Why is there a different result with the *de minimus* doctrine?

6.6 HACCP[37]

The acronym HACCP (pronounced hassip) means hazard analysis and critical control point. Application of HACCP creates a prevention-based food safety system. In this system the inherent risks in ingredients, process, and the final food are analyzed, steps necessary to control the identified risks are established, and those controls are monitored. Thus HACCP provides process control to prevent food safety problems before they happen.

The Seven HACCP Principles[38]

1. *Hazard and risk assessment*.
2. *Determine the critical control points* (CCPs) to control the identified hazards.
3. Establish *critical limits* for the preventative measures.
4. Establish procedures to *monitor the critical control points* (CCPs).
5. Establish *corrective actions* to be taken when monitoring shows that a critical limit has been exceeded (or other deviation occurs in CCP monitoring).
6. Establish *effective record-keeping systems* that document the HACCP system is working correctly.
7. Establish procedures for *verification* that the HACCP system is working correctly.

[37] Adapted from: Neal D. Fortin, *The Hang-up with HACCP*, 58 FOOD AND DRUG LAW JOURNAL 565 (2004).

[38] The seven principles are variously described, but the International Commission on Microbiological Specifications for Food (ICMSF), the National Advisory Committee on Microbiological Criteria of Foods (NACMCF), and the FDA define HACCP as consisting of these seven principles. FDA 2001 FOOD CODE 421–457.

6.6.1 HACCP's History

HACCP was developed in the late 1950s and pioneered in the early 1960s by the Pillsbury Company with participation of the National Aeronautics and Space Administration (NASA), the Natick Laboratories of the U.S. Army, and the U.S. Air Force Space Laboratory Project Group. NASA's impetus for safe food is clear. Typical foodborne illness symptoms—nausea, diarrhea, and vomiting—could be catastrophic in space. Unfortunately, conventional end product testing was (and is) incapable of providing the desired 100 percent assurance against contamination by bacteria, viruses, toxins, and chemical and physical hazards. HACCP was essential to creating food for the space program that approached as near as possible complete assurance against contamination.

HACCP was first described in detail to a larger audience at the Conference for Food Protection in 1971. Then it was applied with great success to low-acid canned foods in 1974. In the decades since development, HACCP has become widely recognized as the best approach for improving food safety.

One of the strongest recommendations came in 1985 from the National Academy of Sciences, which recommended, "[G]overnment agencies responsible for control of microbiological hazards in foods should promulgate appropriate regulations that would require industry to utilize the HACCP system in their food protection programs." In addition HACCP has been endorsed by the U.S. National Advisory Committee on the Microbiological Criteria for Foods, the International Commission on Microbiological Specifications for Foods, the U.S. Food and Drug Administration, the Conference on Food Protection, and the Codex Alimentarius Commission of the United Nations.

6.6.2 The Advantages of HACCP

The traditional inspection works well at accomplishing what it was designed to achieve—cleaner food produced under more sanitary conditions—however, it is inadequate in preventing many foodborne illnesses. Whereas traditional food safety assurance programs rely on general sanitation inspections and end product testing, HACCP identifies the risks and then applies preventative control measures.

HACCP's preventative nature may be its most significant design achievement. Reliance on classical end product testing and inspections is relatively resource intensive and inefficient because it is reactive rather than preventative. "It's much easier to keep all the needles out of the barn than to find the needle in the haystack," one food safety educator noted, "An ounce of prevention is worth several million pounds of recalled product." Recent outbreaks provide dramatic examples of the economics of failed prevention; for example, $12.5 million of apple juice were recalled following contamination of the product with *E. coli*, and the firm, Odwalla, Inc., paid a $1.5 million federal fine.

Three additional benefits of HACCP are worth noting briefly. HACCP creates a complete system to ensure safety, which includes plans for corrective actions, record-keeping systems, and verification steps to ensure that potential risks are controlled. HACCP also clearly recognizes that responsibility for ensuring safe food rests on the food industry. "A HACCP system will emphasize the industry's role in continuous problem solving and prevention rather than relying solely on periodic facility inspections by regulatory agencies." Clearly, the food industry is in the best position to proactively ensure safe food. Third, HACCP allows the traditional inspection methods to be more productive. Traditional inspections and end product testing can achieve clean food produced under sanitary conditions, but they produce only a snapshot of time, rather than a continuous method. HACCP's record keeping improves the ability of food managers and regulators to ensure that the food workers consistently implement traditional sanitary practices.

HACCP's preventative system of process control can and does prevent hazards that traditional reactive methods could not. For example, after a number of botulism food poisonings, the U.S. Food and Drug Administration (FDA) promulgated the federal low-acid canning regulations in 1974. Although not called HACCP, these regulations essentially mandated HACCP for low-acid canning and nearly eliminated the incidence of botulism associated with canned food.

To sum up the primary benefits of HACCP, it is a science-based, preventative, risk control system. HACCP prevents foodborne illness by applying science to identify the risks in a method of food handling or processing. It controls those risks through preventative controls. Finally, HACCP is a complete system that includes corrective actions, record keeping, and verification, which increase the effectiveness and efficiency of both HACCP and conventional sanitation methods.

6.6.3   The Nature and Cost of Foodborne Illness

To fully appreciate the benefits of HACCP, it is necessary to understand the burden of foodborne illness. In excess of 200 known diseases are transmitted through food. These diseases include infections, intoxications, and chronic sequelae. The foodborne infectious agents include bacteria, viruses, and parasites. The intoxications (commonly called poisonings) include bacterial toxins, heavy metals, insecticides, and other chemical contaminants. Disease symptoms range from mild gastrointestinal distress to life-threatening neurological, hepatic, and renal syndromes, and death.

Over the past 10 years, science has begun to reveal the grim potential of foodborne pathogens to cause chronic sequelae, secondary complications that may develop months, even years, after the first unpleasant bout of symptoms. Growing evidence exists of a multitude of chronic illnesses resulting from an attack of foodborne disease, such as "arthropathies, renal disease, cardiac and neurological disorders, and nutritional and other malabsorbtive disorders (incapacitating diarrhea)." Sequelae include the immediate aftereffects of foodborne disease, toxins with long delay in onset, antigenic and autoimmune effects, and intracellular sequestration. It is estimated that chronic sequelae may occur in 2 to 3 percent of foodborne illness cases.

The burden of foodborne illness is estimated as high as 300 million cases per year and patient-related costs in the billions of dollars per year. Each year in the United States, foodborne illness causes an estimated 76 million illnesses, 320,000 hospitalizations, and 5,000 deaths. Contaminated food results in one of every 100 hospitalizations, and one of every 500 deaths in the United States.

This lack of certainty in the identification of foodborne illness is a principal reason why the range of estimates of the costs of foodborne illness is wide. One estimate places the cost for just direct, patient-related costs of foodborne illness at $164 billion per year. Other estimates attempt to calculate the costs for a limited number of foodborne pathogens (typically five to seven major pathogens), and in these, the estimated annual cost of medical treatment and lost productivity varies from $5.6 billion to $37.1 billion.

Many people casually reference the available aggregate estimates as the total cost of foodborne illness. However, these estimates—by design—are partial estimates of the burden of foodborne illness. For example, the USDA's Economic Research Service (ERS) estimated the medical costs and losses in productivity of five major foodborne pathogens at between $5.6 billion and $9.4 billion. However, this estimate does not include hepatitis A virus and other significant pathogens. In addition the ERS estimate and other available estimates do not include difficult to quantify costs, such as the public health expenditures on foodborne illness. Further these aggregate estimates of cost do not include the loss of food (i.e., recall and destruction), lost production, lost sales, or pain and suffering. The aggregate estimates also do not encompass foodborne illness that is too mild to require medical treatment, and they do not include the amount consumers are willing to pay to avoid mild diarrhea and nausea. Finally, none of the aggregate estimates includes the costs of the chronic sequelae of foodborne illness. Estimates of health consequences of chronic sequelae indicate that the economic costs may be higher than those of the acute diseases. In this light, even the highest estimate, $164 billion per year for direct medical costs, may be far below the total burden of foodborne illness.

The significant burden of foodborne illness highlights the importance of an effective and efficient food safety system, and the potential gains from the application of HACCP. Safe food is a goal shared by all. Consumers, obviously, benefit by having fewer illnesses. Society benefits from lower health care costs and lost productivity. Food businesses profit from lower liability, fewer production losses (such as recalls), and improved marketability of their product.

HACCP Implementation HACCP has been adopted, in principle, by the FDA-regulated low-acid canned food industry. FDA further established

HACCP for the seafood industry in a final rule December 18, 1995, and for the juice industry in a final rule with a phased in effective date through January 20, 2004, for very small businesses.

FDA considered developing regulations that would establish HACCP as the food safety standard throughout other areas of the food industry, including both domestic and imported food products. However, no other industries have specifically been proposed.

In 1993, after over 700 people were sickened and a number died from *E. coli* O157:H7-contaminated hamburgers in the Jack-In-The-Box outbreak, USDA received a firestorm of pressure from consumer groups and the media for an improved meat safety system. In 1998, the U.S. Department of Agriculture established HACCP for meat and poultry processing plants. Most of these establishments were required to start using HACCP by January 1999. Very small plants had until January 25, 2000.

USDA's implementation of HACCP faced criticism, both that it was watered down HACCP and (by some in the meat industry) that the plan went beyond USDA's authority. An important challenge came from Supreme Beef Processors.

* * * * *

Protecting the Public from Foodborne Illness

FSIS Backgrounder, the Food Safety and Inspection Service, April 2001

. . .

In 2000, FSIS completed implementation of its landmark rule on Pathogen Reduction and Hazard Analysis and Critical Control Point (HACCP) systems, which published in the *Federal Register* on July 25, 1996. . . . Under the regulations each meat and poultry plant must develop and implement a written plan for meeting its sanitation responsibilities and develop and implement a HACCP plan that systematically addresses all significant hazards associated with its products. In addition all slaughter plants must regularly test for generic *E. coli* to verify their procedures for preventing and reducing fecal contamination—the main source of bacteria that cause human foodborne illness. Raw products from slaughter plants and plants that grind meat and poultry are subject to *Salmonella* testing by FSIS. These efforts are directed at reducing microbial contamination over time.

With the Pathogen Reduction and HACCP final rule, FSIS has shifted its regulatory approach for meat and poultry. The expanded approach includes not only the product but also the process. A system under which potential food safety problems are identified and prevented is replacing a system that focused largely on detecting problems at the end of the production line.

* * * * *

Supreme Beef v. USDA Some believed that the *Supreme Beef v. USDA* decision dealt a serious blow to USDA's food safety reforms.[39] The decision upheld a ruling that USDA lacks the authority to shut down a meat-processing plant for insanitary condition based on repeated failures of performance tests for *Salmonella* in product. The decision prompted vociferous protest from food-safety advocates. Former Agriculture Secretary Dan Glickman said that he believed the decision was "a serious blow" to food safety.[40]

* * * * *

Supreme Beef Processors, Inc. v. USDA

275 F.3d 432 (5th Cir. 2001)

Before: REAVLEY, HIGGINBOTHAM, and PARKER, Circuit Judges
PATRICK E. HIGGINBOTHAM, Circuit Judge

Certain meat inspection regulations promulgated by the Secretary of Agriculture, which deal with the levels of *Salmonella* in raw meat product, were challenged as beyond the statutory authority granted to the Secretary by the Federal Meat Inspection Act. The district court struck down the regulations. We hold that the regulations fall outside of the statutory grant of rulemaking authority and affirm.

I

The Federal Meat Inspection Act authorizes the Secretary of Agriculture to "prescribe the rules and regulations of sanitation":
"covering slaughtering, meat canning, salting, packing, rendering, or similar establishments in which cattle, sheep, swine, goats, horses, mules, and other equines are slaughtered and the meat and meat food products thereof are prepared for commerce."...

Further, the Secretary is commanded to:
"where the sanitary conditions of any such establishment are such that the meat or meat food products are rendered adulterated, . . . refuse to allow said meat or meat food products to be labeled, marked, stamped, or tagged as "inspected and passed."

In sum, the FMIA instructs the Secretary to ensure that no adulterated meat products pass USDA inspection, which they must in order to be legally sold to consumers. The FMIA requires that adulterated meat products be stamped "inspected and condemned" and destroyed. 21 U.S.C. § 606.

The FMIA contains several definitions of "adulterated," including 21 U.S.C. § 601(m)(4), which classifies a meat product as adulterated if "it

---

[39] *See, e.g., Modern Meat,* PBS Frontline, *available at*: http://www.pbs.org/wgbh/pages/frontline/shows/meat/evaluating/supremebeef.html (Apr. 23, 2004).
[40] *Id.*

has been prepared, packed, or held under insanitary conditions whereby it may have become contaminated with filth, or whereby it may have been rendered injurious to health." Thus the FMIA gives the Secretary the power to create sanitation regulations and commands him to withhold meat approval where the meat is processed under insanitary conditions. The Secretary has delegated the authority under the FMIA to the Food Safety and Inspection Service.

In 1996 FSIS, after informal notice and comment rulemaking, adopted regulations requiring all meat and poultry establishments to adopt preventative controls to assure product safety. These are known as Pathogen Reduction, Hazard Analysis and Critical Control Point Systems or "HACCP." HACCP requires, inter alia, that meat and poultry establishments institute a hazard control plan for reducing and controlling harmful bacteria on raw meat and poultry products. In order to enforce HACCP, FSIS performs tests for the presence of *Salmonella* in a plant's finished meat products.

The *Salmonella* performance standards set out a regime under which inspection services will be denied to an establishment if it fails to meet the standard on three consecutive series of tests. The regulations declare that the third failure of the performance standard "constitutes failure to maintain sanitary conditions and failure to maintain an adequate HACCP plan . . . for that product, and will cause FSIS to suspend inspection services." The performance standard, or "passing mark," is determined based on FSIS's "calculation of the national prevalence of *Salmonella* on the indicated raw product."

In June 1998 plaintiff-appellee Supreme Beef Processors, Inc., a meat processor and grinder, implemented an HACCP pathogen control plan, and on November 2, 1998, FSIS began its evaluation of that plan by testing Supreme's finished product for *Salmonella*. After four weeks of testing, FSIS notified Supreme that it would likely fail the *Salmonella* tests. Pursuant to the final test results, which found 47 percent of the samples taken from Supreme contaminated with *Salmonella*, FSIS issued a Noncompliance Report, advising Supreme that it had not met the performance standard. Included in the report was FSIS's warning to Supreme to take "immediate action to meet the performance standards." Supreme responded to FSIS's directive on March 5, 1999, summarizing the measures it had taken to meet the performance standard and requesting that the second round of testing be postponed until mid-April to afford the company sufficient time to evaluate its laboratory data. FSIS agreed to the request and began its second round of tests on April 12, 1999.

On June 2, 1999, FSIS again informed Supreme that it would likely fail the *Salmonella* tests and, on July 20, issued another Noncompliance Report—this time informing Supreme that 20.8 percent of its samples had tested positive for *Salmonella*. Supreme appealed the Noncompliance Report, citing differences between the results obtained by FSIS and Supreme's own tests conducted on "companion parallel samples." Those private tests, Supreme asserted, had produced only a 7.5 percent *Salmonella* infection level, satisfying the performance standard. FSIS denied the appeal; but based on Supreme's commitment to install 180 degree water source on all boning and trimming lines, granted the company's request to postpone the next round of *Salmonella* testing for 60 days. FSIS later withdrew the extension, however, after learning that Supreme was merely considering installation of the water source.

The third set of tests began on August 27, 1999, and after only five weeks, FSIS advised Supreme that it would again fall short of the ground beef performance standard. On October 19, 1999, FSIS issued a Notice of Intended Enforcement Action, which notified Supreme of the agency's intention to suspend inspection activities. The Notice gave Supreme Beef until October 25, 1999, to demonstrate that its HACCP pathogen controls were adequate or to show that it had achieved regulatory compliance. Although Supreme Beef promised to achieve the 7.5 percent performance standard in 180 days, it failed to provide any specific information explaining how it would accomplish that goal, and FSIS decided to suspend inspection of Supreme's plant.

On the day FSIS planned to withdraw its inspectors, Supreme brought this suit against FSIS's parent agency, the USDA, alleging that in creating the *Salmonella* tests, FSIS had overstepped the authority given to it by the FMIA. Along with its complaint, Supreme moved to temporarily restrain the USDA from withdrawing its inspectors. The district court granted Supreme's motion and, after a subsequent hearing, also granted Supreme's motion for a preliminary injunction.

The National Meat Association filed a motion to intervene as a plaintiff in the district court. The district court denied the motion on the ground that NMA was adequately represented by Supreme in this litigation. The district court allowed NMA and other industry groups, as well as various consumer advocacy groups, to file briefs.

On cross-motions for summary judgment, the district court granted summary judgment in favor of Supreme, finding that the *Salmonella* performance standard exceeded the USDA's statutory authority and entering a permanent injunction against enforcement of that standard against Supreme. The USDA now appeals. . . .

Having concluded that this case is not moot, we now turn to the question of whether the *Salmonella* performance standard represents a valid exercise of rulemaking authority under the FMIA.

III

Our analysis in this case is governed by the approach first enunciated by the Supreme Court in *Chevron U.S.A., Inc. v. Natural Resources Defense Council, Inc.* The Chevron inquiry proceeds in two steps. First, the court should look to the plain language of the statute and determine whether the agency construction conflicts with the text. Then, "[i]f the agency interpretation is not in conflict with the plain language of the statute, deference is due." The district court held the *Salmonella* performance standard invalid as exceeding the statutory authority of the USDA under the first step of the Chevron inquiry.

A

Following *Chevron*, we first repair to the text of the statute that the USDA relies upon for its authority to impose the *Salmonella* performance standard. The USDA directs us to 21 U.S.C. § 601(m)(4), which provides that a meat product is adulterated

"if it has been prepared, packed, or held under insanitary conditions whereby it may have become contaminated with filth, or whereby it may have been rendered injurious to health."

This statutory definition is broader than that provided in 21 U.S.C. § 601(m)(1), which provides that a meat product is adulterated

"if it bears or contains any poisonous or deleterious substance which may render it injurious to health; but in case the substance is not an added substance, such article shall not be considered adulterated under this clause if the quantity of such substance in or on such article does not ordinarily render it injurious to health."

Thus if a meat product is "prepared, packed, or held under insanitary conditions" such that it may be adulterated for purposes of § 601(m)(1), then it is, by definition, adulterated for purposes of § 601(m)(4). The USDA is then commanded to refuse to stamp the meat products "inspected and passed."

The difficulty in this case arises, in part, because *Salmonella*, present in a substantial proportion of meat and poultry products, is not an adulterant per se, meaning its presence does not require the USDA to refuse to stamp such meat "inspected and passed." This is because normal cooking practices for meat and poultry destroy the *Salmonella* organism, and therefore the presence of *Salmonella* in meat products does not render them "injurious to health" for purposes of § 601(m)(1). *Salmonella*-infected beef is thus routinely labeled "inspected and passed" by USDA inspectors and is legal to sell to the consumer.

Supreme maintains that since *Salmonella*-infected meat is not adulterated under § 601(m)(1), the presence or absence of *Salmonella* in a plant cannot, by definition, be "insanitary conditions" such that the product "may have been rendered injurious to health," as required by § 601(m)(4). The USDA, however, argues that *Salmonella*'s status as a non-adulterant is not relevant to its power to regulate *Salmonella* levels in end product. This is because the USDA believes that *Salmonella* levels can be a proxy for the presence or absence of means of pathogen controls that are required for sanitary conditions under § 601(m)(4). However, as we discuss, and as the USDA admits, the *Salmonella* performance standard, whether or not it acts as a proxy, regulates more than just the presence of pathogen controls.

The district court agreed with Supreme and reasoned that "[b]ecause the USDA's performance standards and *Salmonella* tests do not necessarily evaluate the conditions of a meat processor's establishment, they cannot serve as the basis for finding a plant's meat adulterated under § 601(m)(4)." The district court therefore held that the examination of a plant's end product is distinct from "conditions" within the plant for purposes of § 601(m)(4) because *Salmonella* may have come in with the raw material.

We must decide two issues in order to determine whether the *Salmonella* performance standard is authorized rulemaking under the FMIA: (a) whether the statute allows the USDA to regulate characteristics of raw materials that are "prepared, packed or held" at the plant, such as *Salmonella* infection; and (b) whether § 601(m)(4)'s "insanitary conditions" such that product "may have been rendered injurious to health" includes the presence of *Salmonella*-infected beef in a plant or the increased likelihood of cross-contamination with *Salmonella* that results from grinding such infected beef. Since we are persuaded that the *Salmonella* performance standard improperly regulates the *Salmonella* levels of incoming meat and that *Salmonella* cross-contamination cannot be an insanitary condition such that product may be rendered "injurious to health," we conclude that the *Salmonella* performance standard falls outside of the ambit of § 601(m)(4).

B

1

In order for a product to be adulterated under § 601(m)(4), as the USDA relies on it here, it must be "prepared, packed, or held under insanitary conditions ... whereby it may have been rendered injurious to health." The use of the word "rendered" in the statute indicates that a deleterious change in the product must occur while it is being "prepared, packed, or held" owing to insanitary conditions. Thus a characteristic of the raw materials that exists before the product is "prepared, packed, or held" in the grinder's establishment cannot be regulated by the USDA under § 601(m)(4). The USDA's interpretation ignores the plain language of the statute, which includes the word "rendered." Were we to adopt this interpretation, we would be ignoring the Court's repeated admonition that, when interpreting a statute, we are to "give effect, if possible, to every clause and word of a statute."

The USDA claims, however, that the *Salmonella* performance standard serves as a proxy for the presence or absence of pathogen controls, such that a high level of *Salmonella* indicates § 601(m)(4) adulteration. Supreme oversimplifies its argument by claiming, essentially, that the USDA can never use testing of final product for a non-adulterant, such as *Salmonella*, as a proxy for conditions within a plant.

We find a similar, but distinct, defect in the *Salmonella* performance standard. The USDA admits that the *Salmonella* performance standard provides evidence of (1) whether or not the grinder has adequate pathogen controls, and (2) whether or not the grinder uses raw materials that are disproportionately infected with *Salmonella*. Supreme has, at all points in this litigation, argued that it failed the performance standard not because of any condition of its facility, but because it purchased beef "trimmings" that had higher levels of *Salmonella* than other cuts of meat. The USDA has not disputed this argument, and has merely argued that this explanation does not exonerate Supreme, because the *Salmonella* levels of incoming meat are fairly regulated under § 601(m)(4). Our textual analysis of § 601(m)(4)[41] shows that it cannot be used to regulate characteristics of the raw materials that exist before the meat product is "prepared, packed or held." Thus the regulation fails, but not because it measures *Salmonella* levels and *Salmonella* is a non-adulterant. The performance standard is invalid because it regulates the procurement of raw materials.

2

Our determination here is not in tension with the Second Circuit's decision interpreting identical language under the Food, Drug, and Cosmetic Act in *United States v. Nova Scotia Food Products Corp.* In *Nova Scotia* the defendant challenged an FDA regulation requiring the heating of smoked fish to combat the toxin formation of *Clostridium botulinum* spores, which cause botulism. The defendant argued that "the prohibition against 'insanitary conditions' embraces conditions only in the plant itself, but does not include conditions which merely inhibit the growth of organisms already in the food when it enters the plant in its raw state." The court gave "insanitary conditions" a broad reading and upheld the regulation. Nevertheless, it conceded that "a plausible argument can, indeed, be made that the references are to insanitary conditions in the plant itself, such as the presence of rodents or insects. . . ."

While this may appear to conflict with our determination that preexisting characteristics of raw materials before they are "prepared, packed, or held" are not within the regulatory reach of § 601(m)(4), the regulations at issue in *Nova Scotia* did not attempt to control the levels of *Clostridium botulinum* spores in incoming fish, as the performance standard does to *Salmonella* in

---

[41] The USDA repeatedly asserts that it has the power to regulate the *Salmonella* levels of incoming raw materials used in grinding establishments. See, e.g., Appellant's Reply Brief at 12 ("To operate in a sanitary manner, a plant must match the level of its pathogen controls to the nature of the meat it purchases. The greater the risk of contamination in the incoming product, the greater the need for strategies to reduce microbial contamination."); 61 Fed. Reg. at 38846 ("Establishments producing raw ground product from raw meat or poultry supplied by other establishments cannot use technologies for reducing pathogens that are designed for use on the surfaces of whole carcasses at the time of slaughter. Such establishments may require more control over incoming raw product, including contractual specifications to ensure that they begin their process with product that meets the standard. . . .") [emphasis added].

incoming raw meat. Instead, the regulations in *Nova Scotia* required the use of certain heating and salination procedures to inhibit growth of the spores.

Nova Scotia did not consider the argument before us today, which is that the statute does not authorize regulation of the levels of bacterial infection in incoming raw materials. The argument that *Nova Scotia* entertained was that "Congress did not mean to go so far as to require sterilization sufficient to kill bacteria that may be in the food itself rather than bacteria which accreted in the factory through the use of insanitary equipment." The required sterilization under the regulations at issue in *Nova Scotia* obviously occurred within the plant and did not regulate the quality of incoming fish.

3

The USDA and its amicus supporters argue that there is no real distinction between contamination that arrives in raw materials and contamination that arises from other conditions of the plant. This is because *Salmonella* can be transferred from infected meat to non-infected meat through the grinding process. The *Salmonella* performance standard, however, does not purport to measure the differential between incoming and outgoing meat products in terms of the *Salmonella* infection rate. Rather, it measures final meat product for *Salmonella* infection. Thus the performance standard, of itself, cannot serve as a proxy for cross-contamination because there is no determination of the incoming *Salmonella* baseline.

Moreover the USDA has not asserted that there is any correlation between the presence of *Salmonella* and the presence of § 601(m)(1) adulterant pathogens. The rationale offered by the USDA for the *Salmonella* performance standard—that "intervention strategies aimed at reducing fecal contamination and other sources of *Salmonella* on raw product should be effective against other pathogens"—does not imply that the presence of *Salmonella* indicates the presence of these other, presumably § 601(m)(1) adulterant, pathogens.[42] Cross-contamination of *Salmonella* alone cannot form the basis of a determination that a plant's products are § 601(m)(4) adulterated, because *Salmonella* itself does not render a product "injurious to health" for purposes of both §§ 601(m)(1) and 601(m)(4).

Not once does the USDA assert that *Salmonella* infection indicates infection with § 601(m)(1) adulterant pathogens. Instead, the USDA argues that the *Salmonella* infection rate of meat product correlates with the use of pathogen control mechanisms and the quality of the incoming raw materials. The former is within the reach of § 601(m)(4), the latter is not.

---

[42] One might speculate that such a conclusion would create problems for the USDA, because a statement that *Salmonella* was a proxy for, for example, pathogenic *E. coli* could arguably require the determination that the presence of *Salmonella* rendered a product § 601(m)(1) adulterated. This would prevent *Salmonella*-infected meat from being sold in the United States to consumers.

IV

Because we find that the *Salmonella* performance standard conflicts with the plain language of 21 U.S.C. § 601(m)(4), we need not reach Supreme's numerous alternative arguments for invalidating the standard, which were not addressed by the district court.

V

We AFFIRM and REMAND with instructions that the final judgment of the district court be amended to include the National Meat Association.

* * * * *

NOTES AND QUESTIONS

6.6. Contrast *Nova Scotia Food Products*. Contrast the result in *United States v. Nova Scotia Food Products Corp,* 568 F.2d 240 (2d. Cir. 1977), *supra* Chapter 5, with the *Supreme Beef* decision. Note that the language for "insanitary conditions" in the FMIA and the FD&C Act are identical. How can you reconcile the different conclusions?

6.7. Problem exercise. In this exercise, put yourself in the shoes of USDA officials upon finding Supreme Beef was in violation of the *Salmonella* performance standard for the third time. USDA has withdrawn inspectors. However, Supreme Beef's attorneys are seeking an injunction against this enforcement.

Supreme Beef's lawyers have argued that *Salmonella* is not an "added" substance under the statute. In addition ordinarily consumers cook their beef adequately, and the pathogens are thus killed. Thus the product is not "injurious to health." Although they admit the ground beef it produced contained *Salmonella* in excess of the allowance in the performance standard, Supreme argues that the *Salmonella* entered Superior's plant on the beef it purchased, and if anyone should be shut down, it should be those packing plants.

Questions. Explain in your most persuasive reasoning why the court should deny the injunction and allow USDA to shut down the Supreme Beef plant. How would you recommend revising the statute?

CHAPTER 7

Food Additives, Food Colorings, Irradiation

7.1 INTRODUCTION

This chapter examines the regulation of food additives and food colorings. It also covers generally recognized as safe (GRAS) status of food ingredients, as this is one of three methods for establishing their acceptability.

Color additives, by definition, are excluded from the definition of food additive. However, colors are included here to compare and contrast their regulation with that of food additives.

The topic of food additives encompasses another interesting, and perhaps surprising, subject. Treatment by irradiation is defined by U.S. law as a food additive. Thus that topic is included in this chapter as well.

7.2 BACKGROUND

* * * * *

Food Additives

U.S. Food and Drug Administration, FDA/IFIC Brochure (Jan. 1992)[1]

Q. What keeps bread mold-free and salad dressings from separating?

Q. What helps cake batters rise reliably during baking and keeps cured meats safe to eat?

Q. What improves the nutritional value of biscuits and pasta, and gives gingerbread its distinctive flavor?

---

[1] Food and Drug Administration/ International Food Information Council, Food Additives Brochure (Jan. 1992) *available at:* http://www.cfsan.fda.gov/~lrd/foodaddi.html (last accessed Oct. 2, 2007).

Q. What gives margarine its pleasing yellow color and prevents salt from becoming lumpy in its shaker?

Q. What allows many foods to be available year-round, in great quantity and the best quality?

A. Food Additives

Food additives play a vital role in today's bountiful and nutritious food supply. They allow our growing urban population to enjoy a variety of safe, wholesome, and tasty foods year-round. And they make possible an array of convenience foods without the inconvenience of daily shopping.

Although salt, baking soda, vanilla, and yeast are commonly used in foods today, many people tend to think of any additive to foods as complex chemical compounds. All food additives are carefully regulated by federal authorities and various international organizations to ensure that foods are safe to eat and are accurately labeled. The purpose of this brochure is to provide helpful background information about food additives, why they are used in foods and how regulations govern their safe use in the food supply.

Why Are Additives Used in Foods?

Additives perform a variety of useful functions in foods that are often taken for granted. Since most people no longer live on farms, additives help keep food wholesome and appealing while en route to markets sometimes thousands of miles away from where it is grown or manufactured. Additives also improve the nutritional value of certain foods and can make them more appealing by improving their taste, texture, consistency, or color.

Some additives could be eliminated if we were willing to grow our own food, harvest and grind it, spend many hours cooking and canning, or accept increased risks of food spoilage. But most people today have come to rely on the many technological, aesthetic, and convenience benefits that additives provide in food.

Additives are used in foods for five main reasons:

- To maintain product consistency. Emulsifiers give products a consistent texture and prevent them from separating. Stabilizers and thickeners give smooth uniform texture. Anti-caking agents help substances such as salt flow freely.
- To improve or maintain nutritional value. Vitamins and minerals are added to many common foods such as milk, flour, cereal, and margarine to make up for those likely to be lacking in a person's diet or lost in processing. Such fortification and enrichment has helped reduce malnutrition among the U.S. population. All products containing added nutrients must be appropriately labeled.

- To maintain palatability and wholesomeness. Preservatives retard product spoilage caused by mold, air, bacteria, fungi, or yeast. Bacterial contamination can cause foodborne illness, including life-threatening botulism. Antioxidants are preservatives that prevent fats and oils in baked goods and other foods from becoming rancid or developing an off-flavor. They also prevent cut fresh fruits such as apples from turning brown when exposed to air.
- To provide leavening or control acidity/alkalinity. Leavening agents that release acids when heated can react with baking soda to help cakes, biscuits, and other baked goods rise during baking. Other additives help modify the acidity and alkalinity of foods for proper flavor, taste, and color.
- To enhance flavor or impart desired color. Many spices and natural and synthetic flavors enhance the taste of foods. Colors, likewise, enhance the appearance of certain foods to meet consumer expectations. Examples of substances that perform each of these functions are provided in the chart "Common Uses of Additives."

Many substances added to food may seem foreign when listed on the ingredient label but are actually quite familiar. For example, ascorbic acid is another name for vitamin C; alphatocopherol is another name for vitamin E; and beta-carotene is a source of vitamin A. Although there are no easy synonyms for all additives, it is helpful to remember that all food is made up of chemicals. Carbon, hydrogen, and other chemical elements provide the basic building blocks for everything in life.

What Is a Food Additive?

In its broadest sense, a food additive is any substance added to food. Legally, the term refers to "any substance the intended use which results or may reasonably be expected to result—directly or indirectly—in its becoming a component or otherwise affecting the characteristics of any food." This definition includes any substance used in the production, processing, treatment, packaging, transportation, or storage of food.

If a substance is added to a food for a specific purpose in that food, it is referred to as a direct additive. For example, the low-calorie sweetener aspartame, which is used in beverages, puddings, yogurt, chewing gum, and other foods, is considered a direct additive. Many direct additives are identified on the ingredient label of foods.

Indirect food additives are those that become part of the food in trace amounts due to its packaging, storage, or other handling. For instance, minute amounts of packaging substances may find their way into foods during storage. Food packaging manufacturers must prove to the U.S. Food and Drug Administration (FDA) that all materials coming in contact with food are safe, before they are permitted for use in such a manner.

What Is a Color Additive?

A color additive is any dye, pigment, or substance that can impart color when added or applied to a food, drug, or cosmetic, or to the human body. Color additives may be used in foods, drugs, cosmetics, and certain medical devices such as contact lenses. Color additives are used in foods for many reasons, including to offset color loss due to storage or processing of foods and to correct natural variations in food color.

Colors permitted for use in foods are classified as certified or exempt from certification. Certified colors are man-made, with each batch being tested by the manufacturer and FDA to ensure that they meet strict specifications for purity. There are nine certified colors approved for use in the United States. One example is FD&C Yellow No. 6, which is used in cereals, bakery goods, snack foods, and other foods.

Color additives that are exempt from certification include pigments derived from natural sources such as vegetables, minerals, or animals. For example, caramel color is produced commercially by heating sugar and other carbohydrates under strictly controlled conditions for use in sauces, gravies, soft drinks, baked goods, and other foods. Most colors exempt from certification also must meet certain legal criteria for specifications and purity.

How Are Additives Regulated?

Additives are not always by-products of twentieth-century technology or modern know-how. Our ancestors used salt to preserve meats and fish; added herbs and spices to improve the flavor of foods; preserved fruit with sugar; and pickled cucumbers in a vinegar solution.

Over the years, however, improvements have been made in increasing the efficiency and ensuring the safety of all additives. Today food and color additives are more strictly regulated that at any other time in history. The basis of modern food law is the Federal Food, Drug, and Cosmetic (FD&C) Act of 1938, which gives the Food and Drug Administration (FDA) authority over food and food ingredients and defines requirements for truthful labeling of ingredients.

The Food Additives Amendment to the FD&C Act, passed in 1958, requires FDA approval for the use of an additive prior to its inclusion in food. It also requires the manufacturer to prove an additive's safety for the ways it will be used.

The Food Additives Amendment exempted two groups of substances from the food additive regulation process. All substances that FDA or the U.S. Department of Agriculture (USDA) had determined were safe for use in specific food prior to the 1958 amendment were designated as prior-sanctioned substances. Examples of prior-sanctioned substances are sodium nitrite and potassium nitrite used to preserve luncheon meats.

A second category of substances excluded from the food additive regulation process are generally recognized as safe or GRAS substances. GRAS substances are those whose use is generally recognized by experts as safe, based on their extensive history of use in food before 1958 or based on published scientific evidence. Salt, sugar, spices, vitamins, and monosodium glutamate are classified as GRAS substances, along with several hundred other substances. Manufacturers may also request FDA to review the use of a substance to determine if it is GRAS.

Since 1958, FDA and USDA have continued to monitor all prior sanctioned and GRAS substances in light of new scientific information. If new evidence suggests that a GRAS or prior sanctioned substance may be unsafe, federal authorities can prohibit its use or require further studies to determine its safety.

In 1960, Congress passed similar legislation governing color additives. The Color Additives Amendments to the FD&C Act require dyes used in foods, drugs, cosmetics, and certain medical devices to be approved by FDA prior to their marketing.

In contrast to food additives, colors in use before the legislation were allowed continued use only if they underwent further testing to confirm their safety. Of the original 200 provisionally listed color additives, 90 have been listed as safe and the remainder have either been removed from use by FDA or withdrawn by industry.

Both the Food Additives and Color Additives Amendments include a provision which prohibits the approval of an additive if it is found to cause cancer in humans or animals. This clause is often referred to as the Delaney Clause, named for its congressional sponsor, Rep. James Delaney (D–N.Y.).

Regulations known as Good Manufacturing Practices (GMP) limit the amount of food and color additives used in foods. Manufacturers use only the amount of an additive necessary to achieve the desired effect.

How Are Additives Approved for Use in Foods?

To market a new food or color additive, a manufacturer must first petition FDA for its approval. Approximately 100 new food and color additives petitions are submitted to FDA annually. Most of these petitions are for indirect additives such as packaging materials.

A food or color additive petition must provide convincing evidence that the proposed additive performs as it is intended. Animal studies using large doses of the additive for long periods are often necessary to show that the substance would not cause harmful effects at expected levels of human consumption. Studies of the additive in humans also may be submitted to FDA.

In deciding whether an additive should be approved, the agency considers the composition and properties of the substance, the amount likely to be consumed, its probable long-term effects and various safety factors. Absolute safety of any substance can never be proved. Therefore FDA must determine if the additive is safe under the proposed conditions of use, based on the best scientific knowledge available.

If an additive is approved FDA issues regulations that may include the types of foods in which it can be used, the maximum amounts to be used, and how it should be identified on food labels. Additives proposed for use in meat and poultry products also must receive specific authorization by USDA. Federal officials then carefully monitor the extent of Americans' consumption of the new additive and results of any new research on its safety to assure its use continues to be within safe limits.

In addition, FDA operates an Adverse Reaction Monitoring System (ARMS) to help serve as an ongoing safety check of all additives. The system monitors and investigates all complaints by individuals or their physicians that are believed to be related to specific foods; food and color additives; or vitamin and mineral supplements. The ARMS computerized database helps officials decide whether reported adverse reactions represent a real public health hazard associated with food, so that appropriate action can be taken.

Summary

Additives have been used for many years to preserve, flavor, blend, thicken, and color foods, and have played an important role in reducing serious nutritional deficiencies among Americans. Additives help assure the availability of wholesome, appetizing, and affordable foods that meet consumer demands from season to season.

Today, food and color additives are more strictly regulated than at any time in history. Federal regulations require evidence that each substance is safe at its intended levels of use before it may be added to foods. All additives are subject to ongoing safety review as scientific understanding and methods of testing continue to improve.

Additional Information about Additives

Q. What is the difference between "natural" and "artificial" additives?

A. Some additives are manufactured from natural sources such as soybeans and corn, which provide lecithin to maintain product consistency, or beets, which provide beet powder used as food coloring. Other useful additives are not found in nature and must be man-made. Artificial additives can be produced more economically, with greater purity and more consistent quality than some of their natural counterparts. Whether an additive is natural or artificial has no bearing on its safety.

Q. Is a natural additive safer because it is chemical-free?

A. No. All foods, whether picked from your garden or your supermarket shelf, are made up of chemicals. For example, the vitamin C or ascorbic acid found in an orange is identical to that produced in a laboratory. Indeed all things in the world consist of the chemical building blocks of carbon, hydrogen, nitrogen, oxygen and other elements. These elements are combined in various ways to produce starches, proteins, fats, water, and vitamins found in foods.

Q. Are sulfites safe?

A. Sulfites added to baked goods, condiments, snack foods, and other products are safe for most people. A small segment of the population, however, has been found to develop hives, nausea, diarrhea, shortness of breath, or even fatal shock after consuming sulfites. For that reason, in 1986 FDA banned the use of sulfites on fresh fruits and vegetables intended to be sold or served raw to consumers. Sulfites added as a preservative in all other packaged and processed foods must be listed on the product label.

Q. Does FD&C Yellow No. 5 cause allergic reactions?

A. FD&C Yellow No. 5, or tartrazine, is used to color beverages, desert powders, candy ice cream, custards, and other foods. The color additive may cause hives in fewer than one out of 10,000 people. By law, whenever the color is added to foods or taken internally, it must be listed on the label. This allows the small portion of people who may be sensitive to FD&C Yellow No. 5 to avoid it.

Q. Does the low calorie sweetener aspartame carry adverse reactions?

A. There is no scientific evidence that aspartame causes adverse reactions in people. All consumer complaints related to the sweetener have been investigated as thoroughly as possible by federal authorities for more than five years, in part under FDA's Adverse Reaction Monitoring System. In addition scientific studies conducted during aspartame's pre-approval phase failed to show that it causes any adverse reactions in adults or children. Individuals who have concerns about possible adverse reactions to aspartame or other substances should contact their physicians.

Q. Do additives cause childhood hyperactivity?

A. No. Although this theory was popularized in the 1970s, well-controlled studies conducted since that time have produced no evidence that food additives cause hyperactivity or learning disabilities in children. A Consensus Development Panel of the National Institutes of Health concluded in 1982 that there was no scientific evidence to support the claim that additives or colorings cause hyperactivity.

Q. Why are decisions sometimes changed about the safety of food ingredients?

A. Since absolute safety of any substance can never be proved, decisions about the safety of food ingredients are made on the best scientific evidence available. Scientific knowledge is constantly evolving. Therefore federal officials often review earlier decisions to assure that the safety assessment of a food substance remains up to date. Any change made in previous clearances should be recognized as an assurance that the latest and best scientific knowledge is being applied to enhance the safety of the food supply.

Q. What are some other food additives that may be used in the future?

A. Among other petitions, FDA is carefully evaluating requests to use ingredients that would replace either sugar or fat in food. In 1990, FDA confirmed the GRAS status of Simplesse, (registered trademark) a fat replacement made from milk or egg white protein, for use in frozen desserts. The agency also is evaluating a food additive petition for olestra, which would partially replace the fat in oils and shortenings.

Q. What is the role of modern technology in producing food additives?

A. Many new techniques are being researched that will allow the production of additives in ways not previously possible. One approach, known as biotechnology, uses simple organisms to produce additives that are the same food components found in nature. In 1990, FDA approved the first bioengineered enzyme, rennin, which traditionally has been extracted from calves' stomachs for use in making cheese.. . .

* * * * *

7.3 FOOD ADDITIVES

* * * * *

Food and Drug Administration, Second Edition

James T. O'Reilly Chapter 11, Food Additives, 11.2 (2004)

Leave the baggage of colloquial understanding behind you, discard the mathematician's view of the concept of adding, and suspend disbelief about what is "food," and you are in the proper mental state for parsing the term "food additive" under the 1958 amendments to the Food Drug and Cosmetic Act and its subsequent court interpretations. Perhaps one should first grapple intellectually with "new drug," that metaphysical concept of 1938, and only then delve into food additive.

Counsel who wishes to write an opinion or brief on "food additive" status must be extremely careful. The reader, client, or judge will need to be told that this term of art does not mean what the simple terms mean. Counsel should read, at least twice, the definition for "food additive" found in § 201(s) of the Act. Its history is found in a parallel to the new drug definition, since the concept of "general recognition of safety" is involved in both definitions, and food additives definitions also have the politically necessary exclusions for "prior sanction" materials. . . .

* * * * *

7.3.1 The Food Additive Amendment of 1958

Before World War II there were relatively few food additives for functional purposes. The revolution in food technology in the 1940s and 1950s brought a proliferation of new additives. Moreover changes in demographics, particularly the migration of the population from farms to the cities, fueled a growing need for additives, such as preservatives.

In the early 1950s, Representative James Delaney (D–N.Y.) chaired a committee to investigate the use of chemicals in food. These Delaney Committee hearings helped spur Congress into enacting the Food Additives Amendment of 1958.[2] The Food Additives Amendment of 1958 revised the FD&C Act to require the FDA approval for the use of an additive before its use in food. The amendment also put the burden on the manufacturer to prove an additive's safety for the intended use.

In the broadest meaning, a food additive is any substance added to food. Similarly, under the FD&C Act, the term broadly encompasses any substance that may reasonably be expected to result—directly or indirectly—in its becoming a component any food. This definition includes any substance used in the production, processing, treatment, packaging, transportation, or storage of food. "Food additive" also means any substance that may reasonably be expected to *affect the characteristics* of a food. Thus irradiation falls under the definition of food additive.

While the definition of food additive strictly includes almost everything entering a food or affecting a food, the Food Additives Amendment exempts two groups of substances. Congress recognized that many substances added to foods would not require a formal premarket review by FDA to ensure their safety. The safety of some additives had been established by a long history of safe use in food. Other additives could be deemed safe by the nature of the additive and the information generally available to scientists. Congress thus adopted a three-part definition of "food additive."[3]

The first part broadly includes any substance the intended use of which results or may reasonably be expected to result, directly or indirectly, in its becoming a component or otherwise affecting the characteristics of food. The second part excludes from the definition of food additive substances that are generally recognized—among experts qualified by scientific training and experience to evaluate their safety—as having been shown to be safe under

---

[2] 72 Stat. 1784 (1958).
[3] 21 U.S.C. § 321(s).

the conditions of their intended use. This is the generally recognized as safe (GRAS) provision. GRAS determination is usually made by recognized experts based on an extensive history of safe use in food before 1958 or based on published scientific evidence. Salt, sugar, spices, vitamins, and monosodium glutamate are GRAS substances. Several hundred substances are listed as GRAS, but the listing is not intended to be comprehensive.

The third part of the definition is that all substances that FDA or the USDA had determined were safe for use in a specific food before the 1958 Food Additive Amendment are designated as prior-sanctioned substances. For example, sodium nitrite and potassium nitrite used to preserve lunch meats are prior-sanctioned substances.

Another important limitation on use of food additives is that the good manufacturing practices (GMPs) regulations limit the amount of food additives and color additives used in foods. Manufacturers may use only the amount of an additive necessary to achieve the desired effect.

7.3.2 FD&C Act Definition

* * * * *

Federal Food, Drug, and Cosmetic Act § 201 (21 U.S.C. § 321)

(s) The term "food additive" means any substance the intended use of which results or may reasonably be expected to result, directly or indirectly, in its becoming a component or otherwise affecting the characteristics of any food (including any substance intended for use in producing, manufacturing, packing, processing, preparing, treating, packaging, transporting, or holding food; and including any source of radiation intended for any such use), if such substance is not generally recognized, among experts qualified by scientific training and experience to evaluate its safety, as having been adequately shown through scientific procedures (or, in the case as a substance used in food prior to January 1, 1958, through either scientific procedures or experience based on common use in food) to be safe under the conditions of its intended use; except that such term does not include—

(1) a pesticide chemical residue in or on a raw agricultural commodity or processed food; or

(2) a pesticide chemical; or

(3) a color additive; or

(4) any substance used in accordance with a sanction or approval granted prior to September 6, 1958, pursuant to this chapter, the Poultry Products Inspection Act (21 U.S.C. § 451 *et seq.*) or the Meat Inspection Act of March 4, 1907, as amended and extended (21 U.S.C. § 601 *et seq.*);

(5) a new animal drug; or

(6) an ingredient described in paragraph (ff) in, or intended for use in, a dietary supplement.

* * * * *

21 C.F.R. Part 170—Food Additives

Sec. 170.3 Definitions.

For the purposes of this subchapter, the following definitions apply: . . .

(e)(1) **Food additives** includes all substances not exempted by section 201(s) of the act, the intended use of which results or may reasonably be expected to result, directly or indirectly, either in their becoming a component of food or otherwise affecting the characteristics of food. A material used in the production of containers and packages is subject to the definition if it may reasonably be expected to become a component, or to affect the characteristics, directly or indirectly, of food packed in the container. "Affecting the characteristics of food" does not include such physical effects as protecting contents of packages, preserving shape, and preventing moisture loss. If there is no migration of a packaging component from the package to the food, it does not become a component of the food and thus is not a food additive. A substance that does not become a component of food, but that is used, for example, in preparing an ingredient of the food to give a different flavor, texture, or other characteristic in the food, may be a food additive.

   (2) Uses of food additives not requiring a listing regulation. Substances used in food-contact articles (e.g., food-packaging and food-processing equipment) that migrate, or may be expected to migrate, into food at such negligible levels that they have been exempted from regulation as food additives under Sec. 170.39.

(f)   Common use in food means a substantial history of consumption of a substance for food use by a significant number of consumers.

(g)   The word substance in the definition of the term "food additive" includes a food or food component consisting of one or more ingredients.

(h)   Scientific procedures include those human, animal, analytical, and other scientific studies, whether published or unpublished, appropriate to establish the safety of a substance.

(i)   **Safe or safety** means that there is a *reasonable certainty in the minds of competent scientists* that the substance is not harmful under the intended conditions of use. It is impossible in the present state of scientific knowledge to establish with complete certainty the absolute harmlessness of the use of any substance. Safety may be determined by scientific procedures or by general recognition of safety. In determining safety, the following factors shall be considered:

(1) The probable consumption of the substance and of any substance formed in or on food because of its use.

(2) The cumulative effect of the substance in the diet, taking into account any chemically or pharmacologically related substance or substances in such diet.

(3) Safety factors which, in the opinion of experts qualified by scientific training and experience to evaluate the safety of food and food ingredients, are generally recognized as appropriate.

(j)   The term nonperishable processed food means any processed food not subject to rapid decay or deterioration that would render it unfit for consumption. Examples are flour, sugar, cereals, packaged cookies, and crackers. Not included are hermetically sealed foods or manufactured dairy products and other processed foods requiring refrigeration.

(k)   General recognition of safety shall be determined in accordance with Sec. 170.30.

(l)   Prior sanction means an explicit approval granted with respect to use of a substance in food prior to September 6, 1958, by the Food and Drug Administration or the United States Department of Agriculture pursuant to the Federal Food, Drug, and Cosmetic Act, the Poultry Products Inspection Act, or the Meat Inspection Act.

(m)   Food includes human food, substances migrating to food from food-contact articles, pet food, and animal feed.

* * * * *

7.3.3 Additive Safety and Approval

The FD&C Act begins with the *presumption that new food additives are unsafe* unless proven otherwise.[4] The Act requires pre-market review of a new food additive and states that no food additive can be approved for use by the FDA absent a showing that the proposed additive will be safe under the specific conditions of its use.

New Food Additive Petition   The process for approval of a new food additive begins with the submission of a food additive petition. The FD&C Act section 409(b)(2) lists the statutory requirements for food additive petitions, and 21 C.F.R. § 171.1(c) describes these requirements in greater detail. Briefly stated, these petition requirements encompass five general areas of information:

---

[4] *See* FD&C Act § 409 [348].

1. Identity of the additive
2. Proposed use of the additive
3. Intended technical effect of the additive
4. Method of analysis for the additive in food
5. Full reports of all safety investigations with the additive (e.g., animal and toxicological studies)

The petitioner generally must also furnish a description of the methods, facilities, and controls used in or for the production of the additive. The petitioner must also submit, upon request, samples of the additive and the food in which the additive will be used.[5] Applicants need to be mindful that much of the information provided in a food additive petition are available for public review.[6]

The FDA defined safe to mean "reasonable certainty in the minds of competent scientists that the substance is not harmful under the intended conditions of use," and in determining safety, the agency must consider, among other relevant factors, the probable consumption of the additive, the cumulative effect of the additive in the diet, and appropriate safety factors. Additionally FDA requires details regarding the method of manufacture, facilities, and the type of analytical controls used to establish that the additive is "a substance of reproducible composition."

Approximately 100 new food and color additives petitions are submitted to FDA annually. Most of these petitions are for indirect additives such as packaging materials. If approved, FDA issues regulations that may include the types of foods in which it can be used, the maximum amounts to be used, and how it should be identified on food labels.

The Food Additive Two-Step for Approved Use in Meats   The USDA Food Safety and Inspection Service (FSIS) shares responsibility with FDA for the safety of food additives used in meat, poultry, and egg products. Therefore approval of food, poultry, and egg products additives in meat follows a two-step process. First all additives are evaluated for safety by FDA under the authority of FD&C Act section 409,[7] and then additives proposed for use in meat must also receive specific authorization by FSIS. FSIS determines the suitability of the use of food additive in accordance with various FSIS laws, regulations, and policies.

Although the FDA has overriding authority on food additive safety, the FSIS may apply stricter standards that take into account the unique characteristics of meat, poultry, and egg products. A number of years ago, for example, permission was sought to use sorbic acid in meat salads. Sorbic acid had already been approved as a food additive; nevertheless, permission for use in

[5] See 21 C.F.R. § 409(b)(3) and (4).
[6] See 21 C.F.R. § 171.1(h).
[7] 21 U.S.C. § 348.

meat salad was denied because such usage could mask spoilage caused by organisms that cause foodborne illness.

No Permanent Approval Food additives are never given permanent approval. FDA continually reviews the safety of approved additives based on the latest scientific knowledge to determine if approvals should be modified or withdrawn.

Timeline The new food additive approval process can take years to complete. Therefore it is not surprising that some firms seek inventive ways around the new food additive petition process.

DISCUSSION QUESTIONS AND NOTES

7.1. The substances added to dietary supplements are exempt from the definition of food additives (FD&C Act § 201(s)(6)). Do you have any thoughts on why this is?

7.2. Why do we require premarket approval of food additives, but not whole foods?

7.3. Why should the GMP regulations limit an additive to the amount necessary to achieve the desired effect?

7.4. On a scale of 1 to 5, with 5 being the most precautionary, how precautionary would you rate the food additive approval process?

7.5. Why did FDA define safe or safety of a food additive to mean a "reasonable certainty" that the substance is not harmful.

7.6. Further Information. Questions on new food additive petitions for human food can be directed to the FDA, CFSAN, Office of Premarket Approval http://www.cfsan.fda.gov/~dms/opa-help.html. Petitions for food additive approval for animal feed are handled by the FDA's Center for Veterinary Medicine, Division of Animal Feeds, see http://www.fda.gov/cvm/default. html.

7.3.4 Prior Sanctioned Substances

Prior sanctioned substances are additives that were sanctioned or approved by FDA or USDA before the 1958 Food Additive Amendment. Substances subject to prior sanction are listed in 21 C.F.R. part 181. Although exempt from food additive approval, these substances must still comply with other requirements for food ingredients.

7.3.5 GRAS (Generally Recognized as Safe)

GRAS substances are exempt them from the food additive requirements:

> [I]f such substance is generally recognized among experts qualified by scientific training and experience to evaluate its safety, as having been adequately shown through scientific procedures (or, in the case of a substance used in food before January 1, 1958, through either scientific procedures or experience based on common use in food) to be safe under the conditions of its intended use.[8]

The GRAS exemption was intended allow the classification of known safe food additives without protracted testing and government review. For example, salt, sugar, and other such common substances had long been used in foods without evidence of harm.

The FD&C Act provides no specific requirements for GRAS affirmation petitions. Section 201(s) exempts GRAS substances from the meaning of "food additive," and therefore (technically) from the statutory requirements for food additive petitions. As a practical matter, this is really only applicable for substances found safe based on common use in America before 1958. To prove the safety of the substance in food, the historical information, such as reports describing and documenting the past uses of the substance in food, must be submitted with a petition.

FD&C Act section 201(s) provides that general recognition of safety must be established either through (1) scientific procedures or (2) experience based on common use in food before January 1, 1958. If the substance did not have a common history of use in food before 1958, then the substance can be considered as GRAS only based on scientific procedures, as set forth in 21 C.F.R. § 170.30(b). As a practical matter, GRAS affirmation based on scientific procedures works out to be nearly as involved as a new food additive petition. GRAS affirmation based on scientific procedures requires "the same quantity and quality of scientific evidence as is required to obtain approval of a food additive."[9] Therefore, in preparing a petition for GRAS affirmation based on scientific procedures,[10] the requirements for food additive petitions[11]—while not technically required—would almost certainly need to be considered.

However, GRAS status does offer one potential benefit over new additive approval: firms are permitted to make a self-determination that the substance is safe. Self-determination of GRAS status permits the immediate marketing of a food additive and avoids the costly and protracted food additive approval process. Remember, only new food additives must obtain premarket approval.

---

[8]FD&C Act § 201(s).
[9]21 C.F.R. § 170.30(b).
[10]21 C.F.R. § 170.35(c)(1).
[11]21 C.F.R. § 171.1(c).

A firm may notify FDA of its determination that a particular use of a substance is GRAS through FDA's proposed notification procedure (although the procedure is not yet finalized, FDA receives the notices).[12]

High-fructose corn syrup provides a example of GRAS self-determination. After safety evaluation and industry determination of safety, high-fructose corn syrup began to be marketed in the mid-1960s. The product was marketed without FDA approval, but a GRAS affirmation petition was filed in 1974. The FDA took nine years to affirm GRAS status and to publish separate regulations for the use of HFCS.

FDA publishes recognized GRAS substances in 21 C.F.R. parts 182 and 582. However, not all GRAS substances are listed because of practicality limitation. You can find out more on GRAS affirmation procedures from the FDA CFSAN Office of Food Additive Safety at: http://www.cfsan.fda.gov/~lrd/foodadd.html.

7.3.6 Proving GRAS Status

In the next case, Coco Rico, Inc., manufactured a coconut concentrate called Coco Rico for use as an ingredient in soft drinks. The product contained potassium nitrate. When FDA charged the company with use of an unsafe food additive, the manufacturer claimed potassium nitrate was GRAS because of prior use by the corporation, and because the substance was approved for use in meats. In addition an industry expert testified that there was no evidence that the substance was unsafe.

The Court ruling illustrates an important point: general acceptance of additive in one food does not make the substance GRAS for another food or use. Additionally the burden is on the proponent of GRAS status to prove its safety. (Lack of evidence of being unsafe does not prove safety.) The proponent must show that there is a consensus of expert opinion regarding the safety of the use of the substance.

* * * * *

United States v. An Article of Food, Coco Rico, Inc.

752 F.2d 11 (1985)

JUDGES: COFFIN and BOWNES, Circuit Judges, and WEIGEL, Senior District Judge
Opinion: WEIGEL, Senior District Judge

This is an appeal from the district court's grant of summary judgment. Appellant Coco Rico, Inc., manufactures in Puerto Rico a coconut concentrate called Coco Rico for use as an ingredient in soft drinks. The Coco Rico concentrate

---

[12] *See* the GRAS proposal, 62 FR 18938.

sold to beverage bottlers in Puerto Rico contains potassium nitrate, added for the purpose of developing and fixing a desirable color and flavor.[13] On March 10, 1982, the United States instituted in rem proceedings against three lots of bottled soft drinks located on the premises of Puerto Rican bottlers. The soft drinks contained Coco Rico concentrate. The government charged that potassium nitrate constitutes an "unsafe" food additive, making the beverages "adulterated" and subject to forfeiture under the Food, Drug, and Cosmetic Act ("the Act"), 21 U.S.C. section 301 et seq. On March 24, 1982, the government seized the three lots of soft drinks pursuant to warrants issued by the district court.

The forfeiture complaints were answered by Coco Rico, Inc. as claimant. Coco Rico did not dispute that the beverages in question contained potassium nitrate. Rather, it alleged that (1) as neither the beverages nor the concentrate they contained had been shipped outside Puerto Rico, they had not traveled in interstate commerce and were therefore not subject to forfeiture, and that (2) the beverages were not "adulterated" within the meaning of the Act.

The three cases were consolidated and the government moved for summary judgment. In support of its motion, the government submitted affidavits of two food chemists, Dr. Shibko and Dr. Wade. These affidavits stated, in summary, that (1) both Dr. Shibko and Dr. Wade know of no scientific studies showing that potassium nitrate is safe for use in beverages; (2) both believe, based on their training and study of the scientific literature, that potassium nitrate is not generally recognized as safe for use in beverages; and (3) the levels of potassium nitrate contained in the beverages in question approach those feared toxic to infants.

Coco Rico submitted one affidavit in opposition to the motion for summary judgment, that of food chemist Algeria B. Caragay. Caragay's affidavit makes the following points:

1. In her opinion, nitrates and nitrites are not "food additives" within the meaning of the Act because they are "prior sanctioned";
2. She believes that on August 19, 1980, a Food and Drug Administration (FDA) Commissioner and an Assistant Secretary of Agriculture stated publicly that there was "no basis for the FDA or USDA to initiate any action to remove nitrite from foods at this time";
3. Although some studies have cast suspicion on nitrates and nitrites as possible carcinogens, she knows of no conclusive scientific evidence that the use of potassium nitrate in beverages is unsafe;

---

[13] In addition to selling its product to Puerto Rican bottlers, Coco Rico ships its concentrate to soft drink bottlers in the mainland United States. Prior to 1978, the concentrate shipped to the continental United States also contained potassium nitrate. Coco Rico subsequently developed a different concentrate formula for interstate sale; since 1978, the concentrate shipped to the continental United States contains no potassium nitrate.

4. Nitrates have been approved by the FDA for use in curing meat; and
5. She knows of no difference in health effects between potassium nitrate as used in meat and as used in beverages.

It was not disputed that all of the potassium nitrate used by Coco Rico originated in New York.

Holding that the interstate shipment of the potassium nitrate was sufficient to bring the beverages under the jurisdiction of the Act, the district court granted summary judgment for the government. The district court also found that because there was no dispute of the material fact that potassium nitrate constituted an "unsafe" food additive, the beverages were adulterated and subject to seizure as a matter of law. We affirm.

1. Jurisdiction

Appellant contends that the seized beverages were not subject to forfeiture because they were to be sold only in Puerto Rico and not shipped in interstate commerce. The governing statute is 21 U.S.C. section 334(a)(1), which provides in pertinent part that:

"any article of food . . . that is adulterated . . . when introduced into or while in interstate commerce or while held for sale (whether or not the first sale) after shipment in interstate commerce . . . shall be liable to be proceeded against while in interstate commerce, or at any time thereafter . . ." (emphasis added).

Commerce between any state and Puerto Rico is "interstate" commerce for purposes of this statute. 21 U.S.C. § 321(a)–(b). This court has held that the "shipment in interstate commerce" requirement is satisfied when adulterated articles held for in-state sale contain ingredients shipped in interstate commerce. Because it is undisputed that the potassium nitrate added to the seized beverages was shipped in interstate commerce, those beverages clearly fall within the scope of statutory forfeiture jurisdiction.

2. Adulteration

21 U.S.C. section 348 provides in part that:

(a) A food additive shall, with respect to any particular use or intended use . . . be deemed to be unsafe for the purposes of [21 U.S.C. § 342(a)(2)(C)] unless [for purposes relevant here]. . . .
(b) there is in effect . . . a regulation issued under this section prescribing the conditions under which such additive may be safely used.

No such regulation authorizes the use of potassium nitrate in beverages. Therefore, if potassium nitrate is a "food additive," it is presumed to be "unsafe" under section 348(a). Any food product containing an "unsafe food additive" is "adulterated" for purposes of a forfeiture proceeding. Thus, if potassium nitrate as used in Coco Rico is a "food additive," the seized beverages were adulterated and subject to forfeiture.

21 U.S.C. section 321(s) defines a "food additive" as:

"any substance the intended use of which results . . . in its becoming a component . . . of any food . . . if such substance is not generally recognized, among experts qualified by scientific training and experience to evaluate its safety, as having been adequately shown through scientific procedures (or, in the case of a substance used in food prior to January 1, 1958, through either scientific procedures or experience based on common use in food) to be safe under the conditions of its intended use; except that such term does not include—. . ."

"(4) any substance used in accordance with a sanction or approval granted prior to September 6, 1958, pursuant to this chapter . . ."

Appellant contends that summary judgment was improper because three genuine issues of fact were presented below as to why the potassium nitrate added to its concentrate is not a "food additive" within the meaning of section 321(s).

First, Coco Rico claims that Caragay's affidavit is sufficient to show the existence of a factual issue as to whether potassium nitrate is "generally recognized" by qualified experts as having been scientifically shown to be safe. To fall within this exception, the substance must be generally recognized as safe under the conditions of its intended use. 21 U.S.C. § 321(s); *United States v. Articles of Food . . . Buffalo Jerky* (general acceptance of sodium nitrate and sodium nitrite as safe for other uses did not establish that they were not "food additives" when used to cure buffalo meats). The burden of proving general recognition of safe use is placed on the proponent of the food substance in question. Caragay's affidavit contained only statements to the effect that she knows of no conclusive scientific evidence that the use of potassium nitrate in beverages is unsafe, or that the health effects of potassium nitrate when used in beverages differ from those caused by its use in meats. Even if these allegations are true, they are insufficient to meet Coco Rico's burden of proving that the use of potassium nitrate in beverages is generally recognized by experts as safe based on scientific evidence. S*ee United States v. Articles of Food and Drug, etc.* (a lack of any studies concerning effects of intended use cannot establish "general recognition" of safety).

For similar reasons, Coco Rico's second argument based on "common use" of potassium nitrate must also fail. Again, a substance may be excluded from classification as a "food additive" only if experience based on common use provides a basis for general recognition by scientists that the substance is safe under the conditions of its intended use. The evidence submitted by Coco Rico tends to show that nitrates are naturally present in many foodstuffs, particularly vegetables, and that they have been used for many centuries to cure meats. No evidence was submitted to show that potassium nitrate has long been added to beverages.[14] Consequently there is no issue of fact as to whether common experience could show that potassium nitrate is not a "food additive" when used in beverages.

The third and final exception to the "food additive" definition invoked by Coco Rico is the one applicable to substances used in accordance with a sanction issued by the FDA prior to 1958. The evidence adduced by appellant shows only that potassium nitrate continues to be sanctioned by the FDA for use in curing meat. The sanction permitting very limited use of potassium nitrate in meats cannot be construed to sanction use of the same substance for an altogether different purpose in beverages. "A prior sanction shall exist only for a specific (uses) of a substance in food, i.e., the (levels), (conditions), (products), etc. for which there was explicit approval. . . ." 21 C.F.R. § 181.5 (1984).

In sum, we conclude that the seized beverages were "held for sale . . . after shipment in interstate commerce" under 21 U.S.C. section 334(a)(1) and that the FDA properly exercised jurisdiction over them. We also hold that, as a matter of law, the potassium nitrate found in the beverages constitutes an unsafe food additive under 21 U.S.C. section 321(s), making the beverages subject to forfeiture as an adulterated food.

The district court's grant of summary judgment is affirmed.

* * * * *

7.3.7   Indirect Additives

A substance added to a food for a specific purpose in that food is a direct additive. For example, the low-calorie sweetener aspartame, which is used in beverages, puddings, yogurt, chewing gum, and other foods, is considered a direct additive. Many direct additives are identified on the ingredient label of foods.

Indirect food additives are those that become part of the food in trace amounts due to its packaging, storage, or other handling. For instance, minute amounts of packaging substances may find their way into foods during storage. Food packaging manufacturers must prove to the FDA that all materials coming in contact with food are safe before they are permitted for use in such a manner.

One issue from the 1960s that FDA wrestled with was the widespread contamination of food packaging paper with PCBs. FDA took the position that food packaging materials were food additives and thus could be regulated as such. Others disputed FDA position and argued that food

---

[14] Coco Rico contends that, as it has continuously used potassium nitrate in the concentrate it sells to Puerto Rican bottlers since 1935 without any complaint of ill effects, its own experience and use of the ingredient indicates that it is safe. Use in one manufacturer's product does not constitute "common use" in that food.

packaging materials—before the materials were actually used to package food—were outside the jurisdiction of the Food Additive Amendments and the FD&C Act.[15]

In *Natick Paperboard Corp. v. Casper Weinberger and FDA*[16] the FDA seized a quantity of paper food packaging materials that contained polychlorinated biphenyls (PCBs). The toxicity of PCBs was not challenged. However, the paper manufacturers challenged the seizure because the FDA seized the paper as being "adulterated food." The Court held that "food additive" includes any substance that may reasonably be expected to indirectly become a component of food. "Unsafe food additives," whether intentional or incidental, are "adulterated food" under FD&C Act section 342 (a)(2)(C), and therefore may be seized.

* * * * *

Natick Paperboard Corp. v. Casper Weinberger and FDA

525 F.2d 1103 (1st Cir. 1975)

Opinion: THOMPSEN, Senior District Judge

Plaintiffs appeal from a summary judgment for defendants, the Secretary of HEW and the Commissioner of Food and Drugs (collectively, FDA), which declared that they have the authority under the Federal Food, Drug and Cosmetic Act, 21 U.S.C. 301 et seq. (the Act), to recommend seizure of paper food packaging material containing polychlorinated biphenyls (PCBs) in excess of 10 parts per million (ppm) as adulterated food.[17]

PCBs are a group of toxic chemical compounds, which find their way into industrial waste, and thence into various products, including recycled paper products. If such a product is used for packaging food, PCBs are likely to migrate into the food unless the food is protected from such migration by an impermeable barrier.

Both plaintiffs manufacture paper and paper products, including paper packaging material from waste paper; they sell such material in interstate commerce, and some of it is used by their customers to make containers for packaging food. Plaintiffs argue that food packaging material is not "food" within the meaning of the Act, and therefore is not subject to seizure as "adulterated food," and that the notice of intended seizure is overbroad.

I

On July 6, 1973, FDA published a proposed regulation, intended to limit the presence of PCBs in human and animal foods by prohibiting, inter alia, PCB

---

[15] *See, e.g.,* Kunkholm, *Are Empty Containers Food?* 15 FOOD DRUG COSM. L. J. 637 (1960).
[16] 525 F.2d 1103 (1st Cir. 1975).
[17] *See also* 367 F. Supp. 885 (D. Mass. 1973) and 498 F.2d 125 (1 Cir. 1974).

residues of more than 10 ppm in paper food packaging material intended for or used with human food, finished animal feed, and any components intended for animal feeds, unless the paper food packaging material is separated from the food therein by a functional barrier which is impermeable to migration of PCBs. 21 C.F.R. § 122.10 (a)(9).[18]

Plaintiffs and others filed objections to subsection (a)(9) of the proposed regulation, and its effectiveness was thereby stayed pending a hearing,[19] which has not yet been scheduled. However, on August 24, 1973, FDA announced that in the interim any paper food packaging material shipped in interstate commerce after September 4, 1973, containing PCBs in excess of 10 ppm, would be seized as "adulterated" in violation of section 402 of the Act, 21 U.S.C. section 342, which defines "adulterated food".

Plaintiffs' complaint herein sought both injunctive and declaratory relief against such seizures. Both were originally denied by the district court because it felt that it lacked authority to grant any relief. 367 F. Supp. 885 (D. Mass. 1973). We affirmed the denial of injunctive relief, but reversed the district court's judgment with respect to declaratory relief jurisdiction and remanded the case for further proceedings.[20]

After a further hearing, the district court granted summary judgment for defendants (FDA), declaring "that they have the authority under the Federal Food, Drug and Cosmetic Act, 21 U.S.C.A. § 301 et seq., to recommend seizure of paper food-packaging material containing polychlorinated biphenyls (PCBs) in excess of 10 parts per million as adulterated food." Plaintiffs appeal from that judgment.

II

The following sections of the Act relating to adulterated food and food additives are material to the issues presented. . . .

Section 342(a)(2)(C) states that a food is adulterated "if it is, or it bears or contains, any food additive which is unsafe within the meaning of section 348."

Section 321(s) defines a "food additive" as "any substance the intended use of which results or may reasonably be expected to result, directly or indirectly, in its becoming a component or otherwise affecting the characteristics of any food (including any substance intended for use in * * * packaging * * * or holding food; * * *), if such substance is not generally recognized, among experts qualified by scientific training and experience to evaluate its safety, as having been adequately shown through scientific procedures * * * to be safe under the conditions of its intended use; * * *."

---

[18] The regulation is quoted in 498 F.2d at 126.

[19] See 21 U.S.C. § 371 (e)(2).

[20] We said, at p. 129: "Nothing in this opinion shall be deemed to bar the institution of seizures in the interim under § 334."

Section 348(a) provides: "A food additive shall, with respect to any particular use or intended use of such additives, be deemed to be unsafe for the purposes of the application of clause (2)(C) of section 342(a) of this title, unless—(1) * * *; or (2) there is in effect, and it and its use or intended use are in conformity with, a regulation issued under this section prescribing the conditions under which such additive may be safely used." No such regulation upon which plaintiffs might rely is in effect. Therefore, if paper food packaging material containing PCBs in excess of 10 ppm is a food additive, it is unsafe within the meaning of section 348.

III

The affidavits before the district court justify the conclusions that PCBs are toxic, that they tend to migrate from paper packaging material to the contained food by a vapor phase phenomenon, that paper packaging material containing PCBs in excess of 10 ppm is not generally recognized as safe for packaging food for human consumption unless the food is protected from such migration by an impermeable barrier,[21] and that if so used, without such barrier, paper food packaging containing PCBs "may reasonably be expected to result, directly or indirectly, in its becoming a component or otherwise affecting the characteristics of * * * food" within the meaning of section 321(s).[22]

Since, therefore, paper food packaging material containing PCBs in excess of 10 ppm will in many instances be an "unsafe food additive" within the meaning of the Act, we proceed to the central issue of this case: whether such material is "adulterated food" under section 342 and thus, under section 334 (a)(1) and (b), subject to seizure by FDA.

Section 342 (a)(2)(C) provides: "A food shall be deemed to be adulterated—* * if it is, or it bears or contains, any food additive which is unsafe within the meaning of section 348 of this title." Plaintiffs argue that, although PCBs may be introduced into food by migration from the packaging, such introduction is not intentional and therefore the packaging is not "used for components" of food within section 321(f)(3). FDA replies that intentional introduction is not required to meet the definition, and refers to the "food additive" definition in section 321 (s), quoted above in Part II of this opinion, and to the legislative history.

The food additive provisions of the Act were added by the Food Additives Amendment of 1958. Its basic purpose is to permit FDA to regulate the use of substances affecting food without first determining that they are in fact

---

[21] *Cf.* United States v. Articles of Food and Drug, . . . Coli-Trol 80, etc., 518 F.2d 743, 746 (5 Cir. 1975); United States v. An Article of Drug "Bentex Ulcerine", 469 F.2d 875, 878–79 (5 Cir. 1972), *cert. denied*, 412 U.S. 938 (1973).

[22] *See* United States v. Articles of Food . . . Pottery . . . Cathy Rose, 370 F. Supp. 371 (E.D. Mich. 1974).

dangerous; the method is to require that such substances be established as safe before being used. A new section 348 was added, establishing a procedure for approval by FDA and permitting the agency to establish tolerances and other regulations to insure that these substances will be used safely. Until the FDA has acted, section 348 (a) provides that substances which meet the definition of "food additive" are deemed "unsafe."[23]

The protection of the public from unsafe food additives was accomplished by amending section 342(a), defining "adulterated food." Among other provisions, a new clause (2)(C) was added to section 342(a), stating that a food shall be deemed adulterated "if it is, or it bears or contains, any food additive which is unsafe within the meaning of section 409 [codified as 21 U.S.C. § 348]" (emphasis added). No other means of prohibiting the unauthorized use of unsafe food additives was provided for in the Amendment; none was needed.[24] We conclude that "unsafe food additives", whether intentional or incidental,[25] are "adulterated food" under section 342(a)(2)(C), and, therefore, may be seized, subject to the provisions of section 334(a)(1) and (b).[26]

---

[23] This purpose of the Amendment was further elucidated in the Senate Report on the bill:"[W]e would point out first that under existing law the Federal Government is unable to prevent the use in foods of a poisonous or deleterious substance until it first proves that the additive is poisonous or deleterious. To establish this proof through experimentation with generations of mice or other animals may require 2 years or even more on the part of the relatively few scientists the Food and Drug Administration is able to assign to a particular problem. Yet, until that proof is forthcoming, an unscrupulous processor of foodstuffs is perfectly free to purvey to millions of our people foodstuffs containing additives which may or may not be capable of producing illness, debility, or death." Report of the Senate Committee on Labor and Public Welfare, S. Rep. No. 2422, 85th Cong., 2d Sess. 1, 2 (1958).

[24] Plaintiffs make the argument from syntax: "it," as used in § 342(a)(2)(C), must first be food before it can be adulterated food. However, plaintiffs do not contend that unsafe food additives intended to be introduced into food may not be seized. We see no sound reason to believe that Congress intended to subject to seizure an unsafe substance reasonably expected to become a component of food through intentional mixing but to exempt from seizure an unsafe substance (in this case packaging material containing PCBs) which is likely to affect the characteristics of food by means of migration when such unsafe substance is put to its intended use.

[25] Both the House and Senate Reports on the Food Additives Amendment contain the following (H. Rep. No. 2284, 85th Cong., 2d Sess. 3 (1958); S. Rep. No. 2422, 85th Cong., 2d Sess. 4, 5 (1958)):

"The legislation covers substances which are added intentionally to food. These additives are generally referred to as 'intentional additives.'

"The legislation also covers substances which may reasonably be expected to become a component of any food or to affect the characteristics of any food. These substances are generally referred to as 'incidental additives.'

"The principal example of both intentional and incidental additives are substances intended for use in producing, manufacturing, packing, processing, preparing, treating, packaging, transporting, or holding food."

[26] Prior court approval of a seizure by the FDA is not required, and, as we held on the previous appeal in this case, 498 F.2d at 127, no court may restrain a contemplated seizure. *Ewing v. Mytinger & Casselberry, Inc.*, 339 U.S. 594 (1950). The seizure is by process pursuant to a libel for condemnation filed in a district court against "the article, equipment or other thing proceeded against," second 334 (b). The owner or other appropriate person may contest the condemnation, and recover the articles or their value if they were seized unlawfully.

It would defeat the policy of the Act to require, as plaintiffs contend, that FDA must wait until the unsafe food additive has actually entered or come in contact with food before it can be seized; it is enough that FDA has reasonable cause to expect that the additive will be used in such a way as to enter or otherwise come in contact with food. To wait until actual contamination occurs, in the warehouse of the food processor, on the shelf of a grocery store, or in a family kitchen would effectively deny FDA the means to protect the public from adulterated food.

IV

We do not hold, however, that FDA can properly take steps to seize any and all paperboard containing PCBs in excess of 10 ppm wherever it is located and whatever its intended use may be. The district court properly limited its judgment to paper food packaging material. We interpret this to mean that the FDA must be able to prove that any paperboard intended to be seized before it has actually been used as a container for food is either in the hands of a packager of food or in transit to, ordered by, or being produced with the intention that it be sold to a packager of food, or that its intended use otherwise meets the test of section 321 (s). If the packager or other claimant can show that the food placed in or to be placed in the paper container is or will be insulated from PCB migration by a barrier impermeable to such migration, so that contamination cannot reasonably be expected to occur, the paperboard would not be a food additive and would not be subject to seizure under the Act. So interpreted, the notice of intended seizure is not overbroad.

The judgment of the district court, as interpreted in this opinion, is Affirmed.

* * * * *

DISCUSSION QUESTIONS AND NOTES

7.7. What would have happened if Natick Paperboard Corporation had shown that the paper was not intended for food packaging?

7.3.8 The FDA Modernization Act of 1997 (FDAMA)

In 1997 FDAMA provided for a notification process, rather than a petition, as a lawful way of establishing the safety of new packaging substances that would otherwise require petitioning as a new food additive. This notification process is contingent on appropriation of funding in any given year.

Packaging used to hold foods need not comply with 21 C.F.R. § 179.45 for new additive approval if the substance's safety is established by the manufacturer by testing. Manufacturers must notify FDA that they intend to use the new substance 120 days before use, and if FDA does not veto the notification, then the packaging may be used.[27]

7.3.9 MSG

Monosodium glutamate (MSG) was classified by the FDA as GRAS in 1959. Nevertheless, MSG remained a controversial food additive. A number of reports have found MSG to be safe when consumed at levels typically used in cooking and food manufacturing. For example, the Federation of American Societies for Experimental Biology (FASEB) completed a comprehensive review of available scientific data on glutamate and, in 1995, reaffirmed the safety of MSG when it is consumed at usual levels by the general population. However, short-term reactions—MSG symptom complex—can occur. The MSG symptom complex may include numbness, burning sensation, tingling, facial pressure, tightness, chest pain, headache, nausea, rapid heartbeat, drowsiness, and weakness.

Special Labeling Requirement When MSG is an added ingredient to food, "monosodium glutamate" must be listed in the ingredient statement. The ingredient cannot be generically listed under "spices" or "flavorings." Other salts of glutamic acid, such as monopotassium glutamate and monoammonium glutamate, similarly have to be specifically declared on the labels.

* * * * *

FDA and Monosodium Glutamate (MSG)

U.S. Food and Drug Administration, FDA Backgrounder (August 31, 1995)[28] Monosodium Glutamate (MSG)

Monosodium glutamate (MSG) is used as a flavor enhancer in a variety of foods prepared at home, in restaurants, and by food processors. Its use has become controversial in the past 30 years because of reports of adverse reactions in people who've eaten foods that contain MSG. Research on the role of glutamate—a group of chemicals that includes MSG—in the nervous system also has raised questions about the chemical's safety.

Studies have shown that the body uses glutamate, an amino acid, as a nerve impulse transmitter in the brain and that there are glutamate-responsive tissues in other parts of the body, as well. Abnormal function of glutamate receptors has been linked with certain neurological diseases, such as Alzheimer's disease and Huntington's chorea. Injections of glutamate in laboratory animals have resulted in damage to nerve cells in the brain. Consumption of glutamate in food, however, does not cause this effect. While people normally

---

[27] *See* 21 U.S.C. § 348.
[28] *Available at:* http://www.cfsan.fda.gov/~lrd/msg.html (last accessed Oct. 2, 2007).

consume dietary glutamate in large amounts and the body can make and metabolize glutamate efficiently, the results of animal studies conducted in the 1980s raised a significant question: Can MSG and possibly some other glutamates harm the nervous system?

A 1995 report from the Federation of American Societies for Experimental Biology (FASEB), an independent body of scientists, helps put these safety concerns into perspective and reaffirms the Food and Drug Administration's belief that MSG and related substances are safe food ingredients for most people when eaten at customary levels.

The FASEB report identifies two groups of people who may develop a condition the report refers to as "MSG symptom complex." One group is those who may be intolerant to MSG when eaten in a large quantity. The second is a group of people with severe, poorly controlled asthma. These people, in addition to being prone to MSG symptom complex, may suffer temporary worsening of asthmatic symptoms after consuming MSG. The MSG dosage that produced reactions in these people ranged from 0.5 grams to 2.5 grams.

Although FDA has not fully analyzed the FASEB report, the agency believes that the report provides the basis to require glutamate labeling. FDA will propose that foods containing significant amounts of free glutamate (not bound in protein along with other amino acids) declare glutamate on the label. This would allow consumers to distinguish between foods with insignificant free glutamate levels and those that might contribute to a reaction.

What Is MSG?

MSG is the sodium salt of the amino acid glutamic acid and a form of glutamate. It is sold as a fine white crystal substance, similar in appearance to salt or sugar. It does not have a distinct taste of its own, and how it adds flavor to other foods is not fully understood. Many scientists believe that MSG stimulates glutamate receptors in the tongue to augment meat-like flavors.

Asians originally used a seaweed broth to obtain the flavor-enhancing effects of MSG, but today MSG is made by a fermenting process using starch, sugar beets, sugar cane, or molasses.

Glutamate itself is in many living things: It is found naturally in our bodies and in protein-containing foods, such as cheese, milk, meat, peas, and mushrooms.

Some glutamate is in foods in a "free" form. It is only in this free form that glutamate can enhance a food's flavor. Part of the flavor-enhancing effect of tomatoes, certain cheeses, and fermented or hydrolyzed protein products (such as soy sauce) is due to the presence of free glutamate.

Hydrolyzed proteins, or protein hydrolysates, are acid- treated or enzymatically treated proteins from certain foods. They contain salts of free amino acids, such as glutamate, at levels of 5 to 20 percent. Hydrolyzed proteins are used in the same manner as MSG in many foods, such as canned vegetables, soups, and processed meats.

Scientific Review

In 1959, FDA classified MSG as a "generally recognized as safe," or GRAS, substance, along with many other common food ingredients, such as salt, vinegar, and baking powder. This action stemmed from the 1958 Food Additives Amendment to the Federal Food, Drug, and Cosmetic Act, which required premarket approval for new food additives and led FDA to promulgate regulations listing substances, such as MSG, which have a history of safe use or are otherwise GRAS.

Since 1970, FDA has sponsored extensive reviews on the safety of MSG, other glutamates and hydrolyzed proteins, as part of an ongoing review of safety data on GRAS substances used in processed foods.

One such review was by the FASEB Select Committee on GRAS Substances. In 1980, the committee concluded that MSG was safe at current levels of use but recommended additional evaluation to determine MSG's safety at significantly higher levels of consumption. Additional reports attempted to look at this.

In 1986, FDA's Advisory Committee on Hypersensitivity to Food Constituents concluded that MSG poses no threat to the general public but that reactions of brief duration might occur in some people.

Other reports gave similar findings. A 1991 report by the European Communities' (EC) Scientific Committee for Foods reaffirmed MSG's safety and classified its "acceptable daily intake" as "not specified," the most favorable designation for a food ingredient. In addition, the EC Committee said, "Infants, including prematures, have been shown to metabolize glutamate as efficiently as adults and therefore do not display any special susceptibility to elevated oral intakes of glutamate."

A 1992 report from the Council on Scientific Affairs of the American Medical Association stated that glutamate in any form has not been shown to be a "significant health hazard."

Also the 1987 Joint Expert Committee on Food Additives of the United Nations Food and Agriculture Organization and the World Health Organization have placed MSG in the safest category of food ingredients.Scientific knowledge about how the body metabolizes glutamate developed rapidly during the 1980s.

Studies showed that glutamate in the body plays an important role in normal functioning of the nervous system. Questions then arose on the role glutamate in food plays in these functions and whether or not glutamate in food contributes to certain neurological diseases.

Anecdotal Evidence

Many of these safety assessments were prompted by unconfirmed reports of MSG-related adverse reactions. Between 1980 and 1994, the Adverse Reaction Monitoring System in FDA's Center for Food Safety and Applied Nutrition received 622 reports of complaints about MSG. Headache was the most frequently reported symptom. No severe reactions were documented, but some reports indicated that people with asthma got worse after they consumed MSG. In some of those cases, the asthma didn't get worse until many hours later.

Also several books and a TV news show have reported widespread and sometimes life-threatening adverse reactions to MSG, claiming that even small amounts of manufactured glutamates may cause adverse reactions.

A problem with these unconfirmed reports is that it is difficult to link the reactions specifically to MSG. Most are cases in which people have had reactions after, but not necessarily because of, eating certain foods containing MSG.

While such reports are helpful in raising issues of concern, they do not provide the kind of information necessary to describe who is most likely to be affected, under what conditions they'll be affected, and with what amounts of MSG. They are not controlled studies done in a scientifically credible manner.

1995 FASEB Report

Prompted by continuing public interest and a flurry of glutamate-related studies in the late 1980s, FDA contracted with FASEB in 1992 to review the available scientific data. The agency asked FASEB to address 18 questions dealing with:

- the possible role of MSG in eliciting MSG symptom complex,
- the possible role of dietary glutamates in forming brain lesions and damaging nerve cells in humans,
- underlying conditions that may predispose a person to adverse effects from MSG,
- the amount consumed and other factors that may affect a person's response to MSG, and
- the quality of scientific data and previous safety reviews.

FASEB held a two-day meeting and convened an expert panel that thoroughly reviewed all the available scientific literature on this issue.

FASEB completed the final report, over 350 pages long, and delivered it to FDA on July 31, 1995. While not a new study, the report offers a new safety assessment based on the most comprehensive existing evaluation to date of glutamate safety.

Among the report's key findings:

An unknown percentage of the population may react to MSG and develop MSG symptom complex, a condition characterized by one or more of the following symptoms:

burning sensation in the back of the neck, forearms, and chest
numbness in the back of the neck, radiating to the arms and back

tingling, warmth, and weakness in the face, temples, upper back, neck, and arms

facial pressure or tightness

chest pain

headache

nausea

rapid heartbeat

bronchospasm (difficulty breathing) in MSG-intolerant people with asthma

drowsiness

weakness

In otherwise healthy MSG-intolerant people, the MSG symptom complex tends to occur within one hour after eating 3 grams or more of MSG on an empty stomach or without other food. A typical serving of glutamate-treated food contains less than 0.5 grams of MSG. A reaction is most likely if the MSG is eaten in a large quantity or in a liquid, such as a clear soup.

Severe, poorly controlled asthma may be a predisposing medical condition for MSG symptom complex.

No evidence exists to suggest that dietary MSG or glutamate contributes to Alzheimer's disease, Huntington's chorea, amyotrophic lateral sclerosis, AIDS dementia complex, or any other long-term or chronic diseases.

No evidence exists to suggest that dietary MSG causes brain lesions or damages nerve cells in humans.

The level of vitamin B6 in a person's body plays a role in glutamate metabolism, and the possible impact of marginal B6 intake should be considered in future research.

There is no scientific evidence that the levels of glutamate in hydrolyzed proteins causes adverse effects or that other manufactured glutamate has effects different from glutamate normally found in foods.

Ingredient Listing

Under current FDA regulations, when MSG is added to a food, it must be identified as "monosodium glutamate" in the label's ingredient list. Each ingredient used to make a food must be declared by its name in this list.

While technically MSG is only one of several forms of free glutamate used in foods, consumers frequently use the term MSG to mean all free glutamate. For this reason FDA considers foods whose labels say "No MSG" or "No Added MSG" to be misleading if the food contains ingredients that are sources of free glutamates, such as hydrolyzed protein.

In 1993, FDA proposed adding the phrase "(contains glutamate)" to the common or usual names of certain protein hydrolysates that contain substantial amounts of glutamate. For example, if the proposal were adopted, hydrolyzed soy protein would have to be declared on food labels as "hydrolyzed soy protein (contains glutamate)." However, if FDA issues a new proposal, it would probably supersede this 1993 one.

In 1994, FDA received a citizen's petition requesting changes in labeling requirements for foods that contain MSG or related substances. The petition asks for mandatory listing of MSG as an ingredient on labels of manufactured and processed foods that contain manufactured free glutamic acid. It further asks that the amount of free glutamic acid or MSG in such products be stated on the label, along with a warning that MSG may be harmful to certain groups of people. FDA has not yet taken action on the petition.

* * * * *

NOTE

7.8. For more information on MSG, see *MSG: A Common Flavor Enhancer* by Michelle Meadows, FDA, FDA CONSUMER (Jan.–Feb. 2003) *available at:* http://www.fda.gov/fdac/features/2003/103_msg.html (last accessed Oct. 2, 2007).

7.3.10 Preservatives

Although food preservatives are ubiquitous—it is almost impossible to eat food without consuming some preservatives—many consumers eye preservatives with skepticism. This attitude is reflected in the requirement that preservatives must include a description of their function, as well as be listed by their common or usual names, in the ingredient statements of all foods that contain them.

Other regulations of preservatives prohibit their use to deceive consumers by changing a food to make it appear other than it is. For example, preservatives that contain sulfites are prohibited on meats because they restore the red color, giving meat a false appearance of freshness. (The USDA regulates meats, but the FDA regulates food additives, including the prohibition of sulfites in meats.) Food preservatives, like other ingredient, must also be of food grade and be prepared and handled as a food ingredient. In addition the quantity used in a food must not exceed the amount needed to achieve the intended effect.

Sulfites Sulfites come under special restrictions. The FDA prohibits the use of sulfites in foods that are important sources of thiamin (vitamin B1), such as enriched flour, because sulfites destroy the nutrient. In addition some people have severe—possibly life-threatening—allergic-type[29] reactions to sulfites.

---

[29] Sulfite sensitivity can result in an allergic-type reaction, which is not fully understood.

Difficulty breathing is the most common symptom, but other problems range from hives to anaphylactic shock. The FDA estimated that more than one million asthmatics are sensitive or allergic to the substance. In 1986, FDA required that sulfites used specifically as preservatives must be listed on the label, regardless of the amount in the finished product. Sulfites used in food processing but not serving as preservatives in the final food must be listed on the label if present at levels of 10 parts per million or higher.

In addition, in 1986, FDA banned the use of sulfites on fruits and vegetables intended to be eaten raw, such as in salad bars and grocery store produce sections. Grocers and restaurateurs were using them to maintain the color and crispness of fresh produce. (Even before the FDA ban, industry trade groups had persuaded many of their members to stop using sulfites on fresh produce.)

There are six sulfiting agents allowed in packaged foods. They are listed on food labels are by the following names:

· Sulfur dioxide
· Sodium sulfite
· Sodium and potassium bisulfite
· Sodium and potassium metabisulfite

* * * * *

A Fresh Look at Food Preservatives

Judith E. Foulke, FDA CONSUMER (October 1993)[30]

Unless you grow all your food in your own garden and prepare all your meals from scratch, it's almost impossible to eat food without preservatives added by manufacturers during processing. Without such preservatives, food safety problems would get out of hand, to say nothing of the grocery bills. Bread would get moldy, and salad oil would go rancid before it's used up.

Food law says preservatives must be listed by their common or usual names on ingredient labels of all foods that contain them—which is most processed food. You'll see calcium propionate on most bread labels, disodium EDTA on canned kidney beans, and BHA on shortening, just to name a few. Even snack foods—dried fruit, potato chips, and trail mix—contain sulfur-based preservatives.

Manufacturers add preservatives mostly to prevent spoilage during the time it takes to transport foods over long distances to stores and then our kitchens. It's not unusual for sourdough bread manufactured in California to be eaten in Maine, or for olive oil manufactured in Spain to be used on a California salad. Rapid transport systems and ideal storage conditions help keep foods

[30] *Available at*: http://www.cfsan.fda.gov/~dms/fdpreser.html (last accessed Oct. 2, 2007).

fresh and nutritionally stable. But breads, cooking oils, and other foods, including the complex, high-quality convenience products consumers and food services have come to expect, usually need more help.

Preservatives serve as either antimicrobials or antioxidants—or both. As antimicrobials, they prevent the growth of molds, yeasts, and bacteria. As antioxidants, they keep foods from becoming rancid, browning, or developing black spots. Rancid foods may not make you sick, but they smell and taste bad. Antioxidants suppress the reaction that occurs when foods combine with oxygen in the presence of light, heat, and some metals. Antioxidants also minimize the damage to some essential amino acids—the building blocks of proteins—and the loss of some vitamins.

Safety Questions

Consumers often ask the Food and Drug Administration about the safety of preservatives, and if there's a system in place to make sure preservatives are safe.

Many preservatives are regulated under the food additives amendment, added to the Federal Food, Drug, and Cosmetic Act in 1958. The amendment strengthened the law to ensure the safety of all new ingredients that manufacturers add to foods. Under these rules a food manufacturer must get FDA approval before using a new preservative, or before using a previously approved preservative in a new way or in a different amount. In its petition for approval the manufacturer must demonstrate to FDA that the preservative is safe for consumers, considering:

- the probable amount of the preservative that will be consumed with the food product, or the amount of any substance formed in or on the food resulting from use of the preservative;
- the cumulative effect of the preservative in the diet; and
- the potential toxicity (including cancer-causing) of the preservative when ingested by humans or animals.

Also a preservative may not be used to deceive a consumer by changing the food to make it appear other than it is. For example, preservatives that contain sulfites are prohibited on meats because they restore the red color, giving meat a false appearance of freshness. (The U.S. Department of Agriculture regulates meats, but depends on the FDA regulation to prohibit sulfites in meats.)

The food additive regulations require the preservative to be of food grade and be prepared and handled as a food ingredient. Also the quantity added to food must not exceed the amount needed to achieve the manufacturer's intended effect.

Regulations about the use of nitrites demonstrate the scrutiny given to the use of additives. Nitrites, used in combination with salt, serve as antimicrobials in meat to inhibit the growth of bacterial spores that cause botulism, a deadly foodborne illness. Nitrites are also used as preservatives and for flavoring and fixing color in a number of red meat, poultry, and fish products.

Since the original approvals were granted for specific uses of sodium nitrite, safety concerns have arisen. Nitrite salts can react with certain amines (derivatives of ammonia) in food to produce nitrosamines, many of which are known to cause cancer. A food manufacturer wanting to use sodium nitrites must show that nitrosamines will not form in hazardous amounts in the product under the additive's intended conditions of use. For example, regulations specify that sodium nitrite, used as an antimicrobial against the formation of botulinum toxin in smoked fish, must be present in 100 to 200 parts per million. In addition other antioxidants, such as sodium ascorbate or sodium erythorbate, may be added to inhibit the formation of nitrosamines.

As scientists learn more about the action of certain chemicals in our bodies, FDA uses the new data to reevaluate the permitted uses of preservatives. Two examples are the commonly used preservatives butylated hydroxyanisole (BHA) and sulfites.

BHA

BHA and the related compound butylated hydroxytoluene (BHT) have been used for years, mostly in foods that are high in fats and oils. They slow the development of off-flavors, odors, and color changes caused by oxidation. When the food additives amendment was enacted, BHA and BHT were listed as common preservatives considered generally recognized as safe (GRAS). GRAS regulations limit BHA and BHT to 0.02 percent or 200 parts per million (ppm) of the fat or oil content of the food product.

Lawrence Lin, Ph.D., of FDA's Center for Food Safety and Applied Nutrition, explains, "The 0.02 percent allowed relates only to the product's fat content. For example, if a product weighs 100 grams and one of those grams is fat, the quantity of BHA in the product cannot exceed 0.02 percent of that one gram of fat."

BHA is also used as a preservative for dry foods, such as cereals. But because such foods contain so little fat, the amount of BHA allowed cannot be measured against the percentage of fat, explains Lin. Therefore, as manufacturers petitioned FDA for approvals for this use, the agency set limits for each type of food. On cereals, for example, FDA limited BHA to 50 ppm of the total product.

In 1978, under contract with FDA, the Life Sciences Research Office of the Federation of American Societies for Experimental Biology (FASEB) examined the health aspects of BHA as part of FDA's comprehensive review of GRAS safety assessments. FASEB concluded that although BHA was safe at permitted levels, additional studies were needed.

Since that evaluation, other studies suggested that at very high levels in the diets of laboratory animals, BHA could cause tumors in the forestomach of rats, mice, and hamsters, and liver tumors in fish. Many experts examined the data and concluded the tests did not establish that such problems could exist in humans, mostly because humans do not have forestomachs. Other studies showed that BHA was protective, inhibiting the effect of some chemical carcinogens, depending on the conditions of the tests.

Studies on BHA were reviewed by scientists from the United Kingdom, Canada, Japan, and the United States. Their findings were published in 1983 in the Report of the Working Group on the Toxicology and Metabolism of Antioxidants and reviewed in the 1990 Annual Review of Pharmacology and Toxicology. The 1983 report stated that data from a Japanese study showed a high incidence of cancerous tumors and papillomas (benign tumors of the skin or mucous membranes) of the forestomach of treated rats and that the effect was dose-related. The report also mentioned the possible existence of a no-effect level, based on dose response, and noted that the level which produced cancer in this study was many thousands of times higher than the level to which humans are exposed.

In November 1990, Glenn Scott, M.D., a physician then living in New York who has since moved to Cincinnati, filed a petition with FDA, asking the agency to prohibit the use of BHA in food. Scott cited animal studies to support his request. Before acting on Scott's petition, however, FDA asked FASEB to reexamine the scientific data on BHA. By March 1994, FASEB is scheduled to provide FDA with a report on the most current scientific information bearing on the relationship of BHA ingestion to cancer in animals.

Sulfites

Sulfites are used primarily as antioxidants to prevent or reduce discoloration of light-colored fruits and vegetables, such as dried apples and dehydrated potatoes. They are also used in wine-making because they inhibit bacterial growth but do not interfere with the desired development of yeast.

Sulfites are also used in other ways, such as for bleaching food starches and as preventives against rust and scale in boiler water used in making steam that will come in contact with food. Some sulfites are used in the production of cellophane for food packaging.

FDA prohibits the use of sulfites in foods that are important sources of thiamin (vitamin B1), such as enriched flour, because sulfites destroy the nutrient.

Though most people don't have a problem with sulfites, some do. FDA's sulfite specialist, consumer safety officer Joann Ziyad, Ph.D., points to a bookcase full of binders and says, "Those are the case histories of adverse reactions to sulfites that have been reported to FDA. Since 1985, when the agency started reporting on sulfites through the Adverse Reaction Monitoring System, over 1,000 adverse reactions have been recorded."

As reports of adverse reactions mounted, FDA asked FASEB to reexamine the use of sulfites. FASEB's report, released in 1985, concluded that sulfites posed no hazard to most Americans, but that they were a hazard of unpredictable severity to people who were sensitive to the substance. Based on the FASEB study, FDA estimated that more than 1 million asthmatics are sensitive or allergic to the substance.

In 1986, FDA ruled that sulfites used specifically as preservatives must be listed on the label, regardless of the amount in the finished product. Sulfites used in food processing but not serving as preservatives in the final food must be listed on the label if present at levels of 10 parts per million or higher. Regulations issued in 1990 extended these required listings to standardized foods.

Also in 1986, FDA banned the use of sulfites on fruits and vegetables intended to be eaten raw, such as in salad bars and grocery store produce sections. Grocers and restaurateurs were using them to maintain the color and crispness of fresh produce. (Even before the FDA ban, industry trade groups had persuaded many of their members to stop using sulfites on fresh produce.)

FDA plans to repropose a ban for sulfites on fresh, peeled potatoes served or sold unpackaged and unlabeled, such as for french fries in restaurants. An earlier FDA rule dealing with sulfites on potatoes was invalidated by the court in 1990 on procedural grounds.

In addition sulfite-sensitive consumers are learning how to avoid sulfites. Consumer awareness combined with FDA actions have slowed the number of adverse reaction reports. Ziyad says that from 1990 to 1992, fewer than 40 were reported, and at press time, there had been only three reports in 1993.

Ziyad says the only way FDA can know about sulfite-sensitivity problems is through consumer and physician reports. Adverse reaction reporting is totally voluntary, and FDA encourages physicians to report patients' reactions to sulfites. But there are times when such reactions are not medically treated because the individual doesn't go to the doctor with the condition or the symptoms are not recognized. Such information would help FDA evaluate the current status of problems with foods among sulfite-sensitive individuals.

The agency's Adverse Reaction Monitoring System collects and acts on complaints concerning all food ingredients, including preservatives. If you experience an adverse reaction from eating a food that contains sulfites, describe the circumstances and your reaction to the FDA district office in your area (see local phone directory) and send your report in writing to:

Adverse Reaction Monitoring System (HFS-636)
200 C St., SW
Washington, DC 20204

Puzzling It Out

Preservatives are a puzzle for many consumers that can sometimes raise safety concerns. Even though these concerns are usually unfounded, some industry publications are reporting attempts to find naturally occurring substitutes for synthetic antioxidants. In a 1990 article, one such publication, INFORM, says alternatives to synthetics are commercially available in the United States, although most are generally more costly or have other drawbacks. For example, tocopherol (vitamin E) generally is not as effective in vegetable fats and oils as it is in animal fats. Also some herbs and herb extracts, such as rosemary and sage, can do the work of antioxidants, but they impart strong color or flavors. And just because these are plant-derived doesn't necessarily mean they are always safe. Inform points to the FDA rule that newly identified natural antioxidants, like other new food additives, must undergo rigorous toxicological tests before they can be approved.

As an additional alternative to synthetic antioxidants, the edible oil industry is increasingly using ultraviolet-barrier packaging and filling under nitrogen to protect the product's stability.

FDA scientists will continue to carefully evaluate all research presented to the agency on new preservatives to ensure that substances added to food to preserve quality and safety are themselves safe.

* * * * *

NOTES

7.9. **Additional information** on sulfites may be obtained from *Sulfites: Safe for Most, Dangerous for Some*, FDA CONSUMER (Dec. 1996), *available at*: http://www.cfsan.fda.gov/~dms/fdsulfit.html (last accessed Oct. 2, 2007).

7.10. **Additional information** about food additives and premarket approval may be obtained at http://www.cfsan.fda.gov/~lrd/foodadd.html (last accessed Oct. 2, 2007).

7.3.11 Botanicals and Other Novel Ingredients

It should be noted that the Dietary Supplement Health and Education Act (DSHEA) of 1994 places dietary supplements under a different regulatory scheme than other foods. The regulation of dietary supplements will be covered in Chapter 8. However, dietary supplements are raised here because botanicals and other novel ingredients allowed in dietary supplements are considered food additives (if not GRAS) that require premarket approval before they are added to a conventional food.

With the growth of the dietary supplement industry there has been significant growth in the marketing of foods containing these novel ingredients, such as botanicals. With the growth of the dietary supplement industry there has also been a blurring in the marketing distinction among foods, supplements, and drugs. It is important to bear in mind that "dietary supplement" has a definition under the law, and—regardless of how a novel food is described—it will be regulated based on how it falls under the legal definitions.

Some of the names for these novel foods are *designer foods, functional foods, nutraceuticals, pharmafood, and techno-foods*. These terms have no definition or status under the law. "Functional food" often is the term applied to conventional food products that are enhanced to provide specific health benefits (as opposed to pills, liquids, and powders). For example, a calcium-enriched orange juice and higher calcium yogurt marketed to help build strong bones would be functional foods. These products are regulated as conventional foods.

The FDA recently issued a blanket letter to the food industry restating the requirements of the Food, Drug, and Cosmetic Act regarding the marketing of conventional foods that containing novel ingredients.[31] The FDA was concerned that some botanical and other novel ingredients that are being added to conventional foods are neither approved food additives nor GRAS.

* * * * *

FDA Issues Advisory on Star Anise "Teas"

U.S. FDA Press Release, *FDA Issues Advisory on Star Anise "Teas"*[32]

The Food and Drug Administration (FDA) today is advising consumers not to consume "teas" brewed from star anise. It has come to FDA's attention that brewed "teas" containing star anise have been associated with illnesses affecting about 40 individuals, including approximately 15 infants. The illnesses, which occurred over the last two years, ranged from serious neurological effects, such as seizures, to vomiting, jitteriness, and rapid eye movement.

Although the labeling of "teas" that contain star anise does not make claims for the product, FDA understands that these products are popularly believed to help against colic in infants. FDA is unaware of scientific evidence to support benefits from "teas" brewed from star anise. Given that fact, consumers should not use them or give them to infants and children. . . .

"One of FDA's highest priorities is to make sure that consumers have accurate information about the products FDA regulates," said FDA Commissioner Mark B. McClellan. "This case illustrates that FDA will take action to protect consumers against products that may pose health risks."

FDA is concerned that commonly available Chinese star anise (*Illicium verum*), a product considered by FDA to be generally recognized as safe (GRAS), may contain Japanese star anise (*Illicium anisatum*), which has long been recognized as toxic in many countries and which should be used for

---

[31] FDA, *Letter to Manufacturers Regarding Botanicals and Other Novel Ingredients in Conventional Foods*" (Jan. 30, 2001) available at http://www.cfsan.fda.gov/~dms/ds-ltr15.html.
[32] FDA Press Release, FDA Issues Advisory on Star Anise "Teas," http://www.fda.gov:80/bbs/topics/NEWS/2003/NEW00941.html (Sept. 11, 2003).

decorative purposes only. At this time FDA cannot determine if the star anise associated with the illnesses was associated with Japanese star anise or a mixture of Chinese and Japanese star anise. . . .

FDA considers Chinese star anise to be GRAS when used as a spice or flavoring; Japanese star anise is not GRAS. GRAS status means that a food substance is considered by qualified experts to be safe for is intended use. Safety must be adequately shown through scientific procedures and/or experience based on a common history of use in food, depending on the substance.

The initial reported illnesses were identified retrospectively through a record review after a resident physician from Miami Children's Hospital treating an infant with seizures associated with the ingestion of a star anise-containing tea reported his findings to the Florida Poison Information Center (FPIC). FPIC then reported the findings to the FDA. . . .

* * * * *

Improvements Needed in Overseeing the Safety of Dietary Supplements and "Functional Foods"

U.S. General Accounting Office, FOOD SAFETY: IMPROVEMENTS NEEDED IN OVERSEEING THE SAFETY OF DIETARY SUPPLEMENTS AND "FUNCTIONAL FOODS" (GAO, RCED-00-156, JULY 11, 2000).

New, so-called functional foods are entering the market that provide the basic attributes of traditional foods—taste, aroma, or nutritive value—and that claim to provide an additional health benefit. For example, recently marketed butter-like spreads include an added ingredient designed to reduce cholesterol levels in the bloodstream. In contrast, dietary supplements generally are available in pill, capsule, tablet, or liquid form; are not used primarily for their taste or aroma; and cannot be represented as a conventional food. Supplements include vitamins, minerals, herbs, amino acids, and other dietary substances that are used to enhance the normal dietary intake of nutrients or for more specialized purposes, such as relaxation or stimulation. On their labels, both functional foods and dietary supplements can make health claims and/or so-called structure/function claims. . . .

FDA's efforts and federal laws provide limited assurances of the safety of functional foods and dietary supplements. While the extent to which unsafe products reach consumers is unknown, we believe weaknesses in three areas of the regulatory system increase the likelihood of such occurrences. First, potentially unsafe products may reach consumers for a variety of reasons, including the lack of a clearly defined safety standard for new dietary ingredients in dietary supplements. Second, some products do not have safety-related information on their labels, which could endanger some consumers. This occurs because FDA has not issued regulations or guidance on the information required. For example, according to the National Institutes of Health, St. John's Wort may decrease the efficacy of a drug used to treat HIV infection, but consumers may not be able to determine this from the dietary supplement label. Finally, FDA cannot effectively assess whether a functional food or dietary supplement is adversely affecting consumers' health because, among other things, it does not investigate most reports it receives of health problems potentially caused by these products. FDA officials recognize these weaknesses but say a lack of resources has precluded them from taking actions to correct them.

We also found that agencies' efforts and federal laws concerning health-related claims on product labels and in advertising provide limited assistance to consumers in making informed choices and do little to protect them against inaccurate and misleading claims. FDA has not clearly established the nature and extent of evidence companies need to adequately support structure/function claims and has taken no actions against companies making claims that the agency believes to be questionable. According to an FDA official, the agency has chosen to use its limited resources on regulating product safety rather than on taking enforcement actions against problematic label claims. Furthermore federal agencies operate under different statutes for regulating claims on product labels and in advertising, which has led to claims being made in products' advertisements that were not allowed on product labels. For example, a product that FDA does not allow to claim to lower cholesterol on its label is permitted by FTC to make this claim in its advertising, provided the claim is truthful, not misleading, and supported by reliable scientific evidence. Finally, consumers may not understand the different purposes of health claims and structure-function claims. As a result they may incorrectly view structure/function claims as claims to reduce the risk of or treat a disease. . . .

* * * * *

DISCUSSION QUESTIONS AND NOTES

7.11. Fortification—too much of a good thing? Regulation of food additives includes the control of the safe use or safety of the substance under the intended conditions of use. Recently concerns have been raised that fortified foods may be result in excess consumption and resulting problems. The Institute of Medicine daily reference intakes (DRIs) include "tolerable upper intake" for most vitamins and minerals, but these are only recommendations. Exceeding recommended limits may have adverse health effects. Chronic excess vitamin A consumption may increase the risk of hip fractures in the elderly. See K.L. Penniston & S.A. Tanumihardjo, *Vitamin A in dietary supplements and fortified foods: Too much of a good thing?* 103(9) J. AM. DIET. ASSOC. 1185–1187. (Sept. 2003). Higher iron intake by males may be associated with an increased risk for cancer and heart disease. Individuals eating fortified foods can easily consume as much as 400 percent of recommended levels of iron. See *Fortified Cereals—Too Much of a Good Thing?* REUTERS (July 11, 2001) (citing JOURNAL OF THE AMERICAN COLLEGE OF NUTRITION 2001; 20:247–254), *available at*: http://chem.chem.rochester.edu/~chemlab/health-cereal.html (Mar. 15, 2008).

7.4 COLOR ADDITIVES

A color additive is any dye, pigment, or substance that can impart color when added or applied to a food. Colors permitted for use in foods are classified as (1) certified or (2) exempt from certification.

Certified colors are synthetic, with each batch being tested by the manufacturer and FDA to ensure that they meet strict specifications for purity. There are nine certified colors approved for use in the United States. For example, FD&C Yellow No. 6 is used in cereals, bakery goods, snack foods, and other foods.

Color additives that are **exempt from certification** include pigments derived from natural sources such as vegetables, minerals, or animals. For example, caramel color is produced commercially by heating sugar under controlled conditions. Even though exempt from certification, these colors must also meet legal requirements for specifications and purity.

In vernacular use, we might include colorings as food additives. However, the legal definition of food additives excludes color additives. Color additives fall under similar requirements to food additives, but with a few differences.

7.4.1 Color Additive Amendments

In 1960 Congress enacted legislation governing color additives. The Color Additive Amendments to the FD&C Act require dyes used in foods, drugs, cosmetics, and certain medical devices to be approved by FDA.

Both the Food Additives and Color Additives Amendments include a provision that prohibits the approval of an additive if it is found to cause cancer in humans or animals. This clause is often referred to as the Delaney Clause, named for its congressional sponsor, Representative James Delaney (D–N.Y.)

Before a color additive may be approved, FDA must find that the additive is suitable and safe for the intended use. In addition FDA must find that the proposed use would not "promote deception of the consumer in violation of this chapter or would otherwise result in misbranding or adulteration within the meaning of this chapter."[33]

---

[33] FD&C Act § 721(b)(6) [21 U.S.C. § 379e(b)(6)].

7.4.2 Provisional Listing

However, in contrast to food additives, no GRAS or prior-sanctioned exceptions exist for colorings. Congress did provide a "provisional" status for colorings that were in use before the amendments.[34] FDA was permitted to "provisionally" list these color additives to allow time for testing to determine safety under the new standard.

FDA provisionally listed 200 color additives. Frustrated with FDA delays in acting on the provisional list, the Health Research Group prodded the agency to remove the provisionally listed colors from the market unless there were adequate safety data. In 1980 Health Research Group sued FDA, arguing that the provisional list of the 1960 Color Additive Amendments was intended to last no longer than two and one-half years. The court upheld FDA's continued use of the provisional list in *McIlwain v. Hayes*, 690 F.2d 1041 (1982), Judge Bork:

> Undoubtedly, in 1960 many members of Congress anticipated that color additive testing would be completed more rapidly than has been the case with respect to some additives. Just as certainly, however, Congress foresaw that unavoidable delays were possible and provided a statutory mechanism for the Commissioner to cope with such problems. Most significantly, for the issues in this case, Congress provided no limit upon the number of times postponements could be made. The primary reason for the repeated postponements here is that testing technology has evolved and improved so rapidly that, by the time a color additive has been shown to be safe under one series of tests or is still undergoing testing, more sophisticated testing procedures have been devised and the Commissioner orders that the time-consuming testing process begin anew.
>
> It is important to realize what "proving" safety necessarily means. A color additive is subjected to the best tests available, and safety is assumed to be shown if no evidence of harm to health is found. We are informed by the Commissioner that no test data supplied so far indicate any problem with the safety of the twenty-three color additives involved here, and it is only the fact that new, more rigorous tests have since become available that makes it possible to say that these products have not yet been conclusively demonstrated to be safe. Under these circumstances, we think it both reasonable and within the express powers conferred upon him by Congress for the Commissioner to postpone the closing dates for the color additives in question. We therefore affirm the judgment of the district court.

Judge Mikva dissented:

> By 1960, Congress clearly was dissatisfied with the manner in which the Food and Drug Administration (FDA) was carrying out its mandate to

---

[34] Color Additive Amendments § 203, 74 Stat. 397, 404 (1960)(the provisional list was not made part of the FD&C Act).

regulate the use of color additives in foods, drugs, and cosmetics. The assignment of the burden of proof, requiring the FDA to show that an additive was unsafe before it could be removed from the market, had caused the law to become largely a dead letter. And so, in the Color Additive Amendments of 1960 (Amendments), Congress shifted the burden of testing color additives from the FDA to industry, forcing industry to prove each additive safe before the additive could be permanently listed and marketed. To manage the problem arising from those additives already on the market, Congress established a provisional listing program that was to end after the additive was proven safe or unsafe, or after a two and one-half year period, whichever came first. Only under exceptional circumstances was it anticipated that any additives already on the market would take more than two and one-half years to be proven safe or removed from the market. Indeed, the major factor that motivated Congress to enact the Amendments was the possibility that if the burden were not shifted, the FDA might take "as much as twenty years" to determine the safety of the existing additives.

Some twenty-two years later, the majority is willing to let the FDA and industry go some more tortured miles to keep color additives that have not been proved safe on the market. The majority has ignored the fact that Congress has spoken on the subject and allows industry to capture in court a victory that it was denied in the legislative arena. The 1960 Color Additive Amendments have been made inoperative by judicial fiat.

Thirty years later after passage of the Color Additive Amendments, FDA completed its review of the provisional list with the termination of the provisional listing for FD&C Red No. 3 (erythrosine or E127). Of the original 200 color additives, 90 have been listed as safe and the remainder have either been removed from use by FDA or withdrawn by industry.

* * * * *

Federal Food, Drug, and Cosmetic Act § 201(t)

(t)(1) The term "**color additive**" means a material which—

    (A) is a dye, pigment, or other substance made by a process of synthesis or similar artifice, or extracted, isolated, or otherwise derived, with or without intermediate or final change of identity, from a vegetable, animal, mineral, or other source, and

    (B) when added or applied to a food, drug, or cosmetic, or to the human body or any part thereof, is capable (alone or through reaction with other substance) of imparting color thereto; except that such term does not include any material which the Secretary, by regulation, determines is used (or intended to be used) solely for a purpose or purposes other than coloring.

(2)   The term "color" includes black, white, and intermediate grays.

(3)   Nothing in subparagraph (1) of this paragraph shall be construed to apply to any pesticide chemical, soil or plant nutrient, or other agricultural chemical solely because of its effect in aiding, retarding, or otherwise affecting, directly or indirectly, the growth or other natural physiological processes of produce of the soil and thereby affecting its color, whether before or after harvest.

* * * * *

Food Color Facts

Food and Drug Administration & International Food Information Council Foundation Brochure (January 1993)[35]

The color of food is an integral part of our culture and enjoyment of life. Who would deny the mouth-watering appeal of a deep-pink strawberry ice on a hot summer day or a golden Thanksgiving turkey garnished with fresh green parsley?

Even early civilizations such as the Roman recognized that people "eat with their eyes" as well as their palates. Saffron and other spices were often used to provide a rich yellow color to various foods. Butter has been colored yellow as far back as the 1300s.

Today all food color additives are carefully regulated by federal authorities to ensure that foods are safe to eat and accurately labeled. This brochure provides helpful background information about color additives, why they are used in foods, and regulations governing their safe use in the food supply.

What Is a Color Additive?

Technically, a color additive is any dye, pigment or substance that can impart color when added or applied to a food, drug, cosmetic, or to the human body.

The Food and Drug Administration (FDA) is responsible for regulating all color additives used in the United States. All color additives permitted for use in foods are classified as "certifiable" or "exempt from certification" (see Table I).

Certifiable color additives are man-made, with each batch being tested by manufacturer and FDA. This "approval" process, known as color additive certification, assures the safety, quality, consistency and strength of the color additive prior to its use in foods.

---

[35] *Available at*: http://www.foodsafety.gov./~lrd/colorfac.html (last visited Sept. 30, 2008).

TABLE I   Color Additives Permitted for Direct Addition to Human Food in the United States

Certifiable colors

FD&C Blue No. 1 (Dye and Lake), FD&C Blue No. 2 (Dye and Lake), FD&C Green No. 3 (Dye and Lake), FD&C Red No. 3 (Dye), FD&C Red No. 40 (Dye and Lake), FD&C Yellow No. 5 (Dye and Lake), FD&C Yellow No. 6 (Dye and Lake), Orange B,* Citrus Red No. 2*

Colors exempt from certification

Annatto extract, B-Apo-8′-carotenal,* Beta-carotene, Beet powder, Canthaxanthin, Caramel color, Carrot oil, Cochineal extract (carmine); Cottonseed flour, toasted partially defatted, cooked; Ferrous gluconate,* Fruit juice, Grape color extract,* Grape skin extract* (enocianina), Paprika, Paprika oleoresin, Riboflavin, Saffron, Titanium dioxide,* Turmeric, Turmeric oleoresin, Vegetable juice

*These food color additives are restricted to specific uses.

There are nine certified colors approved for use in food in the United States. One example is FD&C Yellow No. 6, which is used in cereals, bakery goods, snack foods, and other foods.

Color additives that are exempt from certification include pigments derived from natural sources, such as vegetables, minerals, or animals, and man-made counterparts of natural derivatives.

For example, caramel color is produced commercially by heating sugar and other carbohydrates under strictly controlled conditions for use in sauces, gravies, soft drinks, baked goods, and other foods.

Whether a color additive is certifiable or exempt from certification has no bearing on its overall safety. Both types of color additives are subject to rigorous standards of safety prior to their approval for use in foods.

Certifiable color additives are used widely because their coloring ability is more intense than most colors derived from natural products; thus they are often added to foods in smaller quantities. In addition, certifiable color additives are more stable, provide better color uniformity and blend together easily to provide a wide range of hues. Certifiable color additives generally do not impart undesirable flavors to foods, while color derived from foods such as beets and cranberries can produce such unintended effects.

Of nine certifiable colors approved for use in the United States, seven color additives are used in food manufacturing (see Table II). Regulations known as Good Manufacturing Practices limit the amount of color added to foods. Too much color would make foods unattractive to consumers, in addition to increasing costs.

TABLE II   Color Additives Certifiable for Food Use

Name/Common Name	Hue	Common Food Uses
FD&C Blue No. 1		
Brilliant Blue FCF	Bright blue	Beverages, dairy products powders, jellies, confections, condiments, icings, syrups, extracts
FD&C Blue No. 2		
Indigotine	Royal Blue	Baked goods, cereals, snack foods, ice cream, confections, cherries
FD&C Green No. 3		
Fast Green FCF	Sea Green	Beverages, puddings, ice cream, sherbert, cherries, confections, baked goods, dairy products
FD&C Red No. 40		
Allura Red AC	Orange-red	Gelatins, puddings, dairy products, confections, beverages, condiments
FD&C Red No. 3		
Erythrosine	Cherry-red	Cherries in fruit cocktail and in canned fruits for salads, confections, baked goods, dairy products, snack foods
FD&C Yellow No. 5		
Tartrazine	Lemon Yellow	Custards, beverages, ice cream, confections, preserves, cereals
FD&C Yellow No. 6		
Sunset Yellow	Orange	Cereals, baked goods, snack foods, ice cream, beverages, dessert powders, confections

What Are Dyes and Lakes?

Certifiable color additives are available for use in food as either "dyes" or "lakes." Dyes dissolve in water and are manufactured as powders, granules, liquids, or other special purpose forms. They can be used in beverages, dry mixes, baked goods, confections, dairy products, pet foods, and a variety of other products.

Lakes are the water-insoluble form of the dye. Lakes are more stable than dyes and are ideal for coloring products containing fats and oils or items lacking sufficient moisture to dissolve dyes. Typical uses include coated tablets, cake and donut mixes, hard candies, and chewing gums.

Why Are Color Additives Used in Foods?

Color is an important property of foods that adds to our enjoyment of eating. Nature teaches us early to expect certain colors in certain foods, and our future acceptance of foods is highly dependent on meeting these expectations.

Color variation in foods throughout the seasons and the effects of food processing and storage often require that manufacturers add color to certain foods to meet consumer expectations. The primary reasons of adding colors to foods include:

- To offset color loss due to exposure to light, air, extremes of temperature, moisture and storage conditions.
- To correct natural variations in color. Off-colored foods are often incorrectly associated with inferior quality. For example, some tree-ripened oranges are often sprayed with Citrus Red No. 2 to correct the natural orangy-brown or mottled green color of their peels (Masking inferior quality, however, is an unacceptable use of colors.)
- To enhance colors that occur naturally but at levels weaker than those usually associated with a given food.
- To provide a colorful identity to foods that would otherwise be virtually colorless. Red colors provide a pleasant identity to strawberry ice while lime sherbet is known by its bright green color.
- To provide a colorful appearance to certain "fun foods." Many candies and holiday treats are colored to create a festive appearance.
- To protect flavors and vitamins that may be affected by sunlight during storage.
- To provide an appealing variety of wholesome and nutritious foods that meet consumers' demands.

How Are Color Additives Regulated?

In 1900, there were about 80 man-made color additives available for use in foods. At that time there were no regulations regarding the purity and uses of these dyes.

Legislation enacted since the turn of the century, however, has greatly improved food color additive safety and stimulated improvements in food color technology.

The Food and Drug Act of 1906 permitted or "listed" seven man-made color additives for use in foods. The Act also established a voluntary certification program, which was administered by the U.S. Department of Agriculture (USDA); hence man-made color additives became known as "certifiable color additives".

The Federal Food, Drug, and Cosmetic (FD&C) Act of 1938 made food color additive certification mandatory and transferred the authority for its testing from USDA to FDA. To avoid confusing color additives used in food with those manufactured for other uses, three categories of certifiable color additives were created:

- Food, Drug, and Cosmetic (FD&C)—Color additives with application in foods, drugs, or cosmetics;
- Drug and Cosmetic (D&C)—Color additives with applications in drugs, or cosmetics;
- External Drug and Cosmetic (External D&C)—Color additives with applications in externally applied drugs (e.g., ointments) and in externally applied cosmetics.

In 1960, the Color Additive Amendments to the FD&C Act placed color additives on a "provisional" list and required further testing using up-to-date procedures. One section of the amendment, known as the Delaney Clause, prohibits adding to any food substance that has been shown to cause cancer in animals or man regardless of the dose. Under the amendments, color additives exempt from certification also are required to meet rigorous safety standards prior to being permitted for use in foods.

According to the Nutrition Labeling and Education Act of 1990, a certifiable color additive used in food must be listed in the ingredient statement by its common or usual name. All labels printed after July 1, 1991, must comply with this requirement.

How Are Color Additives Approved for Use in Foods?

To market a new color additive, a manufacturer must first petition FDA for its approval. The petition must provide convincing evidence that the proposed color additive performs as it is intended. Animal studies using large doses of the color additive for long periods are often necessary to show that the substance would not cause harmful effects at expected levels of human consumption. Studies of the color additive in humans also may be submitted to FDA.

In deciding whether a color additive should be approved, the agency considers the composition and properties of the substance, the amount likely to be consumed, its probable long-term effects and various safety factors. Absolute safety of any substance can never be proved. Therefore FDA must determine if there is a reasonable certainty of no harm from the color additive under its proposed conditions of use.

If the color additive is approved FDA issues regulations that may include the types of foods in which it can be used, the maximum amounts to be used and how it should be identified on food labels. Color additives proposed for use in meat and poultry products also must receive specific authorization by USDA.

Federal officials then carefully monitor the extent of Americans' consumption of the new color additive and results of any new research on its safety.

In addition FDA operates an Adverse Reaction Monitoring System (ARMS) to help serve as an ongoing safety check of all activities. The system monitors and investigates all complaints by individuals or their physicians that are believed to be related to food and color additives, specific foods, or vitamin and mineral supplements. The ARMS computerized database helps officials decide whether reported adverse reactions represent a real public health hazard so that appropriate action can be taken.

Additional Information about Color Additives

Q. Are certain people sensitive to FD&C Yellow No. 5 in foods?

A. FDA's Advisory Committee on Hypersensitivity to Food Constituents concluded in 1986 that FD&C Yellow No. 5 may cause hives in fewer that one out of 10,000 people. The committee found that there was no evidence the color additive in foods provokes asthma attacks nor that aspirin-intolerant individuals may have a cross-sensitivity to the color. As with other color additives certifiable for food use, whenever FD&C Yellow No. 5 is added to foods, it is listed on the product label. This allows the small portion of people who may be sensitive to the color to avoid it.

Q. What is the status of FD&C Red No. 3?

A. In 1990 FDA discontinued the provisional listing of all lake forms of FD&C Red No. 3 and its dye form used in external drugs and cosmetics. The uses were terminated because one study of the color additive in male rats showed an association with thyroid tumors. In announcing the decision, FDA stated that any human risk posed by FD&C Red No. 3 was extremely small and was based less on safety concerns than the legal mandate of the Delaney Clause. FD&C Red No. 3 remains permanently listed for use in food and ingested drugs, although FDA has announced its intent to propose rescinding those listings.

Q. Why are decisions sometimes changed about the safety of food color additives?

A. Since absolute safety of any substance can never be proved, decisions about the safety of color additives or other food ingredients are made on the best scientific evidence available. Because scientific knowledge is constantly evolving, federal officials often review earlier decisions to assure that the safety assessment of a food substance remains up to date. Any change made in previous clearances should be recognized as an assurance that the latest and best scientific knowledge is being applied to enhance the safety of the food supply.

Q. Do food color additives cause hyperactivity?

A. Although this theory was popularized in the 1970s, well-controlled studies conducted since then have produced no evidence that food color additives cause hyperactivity or learning disabilities in children. A Consensus Development Panel of the National Institutes of Health concluded in 1982 that there was no scientific evidence to support the claim that colorings or other food additives cause hyperactivity. The panel said that elimination diets should not be used universally to treat childhood hyperactivity, since there is no scientific evidence to predict which children may benefit.

* * * * *

DISCUSSION QUESTIONS AND NOTES

7.12. Give a rationale for why Congress did not provide a GRAS or prior-sanctioned exemption for color additives.

7.13. Why would Congress place a stricter a requirement on previously used certified colors as opposed to food additives?

7.14. Carbon monoxide coloring in fresh meat. In 2005 Kalsec, Inc., petitioned FDA to ban on carbon monoxide in fresh meat packaging and to terminate the FDA acceptance of Generally Recognized As Safe (GRAS) notifications (GRAS Notice Nos. GRN 000083 and 000143). Kalsec claimed, "The use of carbon monoxide gas in fresh meat packaging produces an artificially intense, persistent red color in meat that can simulate the look of fresh meat and mask the natural signs of aging and spoilage that consumers depend upon in making safe food choices, including browning and tell-tale odors. Consumers have no way to tell the difference between meat packaged with carbon monoxide gas that may merely look fresh and safe, and genuinely fresh and wholesome meat. As a result, carbon monoxide presents serious consumer deception and food safety risks which jeopardize the public health."

* * * * *

United States v. An Article of Food Consisting of 12 Barrels . . . Lumpfish Roe

477 F. Supp. 1185 (1979)

Opinion: BRIEANT

Plaintiff in this action, the United States of America, seized 12 barrels of lumpfish roe, referred to during the trial as "caviar," pursuant to the provisions of the Federal Food, Drug, and Cosmetic Act (hereinafter "the Act"). The complaint, filed July 10, 1978, alleges that the seized lumpfish roe constitutes an adulterated food held for sale after shipment in interstate commerce within the meaning of the Act in that it bears and contains a color additive, FD&C Red No. 2 (hereinafter "Red #2"). Iranian Caviar and Sturgeon Corporation (hereinafter "claimant" or "Iranian Caviar Corp.") intervened in the action and filed a claim to the seized lumpfish roe. . . .

The Iranian Caviar Corp. processes and distributes lumpfish roe for sale to delicatessens and others. On July 10, 1977, it received a shipment of 20 barrels of lumpfish roe. This shipment was first transported from Bergen, Norway, to the United States. . . .

On May 5, 1978, a Food and Drug Administration (hereinafter "FDA") inspector entered the Iranian Caviar Corp.'s premises for the purpose of conducting a scheduled routine inspection. While inspecting the plant, the inspector discovered a container labelled "Black Shade," which listed several ingredients, including Red #2. The inspector also discovered several large wooden barrels in a refrigerator that were marked in part "* * * LUMPFISH ROE 100 KG NET * * *." Affixed to a number of these barrels were white tags with the words "colored black" and a date. Samples were taken from five of these barrels, and . . .

FDA chemists apparently discerned the presence of Red #2 in the samples taken from the five wooden barrels and the "Black Shade" container. On the basis of this analysis, the FDA obtained a warrant from the Clerk of this Court for the arrest of 12 barrels labelled in part "* * * LUMPFISH ROE 100 KG NET * * * COLORED BLACK * * *," . . .

The remaining question is whether the lumpfish roe was adulterated before, during, or after shipment in interstate commerce. 21 U.S.C. § 342(c) provides that "(a) food shall be deemed adulterated (under the Act) if it . . . bears or contains a color additive which is unsafe within the meaning of § 376(a) of this title." . . .

Thus a color additive is unsafe unless a regulation promulgated by the FDA permits its use. The FDA specifically found Red #2 to be unsafe in 1976 and banned its use, and that finding has been upheld in subsequent litigation. In addition a review of the current FDA regulations relating to color additives indicates that the status of Red #2 has not changed.

The evidence offered at trial clearly indicates that 11 of the 12 barrels involved in this case contained Red #2. Tests by the FDA and a qualified private chemist came to this conclusion, and there is no evidence to refute it. It follows that the food was adulterated within the meaning of 21 U.S.C. §§ 342 and 376, and therefore I find that the government established a prima facie case for condemnation, and that the 11 barrels of lumpfish roe that contained Red #2 were properly seized under 21 U.S.C. § 334(a)(1).

Claimant asserted a number of theories during the trial as possible defenses to the government's case. First, claimant implied that the lumpfish roe was adulterated or dyed by some mysterious stranger without its knowledge. This contention, if established, would not affect the validity of the government's seizure nor its right to condemn. The Act only requires that the food be adulterated; it does not require proof of how the food was adulterated. However, the issue of whether the Iranian Caviar Corp. in fact put the Red #2 in the lumpfish roe is important in another respect. Title 21 U.S.C. § 334(d)(1), which directs that any food condemned under § 334 be disposed of by destruction or sale, includes the following:

If the article was imported into the United States and the person seeking its release establishes (A) that the adulteration ... did not occur after the article was imported, and (B) that he had no cause for believing that it was adulterated ... before it was released from customs custody, the court may permit the article to be delivered to the owner for exportation in lieu of destruction upon a showing by the owner that all of the conditions of section 381(d) of this title can be and will be met.

Therefore claimant's expression of the theory that another party adulterated the lumpfish roe was asserted indirectly to provide a basis for the return of the caviar for exportation in lieu of destruction.

This point is of more than academic interest. While Red #2 has been finally determined by the appropriate federal bureaucrats to be a dangerous carcinogen, the use of which in food is absolutely prohibited, it seems that in Canada its use remains both lawful and commonplace. Whether there is a physiological difference of some sort between Canadians and Americans rendering the former less susceptible to cancer, or whether there is merely a difference in bureaucrats between the two countries is a question which need not be resolved here. The evidence at the trial justifies an inference, and I find, that an officer or employee of claimant Iranian Caviar Corp. colored the lumpfish roe right here in New York, and that the dye used for coloring included Red #2. An FDA inspector testified, and I find, that Mr. Ura Fridman, President of the Iranian Caviar Corp., admitted to coloring the lumpfish roe with a dye. Mr. Fridman, in his testimony, denied any knowledge of Red #2, but did admit that his firm colored the roe with a "black jet color." Furthermore pictures of tags on the barrels of lumpfish roe in the Iranian Caviar Corp.'s refrigerator reveal that the roe was "colored black" in August 1977, after the barrels were received by claimant. (This evidence, coupled with the fact that a container of black food coloring containing Red #2 was found on the premises, leads to the conclusion that the Red #2 was in the dye used by claimant to color the roe. Claimant offered no credible evidence to oppose this inference, and I conclude that the government has proved, by a fair preponderance of the evidence, that the adulteration did occur after the lumpfish roe was imported. Therefore 21 U.S.C. § 334(d)(1) does not apply, and the 11 barrels properly seized by the government should not be returned to claimant for exportation.

Claimant also charged at the trial that the actions of the government agents leading up to the removal of the lumpfish roe constituted an unreasonable search and seizure, presumably in violation of the Fourth Amendment to the United States Constitution. This theory appears to be based on the assertion that the FDA inspector who originally sampled the roe had no authority to inspect the Iranian Caviar Corp. premises, and that the FDA agent and U.S. Marshals participating in the seizure of the 12 barrels were unnecessarily abusive and used undue physical force. I find no merit in these charges, either factually or in law. The FDA inspector testified, and I find that he was on a regularly scheduled visit. Mr. Fridman admitted in his testimony to having permitted such an inspection in his presence without reservation or objection. Claimant presented no credible contrary evidence, and I conclude that the inspection was not an unreasonable search. As for the events leading up to the subsequent second seizure and removal of the barrels, the testimony of the FDA agent and the Deputy United States Marshal shows, and I find, that their conduct was proper and that it was Mr. Fridman who was impeding their official duties by his actions and by refusing to cooperate. It is undisputed that the seizure took place during daylight hours, and that the agents and marshals had procured a warrant for the arrest of the barrels and a warrant to inspect the premises. While the government concedes that claimant's lock on its refrigerator was broken in order to remove the 12 barrels, this action was only taken after a search warrant was issued, and after Mr. Fridman refused to unlock the door. Claimant has failed to prove that the government's conduct was improper.

Moreover I conclude that the exclusionary rule applied to evidence seized in violation of the Fourth Amendment does not apply to condemnation proceedings under 21 U.S.C. § 334. . . . Supplemental Rule C provides that the clerk of the district court "shall" issue a warrant for the arrest of an article of food under 21 U.S.C. § 334 on the basis of a complaint verified on oath or by affirmation. In addition it is well established that contraband, though unlawfully found or seized in violation of the Fourth Amendment, and therefore suppressed as evidence in criminal proceedings against the owner or possessor, may be condemned by the government and will not be returned. This is so even if the granting of a suppression motion bars the underlying criminal prosecution. The rationale for this latter rule is that returning the contraband would frustrate "the express public policy against the possession of such objects." Accordingly, the exclusionary rule fashioned by the courts to protect the Fourth Amendment rights of individuals is not available to a claimant in a condemnation proceeding brought In rem concerning the contraband itself. Such a condemnation can be maintained so long as the initial pre-seizure requirements of Supplemental Rule C, Supra, have been met by the government. Claimant would not be entitled to the return of the 12 barrels of lumpfish roe found to be contaminated even if it had proved that these articles were illegally seized. It is arguable that the contamination of the lumpfish roe might not have been established if the exclusionary rule, applicable to the prosecution of crimes were applied in this civil In rem action, and the barrels themselves were suppressed before trial as fruits of an improper search or seizure. However, such an argument ignores the express public policy against the possession of contraband. . . .

The exclusionary rules are judge-made law promulgated beginning with *Weeks v. United States*, 232 U.S. 383 (1914), as a result of a perceived necessity for the courts to "deter" those charged with enforcement of the criminal laws from violating the Constitutional rights of persons.

. . . .

Bound as we are by the policy decision made in *Weeks* and its progeny, we must continue to let the criminal go free where the constable has blundered. But claimant would ask the courts, solely in order to vindicate Fourth Amendment rights, secured to Persons under the Constitution, also to turn loose into the byways of commerce contaminated food for consumption by the unsuspecting public. Why? This poisoned carcinogenic caviar has no rights. If Mr. Fridman's rights, or those of claimant were infringed, they have the adequate remedies envisioned by Judge Cardozo. The Court declines to extend the exclusionary rule to the benefit of inanimate dangerous articles, or to protect the profits which flow from the sale of contraband.

Claimant's assertion that the barrels of lumpfish roe were seized unreasonably in violation of the Fourth Amendment would, if proved, be insufficient as a matter of law to justify or compel release of the caviar for human consumption.

Accordingly, the defendant In rem articles of food are condemned and forfeited to the government . . .

* * * * *

7.5   FOOD IRRADIATION

Radiation was studied as a way to preserve food in the 1930s, but research did not take off until after World War II, when the U.S. Army funded research in their look for a means to lessen dependence on refrigeration for food for troops deployed in the field. Also, in the early 1950s, the Atomic Energy Commission research food irradiation as part of President Eisenhower's "Atoms for Peace" program. This research examined the effects radiation on certain fruits and vegetables to kill or sterilize insects.

The FDA first approved the use of irradiation on a food product in 1963 for wheat and wheat flour. In 1997 the FDA approved treating red meat products with radiation to kill harmful bacteria. In between, there have been a number of other approvals, including poultry, fresh fruits and vegetables, dry spices, seasonings, and enzymes. In approving a use of radiation, FDA sets the maximum radiation dose the product can be exposed to, measured in units called kiloGray (kGy).

7.5.1   Irradiation as a Food Additive under FD&C Act

The Food Additives Amendment of 1958 to the FD&C Act expressly defined a source of radiation as a food additive. Note that it is the "source of radiation" intended for use in processing food that is included as a "food additive." Thus, the radiation is not an additive, but the process of being irradiated is defined as an additive.

The Food Additives Amendment also defined adulterated food to include food that has been intentionally irradiated, unless the irradiation is carried out in conformity with a regulation prescribing safe conditions of use.

When meat is irradiated, FDA and USDA are both involved in regulation. The FD&C Act applies for the additive portion. The Federal Meat Inspection Act and the Poultry Products Inspection Act nevertheless still apply. Thus anyone interested in irradiating meat or poultry is also subject to the regulatory authority of the FSIS.

Safety Issues   Before issuing an authorizing regulation FDA is required to establish the safety of the petitioned use. In the case of irradiated food, safety consideration requires that four broad areas be addressed: radiological safety, toxicological safety, microbiological safety, and nutritional adequacy.

Current Regulations   FDA has found irradiation of food to be safe under several conditions and promulgated authorizing regulations have been issued both in response to petitions and at FDA's initiative.[36]

* * * * *

Irradiation: A Safe Measure for Safer Food

John Henkel, FDA CONSUMER (May–June 1998)[37]

Beef is one of the U.S. food industry's hottest sellers—to the tune of 8 billion pounds a year, according to trade figures. Whether at a fast-food meal, a dinner on the town, or a backyard barbecue, beef is often front and center on America's tables.

But in recent years, beef, especially ground beef, has shown a dark side: it can harbor the bacterium *E. coli* O157:H7, a pathogen that threatens the safety of the domestic food supply. If not properly prepared, beef tainted with *E. coli* O157:H7 can make people ill, and in rare instances, kill them. In 1993, *E. coli* O157:H7-contaminated hamburgers sold by a fast-food chain were linked to the deaths of four children and hundreds of illnesses in the Pacific Northwest.

In 1997, the potential extent of *E. coli* O157:H7 contamination came to light when Arkansas-based Hudson Foods Inc. voluntarily recalled 25 million pounds of hamburger suspected of containing *E. coli* O157:H7. It was the largest recall of meat products in U.S. history.

Nationally, *E. coli* O157:H7 causes about 20,000 illnesses and 500 deaths a year, according to the federal Centers for Disease Control and Prevention. Scientists have only known since 1982 that this form of *E. coli* causes human illness.

---

[36] 21 C.F.R. § 179.

[37] *Available at*: http://www.fda.gov/fdac/features/1998/398_rad.html (last accessed Oct. 6, 2007).

To help combat this public health problem, the Food and Drug Administration, in December 1997, approved treating red meat products with a measured dose of radiation. This process, commonly called irradiation, has drawn praise from many food industry and health organizations because it can control *E. coli* O157:H7 and several other disease-causing microorganisms. As with other regulations governing meat and poultry products, irradiation will be authorized when the U.S. Department of Agriculture completes its implementing regulations.

Though irradiation is the latest step toward curbing food-borne illness, the federal government also is implementing other measures, which include developing new technologies and expanding the use of current technologies.

A Long Safety Record

FDA's red meat approval added another product category to the already lengthy list of foods the agency has approved for irradiation since 1963. These include poultry, fresh fruits and vegetables, dry spices, seasonings, and enzymes.

As part of its approval, FDA requires that irradiated foods include labeling with either the statement "treated with radiation" or "treated by irradiation" and the international symbol for irradiation, the radura. Irradiation labeling requirements apply only to foods sold in stores. For example, irradiated spices or fresh strawberries should be labeled. When used as ingredients in other foods, however, the label of the other food does not need to describe these ingredients as irradiated. Irradiation labeling also does not apply to restaurant foods.

FDA has evaluated irradiation safety for 40 years and found the process safe and effective for many foods. Before approving red meat irradiation, the agency reviewed numerous scientific studies conducted worldwide. These included research on the chemical effects of radiation on meat, the impact the process has on nutrient content, and potential toxicity concerns.

In this most recent review and in previous reviews of the irradiation process, FDA scientists concluded that irradiation reduces or eliminates pathogenic bacteria, insects, and parasites. It reduces spoilage, and in certain fruits and vegetables, it inhibits sprouting and delays the ripening process. Also, it does not make food radioactive, compromise nutritional quality, or noticeably change food taste, texture, or appearance as long as it's applied properly to a suitable product.

Health experts say that in addition to reducing *E. coli* O157:H7 contamination, irradiation can help control the potentially harmful bacteria *Salmonella* and *Campylobacter*, two chief causes of foodborne illness. The Centers for Disease Control and Prevention estimates that *Salmonella*—commonly found in poultry, eggs, meat, and milk—sickens as many as 4 million and kills 1,000 per year nationwide. *Campylobacter*, found mostly in poultry, is responsible for 6 million illnesses and 75 deaths per year in the United States. A May 1997 presidential report, "Food Safety from Farm to Table," estimates that "millions" of Americans are stricken by foodborne illness each year and some 9,000, mostly the very young and elderly, die as a result.

FDA officials emphasize that though irradiation is a useful tool for reducing foodborne disease risk, it complements, but doesn't replace, proper food-handling practices by producers, processors and consumers.

Limited Success So Far

Though irradiation would appear to have much going for it, retail outlets have been slow to carry irradiated foods. This, experts say, is partially because many store owners and food producers fear consumers won't buy the products based on misgivings about radiation in general.

But some stores have plunged in anyway—with limited success. Carrot Top, a Chicago-area grocery market, was one of the first to carry irradiated fruits (see "Berry Successful Irradiation"). Owner Jim Corrigan says the products have been selling steadily since 1992. Other stores—mostly small, independent markets—have followed suit, offering irradiated vegetables, fruits, and poultry to a modest, but loyal, group of irradiation-savvy customers.

Because irradiated red meat is not yet on the market, it remains to be seen if consumers will buy products such as irradiated ground beef—or if large food processors will even offer it. Irradiated products sold to date have cost slightly more than their untreated counterparts because of the extra step irradiation adds to food processing. But in the future, these costs could be offset by improved shelf life and increased consumer demand, according to food trade groups.

Major food companies such as poultry processors, meat packers, and grocery chains have yet to embrace irradiation, not only because of perceived consumer attitudes, but also due to logistics. Food Technology Service Inc., in Mulberry, Florida, is the only irradiating facility dedicated solely to treating agricultural products. More than 40 other facilities nationwide primarily handle sterilization of medical supplies, though these plants also can irradiate food products. In fact, it was a New Jersey–based medical irradiation company, Isomedix Inc., that petitioned FDA to approve red meat irradiation.

Beyond physical distances and lack of facilities, sheer product volume makes it unlikely that irradiation will be widespread anytime soon. The domestic poultry trade, for instance, processes about 25 billion pounds per year, according to industry figures. Says Kenneth May, spokesman for the National Broiler Council, which represents poultry producers: "We think [irradiation is] a process that will work. But for practical purposes, we just don't see anything happening with it in the near future." He adds, however, that if the public really wants an irradiated product, the poultry industry will find a way to deliver it.

Will Consumers Accept It?

Before irradiation can really take off, the public must "warm up" to a method associated with nuclear energy, a source that carries its share of negative perceptions. George Pauli, Ph.D., FDA's food irradiation safety coordinator, compares irradiation to milk pasteurization, another decontaminating process that dramatically curbed disease but took decades before achieving public acceptance. "When the public finally sees a need for irradiation and realizes its value, I think people will accept it, maybe even demand it," Pauli says. "But you have to give them time."

A Louis Harris poll released in 1986 found that 76 percent of Americans considered irradiated food a hazard. But later studies have shown that consumer attitudes can be changed through education.

In 1995, researchers at the University of Georgia reported that 87.5 percent of consumers had heard of irradiation but knew little about it. So the university set up a "simulated supermarket setting" and labeled irradiated products, put posters at the point of sale, and developed a slide show explaining irradiation. "Our goal was to see which one of those techniques was most effective in changing people's attitudes," says Kay McWatters, agricultural research scientist and one of the study authors.

The study found that any kind of education helps convey the benefits of irradiation, McWatters says. "But the one that turned out most effective was the slide show, because visual images and [narration] are much more attention-getting than just a static label or poster."

After the study's education strategy, about 84 percent of participating consumers said irradiation is "somewhat necessary" or "very necessary." Fifty-eight percent said they would always buy irradiated chicken if available, and 27 percent said they would buy it sometimes.

Another study in 1997 by the Food Marketing Institute had similar results. After receiving education about the process, 60 percent of those in the study said they would buy irradiated foods.

Carrot Top owner Corrigan also discovered this on a small scale after sending his regular customers information about irradiation in periodic newsletters.

Luggage and Milk

Other studies, however, show that many consumers still question if irradiation is safe. They wonder if the process transfers radiation to the product or if it causes chemical changes in the food that might be hazardous. Even the word "irradiation" is scary to some, carrying images of atomic explosions or nuclear reactor accidents. But as long as radiation is applied to foods in approved doses, it's safe, says FDA's Pauli. Similar to sending luggage through an airport scanner, the process passes food quickly through a radiation field—typically gamma rays produced from radioactive cobalt-60. That amount of energy is not strong enough to add any radioactive material to the food. The same irradiation process is used to sterilize medical products such as bandages, contact lens solutions, and hospital supplies such as gloves, sutures, and gowns. Many spices sold in this country also are irradiated, which eliminates the need for chemical fumigation to control pests. American astronauts have eaten irradiated foods since 1972.

Irradiation is a "cold" process that gives off little heat, so foods can be irradiated within their packaging and remain protected against contamination until opened by users. Because a few bacteria can survive the process in poultry and meats, it's important, Pauli says, to keep products refrigerated and to cook them properly.

Irradiation interferes with bacterial genetics, so the contaminating organism can no longer survive or multiply. Although chemicals called radiolytic products are created when food is irradiated, FDA has found them to pose no health hazard. In fact, the same kinds of products are formed when food is cooked.

Praises and Protests

Though irradiation has its share of detractors, many prestigious organizations endorse it, including the World Health Organization, the International Atomic Energy Agency, the American Medical Association, and the American Dietetic Association. Trade groups such as the National Meat Association, the Grocery Manufacturers of America, and the National Food Processors Association also support irradiation.

However, some groups have given irradiation a thumbs down. Consumer activist Jeremy Rifkin, president of the Pure Food Campaign, says more attention should be placed on raising healthier livestock, which he says would reduce pathogens and make irradiation unnecessary. The Center for Science in the Public Interest calls irradiation "expensive" and "an end-of-the-line solution to contamination problems that can and should be addressed earlier."

But with so many influential organizations backing irradiation, along with concerns about rising numbers of disease cases, the stage is set for the process to pick up momentum, despite negative sentiments, supporters say. First, however, says FDA's Pauli, the food industry needs to get more irradiated products into the marketplace. "Most people in this country haven't even seen an irradiated food," he says. "When products start appearing, then the public can make up its mind."

Radiation's Positive Side

Scientists first studied radiation as a way to improve food products in the 1930s, but research didn't begin in earnest until just after World War II. At that time, the U.S. Army was seeking a means to lessen dependence on refrigeration and replace K rations and other preserved products that troops used in the field.

In the early 1950s, the Atomic Energy Commission (now part of the U.S. Department of Energy) explored food irradiation as part of President

Eisenhower's "Atoms for Peace" program. This research differed from the Army's in that it examined the effects smaller radiation doses had on certain fruits and vegetables. The end result was not a sterile product but one where insects would be killed or sterilized. Because this produce still could spoil, refrigeration was needed. But at least potentially harmful insects would not cross state or national borders.

Such research, augmented by studies from other countries, established that the most important benefit from irradiation could be the control of disease-causing pathogens and that the maximum practical and effective dose depended on the food and the purpose for irradiating.

Berry Successful Irradiation

The huge sign hanging over the rows of boxed strawberries left little doubt for Chicago-area grocery shoppers that the produce before them was something new and unusual.

Not that the berries looked any different. But the massive poster above them bore a message in mammoth letters that might as well have been neon: "Treated by irradiation for freshness and health." To the store owner's surprise, patrons flocked to the new product, buying nine times more of it than of standard strawberries.

That scene took place in 1992 at Carrot Top, one of the first retail stores to venture into the then-uncharted realm of irradiated foods. The decision to stock radiation-treated berries in the store, however, came slowly. Owner Jim Corrigan spent about a year reading up on the irradiation process and passing details to his regular customers through periodic newsletters. He says informing customers before the store actually stocked the new products helped allay possible fears.

When the Florida-grown strawberries finally arrived, along with irradiated oranges and grapefruits, shoppers were well acquainted with the process and responded with sales.

Today, Corrigan remains enthusiastic. He says irradiation ensures that strawberries will be free of insects and will keep longer—in some cases, up to three weeks, versus three to five days for conventional berries.

"One of our ways of rating the freshness of strawberries is to examine the small hairs that grow by the seed," he says. "If they are standing up and plentiful, the strawberries are still fresh. [With irradiated strawberries] we see a lot of that after three weeks."

The products remain steady sellers, and Corrigan has since added irradiated onions and papayas to his stock.

Approved Uses of Irradiation

FDA approved the first use of irradiation on a food product in 1963 when it allowed radiation-treated wheat and wheat flour to be marketed. In approving a use of radiation, FDA sets the maximum radiation dose the product can be exposed to, measured in units called kiloGray (kGy). The following is a list of all approved uses of radiation on foods to date, the purpose for irradiating them, and the radiation dose allowed:

FOOD	APPROVED USE	DOSE
Spices and dry vegetable seasoning	Decontaminates and controls insects and microorganisms	30 kGy
Dry or dehydrated enzyme preparations	Controls insects and microorganisms	10 kGy
All foods	Controls insects	1 kGy
Fresh foods	Delays maturation	1 kGy
Poultry	Controls disease-causing microorganisms	3 kGy
Red meat (e.g., beef, lamb and pork)	Controls spoilage and disease-causing microorganisms	0.5 kGy (fresh) 7 kGy (frozen)

* * * * *

7.5.2 Labeling

Retail foods treated with ionizing radiation must be labeled with the "Radura" symbol and with the statement "Treated by irradiation" or "Treated with radiation".[38] The symbol was intended to put forth a friendly image (Figure 7.1).

There are a number of important exemptions from labeling. Foods only containing an irradiated ingredient (as opposed to complete foods that have been irradiated) need not be labeled as irradiated. In addition irradiation labeling also does not apply to restaurant foods.

Wholesale shipments must be labeled with the statement "Treated with radiation—do not irradiate again" or the statement "Treated by irradiation—do not irradiate again" when shipped to a food manufacturer or processor for further processing, labeling, or packing.

FDA has been criticized by some for requiring labeling of irradiated food. However, FDA found labeling necessary to inform consumers that an irradiated food has been processed, because irradiation, like other forms of processing, can affect the characteristics of food. This reasoning also explains FDA lack of a labeling requirement when irradiated ingredients are added to foods that have not been irradiated. It should also be noted that food products produced or subject to conventional processing technologies, such as pasteurization, also need to be so labeled when there are two forms of the food.

[38] 21 C.F.R. § 179.26(c).

Figure 7.1   Radura symbol for irradiated food.

In November 1997, Congress reopened the issue of labeling for irradiated food in two ways. First, Congress mandated that FDA could not require a label statement to use print that is larger than that required for ingredients. FDA had not mandated a type size but did require the statement to be "prominent and conspicuous." On August 17, 1998, FDA updated it's regulation to clarify that the prominence requirement did not mean larger than usual type size.

7.5.3   Agricultural Pests

An additional reason for irradiating some foods is to protect U.S. agriculture from the import of exotic pests. USDA's Animal and Plant Health Inspection Service (APHIS) administers the law by quarantining certain crops from transport into the country. If irradiation is to be used for such a purpose, requirements of APHIS must be met.

* * * * *

The Truth about Irradiated Meat

CONSUMER REPORTS 34–37 (August 2003)[39]
CR Quick Take

In our tests of more than 500 meat samples from groceries in 60 cities—the largest test of its kind—we found that irradiated beef and chicken have a slight off-taste and come with the same handling and cooking instructions as regular meat. So they offer no real benefit for the careful cook.

One advantage: Irradiated meat generally has lower bacteria levels than regular meat. As such, it may reduce—but not eliminate—the risk of food-borne illness if your meat is undercooked.

[39] © 2003 by Consumers Union of U.S., Inc. Yonkers, NY 10703-1057, a nonprofit organization. Reprinted with permission from CONSUMER REPORTS® for educational purposes only. No commercial use or reproduction permitted. Also *available at*: www.ConsumerReports.org.

Irradiation has fueled a debate over how best to improve meat safety: by more aggressively preventing contamination in the first place, irradiating possible contaminants in packaged meat, or some combination of both.

In the aftermath of record meat recalls, certain supermarkets and restaurants are touting something new: irradiated chicken and ground beef.

Irradiation "eliminates any bacteria that might exist in food," according to a Food Emporium supermarket flyer. "You can't taste the difference," claims a pamphlet from SureBeam, a leading food irradiator. "Enjoy with confidence!" says a poster advertising irradiated double cheeseburgers at a Minneapolis Dairy Queen. Full-page newspaper ads from Wegmans supermarkets tell customers that they can cook a juicy irradiated burger "the way they like it" and "without worrying about safety."

Consumer Reports put claims like those to the test. Our research, taste tests, and microbial analysis of irradiated and nonirradiated chicken and ground beef—the largest analysis of its kind on meat sold at retail—counter many of the assertions:

- Bacteria levels in the irradiated, uncooked ground beef and skinless chicken tenders were generally much lower than levels in the nonirradiated meat. But the irradiated meat still contained some bacteria. And, like any meat, irradiated meat can become contaminated if it is handled improperly. That's why packages carry the same handling and cooking instructions as nonirradiated meat, including directions to "cook thoroughly."

- Our trained taste testers noted a slight but distinct off-taste and smell in most of the irradiated beef and chicken we cooked and sampled, likening it to singed hair. In the beef, the taste was detectable even with a bun, ketchup, and lettuce. Because it was usually subtle, however, some consumers may not notice it.

- Irradiated food is safe to eat, according to federal and world health officials. It certainly does not become radioactive. But a recent study on the chemical byproducts that irradiation creates in meat has led some researchers and the European Parliament to call for further studies. . . .

Should you buy it? There's no reason to if you cook meat thoroughly. Irradiation actually destroys fewer bacteria than does proper cooking.

Irradiation may offer added protection if meat is undercooked, however. Used in institutions such as cafeterias, irradiated meat could help reduce widespread foodborne illness, some experts predict. That's worth knowing if you are among those, such as the immunocompromised, at greatest risk from foodborne illness or if you want an extra measure of safety.

But other experts worry that the way irradiation is being promoted gives consumers a false sense of security. They say this end-stage fix also takes the focus off preventing contamination in the first place. Clearly, much more could be done to clean up unsanitary conditions at feedlots, slaughterhouses, processors, cafeterias, and other places where meat is prepared.

What It Can and Can't Do

Irradiation is the process by which food is bombarded with high-frequency energy capable of breaking chemical bonds. The energy source is electricity (for electron-beam irradiation) or radioactive cobalt-60 (for gamma-ray irradiation).

Food Technology Service, the nation's largest gamma-ray meat irradiator, says the energy passes through food much as "a ray of light passes through a window." But it is a powerful ray; the typical irradiation dose for meat, 1.5 KiloGrays, is 15 million times the energy involved in a single chest X ray, or 150 times the dose capable of killing an adult. . . .

Irradiation works by damaging the DNA of disease-causing bacteria such as *Salmonella* and the potentially deadly *E. coli* O157:H7, as well as of insects, parasites, and some spoilage organisms. They become "inactive" because they can't reproduce.

At approved doses, however, irradiation doesn't wipe out all bacteria in meat. Much higher doses would be needed to do that, but higher doses are not used because they would significantly degrade the taste of the food. And irradiation is ineffective against prions, the infectious proteins thought to cause mad-cow disease, because prions contain no DNA.

Irradiated meat generally harbors far fewer bacteria than nonirradiated meat, so there is less chance it would make you sick if it were not cooked thoroughly. And experts say there would be fewer germs in drippings that could contaminate other foods from, say, a cutting board. But irradiated meat doesn't protect against other food-handling problems. It offers no added safeguards if it is stored improperly, handled with dirty hands, or tainted from the drippings of some other contaminated food.

Why You Should Care Now

Federal regulators are paying attention to irradiation because the kinds of organisms it targets in meats are the nation's biggest food health threat. Last year, producers recalled a record 57 million pounds of meat, including ground beef, poultry, and deli meats, because of potentially deadly bacterial contamination. The Food and Drug Administration is considering a petition to approve irradiation for seafood such as clams and for ready-to-eat foods like deli meats, precooked beef patties, and hot dogs.

The government considers irradiation so effective that it allows tainted ground beef that otherwise would be unlawful to sell, such as meat containing *E. coli* O157:H7, to be irradiated and sold to consumers.

That meat safety needs improving is a given. But irradiation has stoked the debate over how best to do it.

On the one hand, widespread meat irradiation could appreciably reduce foodborne illnesses, says Dr. Robert Tauxe, a medical epidemiologist and chief of the Foodborne and Diarrheal Diseases Branch at the Centers for Disease Control and Prevention in Atlanta.

In a study published in the journal EMERGING INFECTIOUS DISEASES in June 2001, he estimated that irradiating half of all ground beef, poultry, pork, and processed meat would prevent 900,000 cases of foodborne infection, 8,500 hospital admissions, 6,000 grave illnesses, and 350 deaths in the U.S. each year, assuming that those foods are the source of half of *Campylobacter, E. coli* O157:H7, *Listeria, Salmonella,* and toxoplasma infections.

Such reductions would amount to 6 percent of foodborne illnesses reported each year. The rest of those illnesses can be attributed to other foods, like eggs and seafood, or other problems, such as improper food storage.

By contrast, the CDC says 20 percent of foodborne outbreaks are caused simply by commercial food preparers' poor hygiene, such as failing to wash hands before touching food. The Department of Agriculture reported eliminating 99.9 percent of *E. coli* O157:H7 in spiked beef samples with a low-tech step: spraying beef with lactic acid, a food preservative with antimicrobial properties, before grinding.

"It's better to take steps to avoid contaminating food to begin with than it is to try to clean it up afterwards," says Carol Tucker Foreman, director of the Food Policy Institute of the Consumer Federation of America and former assistant secretary of the USDA. "But I'm afraid it's human nature not to spend money to change the way animals are raised, or have a trained workforce in meatpacking plants, or upgrade facilities if they can just irradiate food at the end of the line."

That debate is being played out throughout the country:

School lunches. Beginning in January, the USDA says each school district will have the option of ordering irradiated ground beef for its school lunch program.

Want a Flyer with That Burger?

Since its introduction in major supermarkets such as Food Emporium, Giant Food, Publix, and ShopRite, along with restaurants including Dairy Queen and Embers America, irradiated ground beef has been the subject of marketing blitzes.

Two years ago, the Bush administration proposed allowing irradiated poultry and ground beef into the federal school lunch program instead of requiring that meat be tested for *Salmonella.* That proposal triggered such resistance that the USDA scrapped the plan and banned irradiated foods from the program, which serves 28 million public school lunches each day. But a provision in the Farm Security and Rural Investment Act of 2002 directed the USDA to drop its restrictions, while continuing *Salmonella* testing.

To garner support, the USDA has awarded $151,000 to the Minnesota Department of Children, Families, and Learning to test the effectiveness of sending irradiation-information kits to parents in several school districts.

Meanwhile, the school board of Berkeley, California, became one of the first to pass a resolution explicitly prohibiting the purchase of irradiated foods for its schools. Its November 6 resolution noted that there had been "no long-term health and side-effect studies on humans."

Foodborne illness in schools has been a recurring problem. Schools reported roughly 24 outbreaks of foodborne illness each year between 1973 and 1997, according to research reported in the Pediatric Infectious Disease Journal in July 2002. During that time 50,000 students were sickened, 1,500 were hospitalized, and 1 child died, says the CDC's Tauxe, who adds that the numbers are probably an underestimate.

"Many of these illnesses should be preventable by making sure the foods are prepared following the usual food-safety guidelines," Tauxe says. "The risk would be further reduced by broader application of irradiation."

The problem is that food safety isn't being adequately addressed. Only one-third of some 800 school food-service directors surveyed in March 2003 by the American School Food Service Association said they have programs that detail where contamination might occur and provide systems to prevent it. That finding prompted the group to ask Congress for money to create such safety systems.

Health

Between 1964 and 1992, three United Nations agencies, including the World Health Organization, convened five expert committees to evaluate studies on the safety of consuming irradiated foods. Each found the foods to be safe.

Recent European research, however, suggests that the substances known as 2-alkylcyclobutanones, unique by-products created by irradiating fat in a food such as ground beef, may act as tumor promoters in laboratory rats. Authors of the report, released last fall by Germany's Federal Research Centre for Nutrition in Karlsruhe, say their findings show the need for further study. Meanwhile, the European Parliament in December halted new approvals of irradiated foods going to member nations of the European Union pending more safety studies.

The research was brought to the attention of the FDA by the Washington, DC–based nonprofit consumer groups Public Citizen and the Center for Food Safety. The FDA is reviewing the research, says George Pauli, the agency's associate director for science and policy.

Will the Market Decide?

Until the recent marketing blitz for irradiated meat, irradiated food was a nonissue for consumers. For decades the government has allowed certain foods to be irradiated, including wheat and flour, to control insects, and white potatoes, to inhibit sprouting. Since 1985, the government has approved irradiation of spices, fruits, vegetables, pork, and poultry. In 1997, irradiation was OK'd for beef, and in 2000 for fresh eggs.

But it has rarely been used on foods in the United States, in part because of concerns that consumers wouldn't buy irradiated products. Indeed, when irradiated beef was introduced three years ago in groceries, it was withdrawn because of poor sales.

Today, irradiated beef accounts for less than 5 percent of the 9 billion pounds of ground beef produced annually in the United States, says the American Meat Institute Foundation, a nonprofit meat industry group. (Irradiated frozen chicken was introduced this winter only in Publix stores; the company would not disclose sales.) In some stores irradiated meat is somewhat more expensive than nonirradiated meat. In others, the prices are comparable.

The labeling of irradiated meat is a subject of debate. Currently, packages of irradiated meat must be marked with the radura, the international symbol of irradiation, and with words such as "Treated with irradiation."

But "Electronic pasteurization," a term favored by some irradiators, might also be allowed because the 2002 Farm Act broadens the definition of pasteurization and allows anyone to petition the FDA for alternative labeling of irradiated food.

Restaurant patrons and parents may find themselves even more confused. While some restaurants are promoting their use of irradiated meat, no federal regulation requires restaurants or school cafeterias to disclose that they serve it. . . .

Consumers Union believes that the best way to improve meat quality is to clean up the food-supply chain and strengthen USDA authority over meat safety.

Schools should be given the resources to assess food handling, preparation, and storage procedures and to fix problems.

CU supports further tests of chemical byproducts created by meat irradiation.

Irradiated foods should continue to be labeled "Irradiated." Calling them "Pasteurized" or anything else is misleading.

More care must be taken to ensure that information consumers receive concerning irradiated foods is accurate.

To learn more about CU's position or to contact the appropriate authorities about food labeling, school lunches, and related issues, visit www.consumersunion.org.

* * * * *

DISCUSSION QUESTIONS AND NOTES

7.15. Further information. U.S. Regulatory Requirements for Irradiating Foods, available at: http://www.cfsan.fda.gov/~dms/opa-rdtk.html (last accessed Mar. 15, 2008).

7.16. Labeling. How do you think irradiated food should be labeled? Explain your reasons.

7.17. Further safety information. You may find more information on food irradiation and its safety in the following publications: *Wholesomeness of Irradiated Food*, Report of a Joint FAO/IAEA/WHO Expert Committee, Technical Report Series 659, World Health Organization (1981). *Food Irradiation: A Technique for Preserving and Improving the Safety of Food*, World Health Organization and Food and Agriculture Organization of the United Nations (1988). *Safety and Nutritional Adequacy of Irradiated Food*, World Health Organization (1994).

7.18. Additional news. One source for news and reporting on this topic is the archives of the Food Safety Network (FSNet). Go to http://archives.foodsafety.ksu.edu/search.html and search for "irradiated food."

7.19. Is heating foods a greater risk than irradiation? Morton Satin has recommended a broad-based review on the relative risks of consuming irradiated foods rather than foods cooked medium. He said, "The principal function of food irradiation in food safety is to minimize the risk of pathogens. In this cold process, authoritative scientific studies have made it clear that no new hazards are introduced. But cooking meat well done to an internal temperature of 160 degrees—the most common method for reducing foodborne hazards—introduces a whole range of new problems whose significance is just becoming known." Cooking foods to a well-done increases the amounts of heterocyclic products produced and consumed. "This increases the risk of a range of conditions including colorectal adenomas, adenocarcinoma of the stomach and esophagus and breast, lung and prostate cancer," Satin says. "Parents, dieticians, nutritionists, and administrators are pouring their hearts into the irradiated ground beef debate without the benefit of all the facts. If they had all the facts and understood the negative effects of high-temperature, well-done cooking, they would consider the matter differently. In all likelihood, they would regard food irradiation to be a godsend for school lunch food safety." Morton Satin, author of FOOD IRRADIATION: A GUIDEBOOK, speaking at the International Meeting on Radiation Processing in Chicago, Illinois (Sept. 10, 2003).

PART IV

SPECIALIZED FOOD REGULATION

CHAPTER 8

Dietary Supplements

8.1 INTRODUCTION

The regulation of dietary supplements is one of the most controversial topics in food law. Without taking a position for or against the current regulatory approach, this text discusses aspects of the controversy in the process of illuminating how dietary supplements are regulated.

The Food and Drug Administration regulates dietary supplements under additional and different regulations than those covering conventional foods and pharmaceutical products. However, the conventional food regulations additionally may also apply to dietary supplements. This chapter covers the key elements of dietary supplement regulation, how the Dietary Supplement Health and Education Act of 1994 (DSHEA) changed the U.S. regulatory approach, the main roles of the government agencies and self-regulatory entities involved in the oversight of dietary supplements, and important recent developments in dietary supplement regulation.

8.2 BACKGROUND

8.2.1 What Are Dietary Supplements?

For many years dietary supplements were simply food items intended to balance the diet or provide for special dietary needs. Vitamins, minerals, and foods for those on medical diets were the main products.

Today, the lines are not so clearly drawn. Herbs and amino acids are added to conventional foods. Traditional foods may be enhanced with functional ingredients. In addition, herbs, extracts, and other supplements have entered the mainstream markets, not just the small health food stores.

Accompanying the new products are new terms. Many of these terms lack legal definitions in the United States, therefore, their usage and meaning often blur.

Food Regulation: Law, Science, Policy, and Practice, by Neal D. Fortin
Copyright © 2009 Published by John Wiley & Sons, Inc.

Nutraceutical combines the words nutrition and pharmaceutical to describe foods that contain a supplement or enhancement to provide a specific health benefit. For example, a combination of herbs, amino acids, and vitamins in a capsule that the manufacturer claims raises energy levels would be a nutraceutical.

Functional food is sometimes used as a synonym for nutraceutical, as are pharmafood and FoSHU (food for specified health use). However, functional food more often refers to conventional food products that are enhanced to provide specific health benefits (as opposed to pills, liquids, and powders). For example, a calcium-enriched orange juice or higher calcium yogurt are functional foods.

These new terms have no legal significance, but they do indicate how the marketplace is changing. The traditional lines between drugs, dietary supplements, and conventional foods are more diffuse than ever. Soft drinks may be enriched with vitamin C, plant sterols added to margarine and yogurt to lower cholesterol, and herbs such as *Echinacea* added to foods to boost the body's immunity.

NOTES AND QUESTIONS

8.1. What does the public think? What does the public think dietary supplements are?

8.2.2   Background to DSHEA

At the time of passage of the **Dietary Supplement Health and Education Act of 1994**[1] (**DSHEA** pronounced "DaShay"), the dietary supplement business was a $3.7 billion industry.[2] Since the passage of DSHEA, the sale of dietary supplements has increased, and by 1996, consumers spent over $6.5 billion on dietary supplements.[3] In 2006, total sales for the dietary supplement in the United States were estimated at $22.1 billion.[4]

DSHEA was the culmination of a legislative and regulatory battle that was fought for more than two decades. Prior to DSHEA in 1994, FDA largely

---

[1] Dietary Supplement Health & Education Act (DSHEA) of 1994, Pub. L. No. 103–417, 108 Stat. 4325 (codified in scattered sections of 21 U.S.C.).
[2] Congressional Research Service, *Dietary Supplements: Use and Regulation* 18, 24 (Rep. 94–208 SPR, Mar. 4, 1994).
[3] Kelly Ann Kaczka, *From Herbal Prozac to Mark McGwire's Tonic: How the Dietary Supplement Health and Education Act Changed the Regulatory Landscape for Health Products*, 16 Journal of Contemporary Health Law and Policy 463, 464 (2000).
[4] Kathleen M. Zelman, *The Truth Behind the Top 10 Dietary Supplements*, WebMD Weight Loss Clinic-Feature (citing Patrick Rea, editor of the market research publication NUTRITION BUSINESS JOURNAL), *available at*: http://www.webmd.com/diet/features/truth-behind-top-10-dietary-supplements (last accessed Apr. 3, 2008).

controlled the therapeutic and health-benefit claims on dietary supplements by applying the drug regulatory standards. FDA evaluates and regulates drugs under a rigorous program to ensure effectiveness and safety.[5] Therapeutic claims for drugs must be substantiated and pre-approved by the FDA. This level of regulatory oversight is costly to the industry.

Brief History

> Sometimes perhaps his Guilded pill prevails,
> But if that fails, the Dead can tell no tales.
> What if his medicines thousands Lives should spill?
> Hangmen and Quacks are authorized to Kill.
> —English Ballad (1691)[6]

Prior to passage of the Pure Food and Drug Act of 1906, many believed that peddlers of quack medicines had easy pickings.[7] In response, Congress passed the 1906 Act to empower FDA to take action against "snake oil" and quack health claims. FDA used these powers to regulate health claims on dietary supplements.

Beginning in the 1970s, supplements began to change and to find a larger consumer market. FDA continued its efforts to control the supplement market but found increasing resistance from the industry. In the 1970s FDA tried to write regulations for mega-dose vitamin products because of their potential health risks.[8] The supplement industry waged a lobbying effort against FDA's attempt, and the Proxmire Vitamin and Mineral Amendment of 1976 expressly prohibited FDA from establishing maximum limits on the potency of any vitamin or mineral.[9]

In the 1970s and 1980s, FDA's approach to regulation of nutrition and health claims on foods and dietary supplements was conservative. This approach to handling nutrition and health claims proved too restrictive for modern marketplace and the state of scientific knowledge on nutrition. The Nutrition Labeling and Educational Act of 1990 (NLEA)[10] opened the door for a more lenient approach.

At the same time, NLEA provided FDA with enhanced authority over the labeling of foods, including dietary supplements. About this time, from 1989 to 1990, several thousand illnesses and 37 deaths occurred from ingestion of

---

[5] *See* 21 U.S.C. § 403.

[6] C.J.S. THOMPSON, Quacks of Old London 46 (2003).

[7] *E.g.,* Peggy Robbins, *A Brief History of Quack Medicines in America,* AM. LEGION MAG., (Mar. 1975), at 6.

[8] *See* Nat'l Nutritional Foods v. Matthews, 557 F.2d 325 (2d Cir. 1977).

[9] 21 U.S.C. § 350(1).

[10] Nutrition Labeling and Education Act (NLEA) of 1990, Pub. L. No. 101–535, 104 Stat. 2353 (codified in 21 U.S.C. § 343–1).

the dietary supplement L-tryptophan.[11] FDA established a task force to examine the problems associated with dietary supplements. The task force recommended regulating dietary supplements as drugs when they are used as anything other than food additives.[12]

Opponents of the task force recommendations and of application of the NLEA requirements to dietary supplements mounted a large-scale lobbying effort. The culmination of this battle was DSHEA in 1994. DSHEA represents a significant departure from FDA institutional history of increasing power. Over its 100 plus year history, Congress granted FDA ever expanding authority. On the other hand, DSHEA had an overall deregulatory affect.

Congress reasoned that the public would be better served through less regulation of dietary supplements. Congress also noted that it saw dietary supplements as an inexpensive means to promote public health and that consumers should be empowered to make choices in preventative health care. Moreover Congress found that safety problems with dietary supplements were relatively rare. Therefore these products could be less regulated than drugs.

NOTE

8.2. For the FDA's overview of its role in dietary supplement regulation, see http://www.cfsan.fda.gov/~dms/ds-oview.html#what.

8.3   THE STATUTORY DEFINITION

DSHEA amended the Food, Drug, and Cosmetic Act (FD&C Act). One of the key amendments was the addition of a definition for "dietary supplement."[13] This definition of dietary supplement is convoluted and hard to follow. For example, "dietary supplements" are "deemed to be food" within the meaning of the FD&C Act.[14] Yet supplements are exempt from significant requirements that apply to foods, such as the food additive requirements.[15] In short, dietary supplements are not regulated the same as foods. Therefore, it is confusing to think of dietary supplements as a subcategory of foods under the FD&C Act. The DSHEA provision, "Except for purposes of paragraph (g), a dietary supplement shall be deemed to be a food within the meaning of this chapter,"[16] does not so much define dietary supplements as exempt them from the food additive requirements and other premarket approval require-

---

[11] General Accounting Office, *Report on FDA Management of Dietary Supplements* (HRD-93-28R, July 2, 1993), *reprinted in* 5 FDLI Food and Drug Report 4, at 566 (1994).
[12] 58 Fed. Reg. 33,697 (1993).
[13] 21 U.S.C. § 321 (ff).
[14] *Id.*
[15] *Id.* § 321 (s).
[16] *Id.* § 321 (ff).

ments unless the supplements fall into the definition for drug (paragraph (g) refers to the drug definition).[17]

Statutory definitions often hold important regulatory implications and even explicit requirements. Therefore it is important to remember that our everyday use of a term, such as dietary supplement, may bear little semblance to the statutory definition. This seeming contradiction in usage can become confusing unless the statutory definitions are treated as a separate language.

* * * * *

Federal Food, Drug, and Cosmetic Act

28 U.S.C. § 321 (emphasis added)

For the purposes of this Chapter—

. . . .

(ff) The term "dietary supplement"—

    (1) means a product (other than tobacco) intended to supplement the diet that bears or contains one or more of the following dietary ingredients:

        (A) a vitamin;

        (B) a mineral;

        (C) an herb or other botanical;

        (D) an amino acid;

        (E) a dietary substance for use by man to supplement the diet by increasing the total dietary intake; or

        (F) a concentrate, metabolite, constituent, extract, or combination of any ingredient described in clause (A), (B), (C), (D), or (E);

    (2) means a product that—

        (A)(i) is intended for ingestion in a form described in section 350(c)(1)(B)(i) of this title; or

          (ii) complies with section 350(c)(1)(B)(ii) of this title;

        (B) is not represented for use as a conventional food or as a sole item of a meal or the diet; and

        (C) is labeled as a dietary supplement; and

    (3) does—

        (A) include an article that is approved as a new drug under section 355 of this title or licensed as a biologic under section 262 of title 42 and was, prior to such approval, certification, or license, marketed as a dietary supplement or as a food unless the Secretary has issued a regulation, after notice and comment, finding that

---

[17] *Id.* § 349.

the article, when used as or in a dietary supplement under the conditions of use and dosages set forth in the labeling for such dietary supplement, is unlawful under section 342(f) of this title; and

(B) not include—

(i) an article that is approved as a new drug under section 355 of this title, certified as an antibiotic under section 357 of this title, or licensed as a biologic under section 262 of title 42, or

(ii) an article authorized for investigation as a new drug, antibiotic, or biological for which substantial clinical investigations have been instituted and for which the existence of such investigations has been made public, which was not before such approval, certification, licensing, or authorization marketed as a dietary supplement or as a food unless the Secretary, in the Secretary's discretion, has issued a regulation, after notice and comment, finding that the article would be lawful under this chapter.

Except for purposes of paragraph (g), a dietary supplement shall be deemed to be a food within the meaning of this chapter.

* * * * *

8.3.1   What Is a Dietary Supplement

To be a considered a "dietary supplement" a product must meet *all* of the following requirements.

Intended use. Be intended to supplement the diet (that is, not intended to be a drug or a conventional food).[18]

Ingredients. Contain one or more of:

a vitamin;

a mineral;

an herb or other botanical;

an amino acid; or

a dietary substance for use by humans to supplement the diet by increasing the total dietary intake.[19]

Form. Be for *ingestion* in capsule, tablet, liquid, powder, softgel, or gelcap form, *or* into products that are in conventional food form, *and* must not

---

[18] *Id.* § 321(ff)(1).
[19] *Id.* § 321(ff)(1)(A)-(E).

be represented as a conventional food *nor* as a sole item of a meal or diet.[20]

Label. Be labeled as a dietary supplement.[21]

Exclusions. Not be a new drug—unless sold as a dietary supplement before 1994—or tobacco.[22]

Although the definition of "dietary supplement" is relatively complicated, the coverage is quite broad. For example, note that the list of eligible ingredients is nearly all-encompassing because almost any ingredient could be considered "a dietary substance for use by man to supplement the diet by increasing the total dietary intake."[23]

8.3.2 What Is Not a Dietary Supplement

Represented as a Conventional Food   The most important limitation in the definition of a dietary supplement is that—when sold in conventional food form—a dietary supplement must not be represented as a conventional food. No definitive rule exists for determining what constitutes representation as a conventional food; however, if the food is consumed primarily for taste or aroma, it probably would be considered a conventional food.

Sole Item of a Meal or Diet   Note that meal replacements also are excluded from the dietary supplement category.

Intended for Ingestion   The definition of dietary supplement also limits the category to those products that are "intended for ingestion." "Ingestion" within definition of "food for special dietary use" (which is excluded from regulation as a "drug") means to take into the stomach and gastrointestinal track by means of enteral administration, and thus does not apply to nasal administration of vitamin preparation.[24]

In the next case, *United States v. Ten Cartons of Ener-B Nasal Gel*, the FDA brought enforcement action against Nature's Bounty and their product, Ener-B Nasal Gel, a vitamin B-12 supplement in gel form designed to be applied to the inside of the nose. FDA charged that Ener-B was an unapproved new drug within meaning of the Food, Drug, and Cosmetic Act. Nature's Bounty argued their product was a dietary supplement, but the court held that a preparation for nasal administration was neither a "food for special dietary use" nor a "dietary supplement" under DSHEA. An injunction against the shipment of Ener-B was issued.

---

[20] *Id.* § 321(ff)(2).
[21] *Id.* § 321(ff)(2)(C).
[22] *Id.* §§ 321(ff)(1), 321(ff)(3).
[23] *Id.* § 321(ff)(1)(E).
[24] Federal Food, Drug, and Cosmetic Act § 411(c)(1)(B), 21 U.S.C. § 350(c)(1)(B).

* * * * *

United States v. Ten Cartons of Ener-B Nasal Gel

888 F. Supp. 381 (1995)

SPATT, District Judge:

Before the Court are the objections of the defendant Nature's Bounty, Inc. ("Nature's Bounty" or "defendant") to the Report and Recommendation of United States Magistrate Judge Allyne R. Ross . . . regarding the defendant's nasally administered vitamin B-12 preparation called Ener-B Nasal Gel ("Ener-B"). This Court referred to Judge Ross the issue of whether Ener-B is a "food" or a "drug" within the meaning given to these terms by the Federal Food, Drug, and Cosmetic Act ("FD&C Act" or the "Act"). Judge Ross found that the Food and Drug Administration ("FDA") reasonably determined that Ener-B was a drug and not a "food" within the meaning of sections 201(f) and 201(g)(1)(C) of the Act, 21, and recommended that the Court defer to the agency's determination. . . .

Background

The defendant markets Ener-B, which is intended to be applied to the inside of one's nose. As intended to be used, the vitamin B-12 contained in Ener-B bypasses digestion through the gastrointestinal tract, where it would be absorbed into the body through the intestines. Instead, Ener-B's vitamin B-12 is absorbed directly into the bloodstream through the nasal mucosa.

On February 26, 1987, the FDA notified Nature's Bounty that the FDA considered Ener-B to be a "drug" under the FD&C Act, and that Ener-B was being marketed illegally because it had not received recognition or approval as a "new drug" under the Act. The FDA also alleged that Ener-B was misbranded and improperly labeled under the Act. The FDA's notice informed Nature's Bounty that the Act provided for the seizure of illegal products, and for an injunction against the distributor of such products.

Nature's Bounty responded to the FDA's letter, and on April 2, 1987, filed a Citizen Petition with the FDA pursuant to 21 C.F.R. § 10.30 (1994). In its petition Nature's Bounty essentially contended that Ener-B was a dietary supplement which was considered a "food" under the Act, and Ener-B's route of administration into the body bypassing digestion through the gastrointestinal tract did not reconstitute it as a "drug" under the Act. The petition requested (i) that the FDA establish and make public its policy regarding whether the method of ingestion of a substance otherwise classified as a food may make it a drug under the Act; (ii) promulgate a rule or guideline subject to notice and comment with respect to its policy; and (iii) refrain from taking any administrative or enforcement action against Ener-B in the absence of any policy delineated by a rule or guideline.

On May 24, 1988, the FDA denied Nature's Bounty's petition. As described in greater detail later in this Opinion, the FDA explained its denial on the grounds that it considered Ener-B to be a "drug" within the meaning of the Act because Ener-B affected the structure of the human body, and that Ener-B could not be a "food" within the meaning of the relevant statutory section because it was not ingested—namely it was not enterally administered into the gastrointestinal tract.

Subsequent to the denial of Nature's Bounty's Citizen Petition, the United States ("government" or "plaintiff"), on behalf of the FDA, instituted an *in rem* proceeding against Ener-B pursuant to 21 U.S.C. § 334 on September 28, 1988, and seized ten cartons of Ener-B from Nature's Bounty. Approximately eighteen months later, on May 11, 1990, the government brought a second action against Nature's Bounty *in personam*, pursuant to 21 U.S.C. § 332(a), seeking to permanently enjoin Nature's Bounty from selling Ener-B.

In October of 1991, the government moved for summary judgment in its favor on the complaints in both of these cases. Relying on the depositions and declarations of two FDA scientists, the government expounded on the rationale for denying Nature's Bounty's Citizen Petition and contended that Ener-B is a drug within the scope of 21 U.S.C. § 321(g)(1)(C) because it is labeled and marketed as a product which "affects the structure or function of the body." That section defines the term "drug" as "articles (other than food) intended to affect the structure or any function of the body of man or other animals."

The government also contended that Ener-B cannot be a "food" within the meaning of the parenthetical exception "(other than food)" found in section 321(g)(1)(C), because the phrase "other than food" is, according to the government, construed to mean food in the conventional sense; namely articles which are ingested through the mouth for the primary purposes of nutrition, taste or aroma, and which are absorbed into the body through the gastrointestinal tract. According to the government, a vitamin nasal gel by which the vitamins are absorbed into the bloodstream through the nasal mucosa hardly fits the conventional meaning of the term "food."

In addition, the government contended that Ener-B is a drug within the meaning of 21 U.S.C. § 321(g)(1)(B), insofar as the defendant's labeling and promotional material claimed that Ener-B mitigates the effects of several medical conditions, including lack of the "intrinsic factor," a substance produced by the stomach which is required for the absorption of vitamin B-12. Section 321(g)(1)(B) defines the term "drug" as "articles intended for use in the diagnosis, cure, mitigation, treatment, or prevention of disease in man or other animals."

In opposing the government's motion, the defendant submitted the declarations of seven experts, including medical doctors, attesting to the fact that Ener-B is a food, as well as the declaration of the executive vice-president of Nature's Bounty attesting to the company's marketing practices. Essentially, the defendant contended that Ener-B is a food because it functions as a "food for special dietary use" within the meaning of section 411 of the FD&C Act, 21 U.S.C. § 350, which is the section governing the regulation of vitamins and minerals.

. . . .

According to Nature's Bounty, Ener-B met the definition of a "food to which [section 350] applies," because it is a "food for special dietary use" which contains a vitamin that is not intended for ingestion, and does not simulate or represent itself to be a conventional food or the sole item of a meal or of a diet. Moreover Nature's Bounty contended that because Ener-B is a "food for special dietary use," Ener-B also met the statutory definition of "food" in 21 U.S.C. § 321(f). Section 321(f) provides that the term "food" means "(1) articles used for food or drink for man or other animals, (2) chewing gum, and (3) articles used for components of any such article."

In addition, Nature's Bounty contended that nowhere in the FD&C Act is the classification of a product as a food or a drug dependent on its route of administration or place of absorption. Rather, the defendant contended that the function of Ener-B as a "food for special dietary use," and not its route of absorption into the body, is the relevant factor for determining whether or not Ener-B is a food under the meaning given to that term by the Act.

. . . .

[T]he Court referred the matter of whether Ener-B was a "drug" or "food" under the Act to Judge Ross for a hearing and report and recommendation. The hearing was to include the testimony of expert witnesses, if offered.

Findings and Conclusions by Judge Ross

. . . .

Judge Ross found that the credible expert testimony offered at the hearing, by both the government's and the defendant's experts, strongly supported the conclusion that the common sense and scientific definitions of "food" entail two elements: (i) nutrient intake, and (ii) ingestion into the gastrointestinal tract of such nutrients, also known as enteral administration of nutrients. According to Judge Ross, ingestion into the gastrointestinal tract was viewed as a necessary element of a food by the medical and scientific communities:

> [The] common sense and scientific definitions of 'food ... for man' that incorporate as a necessary element ingestion into the gastrointestinal tract are both reasonable and accepted by a substantial segment of the medical and scientific community. As evidenced by certain defense expert definitions, other respectable segments of the scientific community apparently adopt a more expansive definition that includes parenterally administered nutrients as well. This showing, however, does not impeach the evidence that a definition requiring enteral administration is reasonable and is also well accepted by credible scientists.

. . . .

According to Judge Ross, the experts also agreed "virtually unanimously" that the mucosa lining the nasal cavity are functionally dissimilar to the membranes lining the gastrointestinal tract. As a result of this difference, Judge Ross found that both parties' experts acknowledged (1) "that parenteral administration bypasses the normal physiological safety mechanisms present in the gastrointestinal tract," and (2) that the route of exposure to the nutrients, namely whether absorption occurs through the nasal mucosa or through ingestion, has an impact on toxicity.

Based on these findings, Judge Ross concluded that the agency's determination that Ener-B is not a food based on its route of administration into the body is reasonable, and in accord with the Second Circuit's interpretation of the term "food" in the parenthetical exception to section 321(g)(1)(C); namely, that Congress intended the term to have "the everyday meaning of food."

With regard to the defendant's contention that Ener-B was a "food for special dietary use" under the Vitamins and Minerals section of the FDA, Judge Ross concluded that the contention was untenable because that section applies to "food for humans which is a food for special dietary use." See 21 U.S.C. § 350(c)(1). Interpreting the statute as requiring that the vitamin or mineral must first be a "food for humans" within the meaning of section 321(f)(1) before it can be subject to section 350(c)'s provisions as a "food for special dietary use," Judge Ross determined that Ener-B was unqualified as a food for special dietary use because it was not a "food" within the meaning of section 321(f)(1).

In addition, Judge Ross concluded that the defendant's construction of section 350(c)(1)(B)(ii) was erroneous. As mentioned earlier, section 350(c)(1)(B) provides that a food for special dietary use must "(i) [be] intended for ingestion in tablet, capsule, or liquid form, or (ii) if not intended for ingestion in such a form, does not simulate and is not represented as conventional food ... [or] for use as a sole item of a meal or of a diet." Nature's Bounty contended that Ener-B is a "food for special dietary use" because under section 350(c)(1)(B)(ii) it is "not intended for ingestion."

Judge Ross determined that the defendant's construction did not accord with the language of section 350(c)(1)(B)(ii) because the defendant selectively interpreted the statute by neglecting the words "in such form" at the end of the phrase "is not intended for ingestion in such form." According to Judge Ross, the words "in such form" are properly interpreted as referring to the forms of ingestion specified in the previous section of the statute, 350(c)(1)(B)(i), namely "tablet, capsule, or liquid form." Thus, Judge Ross interpreted the language of section 350(c)(1)(B)(ii) as supporting the FDA's contention that Ener-B was not a "food," since the forms of ingestion specified in the statute involve enteral administration of the vitamin by swallowing. ...

Based on her findings and conclusions, Judge Ross recommended that this Court (1) defer to the agency's determination that Ener-B is an unapproved "new drug" pursuant to 21 U.S.C. § 321(p), (2) that as such, Ener-B is subject to condemnation under 21 U.S.C. § 334, and (3) that Nature's

Bounty be permanently enjoined from selling Ener-B, pursuant to 21 U.S.C. § 332.

Nature's Bounty's Objections

In response to Judge Ross's Report, Nature's Bounty filed its Objections to the Report and Recommendation of Judge Ross ("Objections") contending that Judge Ross erred both legally and factually. . . .

. . . .

Nature's Bounty contends that the DSHEA substantially affects the posture of this case because the DSHEA has established a new class of products called "dietary supplements," which are defined as foods and are allegedly excluded from regulation as a drug under section 321(g)(1)(C). According to the defendant, all of the testimony and evidence presented by the government before Judge Ross to support its contention that Ener-B is not a food is "now moot." Indeed, the defendant contends that the debate over the "food" exemption under section 321(g)(1)(C) is "totally irrelevant" in light of the DSHEA, and that the Court should disregard Judge Ross's report as well as the parties' objections and response papers. Instead, Nature's Bounty claims that the Court should focus on whether Ener-B is exempt from regulation as a "drug" because it is a "dietary supplement" under the DSHEA that is subject to an entirely new statutory-regulatory framework.

The Court will consider the defendant's objections to Judge Ross's Report, and its contentions regarding the affect of the DSHEA on this case, after reviewing the applicable standard governing review of this matter.

Discussion

Applicable Standard of Review

. . . .

1. Construction of the Statutory Provision at Issue

The relevant statutory provision at issue is the meaning of the term "food" set forth in the parenthetical exception "(other than food)" in section 321(g)(1)(C). That meaning, in turn, is provided by the definition of "food" in section 321(f)(1), which states "food" is: "articles used for food or drink for man or other animals." According to Nature's Bounty, rather than considering whether Ener-B comes within the ambit of the entire relevant definition of "food," namely whether Ener-B is an "article[ ] used for food . . . for man," Judge Ross erroneously formulated the issue to be determined by focusing her inquiry on the latter part of the definition at issue, that is, whether Ener-B is "food . . . for man." As a result, Nature's Bounty objects to Judge Ross's Report because it is allegedly premised on neglecting "key elements" of the statutory definition of food, specifically the phrase "used for food."

The Court disagrees with the defendant's objection for two reasons. First, in the Court's view, whether one concentrates on the phrase "used for food ... for man" or on the phrase "food ... for man," the object of the statutory construction is the same; namely determining what is denoted by the word "food." In this regard, the meaning of the term "food" in the parenthetical exclusion of section 321(g)(1)(C) and in section 321(f)(1) has been construed to refer to that term's "everyday meaning," and a particular article's "common usage" as food. *See Nutrilab, Inc. v. Schweiker*, 713 F.2d 335, 337–38 (7th Cir. 1983) (the term "food" in section 321(f)(1) is to be defined in terms of its function, and means "articles used by people in the ordinary way most people use food—primarily for taste, aroma or nutritive value"). Moreover, the common-use meaning given to the term "food" in the parenthetical exclusion of section 321(g)(1)(C) and in section 321(f)(1) is more circumscribed and not identical to the statutory meaning of "food" under section 321(f).

Judge Ross appropriately relied on this interpretation of the phrase "used for food ... for man" to reach her own conclusions and recommendations.

Second, and more significant, the Court believes that Judge Ross properly concentrated on construing the term "food" by focusing on its context in the phrase "food ... for man" rather than "use for food." The pivotal and novel issue presented in this case is not so much whether Ener-B is a food, as it is whether the FDA may consider the route of administration of a product in determining its classification as a food or a drug under the Act. Given that the definition of a food in section 321(f)(1) already entails consideration of the functional, everyday-use aspect of food, emphasis on the human—"for man"—aspect of the "use of food" seems appropriate and rational when determining whether the route of administration can be a factor in classification. Indeed, with regard to humans it is the Court's view that the route of administration cannot be ignored when considering the everyday, common-use aspect of a particular article as food.

The Court further believes that by contending Ener-B is a "food" because it is "used for food," the defendant has lost sight of the real issue in this case. Nature's Bounty's syllogism is as follows: the phrase in the statute "used for food" means using a product to introduce nutritive substances into the body; vitamin B-12 is an essential nutrient for humans and is used for food; Ener-B provides vitamin B-12 for the body; Ener-B, therefore, meets the statutory definition of food because it is "an article used for food" for man. The problem with this reasoning is that it presumes to answer the very question at issue by substituting the "article" Ener-B for the "food" vitamin B-12. As explained above, the phrase "used for food for man" in section 321(f)(1) is interpreted by the everyday meaning and common usage of the term food. Although vitamin B-12 may commonly be used as a food, gels containing vitamin B-12 that are administered through the nose hardly meet the everyday definition of food and are not commonly used as food, anymore than an enema containing vitamin B-12 meets the everyday definition of food.

The defendant also misconstrues the pertinent case law and this Court's October 9, 1991, Decision by contending that the key in determining whether a product is a food or a drug is the vendor's intent. The vendor's intent is a key element only with respect to the statutory definition of a drug under sections 321(g)(1)(B) and 321(g)(1)(C). That is not the case, however, with regard to determining whether a product is a food. As the court in *American Health Products* stated:

> The ordinary way in which an article is used, therefore, not any marketing claim on the part of the manufacturer or distributor as to specific physiological purpose of that use, should determine whether it is a food for the purpose of the parenthetical exclusion of section 321(g)(1)(C).

Am. Health Prods., 574 F.Supp. at 1505.

This Court's statement in its October 9, 1991, Decision that "where the substance in question is not recognized in official pharmacopoeia, 'the vendor's intent in selling the product to the public is the key element in this statutory definition,'" is not to the contrary. In its Decision the Court was referring to a vendor's intent with regard to the definition of a drug, not of a food. Moreover, the Court notes that although the vendor may not intend to sell an item as a drug, the product may still be regulated as such. *See, e.g., Mathews*, 557 F.2d at 334 (the FDA is not bound by the manufacturer's subjective claims of intent, and can classify an article as a drug based on objective evidence of therapeutic intent).

Accordingly, Judge Ross properly formulated the statutory issue to be construed, and did not err or contravene this Court's September 29, 1991, Decision by discounting the defendant's proclaimed intent that Ener-B is a food because it is to be used for its nutritional value.

2. Ener-B Is Neither a Food to Which Section 350 Applies, nor a Dietary Supplement under the DSHEA

In 1976, Congress amended the FD&C Act by enacting section 411 of the Act, 21 U.S.C. § 350, governing the regulation of vitamins and minerals, commonly known as the Proxmire Amendments. The underlying purpose of this amendment involves the promotion of concentrated vitamins to help supplement peoples' diets. Basically, section 350 precludes the FDA from regulating a vitamin or mineral solely on the basis of its potency or combination with another vitamin or mineral. The provisions of section 350 apply to certain vitamins or minerals which meet the definition in section 350(c) of a "food to which [section 350] applies." Up until the enactment of the DSHEA, Nature's Bounty contended that Ener-B met the criteria set forth in section 350(c) of a "food to which [section 350] applies."

In order to further facilitate the use of vitamins and minerals to combat nutritional deficiencies and disease, portions of the FD&C Act, including section 350, were amended by the DSHEA in October 1994. The basic purpose of the DSHEA amendments to the FD&C Act is to ensure that the public has over-the-counter access to "dietary supplements," which include vitamins, minerals, amino acids, and herbs. In order to accomplish this, the DSHEA precludes the FDA from regulating "dietary supplements" as a "drug" under section 321(g)(1)(C) solely because of any statements on the products' labeling regarding claims that the product can treat or affect a nutritional deficiency or disease, unless the FDA determines that the product is not safe.

Following the enactment of the DSHEA amendments to sections 321(g) and 350, Nature's Bounty now contends that Ener-B is a "dietary supplement" as that term is defined in the DSHEA, and as such is excluded from regulation as a drug under the FD&C Act. The Court disagrees with both of Nature's Bounty's contentions, and believes that Ener-B is neither a food, to which section 350 applies, nor a dietary supplement under the DSHEA.

A. Ener-B Is Not a Food to which Section 350 Applies

In order to be excluded from regulation as a "drug" under the provisions of sections 350(a) and (b)—in other words, in order to be a "food to which [section 350] applies"—a product must, under the definition of that phrase in section 350(c) prior to the DSHEA amendments, be a "food for humans which is a food for special dietary use" which (A) is a vitamin or mineral, and (B) "which is (i) intended for ingestion in tablet, capsule, or liquid form, or (ii) if not intended for ingestion in such a form, does not simulate and is not represented as conventional food . . . [or] for use as a sole item of a meal or of a diet."

As explained earlier, Nature's Bounty contends that Ener-B meets the definition of a "food to which [section 350] applies" because Ener-B is a food for special dietary use which is a vitamin, and which, according to the defendant, is "not intended for ingestion" within the meaning of section 350(c)(1)(B)(ii).

While the defendant is correct to assert that Ener-B may be used for a "special dietary use" as that term is defined in section 350(c)(3)(B), namely "to supply a vitamin for use by man to supplement his [or her] diet," the defendant's contention that Ener-B is subject to the protection of section 350 ultimately fails because the remaining requirements that are necessary for Ener-B to be a "food" to which section 350 applies are not met. Specifically, under section 350(c)(1)(B) Ener-B must be "intended for ingestion" either in a tablet, capsule, or liquid form, or if not in that form, then it must be "intended for ingestion" in some other form. Ener-B, however, is not intended for "ingestion" as that term is meant to be used in the statute. The defendant's construction of section 350(c)(1)(B)(ii) to apply to foods for special dietary use that are "not intended for ingestion" is erroneous because it reads out of the statute the words "in such form."

The basic canon of statutory construction is that interpretation of the statute must "begin with the language of the statute itself." The usual "assumption [is] that the legislative purpose is expressed by the ordinary meaning of the words used," and "the plain meaning of the statute's language should control except in the 'rare cases [in which] the literal application of a statute will produce a result demonstrably at odds with the intentions of the drafters.'"

If the statute is clear and unambiguous that is the end of the matter, for the court must give effect to the unambiguously expressed intent of Congress. "In ascertaining the plain meaning of the statute, the court must look to the particular statutory language at issue, as well as the language and design of the statute as a whole."

The ordinary and plain meaning of the term "ingestion" means to take into the stomach and gastrointestinal tract by means of enteral administration.

The interpretation of the term "ingestion" to mean enteral administration into the stomach and gastrointestinal tract is also supported by the language of the statutory sections immediately preceding and following section 350(c)(1)(B)(ii). Section 350(c)(1)(B)(i) states that the vitamin must be intended for ingestion in tablet, capsule or liquid form. Each of these forms denotes a method of ingestion that involves swallowing into the stomach. Section 350(c)(2) states that a food is intended for ingestion in liquid form under section 350(c)(1)(B)(i) "only if it is formulated in a fluid carrier and is intended for ingestion in daily quantities measured in drops or similar small units of measure." This elaboration of "liquid form" also denotes ingestion by swallowing the fluid.

The legislative history of section 350 further supports this interpretation of "ingestion." Finally, interpretation of "ingestion" in section 350(c)(1)(B) to mean enteral administration into the stomach and gastrointestinal tract is not antithetical to the purposes underlying section 350, which primarily concern precluding the FDA from regulating vitamins and minerals based solely on their level of potency or combination.

Accordingly, the Court agrees with Judge Ross's construction of the phrase "intended for ingestion" in section 350(c)(1)(B), and her determination that Ener-B is not a food to which section 350 applies.

B. Ener-B Is Not a Dietary Supplement

. . . .

Thus, the definition of a "dietary supplement" in the DSHEA incorporates part of the definition of a "food" to which section 350 applies: namely the part which provides that the vitamin must be a product that is (i) "intended for ingestion in tablet, capsule, powder, softgel, gelcap, or liquid form," or (ii) "if not intended for ingestion in such a form, is not represented as a conventional food and is not represented for use as a sole item of a meal or of a diet."

Accordingly, Ener-B cannot be a dietary supplement because, as shown earlier, Ener-B is not "intended for ingestion" and, therefore, does not meet the criteria of section 350(c)(1)(B)(i) or (ii). Nature's Bounty's contention that Ener-B is perforce excluded from regulation as a drug under section

321(g)(1)(C), because it is a dietary supplement that the DSHEA defines as a food not subject to regulation under section 321(g)(1)(C), is a tautology. In order to be outside the reach of section 321(g)(1)(C), Ener-B must first meet the definition of a "dietary supplement." It does not meet the definition, and unless Ener-B falls within the parenthetical exception in the statute—which it does not—the FDA can regulate Ener-B as a drug under section 321(g)(1)(C).

Indeed, contrary to Nature's Bounty's claims, the amendments to section 350(c)(1)(B)(ii) support the interpretation of "intended for ingestion" as referring to enteral administration into the gastrointestinal tract. The DSHEA amended section 350(c)(1)(B)(ii) by adding several more forms by which the product can be ingested: powder, softgel, and gelcap. According to the *ejusdem generis* maxim of statutory construction, which states that items following an enumeration of particular things are to be construed as applying to the same class as those things enumerated, the addition of powder, softgel, and gelcap to the forms of tablet, capsule, or liquid does not indicate any intent by Congress to change the scope of "foods to which section 350 applies" by including products that are to be administered into the body parenterally. If anything, the opposite intent can be inferred because all the items listed, tablet, capsule, powder, softgel, gelcap, and liquid form, involve a form of the product that must be enterally administered into the stomach through swallowing.

If Congress wished to make such a drastic change in the law by allowing vitamins administered through the nose to be regulated as dietary supplements that are intended for ingestion, one would expect the change to have been expressly stated in the delineation of the other methods of delivery added to section 350(c)(1)(B)(i).

However, because Congress expressly precluded this Senate Report and any other report or statement other than the Statement of Agreement released jointly by the House and Senate when the DHSEA was enacted from being considered as legislative history of the DSHEA, the report cannot be relied upon, and this Court does not rely upon it, as indicative of any congressional intent.

Accordingly, it is the Court's view that Ener-B is not a dietary supplement within the meaning given to that term under the DSHEA. Moreover, the Court believes that the DSHEA amendments to section 350(c)(1)(B) do not alter, but rather support, the conclusion that Ener-B is not a "food to which [section 350] applies."

3. The FDA's Determination That Ener-B Is Not a Food is Reasonable

Nature's Bounty contends that the FDA's determination that Ener-B is not a food is irrational and contrary to the holdings in *Nutrilab* and *Am. Health Prods*. Essentially, Nature's Bounty contends that (1) the ingestibility of an article was explicitly excluded by the *Nutrilab* court as a factor by which to determine whether a particular article is food, (2) the fact that Ener-B is not consistent with the traditional notion of a "food" does not preclude its being regulated as a food, and (3) the FDA's determination that Ener-B is not a food is irrational and arbitrary.

A. Ingestibility as a Factor

In the Court's opinion, Nature's Bounty misinterprets the holding of *Nutrilab* with regard to ingestibility. In *Nutrilab*, the Court considered a challenge to the FDA's classification of starch blocker as a drug. The starch blockers at issue in that case were in tablet and capsule form and used to control weight by blocking the digestion of starch. The district court construed section 321(f)(1) to mean that articles were used for food if they were solely used for taste, aroma or nutritive value. The Seventh Circuit disagreed, and held that the district court's interpretation was unduly restrictive, since foods like prune juice and coffee could be used for reasons other than taste, aroma, and nutritive value. Instead, the Seventh Circuit held that the common-sense definition of a food under section 321(f)(1) encompassed "articles used by people in the ordinary way most people use food—primarily for taste, aroma, or nutritive value." (To these uses the Court would also add aphrodisiacal purposes).

In arriving at its decision, the Seventh Circuit rejected the industry's argument that starch blockers were food because they were derived from a food, namely kidney beans. The Seventh Circuit held that the congressional intent underlying the definition of food in section 321(f)(1) indicates that "'food' is to be defined in terms of its function as food, rather than in terms of its source, biochemical composition, or ingestibility." As an example, the court stated that although caffeine and penicillin are derived from foods, they are not foods. The *Nutrilab* court was, thus, emphasizing that neither an article's source nor its ingestibility can be the sole criterion for classifying it as a food.

However, stating that the ingestibility of an article cannot be the sole basis for classifying it as a food does not mean that an article's administration through a route other than ingestion cannot be a basis for determining that it is a nonfood. The two propositions are exclusive. Moreover, this Court does not read the Seventh Circuit's holding as excluding consideration of ingestibility as a functional factor when determining whether a particular article is a food. In the Court's view, ingestion has everything to do with the common place and everyday meaning of "food," and it is completely rational to use non-ingestion as a factor for determining that something is not as food. The mistake Nature's Bounty makes here is in contending that (1) the *Nutrilab* court's preclusion of ingestibility as a basis for determining whether an article is a food also precludes the converse, namely that non-ingestibility can be the basis for determining that an article is not a food, and (2) ingestion has nothing to do with the functional aspects of the everyday meaning of a food.

For these reasons—and especially in light of the very clear holding in *Am. Health Prods.* that the term "food" in the parenthetical exception of section 321(g)(1)(C) refers to the term's common usage—Nature's Bounty's contention that the FDA and Judge Ross erred in determining that Ener-B is not a food because it does not meet the traditional notion of a "food," is untenable.

Equally unconvincing is the defendant's contention that nasal administration of a vitamin is no more unusual than administration by capsules and tablets once was. Assuming that one day in the future administration of vitamins through the nose will be as common as the taking of vitamin capsules and tablets now is, the decision to expedite the process is not for the Court to make in the first instance by declaring that Ener-B is a "food." Rather, in this very nebulous area of the law where an article can be both a food and a drug, the Court believes that the decision is for the FDA to make in the first instance, as it is the agency containing specialized knowledge that is charged with administering the statutory provisions of the FD&C Act.

B. The FDA's Determination Is Not Irrational

The statutory language in sections 321(g)(1)(C) and 321(f)(1) regarding what is meant by the term "food" is ambiguous. Moreover, Congress has not directly addressed whether a nasal gel is a food, or whether route of administration of an article affects its classification as a food or drug. Accordingly, substantial deference to the FDA's interpretation of the statute is warranted, so long as its interpretation is based on a permissible construction of the statute:

> [where] the court determines [that] Congress has not directly addressed the precise question at issue, the court does not simply impose its own construction on the statute as would be necessary in the absence of an administrative interpretation. Rather, if the statute is silent or ambiguous with respect to the specific issue, the question for the court is whether the agency's answer is based on a permissible construction of the statute.

Chevron, 467 U.S. at 843.

In determining whether a construction is permissible, "[t]he court need not conclude that the agency construction was the only one it permissibly could have adopted . . . or even the reading the Court would have reached if the question initially had arisen in a judicial proceeding." Rather, a permissible construction of the statute is one that "reflects a plausible construction of the plain language of the statute and does not otherwise conflict with Congresses' expressed intent." Moreover, the Court "may not substitute its own construction of a statutory provision for a reasonable interpretation" made by the agency.

Here, the FDA's determination that the term "food" in sections 321(g)(1)(C) and 321(f)(1) is to be interpreted according to the ordinary meaning of the term, and is to include foods that provide nutrition by enteral administration into the gastrointestinal tract—in other words, that the food is ingested—is a permissible construction of the statute: it reflects a plausible construction of the plain language of the statute, and does not conflict in any way with the limited congressional intent that can be gleaned regarding sections 321(g)(1)(C) and 321(f)(1).

In addition to contending that the FDA has impermissibly used ingestion as a factor in determining that Ener-B is not a food, Nature's Bounty makes two more claims in support of its contention that the FDA's determination is unreasonable. These are: (i) adopting the FDA's and Judge Ross's interpretation would render all dietary supplements in tablet and capsule form as drugs, because they are not "foods" in the common-sense meaning of the term; and (ii) the FDA's interpretation of the word "food" is in direct contradiction to the agency's determination that nasal, esophageal, and jejunal tubes used to feed people are "food," even though these methods cannot be considered "ingesting food" in the common, everyday meaning of the term.

The Court finds these contentions unconvincing. The first claim is rendered moot by the DSHEA because under the DSHEA vitamins, minerals, and other dietary supplements are expressly deemed to be a "food" within the meaning of the FD&C Act. *See* DSHEA § 3(a), 21 U.S.C. § 321(ff) (1994). Moreover, in order to ensure that a dietary supplement is not regulated as a drug under section 321(g)(1)(C), the DSHEA amends section 321(g)(1) and related labeling provisions of the FD&C Act to expressly provide that a dietary supplement cannot be regulated as a drug under section 321(g)(1)(C) on the basis of any claims in its labeling regarding benefits related to a disease or the process by which the supplement affects the structure of the body, provided the claim is truthful and not misleading.

With regard to the second claim, the Court believes that the defendant is confusing different definitions of "food" under the FD&C Act. The FD&C Act establishes a separate category of "medical foods" in order to accommodate the enteral feeding and dietary management of people who are ill or injured. The definition of a "medical food" provides in relevant part: "[t]he term medical food means a food which is formulated to be consumed or administered enterally under the supervision of a physician[.]" 21 U.S.C.A. § 360ee(b)(3) (1994). In the Court's view, there is no contradiction between Congress's special designation of a food that is formulated for enteral administration through devices such as nasal, esophageal or jejunal tubes to be a "medical food," and the FDA's determination that Ener-B is not a "food" because it is not "ingested." In the former case, Congress simply created a special category of "food," notwithstanding the unconventional method of delivery into the body, in order to serve a special medical purpose.

Moreover, the statute's premising of a "medical food" on enteral administration undercuts the defendant's contention that a parenterally administered nutrient like Ener-B should also be considered a "food." The regulatory interpretation of "medical food" makes it very clear that if the nutrient is not administered enterally, it is not a food: "[p]arenteral nutrients ... are drugs and not medical foods. By definition, medical foods are consumed or administered enterally." 56 Fed. Reg. 60377–78 (Nov. 27, 1991).

Accordingly, the FDA's determination that Ener-B is neither a "food" within the parenthetical exception of section 321(g)(1)(C) nor within the definition of that term in section 321(f)(1) is reasonable. Certainly the FDA's determination is not "so directly in conflict with the statutory definition, [that] it must be invalidated as arbitrary and capricious and not in accordance with law."

4. Other Alleged Improper Bases for Judge Ross's Report

Nature's Bounty raises several other alleged bases for objecting to Judge Ross's Report. These are: (i) Judge Ross based her conclusions on "post hoc rationalizations" reflecting an alleged changed litigation position by the government subsequent to the denial of its summary judgment motion; (ii) Judge Ross erroneously relied on prior agency actions as indicative of consistent FDA policy; and (iii) Judge Ross improperly based her recommendations on alleged safety concerns regarding Ener-B. . . .

. . . .

5. Evidence That Ener-B Is a Food

. . . .

. . . Based on the evidence adduced at trial, Judge Ross determined that the common use meaning of "food" consisted of two elements: delivering nutrition to the body, and ingestion into the gastrointestinal tract. The fact that Ener-B functionally delivers vitamin B-12 just like any other source of vitamin B-12 is, therefore, not enough to categorize it as a "food" within the meaning of section 321(f)(1) or the parenthetical exception to section 321(g)(1)(C). In addition to providing nutrition, Ener-B must be ingested. Ener-B is not ingested. It is delivered parenterally. Accordingly, because it does not meet the common use definition of a "food," it does not fall within the "(other than food)" exception to section 321(g)(1)(C), no matter how closely to other food sources it delivers vitamin B-12 into the body.

. . . .

Conclusion and Order

In the modern technocratic state where regulatory implementation and enforcement of scientifically technical and complex legislation is delegated to administrative agencies, the courts must beware of usurping the prerogative of Congress that permits the agencies in the first instance to interpret legislation concerning the subject matter of their expertise.

The issue raised by the present cases is novel, and has not been squarely addressed by Congress in either the language of the statute or in the deliberations of Congress prior to enacting the legislation. Deference to the FDA's interpretation of the statute is, thus, appropriate. Perhaps someday in the future, as Nature's Bounty predicts, society will cross a threshold where administration of vitamins via a nasal gel will be as commonplace as the taking of vitamin tablets is today. At that time Congress and/or the FDA may wish to regulate Ener-B as a food or dietary supplement. However, until that time the Court believes it is not this Court's function to do so.

Judge Ross correctly concluded that the FDA's interpretations of the statutory provisions in this case were reasonable, and that deference was warranted to the FDA's determination that Ener-B is a drug, and not a "food" within the parenthetical exception to 21 U.S.C. § 321(g)(1)(C) because it is not ingested. The Court finds that in reaching this conclusion, Judge Ross did not err legally or factually, nor did she base her conclusion on any legal or factual error. The Court, therefore, adopts the Report and Recommendation of Judge Ross in its entirety without qualification.

Accordingly, it is the determination of this Court that Ener-B is not a food, but rather is a drug within the meaning of 21 U.S.C. § 321(g)(1)(C). Moreover, Ener-B is a new drug within the meaning of 21 U.S.C. § 321(p) that is unapproved, and therefore subject to condemnation under 21 U.S.C. § 334

. . . .

* * * * *

New Drug Ingredients   Under some circumstances, a dietary supplement may include substances that are approved drugs. This is allowed so long as the substance was marketed as a dietary supplement or as a food *before* the date it was approved as a drug.[25] This provision was designed to provide first-in-time protection for research and development of a new drug.

In *Pharmanex v. Shalala*, the FDA stopped the importation of red yeast rice for encapsulation into a product intended to promote healthy cholesterol levels. Pharmanex, the marketer of the product filed action seeking declaratory and injunctive relief. The Court of Appeals held that Congress did not unambiguously manifest its intent to exclude only finished drug products from definition of "dietary supplement" in DSHEA, and thus, FDA did not act arbitrarily or capriciously in seeking to regulate active ingredients as well as finished drug products.

* * * * *

Pharmanex v. Shalala

221 F.3d 1151 (10th Cir. 2000)

Before: KELLY, PORFILIO, Circuit Judges, and ALLEY, Senior District Judge
PAUL KELLY JR., Circuit Judge

This case requires that we address the scope of 21 U.S.C. § 321(ff)(3)(B) as it relates to the FDA's power to regulate dietary supplements. Appellants (here-

[25] 21 U.S.C. § 321(ff)(3).

inafter, "FDA") appeal from the federal district court's order setting aside the FDA's Administrative Decision of May 20, 1998. Our jurisdiction arises under 28 U.S.C. § 1291, and we reverse and remand for resolution of record-based arguments not reached below.

Background

Plaintiff-Appellee, Pharmanex, markets a product, Cholestin, that is intended to promote healthy cholesterol levels. Cholestin is made from red yeast rice, and contains a natural substance, mevinolin, which is chemically identical to the active ingredient, lovastatin, in the prescription drug, Mevacor. Mevacor was approved by the FDA in 1987 for the treatment of high cholesterol and heart disease. On April 7, 1997, the FDA advised Pharmanex that it considered Cholestin to be a drug, which may not be marketed without FDA approval. While discussions between the parties were ongoing, the FDA issued a Notice of Detention and Hearing that prevented importation of a shipment of red yeast rice for encapsulation into Cholestin. On May 20, 1998, the FDA issued a final decision, holding that Cholestin does not meet the definition of "dietary supplement" provided by 21 U.S.C. § 321(ff)(3)(B)(i), and is thus subject to regulation as a drug. Subsequently, Pharmanex filed an action in district court, seeking declaratory and injunctive relief, and asking the court to hold unlawful and set aside the FDA's decision. The district court granted a preliminary injunction, and ultimately entered a final order setting aside the FDA decision, holding that Cholestin is a "dietary supplement" within the definition set forth by § 321(ff). The district court based its decision on the determination that § 321(ff)(3)(B) refers unambiguously to finished drug products, rather than their individual constituents. . . .

Discussion

As noted at the outset, this case involves an interpretation of 21 U.S.C. § 321(ff)(3)(B) of the Food, Drug, and Cosmetic Act (hereinafter, "FD&C Act"), as amended by the Dietary Supplement Health and Education Act (hereinafter, "DSHEA"). Because we are confronted with conflicting interpretations of the statute that the Food and Drug Administration is charged with administering, the analytic framework set forth in *Chevron, U.S.A. v. Natural Resources Defense Council, Inc.*, governs our analysis. That is, we must decide, using the traditional tools of statutory construction, "whether Congress has directly spoken to the precise question at issue." If so, that is the end of the matter, and Congress' clear intent controls. If the statute is silent or ambiguous as to the specific issue before us, then we must defer to the agency's interpretation, if it is based on a permissible construction. We need not conclude that the agency construction is the only one possible, or even that we would have so construed the statute had the issue arisen in a judicial proceeding. Rather, we will give effect to the agency's interpretation unless it is arbitrary, capricious, or manifestly contrary to the statute. We accord the agency such deference given its special institutional competence regarding the "facts and circumstances surrounding the subjects regulated," particularly those which touch and concern competing views of the public interest.

In evaluating whether Congress has squarely and unambiguously addressed the question before us, we need not limit ourselves to scrutiny of the discrete statutory section in isolation. Rather, we examine the statutory provision in context. We must "interpret the statute 'as a symmetrical and coherent regulatory scheme,' and 'fit, if possible, all parts into a harmonious whole." In this case we must determine whether Congress unambiguously manifested its intent to exclude only finished drug products (rather than ingredients) from the definition of dietary supplement in § 321(ff)(3)(B), which states in relevant part:

The term "dietary supplement" . . . does . . . not include . . . an article that is approved as a new drug under section 355 of this title, . . . which was not before such approval, certification, licensing, or authorization marketed as a dietary supplement or as a food. . . .

The Parties' Contentions

The FDA argues that the phrase "an article that is approved as a new drug" is properly understood to contemplate active ingredients as well as finished drug products. To support this claim, FDA makes what is effectively a textual argument, pointing out that the word "article" is used throughout the FD&C Act to connote both component and finished drug product. The FDA notes that § 321(ff)(1) and (2) refer to a dietary supplement as a "product" with certain qualities, whereas § 321(ff)(3)(B) uses the word "article," a much broader term. Moreover, the FDA contends that the district court erred in finding that the phrase "approved as a new drug" is dispositive evidence of Congress' unambiguous intent to restrict the application of § 321(ff)(3)(B) to finished drug products. Additionally the FDA argues that the district court misconstrued judicial and regulatory authorities to support its finding of clear congressional intent, and its conclusion that in the past, the FDA has endorsed statutory interpretations squarely contrary to those of the instant case. The FDA also asserts that the district court's reliance on legislative history was misplaced. Finally, the FDA argues that the interpretation advanced by the district court and Pharmanex would undercut the broad purposes of the FD&C Act, with respect to orphan drugs, pioneer drugs, and leave a gap in protection for the public.

The essence of Pharmanex's argument is that the plain meaning of 21 U.S.C. § 321(ff)(3)(B)(i) cannot exclude Cholestin or lovastatin from the definition of "dietary supplement." Pharmanex contends that Cholestin cannot be "an article that is approved as a new drug" because it was never approved as a new drug. Additionally, Pharmanex asserts that "an article that is approved as a new drug" cannot apply to lovastatin because an ingredient is never "approved as a new drug." Moreover, Pharmanex claims that the definition of "new drug," § 321(p), itself precludes § 321(ff)(3)(B) from applying to drug components, as they are not approved, are not the subject of investigation, and do not have labeling. Additionally, Pharmanex contends that in the past, the FDA has advanced the very definition of "new drug" that it now resists. Pharmanex also argues that the FDA's interpretation defeats the unambiguously articulated policies enshrined in DSHEA and would produce absurd results. Pharmanex disputes the FDA's claim that limiting § 321(ff)(3)(B) would leave a gap in public protection, arguing that dietary supplements are adequately regulated by other provisions of FD&C Act. Finally, Pharmanex asserts that the FDA's arguments about statutory ambiguity are in error.

Plain Language

a. Statutory Text

The district court resolved this matter by concluding that Congress had clearly and unambiguously expressed its intent to limit the application of § 321(ff)(3)(B) to finished drug products. Having carefully reviewed the text of the provision in question, other relevant statutory materials, legislative history, and the parties' above arguments, we reach the opposite conclusion. We begin, as always, with the language of the provision in question. First, the use of the word "article" creates ambiguity. As the FDA points out, the term has a broad meaning throughout the FD&C Act, alternatively referring both to products and their individual constituents. *See* 21 U.S.C. § 321(g)(1)(A)–(D) (using "article" to refer to both drugs and their components). The use of the broad term "article" in § 321(ff)(3)(B) is especially striking in contrast with the immediately preceding sections, §§ 321(ff)(1) and (2), which use the word "product" to expand on the definition of dietary supplement. The drafters could have clarified their intent by using the words "active ingredient" rather than "article," as is used in other provisions of the FD&C Act. *See generally* 21 U.S.C. §§ 355(c)(3)(D)(i), 355(j)(2)(A)(ii)(I)–(II). Instead of using the more precise terms such as "product" and "active ingredient" or some combination of the terms, the drafters opted for the more general expression "article." This suggests ambiguity.

Further suggesting ambiguity, the previous section, § 321(ff)(3)(A), refers to "the article, when used as or in a dietary supplement. . . ." The clause at issue here, § 321(ff)(3)(B), omits these descriptive phrases. It could be that the omission reflects the drafters' intent to use "article" to comprehend both product and components in § 321(ff)(3)(A), but not for purposes of § 321(ff)(3)(B). Alternatively, the drafters could have omitted the prepositions because they were superfluous, as 321(ff)(3)(A) has already established that "article" contemplates both product and ingredient. The drafters' intent in this respect is altogether unclear.

We reject Pharmanex's contention that the phrase modifying "article," namely "approved as a new drug," sufficiently clarifies the section for purposes of our analysis. Pharmanex argues that this phrase resolves any doubt as to the scope of § 321(ff)(3)(B). That is, the clause could not possibly apply to drug components because components are never "approved as a new drug." Pharmanex argues that approval only attaches to drug products. Additionally, the very definition of "new drug," 21 U.S.C. § 321(p), refers by its terms only to drug products. We do not find these arguments persuasive.

While it is true that the FD&C Act provisions relating to approval of new drugs, 21 U.S.C. § 355, discuss approval in the overarching context of finished product approval, it is too simple to suggest that ingredients are in no sense "approved" in the new drug approval process. *See, e.g.*, 21 U.S.C. § 355(c)(3)(D)(i)–(ii) (referring to a drug, "no active ingredient . . . of which has been approved" as part of the new drug approval process). It is evident from § 355 that approval of active ingredients is integral to the overall new drug approval process. *See, e.g.*, 21 C.F.R. § 314.50(d)(1)(i)–(ii) (requiring a listing and description of drug substance and drug product components as part of the application for new drug approval). The use of the phrase "approved as a new drug" cannot bear the interpretive weight Pharmanex applies to it. It does not clarify the scope of § 321(ff)(3)(B) to the extent Pharmanex suggests.

We likewise reject Pharmanex's contention that because the definition of "new drug" in § 321(p) refers to composition, investigation, and labeling, it clarifies the intent of § 321(ff)(3)(B), and precludes its application to active ingredients. The definition of "new drug" found in § 321(p) provides in relevant part:

(1) Any drug . . . the composition of which is such that such drug is not generally recognized . . . as safe and effective for use under the conditions prescribed . . . in the labeling thereof. . . .

(2) Any drug . . . the composition of which is such that such drug, as a result of investigations . . . has become so recognized, but which has not . . . been used to a material extent or for a material time under such conditions.

As the FDA points out, this definition (including references to composition, labeling, and investigation) modifies the initial phrase "any drug." As stated previously, the term "drug" is defined in § 321(g) to include both finished drug products as well as individual constituents. Thus, the definition of "new drug" is largely colored by the ambiguity that attends the broad term "drug." Moreover, the claim that an ingredient cannot have composition finds no support in the FD&C Act or common sense. Similarly it is not accurate to suggest that drug components are not the subject of investigation in any sense. *See generally* 21 C.F.R. § 312.23(a)(7)(i). ("Therefore the emphasis in an initial phase I submission should generally be placed on the identification and control of the raw materials and the new drug substance [defined in 21 C.F.R. § 314.3(b) as active ingredient].") In view of the preceding, we find that the definition of "new drug" does not itself sufficiently clarify the drafters' intent with respect to § 321(ff)(3)(B).

b. Prior Judicial and Regulatory Authorities

The district court emphasized that a court must assume that Congress drafts legislation with knowledge of relevant preexisting authorities that bear on judicial interpretations of statutory terms. Even so, the judicial and regulatory authorities upon which the district court relied (and upon which Pharmanex now relies) for the proposition that only finished drug products are approved under § 355, do not sufficiently illuminate the meaning of § 321(ff)(3)(B) for our purposes.

. . . .

In *Pfizer, Inc. v. FDA*, the FDA successfully argued that "drug" in § 355(b)(1) and (c)(2) means "drug product," thus requiring Pfizer to get a new NDA for its tablet version of its previously approved soft gelatin capsule version of nifedipine, on the ground that although it contained the same active ingredient, it was nevertheless a different drug. In *Apotex, Inc. v. Shalala*, the FDA successfully argued that the market exclusivity accorded to one drug product did not extend so as to preclude a generic product with the same active ingredient (although of a differing strength) from receiving a 180-day period of market exclusivity pursuant to § 355(j)(5)(B)(iv). In an FDA decision that was part of the *Apotex* litigation, the agency noted that "FDA could not approve an application that requested approval of only the active ingredient. . . . The Agency, therefore, can only award such exclusivity to an [Abbreviated New Drug Application] applicant for a drug product, and a particular strength."

From these authorities Pharmanex invites us to infer that because it is the drug product, not the active ingredient to which approval attaches, it would not make sense for the phrase "article that is approved as a new drug" to connote an ingredient—but rather it can only refer to a finished drug product. We disagree. Because these arguments arose in the specialized context of the Drug Price Competition and Patent Term Restoration Act, Pub. L. No. 98-417 (1984), these cases are of limited relevance to the instant matter. The Hatch–Waxman amendments alter § 355 in certain provisions to establish periods of market exclusivity for pioneer drugs, while streamlining the approval process for less expensive generic drugs. Thus, in these cases, the FDA interpreted the word "drug" to refer only to drug products to advance these very specific policy objectives, and as such, these precedents do not illuminate congressional intent for § 321(ff)(3)(B) in the instant case. Moreover, it bears noting that "it is not impermissible under *Chevron* for an agency to interpret an imprecise term differently in two separate sections of a statute which have different purposes."

In sum, we reject Pharmanex's argument that the plain language of § 321(ff)(3)(B) evinces a clear intent to exclude only finished drug products from the definition of dietary supplement. In *Brown v. Gardner*, the Supreme Court observed that "[a]mbiguity is a creature not of definitional possibilities but of statutory context." A corollary of this principle is that for purposes of *Chevron* analysis, statutory clarity is a creature not of definitional isolation but of statutory context. Pharmanex isolates the discrete phrase "article that is approved as a new drug," and tries to import clarity, ignoring the linguistic ambiguity that attends these words in the larger context of the surrounding DSHEA provisions, and the FD&C Act more generally. As the following section shows, Pharmanex's argument is further undercut by the policies that undergird DSHEA and FD&C Act.

Legislative History and Policies of DSHEA and FD&C Act

a. Legislative History

Turning to the legislative history, we find that the intended application of § 321(ff)(3)(B) is not elucidated but rather becomes less clear. . . . The following statements, however, bear more directly on the present issue:

> During consideration of S. 784, concerns were expressed that manufacturers or importers of drugs could avoid the drug approval process by marketing drug products as dietary supplements. Although current authorities should be adequate to deal with such potential problems, the committee is sensitive to those concerns. Accordingly, Senators Harkin and Hatch agreed to formulate additional language prior to consideration of S. 784 in the Senate.
>
> Under the substitute to S. 784 as approved by committee, a substance which has been marketed as a dietary ingredient in a dietary supplement, or otherwise as a food, does not lose its status as a food . . . just because FDA approves the substance for use as an active ingredient in a new drug. . . .

This passage suggests that the scope of § 321(ff)(3)(B) is not limited to finished drug products, as Pharmanex suggests, . . . The above example demonstrates how the prior market clause would protect a dietary supplement with an ingredient that is subsequently approved as the active ingredient in a new drug. Provided that the dietary ingredient had been previously marketed as such, it would not lose its food status. By extension to the scenario that apparently troubled some legislators, the above language suggests that if a drug manufacturer sought to market a dietary supplement containing a natural substance that is the active ingredient in a previously approved drug product, it would be subject to the strictures of § 321(ff)(3)(B)'s exclusionary clause, unless it could show that prior to approval of the new drug, the natural substance was marketed as a dietary ingredient or food.

b. Policies of DSHEA and FD&C Act

The policies undergirding DSHEA and FD&C Act do not support a finding that Congress clearly intended § 321(ff)(3)(B) to apply only to finished drug products. It is true that DSHEA was enacted to alleviate the regulatory burdens on the dietary supplement industry, allowing consumers greater access to safe dietary supplements in order to promote greater wellness among the American population. However, the clause at issue in the instant case constitutes a limiting principle to this goal. That is, § 321(ff)(3)(B) specifically excludes certain articles from the definition of dietary supplement. To find that this clause only refers to finished drug products would be to restrict this provision so as to render it without practical application. Under the interpretation proposed by Pharmanex, a manufacturer could identify a naturally occurring substance that was identical to or had the same pharmacological effect as the active ingredient in a prescription drug, and market it in a dietary supplement. The manufacturer could evade the strictures of § 321(ff)(3)(A) by arguing, as Pharmanex does, that the naturally occurring ingredient is not a "finished drug product," and that § 321(ff)(3)(B) would apply only if the prescription drug itself were being held out as a dietary supplement.

To permit this result would contravene one of the primary objectives of the FD&C Act, namely "to ensure that any product regulated by the FDA is 'safe' and 'effective' for its intended use." We understand Pharmanex's argument that dietary supplements are already adequately regulated by provisions such as 21 U.S.C. § 342(f)(1) (setting forth provisions governing adulterated dietary supplements), but Pharmanex has not adequately responded to the FDA's strenuous objection that these provisions only empower the FDA to remove unsafe products rather than preclude their entry into the marketplace *ab initio*. Pharmanex represents only that premarket notification would almost surely be required by § 21 U.S.C. § 350b (provision governing new dietary ingredients). This is not a sufficient response to the concerns raised by the FDA. More importantly, it is not sufficient to demonstrate the clarity necessary under *Chevron* to preclude deference to the FDA's interpretation of § 321(ff)(3)(B).

To permit manufacturers to market dietary supplements with components identical to the active ingredients in prescription drugs would, as the FDA points out, contravene the incentive structures in place in the FDA for the development of orphan drugs and pediatric drugs. . . .

Finally, we reject Pharmanex's argument that the FDA's interpretation would produce absurd results by subjecting to regulation all the traditional food substances that are active ingredients in new drugs. The FDA's reading of the prior market clause would protect such substances if they did, in fact, have a history of marketing as a dietary supplement or food substance. As the FDA interprets § 321(ff)(3)(B), the exclusionary clause would reach naturally occurring substances identical or indistinguishable from the active ingredients in new drugs, provided that the substance in question was not previously marketed as a food or dietary substance. This comports with common sense and the overall purposes of the FD&C Act. It bears noting that many prescription drugs are derived from natural substances that are not benign. For example, "digitalis is extracted from purple foxglove, morphine from poppy, and quinine from cinchona bark."

Thus, the policies of DSHEA and FD&C Act do not move this court to the conclusion that § 321(ff)(3)(B) is clearly meant only to apply to finished drug products. In fact, in light of the foregoing, it seems that to so interpret the provision would be to restrict its scope so as to render it a meaningless limitation, and also contravene the fundamental purposes of the FD&C Act.

Conclusion

As stated previously, § 321(ff)(3)(B) represents the limiting principle of DSHEA's general purpose, namely to assuage the regulatory burdens on the dietary supplement industry. The provision balances this goal with the other policies of the FD&C Act, in effect carving out breathing room for dietary supplements while ensuring that drug manufacturers will not exploit this flexibility to make an end-run around the strictures of the new drug approval process. Congress has not specified the exact contours of this balance, choosing to use broad terminology in crafting the provision. Considering the lack of linguistic clarity and the overall statutory context, we hold that § 321(ff)(3)(B) is sufficiently ambiguous to merit *Chevron* deference. While Pharmanex's argument is linguistically possible, we are not compelled to adopt the conclusion that Congress clearly intended to limit § 321(ff)(3)(b)'s application to finished drug products, in light of the statutory context and the surrounding provisions. Deference is particularly appropriate in the instant case, which involves important questions of public health and safety. Additionally, we hold that the FDA's interpretation of § 321(ff)(3)(B) is not arbitrary, capricious, or manifestly contrary to the statute. Accordingly, we REVERSE and REMAND for consideration of the record based issues not reached below.

* * * * *

8.4   APPROVAL AND SAFETY

8.4.1   Presumption of Safety

Under DSHEA, the dietary supplement manufacturer is responsible for ensuring that a dietary supplement is safe before marketing. However, the manufacturer is *not* required to demonstrate safety or efficacy before marketing a dietary supplement ingredient that was marketed before 1994. Therefore manufacturers do generally not need to register with FDA, nor get FDA approval before producing or selling dietary supplements. In passing DSHEA, Congress found that safety problems with dietary supplements were rare, and thus these supplements could be provided a presumption of safety (unlike drugs and food additives).

* * * * *

Dietary Supplement Health and Education Act of 1994

Public Law No. 103-417 , 103rd Congress (1994)

. . . .

Section 2. Findings

Congress finds that—

(1) improving the health status of United States citizens ranks at the top of the national priorities of the federal government;

(2) the importance of nutrition and the benefits of dietary supplements to health promotion and disease prevention have been documented increasingly in scientific studies;

(3)(A) there is a link between the ingestion of certain nutrients or dietary supplements and the prevention of chronic diseases such as cancer, heart disease, and osteoporosis; and

(B) clinical research has shown that several chronic diseases can be prevented simply with a healthful diet, such as a diet that is low in fat, saturated fat, cholesterol, and sodium, with a high proportion of plant-based foods;

(4) healthful diets may mitigate the need for expensive medical procedures, such as coronary bypass surgery or angioplasty;

(5) preventive health measures, including education, good nutrition, and appropriate use of safe nutritional supplements will limit the incidence of chronic diseases, and reduce long-term health care expenditures;

(6)(A) promotion of good health and healthy lifestyles improves and extends lives while reducing health care expenditures; and

(B) reduction in health care expenditures is of paramount importance to the future of the country and the economic well-being of the country;

(7) there is a growing need for emphasis on the dissemination of information linking nutrition and long-term good health;

(8) consumers should be empowered to make choices about preventive health care programs based on data from scientific studies of health benefits related to particular dietary supplements;

(9) national surveys have revealed that almost 50 percent of the 260,000,000 Americans regularly consume dietary supplements of vitamins, minerals, or herbs as a means of improving their nutrition;

(10) studies indicate that consumers are placing increased reliance on the use of nontraditional health care providers to avoid the excessive

costs of traditional medical services and to obtain more holistic consideration of their needs;

(11)   the United States will spend over $1,000,000,000,000 on health care in 1994, which is about 12 percent of the Gross National Product of the United States, and this amount and percentage will continue to increase unless significant efforts are undertaken to reverse the increase;

(12)(A) the nutritional supplement industry is an integral part of the economy of the United States;

(B) the industry consistently projects a positive trade balance; and

(C) the estimated 600 dietary supplement manufacturers in the United States produce approximately 4,000 products, with total annual sales of such products alone reaching at least $4,000,000,000;

(13)   although the federal government should take swift action against products that are unsafe or adulterated, the federal government should not take any actions to impose unreasonable regulatory barriers limiting or slowing the flow of safe products and accurate information to consumers;

(14)   dietary supplements are safe within a broad range of intake, and safety problems with the supplements are relatively rare; and

(15)(A) legislative action that protects the right of access of consumers to safe dietary supplements is necessary in order to promote wellness; and

(B) a rational Federal framework must be established to supersede the current ad hoc, patchwork regulatory policy on dietary supplements.

. . . .

* * * * *

This presumption of safety for dietary supplement ingredients in use before 1994 has been compared to the GRAS provision of the Food Additive Amendments. However, this comparison only goes so far; there are significant differences between the two categories. Unlike GRAS, there is no requirement that dietary supplement ingredients be generally recognized as safe by scientific experts or that the safety be based on information that is published or otherwise generally available to the scientific community.

In addition, GRAS status is based on "reasonable certainty that the substance is not harmful under the intended conditions of use."[26] As discussed below, dietary supplements may not present a "significant or unreasonable risk of illness or injury" under conditions of use recommended or labeled. The difference between these two standards is muddled, but it is clear that the proof needed to support the safety of a dietary supplement is lower than for

---

[26] 62 Fed. Reg. 18,937 (Apr. 17, 1997).

conventional foods. The difference in language also reflects the difference in the burden of proof. The burden is on the proponent of GRAS status to prove safety. On the other hand, dietary supplement manufacturers do not have to prove safety at all, and the FDA must prove a reasonable certainty of harm.

Nonetheless, prudent manufacturers recognize that they should voluntarily substantiate the safety of their dietary supplements in such a way that the scientific community would recognize the safety. A company should document whatever methods and information relied on to conclude a dietary supplement is safe. This documentation is necessary to meet the regulations but also could be vital in minimizing the risk of product liability. In particular, manufacturers must make sure that product label information is truthful and not misleading, and also provides clear warnings and explanations on the conditions of use. If the label does not limit the product's use, the manufacturer is liable for the "ordinary" use by consumers.

8.4.2 No Food Additive Approval

Easy to miss in the cross-references is DSHEA's amendment of the definition of *food additive* to exclude ingredients in or intended for use in dietary supplements. The effect is to exempt dietary supplements from the requirements for food additives.[27] This is of great importance to dietary supplement manufacturers. Perhaps even easier to miss is a second effect of this amendment, which exempts dietary supplement ingredients from the food additive Delaney Clause.

8.4.3 New Dietary Ingredients (Post-1994)

Dietary supplement ingredients used in the United States food supply before October 15, 1994 are exempt from any requirement for FDA approval.[28] Dietary supplement ingredients that were not present in our food supply or used as dietary supplements before 1994 are subject to additional review requirements and must be reported to the FDA prior to marketing. The manufacturer is responsible for determining whether or not their ingredients meet this definition of "new."

Adequate information on safety   For new dietary supplement ingredients, the manufacturer must have adequate "information to provide reasonable assurance that such ingredient does not present a significant or unreasonable risk of illness or injury."

Reporting to the FDA   A manufacturer must report new dietary supplement ingredients to the FDA at least 75 days before being introduced into

---

[27] 21 U.S.C. § 321(s)(6).
[28] 21 U.S.C. § 350b(c); FD&C Act § 413(c).

commerce. The manufacturer must provide its information that supports the conclusion that the dietary supplement is reasonably expected to be safe. This information is kept confidential for 90 days after receipt and then is made public (except for certain trade secrets).

The DSHEA requirement for notice to the FDA is not an FDA approval or clearance. FDA's failure to object to a notice is not a finding of safety.[29]

8.4.4 Adulteration

* * * * *

Federal Food, Drug, and Cosmetic Act

21 U.S.C. § 342

A food shall be deemed to be adulterated—

. . . .

(f)(1) If it is a dietary supplement or contains a dietary ingredient that—

    (A) presents a significant or unreasonable risk of illness or injury under—

        (i) conditions of use recommended or suggested in labeling, or

        (ii) if no conditions of use are suggested or recommended in the labeling, under ordinary conditions of use;

    (B) is a new dietary ingredient for which there is inadequate information to provide reasonable assurance that such ingredient does not present a significant or unreasonable risk of illness or injury;

    (C) the Secretary declares to pose an imminent hazard to public health or safety, except that the authority to make such declaration shall not be delegated and the Secretary shall promptly after such a declaration initiate a proceeding in accordance with sections 554 and 556 of title 5 to affirm or withdraw the declaration; or

    (D) is or contains a dietary ingredient that renders it adulterated under paragraph (a)(1) under the conditions of use recommended or suggested in the labeling of such dietary supplement.

In any proceeding under this subparagraph, the United States shall bear the burden of proof on each element to show that a dietary supplement is adulterated. The court shall decide any issue under this paragraph on a de novo basis.

* * * * *

---

[29] 62 Fed. Reg. 49,892 (Sept. 23, 1997) (codified at 21 C.F.R. § 190.6).

NOTES AND QUESTIONS

8.3. A dietary supplement is considered adulterated if it "presents a *significant or unreasonable risk of illness or injury*" under conditions of use recommended or suggested in the labeling, or under conditions ordinarily used, if there is no recommended use. Contrast and compare this definition with FD&C Act, section 402(a)(1)'s "may render injurious" standard for added components of food; and section 402(a)(1)'s "ordinarily injurious" standard for non-added components of food.

8.5 ENFORCEMENT

The FDA is responsible for taking action against any unsafe dietary supplement product after it reaches the market. The FDA's postmarketing responsibilities include monitoring safety and product information, such as labeling, claims, package inserts, and accompanying literature.

8.5.1 Hurdles for the FDA

Compared to conventional foods, DSHEA places three additional hurdles in front of the FDA before the agency can take enforcement action against the manufacturers of dietary supplements for unsafe products.

Burden of Proof of Adulteration The FDA bears the burden of proof on each element in proving that a dietary supplement is adulterated[30] or misbranded.[31] This is a huge shift considering the historical deference to the FDA's technical expertise.

Advanced Notice of Prosecution The FDA must provide a manufacturer with the opportunity to present its side to the FDA before the FDA proceeds with an adulteration case to the U.S. Attorney.[32] This procedural right is not found with any other product category regulated by the FDA.

De Novo Review DSHEA eliminates deference to the viewpoint of the FDA when an adulteration case is brought to court.[33]

Also significant is that the standard is moved away from the product itself to the intended or recommended use of the product. For example, although a dietary supplement may present serious risks of illness or even death when

---

[30] 21 U.S.C. § 342(f)(1).
[31] *Id.* § 343–2(c).
[32] *Id.* § 342(f)(2).
[33] *Id.* § 342(f)(1).

used before athletic activity, the product may not be considered adulterated if it was not promoted for athletic activity. This places the FDA in the position of carrying the burden of proving both the toxicity of the ingredients and the toxicity under recommended use. Because recommended use supersedes ordinary use in the definition, a supplement might be commonly used in an abusive manner—leading to serious health risks and even death—but nevertheless be allowed to remain on the marketplace if the FDA cannot prove that the recommended use is unsafe.

8.5.2 Adverse Publicity

FDA's authority under Section 705 to bring adverse publicity remains intact. An example of FDA's use of adverse publicity is the issuance of Consumer Advisories.

* * * * *

Kava Linked to Liver Damage

FDA Consumer Advisory (July 23, 2002), http://nccam.nih.gov/health/alerts/kava/index.htm

The U.S. Food and Drug Administration (FDA) is advising consumers of the potential risk of severe liver injury from the use of dietary supplements containing kava (also known as kava kava or *Piper methysticum*). Recent reports from health authorities in Germany, Switzerland, France, Canada, and the United Kingdom have linked kava use to at least 25 cases of liver toxicity, including hepatitis, cirrhosis, and liver failure. Although liver damage appears to be rare, the FDA believes consumers should be informed of this potential risk. Kava, a member of the pepper family, is an herbal supplement. Products containing kava are sold in the United States for a variety of uses, including insomnia and short-term reduction of stress and anxiety. These products are marketed to men, women, children, and the elderly. . . .

* * * * *

The advisory from the FDA Center for Food Safety and Applied Nutrition is available at the FDA Web site: www.cfsan.fda.gov/~dms/addskava.html.

8.5.3 Other Tools

The FDA may still bring traditional actions for misbranding based on a dietary supplement containing ingredients other than those listed on the label. FDA may also bring traditional adulteration actions under section 342(a)(1) when a dietary supplement contains a poisonous or deleterious substance.

8.6   GOOD MANUFACTURING PRACTICES (GMPs)

A dietary supplement is also adulterated if it has been prepared, packed, or held under conditions that do not meet current good manufacturing practice (GMP) regulations.

* * * * *

Federal Food, Drug, and Cosmetic Act

21 U.S.C. § 342

A food shall be deemed to be adulterated—

. . . .

(g)(1)   If it is a dietary supplement and it has been prepared, packed, or held under conditions that do not meet **current good manufacturing practice regulations**, including regulations requiring, when necessary, expiration date labeling, issued by the Secretary under subparagraph (2).

(2)   The Secretary may by regulation prescribe good manufacturing practices for dietary supplements. Such regulations shall be modeled after current good manufacturing practice regulations for food and may not impose standards for which there is no current and generally available analytical methodology. No standard of current good manufacturing practice may be imposed unless such standard is included in a regulation promulgated after notice and opportunity for comment in accordance with chapter 5 of title 5.

* * * * *

In June 2007 the FDA published the final rule for good manufacturing practices (GMPs) for dietary supplements.[34] As described by the FDA, this rule addresses:[35]

- Minimum requirements for personnel, physical plant and grounds, and equipment and utensils.
- Establishment and use of written procedures for certain operations, including those related to equipment, physical plant sanitation, certain manufacturing operations, quality control, laboratory testing, packaging and labeling, and product complaints.
- Establishment of specifications in the production and process control system that will ensure dietary supplements meet the identity, purity, strength, and composition established in specifications and are properly packaged and labeled as specified in the master manufacturing record.

---

[34] 72 Fed. Reg. 34751–34958 (June 25, 2007) *available at*: http://www.cfsan.fda.gov/~lrd/fr07625a. html (last accessed Mar. 1, 2008).
[35] *Id*. at 34764–34765.

- Testing of a subset of finished batches of dietary supplements based on a sound statistical sampling or, alternatively, testing all finished batches.
- Implementation of quality control operations to ensure the quality of a dietary supplement.
- Preparation and use of a written master manufacturing record for each unique formulation of manufactured dietary supplement, and for each batch size, to ensure that the manufacturing process is performed consistently and there is uniformity in the finished batch from batch to batch.
- Preparation of a batch production record every time a dietary supplement batch is made to ensure that batch production record accurately follows the appropriate master manufacturing record.
- Establishment and use of laboratory control processes related to specifications and to the selection and use of testing and examination methods.
- Identification and quarantine of returned dietary supplements until quality control personnel conduct a material review and make a disposition decision.
- A qualified person to investigate any "product complaint" that involves a possible failure of a dietary supplement to meet any CGMP requirement, with oversight by quality control personnel.
- Records associated with the manufacture, packaging, labeling, or holding of a dietary supplement to be kept for one year beyond the shelf life dating (e.g., expiration dating, shelf life dating, or "best if used by" dating), or if shelf life dating is not used, for two years beyond the date of distribution of the last batch of dietary supplements associated with those records.

8.7   LABELS

8.7.1   Basic Labeling Requirements

The basic labeling requirements for conventional foods also apply to dietary supplements: statement of identity, ingredient statement, nutrition information, the name and place of business of the manufacturer, packer, or distributor, and accurate net contents. A dietary supplement generally must have the term "dietary supplement" in its statement of identity.[36] However, the word "dietary" may be replaced by the name of the dietary ingredient(s) in the product, for example, "vitamin C supplement."[37]

Most dietary ingredients are to be listed by their common names. Botanicals must be listed in accordance with the terminology in the book, HERBS OF COMMERCE. When the Latin name cannot be obtained from this reference, the label must state the Latin name of the botanical. Additionally, when a supple-

[36] FD&C Act § 403(s)(2)(B).
[37] 21 C.F.R. § 103(g).

ment contains any material from a plant, the label must identify any part of a plant used.

DSHEA eliminated the NLEA rules for labeling dietary supplements, but directed the FDA to establish new rules. DSHEA specifies that the labels of dietary supplements must bear nutrition information followed by ingredient information. The nutrition information is to list the dietary ingredients present in a supplement and to state the amounts of these ingredients. With conventional foods, only nutrients that have daily recommendations, such as 1,000 mg for calcium, may be listed in the nutrition information. DSHEA allows dietary ingredients for which recommendations have not been established to be listed as long as the label indicates this fact by an asterisk in the "% Daily Value" column that refers to the footnote "Daily Value not established." The nutrition information is titled "Supplement Facts."

Supplement nutrition information may include ingredient sources. For conventional foods, this information is given in the ingredient list (see Figure 8.1). For example, if calcium is from calcium carbonate, a supplement's nutrition information can state "calcium (as calcium carbonate)." When a source is listed in this manner, it need not be listed again in the ingredient information.

The labels of dietary supplements are to list the names and amounts for a serving of the dietary ingredients. A serving of a dietary supplement is the amount recommended in one eating occasion. This information is stated at the top of the nutrition information following the words "serving size." Similar to the labels of conventional foods, FDA requires that the nutrients most important to public health be listed. Other vitamins and minerals may be declared, but they must be declared when they are added for purposes of supplementation or when a claim is made about them. For uniformity, amounts are to be declared using the units of measurement specified in the regulation.

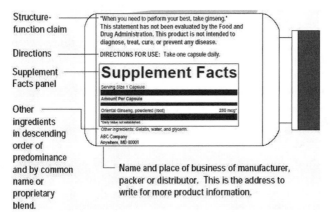

Figure 8.1 Supplement label (image courtesy CFSAN, FDA)

DSHEA states that when the ingredients in a supplement are considered to be a proprietary blend, just the total amount of the blend needs to be stated. In the absence of individual amounts, FDA requires that the dietary ingredients be listed in order of predominance by weight. These blends also are to be identified by the term "proprietary blend" or other appropriately descriptive term or fanciful name.

Supplements should contain the amounts of dietary ingredients that are declared. However, for a dietary ingredient that is naturally occurring, the FDA does not consider a label misbranded when an FDA analysis finds 80 percent of the amount declared on the label. Also, reasonable excesses of most dietary ingredients are acceptable within current good manufacturing practice. No more than a 20 percent excess is allowed for calories, sugars, total fat, saturated fat, cholesterol, or sodium.

8.7.2 Literature and Labeling

Definition of Labeling

* * * * *

Federal Food, Drug, and Cosmetic Act

§ 403B, 21 U.S.C. § 343-2

(a) **In general.** A publication, including an article, a chapter in a book, or an official abstract of a peer-reviewed scientific publication that appears in an article and was prepared by the author or the editors of the publication, which is reprinted in its entirety, shall not be defined as labeling when used in connection with the sale of a dietary supplement to consumers when it—

   (1) is not false or misleading;

   (2) does not promote a particular manufacturer or brand of a dietary supplement;

   (3) is displayed or presented, or is displayed or presented with other such items on the same subject matter, so as to present a balanced view of the available scientific information on a dietary supplement;

   (4) if displayed in an establishment, is physically separate from the dietary supplements; and

   (5) does not have appended to it any information by sticker or any other method.

(b) **Application.** Subsection (a) of this section shall not apply to or restrict a retailer or wholesaler of dietary supplements in any way whatsoever in the sale of books or other publications as a part of the business of such retailer or wholesaler.

(c) **Burden of proof.** In any proceeding brought under subsection (a) of this section, the burden of proof shall be on the United States to establish that an article or other such matter is false or misleading.

* * * * *

Drug Claims on Foods DSHEA also changed how health claims are evaluated for determining whether a product would be regulated as a drug.

* * * * *

Federal Food, Drug, and Cosmetic Act

§ 201, 21 U.S.C. § 321 (emphasis added)

For the purposes of this chapter—

. . . .

(g)(1) The term "drug" means

(A) articles recognized in the official United States Pharmacopoeia, official Homoeopathic Pharmacopoeia of the United States, or official National Formulary, or any supplement to any of them; and

(B) articles intended for use in the diagnosis, cure, mitigation, treatment, or prevention of disease in man or other animals; and

(C) articles (other than food) intended to affect the structure or any function of the body of man or other animals; and

(D) articles intended for use as a component of any article specified in clause (A), (B), or (C). A food or dietary supplement for which a claim, subject to sections 343(r)(1)(B) and 343(r)(3) of this title or sections 343(r)(1)(B) and 343(r)(5)(D) of this title, is made in accordance with the requirements of section 343(r) of this title is not a drug solely because the label or the labeling contains such a claim. *A food, dietary ingredient, or dietary supplement for which a truthful and not misleading statement is made in accordance with section 343(r)(6) of this title is not a drug under clause (C) solely because the label or the labeling contains such a statement.*

. . . .

* * * * *

8.8 HEALTH CLAIMS

St. John's Wort doth charm all witches away,
If gathered on the saint's holy day.

Any devils and witches have no power to harm
Those that gather the plant for a charm.
Rub the lintels with that red juicy flower;
No thunder nor tempest will then have the power
To hurt or hinder your house; and bind
Round your neck a charm of similar kind.[38]

8.8.1 Background

Claims made for dietary supplements are sometimes controversial and confusing.[39] This confusion may be accentuated by the layered nature of the regulation of claims. Once health claims were only permitted on foods if there was "significant scientific agreement" that a claim was valid. Now there are qualified claims, where the scientific support for some claims is weak, but claims are permitted when properly qualified to be nonmisleading. In addition, there are structure-function claims that skirt the edge of health claims and avoid stricter scrutiny. "Manufacturers have, therefore, made the formulation of soft claims into a fine art, creating claims that imply health effects without actually naming a disease."[40]

The claims that can be used on food and dietary supplement labels may be divided into three categories:

1. Dietary guidance
2. Nutrient content claims
3. Health claims

Dietary Guidance   Dietary guidance consists of statements that address a role in good health of general *dietary patterns* or *general* food. For example, "Five servings of fruits and vegetables a day are recommended for good health," is dietary guidance.

Dietary guidance statements are not considered health claims, provided that the guidance and the context of the statements do not suggest that a specific food or substance is the subject of a claim. For example, "Our fruit bar can help maintain a healthy diet," is a health claim rather than dietary guidance.

Dietary guidance statements must be truthful and nonmisleading.

---

[38] Poem AD 1400, from St. John's Wort, Avery 1998, cited by Kelly Ann Kaczka, *From Herbal Prozac to Mark McGwire's Tonic: How the Dietary Supplement Health and Education Act Changed the Regulatory Landscape for Health Products*, 16 JOURNAL OF CONTEMPORARY HEALTH LAW AND POLICY 463 (2000).

[39] *See, e.g.,* Martijn B. Katan, *Health Claims for Functional Foods: Regulations Vary between Countries and Often Permit Vague Claims*, 328 BRITISH MEDICAL JOURNAL 180–81 (2004), *available at* http://bmj.bmjjournals.com/cgi/content/full/328/7433/180 (last accessed Dec. 1, 2005).

[40] *Id.*

Nutrient Content Claims Nutrient content claims are statements that characterize or imply the level of a nutrient in a food. For example, "high in fiber" or "low carb" are nutrient content claims. The Nutrition Labeling and Education Act of 1990 (NLEA) permits the use of label nutrient content claims when the claims are made in accordance with the FDA's authorizing regulations.

Health Claims Health claims describe a relationship between a food, food component, or dietary supplement ingredient and reducing risk of a disease or health-related condition. Health claims may be further subcategorized by determining under what level of regulatory oversight they fall. The FDA exercises its oversight in determining which health claims may be used on a label or in labeling for a food or dietary supplement with:

1. the Nutrition Labeling and Education Act (NLEA), which provides for the FDA to issue regulations authorizing health claims;
2. the 1997 Food and Drug Administration Modernization Act (FDAMA), which provides for health claims based on an authoritative statement of a scientific body of the U.S. government or the National Academy of Sciences; and
3. the 2003 FDA Consumer Health Information for Better Nutrition Initiative provides for qualified health claims where the quality and strength of the scientific evidence falls below that required for the FDA to issue an authorizing regulation. Such health claims must be qualified to assure accuracy and nonmisleading presentation to consumers.
4. Structure-function claims describe the role of an ingredient in affecting or maintaining the normal structure or function in humans; for example, "calcium builds strong bones" and "antioxidants maintain cell integrity." The line between structure/function claims and drug-health claims can be extremely thin. For example, "supports the immune system" would be viewed as a structure-function claim—on the other hand, "supports your body's anti-viral capabilities" would be questioned as a veiled drug-heath claim.

The FDA explained some of the differences in oversight between types of claims as follows:

* * * * *

Claims That Can Be Made for Conventional Foods and Dietary Supplements

FDA/CFSAN, http://www.cfsan.fda.gov/~dms/hclaims.html

. . . .

A **health claim** by definition has two essential components: (1) a substance (whether a food, food component, or dietary ingredient) and (2) a disease or health-related condition. A statement lacking either one of these components does not meet the regulatory definition of a health claim. For example, statements that address a role of dietary patterns or of general categories of foods (e.g., fruits and vegetables) in health are considered to be **dietary guidance** rather than health claims, provided that the context of the statement does not suggest that a specific substance is the subject. Dietary guidance statements used on food labels must be truthful and nonmisleading. Statements that address a role of a specific substance in maintaining normal healthy structures or functions of the body are considered to be **structure-function claims**. Structure-function claims may not explicitly or implicitly link the relationship to a disease or health-related condition. Unlike health claims, dietary guidance statements and structure-function claims are not subject to FDA review and authorization. There are some regulatory requirements associated with the use of structure-function claims; see http://www.cfsan.fda.gov/~dms/labstruc.html.

NLEA Authorized Health Claims

The Nutrition Labeling and Education Act (NLEA) of 1990, the Dietary Supplement Act of 1992, and the Dietary Supplement Health and Education Act of 1994 (DSHEA) provide for health claims used on labels that characterize a relationship between a food, a food component, dietary ingredient, or dietary supplement and risk of a disease (e.g., "diets high in calcium may reduce the risk of osteoporosis"), provided the claims meet certain criteria and are authorized by an FDA regulation. FDA authorizes these types of health claims based on an extensive review of the scientific literature, generally as a result of the submission of a health claim petition, using the significant scientific agreement standard to determine that the nutrient/disease relationship is well established. For an explanation of the significant scientific agreement standard, see: http://www.cfsan.fda.gov/~dms/ssaguide.html.

Health Claims Based on Authoritative Statements

The Food and Drug Administration Modernization Act of 1997 (FDAMA) provides a second way for the use of a health claim on foods to be authorized. FDAMA allows certain health claims to be made as a result of a successful notification to FDA of a health claim based on an "authoritative statement" from a scientific body of the U.S. Government or the National Academy of Sciences. FDA has prepared a guide on how a firm can make use of authoritative statement-based health claims. This guide can be found at: http://www.cfsan.fda.gov/~dms/hclmguid.html. FDAMA does not include dietary supplements in the provisions for health claims based on authoritative statements. Consequently this method of oversight for health claims cannot be used for dietary supplements at this time. Examples of health claims based on authoritative statements may also be found at: http://www.cfsan.fda.gov/~dms/flg-6c.html.

Qualified Health Claims

FDA's 2003 Consumer Health Information for Better Nutrition Initiative provides for the use of qualified health claims when there is emerging evidence for a relationship between a food, food component, or dietary supplement and reduced risk of a disease or health-related condition. In this case the evidence is not well enough established to meet the significant scientific agreement standard required for FDA to issue an authorizing regulation. Qualifying language is included as part of the claim to indicate that the evidence supporting the claim is limited. Both conventional foods and dietary supplements may use qualified health claims. FDA uses its enforcement discretion for qualified health claims after evaluating and ranking the quality and strength of the totality of the scientific evidence. Although FDA's "enforcement discretion" letters are issued to the petitioner requesting the qualified health claim, the qualified claims are available for use on any food or dietary supplement product meeting the enforcement discretion conditions specified in the letter. . . . For more information on Qualified Health Claims, see http://www.cfsan.fda.gov/~dms/lab-qhc.html.

. . . .

* * * * *

8.8.2   DSHEA Authorized Health Claims

DSHEA provided special status to four types of health claims made for dietary supplements. These claims may be made without FDA pre-approval, and the claims—without more—will not make a dietary supplement fall under regulation as a drug. Prior to DSHEA, health claims on a dietary supplement often took the product into drug status with its lengthy pre-approval, premarket evaluation. DSHEA allows certain health claims on a dietary supplement without the product being deemed a drug.

While these claims may be made without FDA approval, the manufacturer is responsible for ensuring that a claim is truthful and not misleading. In addition, manufacturers of a dietary supplement that make these health claims on labels or in labeling must submit a notification to the FDA within 30 days after marketing the dietary supplement.

While a dietary supplement can now make claims that were previously only permitted on drugs, a dietary supplement may not be intended for use as a drug. Specifically, the supplement may not be intended for use in the diagnosis, cure, mitigation, treatment, or prevention of disease, or the FDA may regulate the product as a drug.

8.8.3 Permitted Health Claims

DSHEA added FD&C Act section 403(r)(6) (21 U.S.C. § 343), which defines four types of optional claims that may be made on dietary supplements without pre-approval (i.e., without first obtaining new drug approval or health claim authorization from the FDA). For ease of reference, these health claims under Section 403(r)(6) are often generically called "structure-function" claims. The four types of claims granted this status are as follows:

Classic nutrient deficiency disease. Claims of a benefit related to a classical nutrient deficiency disease may be made if the prevalence of the disease in the United States is disclosed. An example would be vitamin C and scurvy.

Structure and function. There are two types of structure-function claims permitted (both are nondisease claims):

1. Affects. Descriptions of the role of an ingredient intended to affect the structure or function of the body. "Supports the immune system" and "maintains a healthy circulatory system," for example.

2. Maintains. Characterization of the mechanism by which an ingredient maintains the structure or function of the body. "Anti-oxidants maintain cell integrity," for example.

General well-being. Descriptions of general well being from consumption of the ingredient. For example, "helps you relax."

For more information, see 21 C.F.R. § 101.93 and 65 Fed. Reg. 1,034–1,035 (Jan. 6, 2000).

Substantiation These permitted claims are not subject to premarket approval. However, the manufacturer making the claim must have *substantiation* that the claim is truthful and not misleading. "Substantiation" of a health claim made for a dietary supplement must be on file with the business. However, the FDA cannot demand to see this information. The FDA may ask that the information be volunteered, or the FDA may sue for the information if there is a contested issue.[41]

Notification Manufacturers must notify the FDA of the health claims they are making within 30 days of marketing a given dietary supplement. The details of this notification are specified in 21 C.F.R. § 101.93. If the FDA objects to the claim, it will send a letter to the manufacturer. A firm receiving such a letter is well advised to change or remove the claim, or it will face the risk that the FDA will start enforcement action.

---

[41] 21 U.S.C. § 343(r)(6)(B).

Disclaimer for Health Claims   Labels with these health claims must also include a disclaimer that the dietary supplements are not drugs and do not receive FDA premarket approval.[42]

Conventional Foods   Under *Nutrilab v. Schweiker*[43] conventional food claims are limited to structure-function effects that derive from the taste, aroma, or nutritive value of the food. Dietary supplement claims are not subject to that limitation.

Limit on "Labeling"   Section 403B limits the FDA's authority to reach beyond the label and specifically excludes books, articles, abstracts, and the like, used in connection with the sale of a product.

Drug Claims   Labeling a product as a dietary supplement does not shield it from being considered a drug based on the product's claims and marketing of intended use.[44] Drug claims—about disease diagnosis, treatment, cure, or prevention—are still considered drug claims. Drug claims may be express disease claims, such as, "prevents osteoporosis," or implied disease claims, such as, "prevents bone fragility in postmenopausal women."

* * * * *

Federal Food, Drug, and Cosmetic Act

§ 201, 21 U.S.C. § 321 (emphases added)

For the purposes of this chapter—

. . . .

(g)(1) The term "**drug**" means

 (A) articles recognized in the official United States Pharmacopoeia, official Homoeopathic Pharmacopoeia of the United States, or official National Formulary, or any supplement to any of them; and

 (B) articles intended for use in the diagnosis, cure, mitigation, treatment, or prevention of disease in man or other animals; and

 (C) articles (other than food) intended to affect the structure or any function of the body of man or other animals; and

 (D) articles intended for use as a component of any article specified in clause (A), (B), or (C). A food or dietary supplement for which a claim, subject to sections 343(r)(1)(B) and 343(r)(3) of this title or

---

[42] 21 U.S.C. § 343(r)(6)(C).
[43] 713 F.2d 335 (7th Cir. 1983).
[44] United States v. Undetermined Quantities of Articles of Drug, 145 F. Supp. 2d 692, 698–99 (D. Md. 2001).

sections 343(r)(1)(B) and 343(r)(5)(D) of this title, is made in accordance with the requirements of section 343(r) of this title is not a drug solely because the label or the labeling contains such a claim. A food, dietary ingredient, or dietary supplement for which a truthful and not misleading statement is made in accordance with section 343(r)(6) of this title is not a drug under clause (C) solely because the label or the labeling contains such a statement.

. . . .

* * * * *

Express and implied disease claims can be made through the name of a product (e.g., "Carpaltum," "CircuCure"), through a statement about the formulation of a product (contains aspirin), or through the use of pictures, vignettes, or symbols (e.g., electrocardiogram tracings).

Violation of DSHEA through drug claims may be determined by the overall marketing scheme, including the comments of employees. In an unusual case the FDA saw the photograph of the CEO of a dietary supplement corporation on a magazine cover holding up his company's product and promoting it to cure a disease.[45] The FDA determined the product was a new drug under the FD&C Act and forced a major recall and relabeling. The CEO was fired by his board of directors.

In *Whitaker v. Thompson*[46] the court found that the FDA could properly conclude that a dietary supplement's health claims related to the treatment of a disease required approval under the standards for new drugs. The appellants were Julian Whitaker, Pure Encapsulations Inc., Durk Pearson, Sandy Shaw and the American Association for Health Freedom. Specifically, in *Whitaker* the claim reviewed was: "Consumption of 320 mg daily of saw palmetto extract may improve urine flow, reduce nocturia, and reduce voiding urgency associated with mild benign prostatic hyperplasia (BPH)." This claim differs from the structure-function claim allowed by the FDA that consumption of saw palmetto supports good prostate health.

The FDA's position was supported unanimously by a three-judge panel of the U.S. Court of Appeals.

* * * * *

Whitaker v. Thompson

239 F. Supp. 2d 43 (D.D.C. 2003)

KESSLER, District Judge

Plaintiffs are individuals and companies with a direct financial interest in dietary supplements containing saw palmetto extract as well as a nonprofit

---

[45] Paula Kurtweil, *Promo Shot Backfires on Publicity Seeker*, 32 FDA CONSUMER 4 (1998).
[46] 239 F. Supp. 2d 43 (D.D.C. 2003).

therapeutic health organization composed of physician members who sell dietary supplements containing saw palmetto extract. They bring this action against the Food and Drug Administration ("FDA"), Jane E. Henney, Commissioner of the FDA, the Department of Health and Human Services ("HHS"), Tommy G. Thompson, Secretary of the HHS, and the United States of America. Plaintiffs challenge the FDA's denial of a health claim application for saw palmetto. . . .

I. Background

A. *Statutory and Regulatory Framework*

Prior to November 8, 1990, the FD&C Act provided that dietary supplements— including the saw palmetto supplements at issue in this case—would be regulated as a food, unless their intended use was as a drug. In other words, if a dietary supplement's label contained a disease-specific claim, that supplement was subject to the FDA's drug approval and drug labeling requirements.

However, during the mid-1980s companies began making disease-specific claims about foods with increasing frequency and without the approval of the FDA. Passage of the NLEA was intended to address concerns that the FDA had brought "virtually no enforcement actions" against the types of claims it had previously prohibited by clarifying and strengthening "the [FDA's] legal authority . . . to establish the circumstances under which claims may be made about the nutrients in foods."

The NLEA liberalized the FD&C Act to permit health claims to be "made in the label or labeling of [a] food which expressly or by implication . . . characterizes the relationship of any nutrient . . . to a disease or a health-related condition." 21 U.S.C. § 343(r)(1)(B). However, Congress clearly stated that the NLEA and FDA regulatory standards were to concern "only nutrients or substances in foods that 'nourish' and . . . [not] other, non-nutritive substances in foods." Congress delegated to the FDA the task of developing a procedure and standard for approving health claims for dietary supplements, providing that health claims made with respect to a dietary supplement of vitamins, minerals, herbs, or other similar nutritional substances . . . shall be subject to a procedure and standard, respecting the validity of such a claim, established by regulation of the Secretary.

Thus, under the NLEA, a dietary supplement health claim is not automatically subject to the FD&C Act's far more extensive and demanding approval and labeling requirements for drugs so long as the claim is made in accordance with other sections of the statute, including 21 U.S.C. § 343(r)(5)(D).

In 1993, the FDA responded to the NLEA by promulgating 21 C.F.R. §§ 101.14 and 101.70, which explained the standards and procedures for FDA consideration of nutrient-disease claims. The FDA chose the same standard for authorizing dietary supplement health claims as the NLEA prescribed for authorizing food health claims—significant scientific agreement. In requesting authorization for a health claim, a party first submits a petition with the proposed health claim, accompanied by supporting evidence. The FDA must then notify the applicant within 100 days whether the request will be denied or else "filed" for further review. If further review is warranted, within the next 90 days the FDA must either deny the petition or publish a proposed regulation authorizing the health claim. If the FDA publishes a proposed rule authorizing a health claim, the FDA must publish a final regulation approving or denying the claim within 270 days of the date of publication.

In 1994, Congress passed the Dietary Supplement Health and Education Act of 1994 ("DSHEA"),[47] to further recognize "the importance of nutrition and the benefits of dietary supplements to health promotion and disease prevention." The DSHEA clarified the FDA's role in authorizing health claims by creating "a rational Federal framework . . . to supersede the current ad hoc, patchwork regulatory policy on dietary supplements" to protect consumers' right of access to "safe dietary supplements . . . to promote wellness." Passage of the DSHEA attempted to further clarify the authorization of dietary supplement health claims by including within the FD&C Act a dietary supplement definition and an amended drug definition.

B. Procedural History

On May 25, 1999, Plaintiffs filed a health claim petition with the FDA seeking approval for the labels of saw palmetto supplements to include the following health claim:

> Consumption of 320 mg daily of Saw Palmetto extract may improve urine flow, reduce nocturia and reduce voiding urgency associated with mild benign prostatic hyperplasia (BHP).

In accordance with FDA regulations, Plaintiffs included scientific evidence supporting their claim. Plaintiffs also requested that the FDA "approve the claim with such disclaimer or disclaimers as the agency reasonably deemed necessary to avoid any potentially misleading connotation." . . .

The Plaintiffs' petition was denied under operation of law on December 1, 1999, because the FDA allowed 90 days to pass without issuing a decision. The FDA stated that the denial was necessary because the prescribed time frame was insufficient to resolve the "important and novel issue . . . whether health claims for foods (including dietary supplements) may encompass [a claim of an effect on an existing disease] or whether such a claim is appropriate only on a product that has been shown to meet the safety and efficacy requirements for drugs."

On December 7, 1999, the Plaintiffs filed this suit seeking declaratory and injunctive relief in light of the FDA's denial of their health claim petition. The action was stayed pending FDA reconsideration of its decision, and on April

---

[47] Pub.L. No. 103–417, 108 Stat. 4325 (1994).

4, 2000, the FDA held public hearings to determine whether the Plaintiffs' saw palmetto claim was a health claim or a drug claim under the FD&C Act. Upon consideration of the hearings and other comments, the FDA issued a formal letter on May 26, 2000, providing further explanation of its refusal to process the Plaintiffs' saw palmetto petition.

The FDA concluded that "claims about effects on existing diseases do not fall within the scope of the health claims provisions in 21 U.S.C. § 343(r) and therefore may not be the subject of an authorized health claim." With regard to Plaintiffs' proposed health claim, the FDA found that:

> [The] petition clearly identifies the intended use of saw palmetto extract products bearing that proposed claim as the treatment of the urinary symptoms of BPH. The proposed model claim . . . explicitly describes the mitigation of disease by treating its symptoms and establishes the intended use of products bearing the claim as drugs.

The FDA stated that its decision was based on the language and legislative history of the FD&C Act, prior agency interpretations of claims to treat disease, and concern that men would miss early diagnosis of prostate cancer by self-medicating with saw palmetto. . . .

III. Analysis

Plaintiffs argue that the central issue in this case is whether the FDA may deny their petition and indefinitely suppress the saw palmetto health claim. Plaintiffs contend that the FDA's refusal to evaluate the claim under 21 U.S.C. § 343(r)(5)(D) violates both the FD&C Act and the APA and constitutes a blanket ban on commercial speech in violation of the First Amendment. Plaintiffs argue that a claim made in accordance with § 343(r)(5)(D) is a health claim for a dietary supplement and cannot be regulated by the FDA as a drug claim because Congress clearly intended health claims to include any nutrient-disease claim, not just risk reduction claims.

However, Defendants argue that when reading the health claims provision in the context of the entire FD&C Act, Congress never intended to limit the statute's central purpose of drug regulation. Thus, Defendants contend that the statute only allows health claims regarding disease prevention and mandates denial of any claims falling outside that scope—that is, claims containing a nonpreventative intent included in the FD&C Act's drug definition, such as providing treatment for an existing disease. Defendants argue that the FDA's statutory construction violates neither the FD&C Act nor the APA because the FD&C Act's definitions for dietary supplements and drugs are not mutually exclusive. Furthermore, Defendants argue that the FDA's decision to ban Plaintiffs' saw palmetto claim, given its classification as a drug claim, is in accordance with First Amendment principles for government regulation of commercial speech.

A. *The FDA's Determination That It May Authorize Only Those Health Claims Regarding Disease Prevention Violates neither the FD&C Act nor the APA*

Plaintiffs contend that the FDA's classification of the saw palmetto claim as a drug claim violates both the FD&C Act and the APA because the health claims provision in § 343(r)(1)(B) precludes dietary supplement claims from being categorized as drug claims under § 321(g)(1). . . .

The language of § 343(r)(1)(B) provides that a health claim for a dietary supplement is a claim that "characterizes the relationship of any nutrient . . . to a disease or health-related condition." 21 U.S.C. § 343(r)(1)(B). Plaintiffs contend that by enacting § 343(r)(1)(B), Congress clearly intended health claims for dietary supplements to include any nutrient-disease claim and did not intend to limit these claims to disease risk reduction claims. However, Defendants argue that, when read in the context of the whole FD&C Act, it is clear that Congress never intended § 343(r)(1)(B) to limit the statute's core function of drug regulation. Defendants thus argue that Congress intended the FDA to evaluate the validity of a health claim under both the dietary supplement and drug provisions of the FD&C Act.

Accordingly, the Court must examine not only the language of § 343(r)(1)(B) but also the provision regarding FDA health claim approval and other relevant definitions. Approval of a dietary supplement health claim "shall be subject to a procedure and standard, respecting the validity of such claim, established by regulation of the Secretary [of the FDA]." 21 U.S.C. § 343(r)(5)(d). A dietary supplement is a product "intended to supplement the diet that bears or contains . . . an herb or other botanical" that will be deemed a food "[e]xcept for purposes of [the drug definition at § 321(g)]." 21 U.S.C. § 321(ff)(1)(c). A drug is a product "intended for use in the diagnosis, cure, mitigation, treatment, or prevention of disease in man or other animals." 21 U.S.C. § 321(g)(1)(B). While § 343(r)(5)(d) clearly indicates that the FDA has authority to determine the standards regarding health claims for dietary supplements, it is unclear how those standards affect the approval of health claims for products that treat an existing disease.

Plaintiffs argue that if the dietary supplement definition is not read to supersede the drug definition, all health claims could be regulated as drugs given their intended use for "prevention of disease." However, Congress rejected Plaintiffs' view when it refused to add a provision to the DSHEA's drug definition stating that, subject to certain exceptions, "[t]he term 'drug' does not include a dietary supplement as defined in paragraph (ff). . . ." Thus, Plaintiffs' argument has little force.

Plaintiffs also contend that congressional intent to restrict health claims from classification as drug claims is found in other language added to the drug definition by the DSHEA, stating that:

A food or dietary supplement for which a claim, subject to . . . sections 343(r)(1)(B) and 343(r)(5)(D) of this title, is made in accordance with the requirements of section 343(r) of this title is not a drug solely because the label or the labeling contains such a claim. 21 U.S.C. § 321(g)(1).

However, Defendants argue that the term "solely" indicates that Congress still intended the FDA to retain its long-established discretion to classify a claim as a drug claim if it provided adequate grounds for the classification.

In this case, the FDA has provided a further explanation for its decision to classify Plaintiffs' claim as a drug claim—that is, the proposed claim goes beyond risk reduction and purports to treat a disease. The FDA argues that because the FD&C Act's definitions for dietary supplements and drugs are not mutually exclusive, it is authorized to determine that health claims for dietary supplements are actually drug claims when the claim is directed at disease treatment. Congress' intent regarding the scope of health claims is not clear on the face of the FD&C Act, as amended by the NLEA and the DSHEA, given the interconnectedness of the statute's provisions.

Unfortunately, the legislative histories of the FD&C Act, NLEA, and DSHEA do not demonstrate a clear congressional intent with regard to the appropriate scope of health claims. There is no question that the legislative intent behind enactment of the original FD&C Act was to protect the public from unsafe drugs.

In amending the FD&C Act through the NLEA, Congress created a framework for authorization of health claims but also delegated full authority to the FDA to adopt whichever standard the agency deemed most appropriate for approving such claims. While the NLEA provided the statutory authority for authorizing health claims, it clearly gave the FDA wide discretion in approving such claims.

Nor do the DSHEA amendments demonstrate any clear congressional intent with respect to the specific scope of health claims. Here too, the legislative history is ambiguous. It is clear that Congress intended the FDA to establish a more principled regulatory framework for authorizing health claims in order to provide consumers with more access to such information. However, Congress specifically stated that the amendments were added to recognize "the benefits of dietary supplements to health promotion and *disease prevention.*"

Furthermore, Congress issued a Statement of Agreement for the DSHEA that compromised the amendments' "entire legislative history." The sponsors of the bill intended "that no other reports or statements be considered as legislative history for the bill." Because the Statement of Agreement provides explanations for only four DSHEA amendments and does not include any statement regarding the appropriate scope of health claims, the Court finds that the Statement of Agreement further demonstrates an overall lack of specific intent regarding the meaning of the DSHEA amendments in the context of the whole FD&C Act.

In this case, the Court finds an absence of a clear congressional intent with respect to the appropriate scope of health claims. Given the ambiguity inherent in the FD&C Act's intertwined definitions for drugs and dietary supplements, the lack of decisive legislative history, and the FD&C Act's dual function of regulating both drugs and dietary supplements, the Court determines that "Congress has [not] directly spoken to the precise question at issue."

2. Chevron Step Two

Because the Court has determined that the intent of Congress with respect to the scope of health claims is ambiguous, analysis of the FDA's decision under the second stage of *Chevron* is required. In the second stage a court must evaluate the same text, history, and purpose used in the first stage, but instead of determining "whether these convey a plain meaning that requires a certain interpretation," the court will determine "whether these permit the interpretation chosen by the agency." At stage two, a court "need not conclude that the agency construction was the only one it permissibly could have adopted to uphold the construction, or even the reading the court would have reached if the question originally had arisen in a judicial proceeding." The court need merely find that the agency's choice is a rational one.

Courts have long upheld FDA decisions to classify products as drugs based on their intended use.

Furthermore, courts have found that because the FD&C Act definitions of dietary supplements and drugs are not mutually exclusive, FDA regulation may properly focus on intent.

Plaintiffs argue that the FDA has already interpreted the FD&C Act to allow health claims for disease treatment. For example, one of five model health claims suggested by the FDA for the dietary fat and cholesterol–coronary heart disease relationship states that a "healthful diet low in saturated fat, total fat, and cholesterol . . . may lower blood cholesterol levels and may reduce the risk of heart disease." 21 C.F.R. 101.75(e)(3) (2002). Plaintiffs contend that this health claim goes beyond risk prevention because it includes a claim to lower blood cholesterol, which is treating an existing disease. *See also* 21 C.F.R. 101.77(e)(2) (2002). (A model dietary fiber-coronary heart disease health claim also states that eating a diet high in dietary fiber "may lower blood cholesterol levels and reduce your risk of heart disease.") However, Defendants argue that these claims are primarily concerned with risk reduction of heart disease, not the treatment of an existing disease. In fact, the FDA's rulemaking clearly stated that these claims linked "dietary factors to heart disease *risk* via the *intermediate* mechanism of reducing blood LDL-cholesterol levels. . . ." 58 Fed.Reg. 2552, 2573 (1993) (emphasis added).

Moreover, reports relied upon by Congress in enacting the NLEA clearly focused on the role of diet in reducing disease risk, not in treating an existing disease. See House Rep. at 9 (referring to the Surgeon General's "Report on Nutrition and Health" (1988) for the argument that certain diets "can reduce the *risk* of chronic disease") and 13–14 (relying upon the National Research

Council's "Diet and Health Implications for Reducing Chronic Disease *Risk*" (1989)) (emphasis added).

In this case, the FDA concluded that Plaintiffs' health claim indicated that saw palmetto's intended use was purely pharmacological. The proposed claim addresses only BHP symptom treatment and does not include any claim of disease prevention, bolstering FDA concerns that approval of the claim would not provide an adequate level of protection to vulnerable consumers. In fact the FDA previously withdrew over-the-counter approval for saw palmetto by concluding that while "saw palmetto 'probably' provides some 'minimal' [BHP] symptomatic relief," it was concerned that "as long as only the symptoms of the disease [were] relieved, men with BHP may be lulled into a false sense of security" and postpone medical examinations necessary for treatment of BHP, diagnosis of secondary complications, and screening for prostate cancer.

The FD&C Act, as amended by the NLEA and the DSHEA, establishes both the FDA's authority to regulate drugs and dietary supplements and the FDA's responsibility to protect consumers. The FDA has decided that approval of a health claim with a purely pharmacological purpose would not provide an adequate level of consumer protection. As the language, structure, and legislative history of the FD&C Act do not clearly state the appropriate scope of health claims for dietary supplements, the Court finds the FDA's decision to limit approved health claims to those involving disease risk reduction is both permissible and reasonable under the second stage of the *Chevron* analysis.

3. The APA

Under the APA, an agency's action may be set aside only if it is "arbitrary, capricious, an abuse of discretion, or otherwise not in accordance with law." 5 U.S.C. § 706(2)(A). In making this finding, the court "must consider whether the decision was based on a consideration of the relevant factors and whether there has been a clear error of judgment." The court's role is to ensure that the agency's decision was based on relevant factors and not a "clear error of judgment," not substitute its judgment for that of the agency. If the "agency's reasons and policy choices . . . conform to certain minimal standards of rationality . . . the rule is reasonable and must be upheld."

As explained above, the FDA has provided an adequate rationale for its determination that the FD&C Act, as amended by the NLEA and the DSHEA, authorizes the FDA to deny health claims aimed primarily at treatment for an existing disease. Therefore, the FDA's decision to deny the saw palmetto claim as a drug claim, given its intended treatment of BHP, was neither arbitrary nor capricious. In addition, since the FDA's interpretation of the FD&C Act was permissible under the two-step analysis of *Chevron*, its decision was not contrary to law. Accordingly, the FDA did not violate the APA in denying Plaintiffs' saw palmetto health claim petition.

B. The FDA's Decision to Deny Plaintiffs' Health Claim Does Not Violate the First Amendment

Plaintiffs argue that the saw palmetto claim is either scientific or commercial speech protected by the First Amendment. However, it is "undisputed that [ ] restrictions on [ ] health claims are evaluated under the commercial speech doctrine." Therefore, the FDA's refusal to authorize Plaintiffs' proposed claim must be evaluated under the analytical framework established in *Central Hudson Gas & Elec. Corp. v. Public Serv. Comm'n of New York.*[48] . . .

Plaintiffs argue that the FDA's denial of the saw palmetto health claim cannot meet the *Central Hudson* test as articulated by *Pearson* because it impermissibly restricts commercial speech by not allowing "the Plaintiffs' health claims to be made with such disclaimers as are reasonably necessary to avoid a misleading connotation." While the *Pearson* decision did restrict FDA regulation of potentially misleading speech in health claims, Plaintiffs mistakenly construe the FDA's current decision to deny the saw palmetto health claim petition as a decision based on misleadingness. In this case, the Court has determined that the FDA has reasonably interpreted the FD&C Act to conclude that "claims about the effects on existing disease do not fall within the scope of the health claim provisions in 21 U.S.C. § 343(r) and therefore may not be the subject of an authorized health claim." Because the FDA determined that the saw palmetto claim was a drug claim for disease treatment, it concluded that the claim was an unlawful health claim and thus denied Plaintiffs' petition.

As there is no doubt that unlawful speech can be banned under the first step of the *Central Hudson* analysis, the FDA's prohibition of Plaintiffs' saw palmetto claim does not violate the First Amendment.

IV. Conclusion

The FDA's denial of Plaintiffs' saw palmetto claim did not violate the FD&C Act, APA, or First Amendment. The FDA's interpretation of the various provisions of the FD&C Act to permit only disease prevention health claims was reasonable given the ambiguity of the statute; therefore, its decision to deny Plaintiffs' claim based on this interpretation is neither arbitrary nor capricious nor contrary to law. For the reasons discussed above, Plaintiffs' Motion for Summary Judgment is denied and Defendants' Motion to Dismiss is granted. An Order will issue with this Opinion. . . .

ORDERED, that Defendants' Motion to Dismiss is granted; and it is further

ORDERED, that Plaintiffs' Motion for Summary Judgment is denied.

* * * * *

---

[48] 447 U.S. 557 (1980).

Pearson v. Shalala A landmark case in the dietary supplement field is *Pearson v. Shalala*, 164 F.3d 650 (D.C. Cir. 1999). The *Pearson* case began when the FDA rejected four proposed health claims by the *Pearson* plaintiffs. These four claims linked the consumption of a particular food (supplement) to the reduction in risk of a particular disease:

1. "Consumption of antioxidant vitamins may reduce the risk of certain kinds of cancers."
2. "Consumption of fiber may reduce the risk of colorectal cancer."
3. "Consumption of omega-3 fatty acids may reduce the risk of coronary heart disease."
4. ".8 mg of folic acid in a dietary supplement is more effective in reducing the risk of neural tube defects than a lower amount in foods in common form."

Relying on arguments grounded in the First Amendment and the Administrative Procedures Act (APA), the U.S. Court of Appeals invalidated the FDA regulations prohibiting certain health claims on foods. The court required that FDA reconsider its disapproval of the plaintiffs' claims. The court ruled that FDA (1) violated the First Amendment by banning misleading health claims without considering the use of curative disclaimers, and (2) violated the arbitrary and capricious standard of the APA by failing to clarify the standard of "significant scientific agreement."

Health claims are a form of commercial speech, and under First Amendment protections, FDA cannot unnecessarily restrain such speech. FDA argued that health claims lacking "significant scientific agreement" are inherently misleading to consumers and therefore are incapable of being cured by disclaimers. However, the Court of Appeals ruled that the FDA had no basis to reject the health claims without first assessing whether the use of a disclaimer could communicate meaningful, nonmisleading information to the consumer. Where commercial speech is potentially misleading but can be "presented in a way that is not deceptive," the government cannot ban it. For example, a disclaimer might be able to communicate that available scientific evidence is inconclusive regarding the health benefit because the studies performed have been on foods containing those components and not on the dietary substances themselves.

The court also found that FDA had not followed appropriate administrative procedures because it failed to fully explain why the four health claims did not meet the "significant scientific agreement" standard applicable to health claims. FDA had not defined the criteria being applied to determine whether such agreement exists. The court noted the legal and practical need to provide a governing rationale for approving or rejecting proposed health claims based on a lack of "significant scientific agreement." The court concluded that the FDA's denial of these health claims without defining "significant scientific agreement" constituted arbitrary and capricious action under the APA. Accordingly, the court ordered the FDA to explain the meaning of "significant scientific agreement." At a minimum the FDA must make it possible "for the regulated class to perceive the principles which are guiding agency action."

The decision created legal hurdles to the FDA's efforts to reject petitions filed in support of health claims. However, the decision did not permit the *Pearson* plaintiffs to make their health claims. The decision directed the FDA to reconsider the plaintiffs' four proposed claims in light of the possible value of disclaimers. Basically, the decision invalidated the FDA's regulations, but put the *Pearson* plaintiffs back at square one in the FDA pre-clearance process. In addition, the court did not rule out the possibility that "where evidence in support of a claim is outweighed by evidence against the claim," the FDA could deem the claim "incurable" by a disclaimer and, therefore, reject the claim as unlawful.

* * * * *

Pearson v. Shalala

164 F.3d 650 (D.C. Cir. 1999)

Before: WALD, SILBERMAN, and GARLAND, Circuit Judges
Opinion: SILBERMAN, Circuit Judge

Marketers of dietary supplements must, before including on their labels a claim characterizing the relationship of the supplement to a disease or health-related condition, submit the claim to the Food and Drug Administration for preapproval. The FDA authorizes a claim only if it finds "significant scientific agreement" among experts that the claim is supported by the available evidence. Appellants failed to persuade the FDA to authorize four such claims and sought relief in the district court, where their various constitutional and statutory challenges were rejected. We reverse.

I

Dietary supplement marketers Durk Pearson and Sandy Shaw, presumably hoping to bolster sales by increasing the allure of their supplements' labels, asked the FDA to authorize four separate health claims. . . . A "health claim" is a "claim made on the label or in labeling of . . . a dietary supplement that expressly or by implication . . . characterizes the relationship of any substance to a disease or health-related condition." 21 C.F.R. § 101.14(a)(1) (1998). Each of appellants' four claims links the consumption of a particular supplement to the reduction in risk of a particular disease:

(1) "Consumption of antioxidant vitamins may reduce the risk of certain kinds of cancers."

(2) "Consumption of fiber may reduce the risk of colorectal cancer."

(3) "Consumption of omega-3 fatty acids may reduce the risk of coronary heart disease."

(4) ".8 mg of folic acid in a dietary supplement is more effective in reducing the risk of neural tube defects than a lower amount in foods in common form."

. . . .

. . . The NLEA addressed foods and dietary supplements separately. Health claims on foods may be made without FDA approval as a new drug, or the risk of sanctions for issuing a "misbranded" product, if it has been certified by the FDA as supported by "significant scientific agreement." Congress created a similar safe harbor for health claims on dietary supplements, but delegated to the FDA the task of establishing a "procedure and standard respecting the validity of [the health] claim." *Id.* § 343(r)(5)(D).

The FDA has since promulgated 21 C.F.R. § 101.14—the "significant scientific agreement" "standard" (quoted above)—and 21 C.F.R. § 101.70—a "procedure" (not particularly relevant to this case)—for evaluating the validity of health claims on dietary supplements. In doing so, the agency rejected arguments asserted by commenters—including appellants—that the "significant scientific agreement" standard violates the First Amendment because it precludes the approval of less-well supported claims accompanied by a disclaimer and because it is impermissibly vague. The FDA explained that, in its view, the disclaimer approach would be ineffective because "there would be a question as to whether consumers would be able to ascertain which claims were preliminary [and accompanied by a disclaimer] and which were not," and concluded that its prophylactic approach is consistent with applicable commercial speech doctrine. The agency, responding to the comment that "significant scientific agreement" is impermissibly vague, asserted that the standard is "based on objective factors" and that its procedures for approving health claims, including the notice and comment procedure, sufficiently circumscribe its discretion.

Then the FDA rejected the four claims supported by appellants. The problem with these claims, according to the FDA, was not a dearth of supporting evidence; rather, the agency concluded that the evidence was inconclusive for one reason or another and thus failed to give rise to "significant scientific agreement." But the FDA never explained just how it measured "significant" or otherwise defined the phrase. The agency refused to approve the dietary fiber–cancer claim because "a supplement would contain only fiber, and there is no evidence that any specific fiber itself caused the effects that were seen in studies involving fiber-rich [foods]." The FDA gave similar reasons for rejecting the antioxidant vitamins–cancer claim, and the omega-3 fatty acids–coronary heart disease claim. As for the claim that 0.8 mg of folic acid in a dietary supplement is more effective in reducing the risk of neural tube defects than a lower amount in foods in common form, the FDA merely stated that "the scientific literature does not support the superiority of any one source over others." The FDA declined to consider appellants' suggested alternative of permitting the claim while requiring a corrective disclaimer such as "The FDA has determined that the evidence supporting this claim is inconclusive."

A more general folate–neural tube defect claim supported by appellants—that consumption of folate reduces the risk of neural tube defects—was initially rejected but ultimately approved for both dietary supplement and food labels. *See* 21 C.F.R. § 101.79 (1998). The parties disagree on what caused the FDA's change of position on this claim. Appellants contend that political objections—Senator Hatch was one of the complainers—concentrated the agency's mind. The FDA insists that its initial denial of the claim was based on a concern that folate consumption might have harmful effects on persons suffering from anemia, and that its concern was alleviated by new scientific studies published after the initial denial of the claim.

Appellants sought relief in the district court, raising APA and other statutory claims as well as a constitutional challenge, but were rebuffed.

II

Appellants raise a host of challenges to the agency's action. But the most important are that their First Amendment rights have been impaired and that under the Administrative Procedure Act the FDA was obliged, at some point, to articulate a standard a good deal more concrete than the undefined "significant scientific agreement." Normally we would discuss the nonconstitutional argument first, particularly because we believe it has merit. We invert the normal order here to discuss first appellants' most powerful constitutional claim, that the government has violated the First Amendment by declining to employ a less draconian method—the use of disclaimers—to serve the government's interests, because the requested remedy stands apart from appellants' request under the APA that the FDA flesh out its standards. That is to say, even if "significant scientific agreement" were given a more concrete meaning, appellants might be entitled to make health claims that do not meet that standard—with proper disclaimers.

Appellants also claim that the agency's "nondefinition" runs afoul of Fifth Amendment concerns for vagueness. This contention is, however, closely connected to appellants' APA challenge and may well not be implicated if appellants' APA challenge affords ultimate relief. Therefore we will defer it until our APA analysis.

A. Disclaimers

It is undisputed that FDA's restrictions on appellants' health claims are evaluated under the commercial speech doctrine. It seems also undisputed that the FDA has unequivocally rejected the notion of requiring disclaimers to cure "misleading" health claims for dietary supplements. (Although the general regulation does not in *haec verba* preclude authorization of qualified claims, the government implied in its statement of basis and purpose that disclaimers were not adequate, and did not consider their use in the four subregulations before us. *See* 21 C.F.R. § 101.71(a), (c), (e); *id.* § 101.79(c)(2)(i)(G).) The government makes two alternative arguments in response to appellants' claim that it is unconstitutional for the government to refuse to entertain a disclaimer requirement for the proposed health claims: first, that health claims lacking "significant scientific agreement" are inherently misleading and thus entirely outside the protection of the First Amendment; and second, that even if the claims are only potentially misleading, under *Central Hudson*, the government is not obliged to consider requiring disclaimers in lieu of an outright ban on all claims that lack significant scientific agreement.

If such health claims could be thought inherently misleading, that would be the end of the inquiry.

Truthful advertising related to lawful activities is entitled to the protections of the First Amendment. But when the particular content or method of the advertising suggests that it is inherently misleading or when experience has proved that in fact such advertising is subject to abuse, the States may impose appropriate restrictions. Inherently misleading advertising may be prohibited entirely. But the States may not place an absolute prohibition on . . . potentially misleading information . . . if the information also may be presented in a way that is not deceptive.

As best we understand the government, its first argument runs along the following lines: that health claims lacking "significant scientific agreement" are inherently misleading because they have such an awesome impact on consumers as to make it virtually impossible for them to exercise any judgment at the point of sale. It would be as if the consumers were asked to buy something while hypnotized, and therefore they are bound to be misled. We think this contention is almost frivolous. We reject it. But the government's alternative argument is more substantial. It is asserted that health claims on dietary supplements should be thought at least potentially misleading because the consumer would have difficulty in independently verifying these claims. We are told, in addition, that consumers might actually assume that the government has approved such claims.

Under *Central Hudson*, we are obliged to evaluate a government scheme to regulate potentially misleading commercial speech by applying a three-part test. First, we ask whether the asserted government interest is substantial. The FDA advanced two general concerns: protection of public health and prevention of consumer fraud. The Supreme Court has said "there is no question that [the government's] interest in ensuring the accuracy of commercial information in the marketplace is substantial," and that government has a substantial interest in "promoting the health, safety, and welfare of its citizens." At this level of generality, therefore, a substantial governmental interest is undeniable.

The more significant questions under *Central Hudson* are the next two factors: "whether the regulation directly advances the governmental interest asserted," and whether the fit between the government's ends and the means chosen to accomplish those ends "is not necessarily perfect, but reasonable." We think that the government's regulatory approach encounters difficulty with both factors.

It is important to recognize that the government does not assert that appellants' dietary supplements in any fashion threaten consumer's health and safety. The government simply asserts its "commonsense judgment" that the health of consumers is advanced directly by barring any health claims not approved by the FDA. Because it is not claimed that the product is harmful, the government's underlying—if unarticulated—premise must be that consumers have a limited amount of either attention or dollars that could be devoted to pursuing health through nutrition, and therefore products that are not indisputably health enhancing should be discouraged as threatening to crowd out more worthy expenditures. We are rather dubious that this simplistic view of human nature or market behavior is sound, but, in any event, it surely cannot be said that this notion—which the government does not even dare openly to set forth—is a direct pursuit of consumer health; it would seem a rather indirect route, to say the least.

On the other hand, the government would appear to advance directly its interest in protecting against consumer fraud through its regulatory scheme. If it can be assumed—and we think it can—that some health claims on dietary supplements will mislead consumers, it cannot be denied that requiring FDA preapproval, and setting the standard extremely, perhaps even impossibly, high will surely prevent any confusion among consumers. We also recognize that the government's interest in preventing consumer fraud/confusion may well take on added importance in the context of a product, such as dietary supplements, that can affect the public's health.

The difficulty with the government's consumer fraud justification comes at the final *Central Hudson* factor: Is there a "reasonable" fit between the government's goals and the means chosen to advance those goals? The government insists that it is never obliged to utilize the disclaimer approach, because the commercial speech doctrine does not embody a preference for disclosure over outright suppression. Our understanding of the doctrine is otherwise. In *Bates v. State Bar of Arizona*, the Supreme Court addressed an argument similar to the one the government advances. The State Bar had disciplined several attorneys who advertised their fees for certain legal services in violation of the Bar's rule, and sought to justify the rule on the ground that such advertising is inherently misleading "because advertising by attorneys will highlight irrelevant factors and fail to show the relevant factor of skill." *Id.* at 372. The Court observed that the Bar's concern was "not without merit" but refused to credit the notion that "the public is not sophisticated enough to realize the limitations of advertising, and that the public is better kept in ignorance than trusted with correct but incomplete information." Accordingly, the Court held that the "incomplete" attorney advertising was not inherently misleading and that "the preferred remedy is more disclosure, rather than less." In more recent cases, the Court has reaffirmed this principle, repeatedly pointing to disclaimers as constitutionally preferable to outright suppression.

The government suggests that the Supreme Court's guidance on this issue is not so consistent (or coherent?). It points to *Friedman v. Rogers*, where the Court, in the course of upholding a ban on the use of trade names by optometrists, stated that "there is no First Amendment rule . . . requiring a State to allow deceptive or misleading commercial speech whenever the publication of additional information can clarify or offset the effects of the spurious communication." To be sure, this language cuts against the notion that government must, where possible, regulate misleading commercial speech by requiring disclaimers rather than by imposing an outright ban. But the Court in Friedman made clear the narrowness of its holding as limited to the special status of trade names:

> We emphasize . . . that the restriction on the use of trade names has only the most incidental effect on the content of the commercial speech of Texas optometrists. . . . [A] trade name conveys information only because of the associations that grow up over time between the name and a certain level of price and quality of service. . . . Since the Act does not prohibit or limit the type of informational advertising held to be protected in . . . *Bates*, the factual information associated with trade names may be communicated freely and explicitly to the public.

The government does not assert here that appellants' health claims convey no factual information, only that the factual information conveyed is misleading. *Friedman* is thus not at odds with the relevant First Amendment principles established in *Bates*, which in any event the Supreme Court has reaffirmed— post-*Friedman*—in *R.M.J., Shapero,* and *Peel.*

Nor do we agree with the FDA's suggestion that the Supreme Court's decision in *Fox*—a case that did not involve assertedly misleading commercial speech—mandates a more deferential review of government regulations on potentially misleading commercial speech. In *Fox*, the Court elaborated on the degree of scrutiny appropriate under the *Central Hudson* test, making clear that the final step does not require that "the manner of restriction is absolutely the least severe that will achieve the desired end," but only that the fit between the legislature's ends and means is a "reasonable" one. It is clear, then, that when government chooses a policy of suppression over disclosure—at least where there is no showing that disclosure would not suffice to cure misleadingness—government disregards a "far less restrictive" means. . . .

Our rejection of the government's position that there is no general First Amendment preference for disclosure over suppression, of course, does not determine that any supposed weaknesses in the claims at issue can be remedied by disclaimers and thus does not answer whether the subregulations, 21 C.F.R. § 101.71(a), (c), (e); *id.* § 101–79(c)(2)(i)(G), are valid. The FDA deemed the first three claims—(1) "Consumption of antioxidant vitamins may reduce the risk of certain kinds of cancers," (2) "Consumption of fiber may reduce the risk of colorectal cancer," and (3) "Consumption of omega-3 fatty acids may reduce the risk of coronary heart disease"—to lack significant scientific agreement because existing research had examined only the relationship between consumption of foods containing these components and the risk of these diseases. The FDA logically determined that the specific effect of the component of the food constituting the dietary supplement could not be determined with certainty. (The FDA has approved similar health claims on foods containing these components. *See, e.g.*, 21 C.F.R. § 101.79 (folate–neural tube defects).) But certainly this concern could be accommodated, in the first claim, for example, by adding a prominent disclaimer to the label along the following lines: "The evidence is inconclusive because existing studies have been performed with foods containing antioxidant vitamins, and the effect of those foods on reducing the risk of cancer may result from other components in those foods." A similar disclaimer would be equally effective for the latter two claims.

The FDA's concern regarding the fourth claim—0.8 mg of folic acid in a dietary supplement is more effective in reducing the risk of neural tube defects than a lower amount in foods in common form"—is different from its reservations regarding the first three claims; the agency simply concluded that "the scientific literature does not support the superiority of (concluding that "losses [of folic acid] in cooking and canning [foods] can be very high due to heat destruction"), and we suspect that a clarifying disclaimer could be added to the effect that "The evidence in support of this claim is inconclusive."

The government's general concern that, given the extensiveness of government regulation of the sale of drugs, consumers might assume that a claim on a supplement's label is approved by the government, suggests an obvious answer: The agency could require the label to state that "The FDA does not approve this claim." Similarly, the government's interest in preventing the use of labels that are true but do not mention adverse effects would seem to be satisfied—at least ordinarily—by inclusion of a prominent disclaimer setting forth those adverse effects.

The government disputes that consumers would be able to comprehend appellants' proposed health claims in conjunction with the disclaimers we have suggested—this mix of information would, in the government's view, create confusion among consumers. But all the government offers in support is the FDA's pronouncement that "consumers would be considerably confused by a multitude of claims with differing degrees of reliability." Although the government may have more leeway in choosing suppression over disclosure as a response to the problem of consumer confusion where the product affects health, it must still meet its burden of justifying a restriction on speech—here the FDA's conclusory assertion falls far short. See *Ibanez*, 512 U.S. at 146 ("If the protections afforded commercial speech are to retain their force, we cannot allow rote invocation of the words 'potentially misleading' to supplant the [government's] burden to demonstrate that the harms it recites are real and that its restriction will in fact alleviate them to a material degree."); *Edenfield*,

507 U.S. at 771 (invalidating a ban on in-person solicitation by accountants where the government failed to present "studies" or "anecdotal evidence" showing that such solicitation posed dangers of fraud, overreaching, or compromised independence).

We do not presume to draft precise disclaimers for each of appellants' four claims; we leave that task to the agency in the first instance. Nor do we rule out the possibility that where evidence in support of a claim is outweighed by evidence against the claim, the FDA could deem it incurable by a disclaimer and ban it outright. For example, if the weight of the evidence were against the hypothetical claim that "Consumption of vitamin E reduces the risk of Alzheimer's disease," the agency might reasonably determine that adding a disclaimer such as "The FDA has determined that no evidence supports this claim" would not suffice to mitigate the claim's misleadingness. Finally, while we are skeptical that the government could demonstrate with empirical evidence that disclaimers similar to the ones we suggested above would bewilder consumers and fail to correct for deceptiveness, we do not rule out that possibility.

B. The Unarticulated Standard

Wholly apart from the question whether the FDA is obliged to consider appropriate disclaimers is appellants' claim that the agency is obliged to give some content to the phrase "significant scientific agreement." Appellants contend that the agency's failure to do so independently violates their constitutional rights under the First and Fifth Amendments. The First, because producers of dietary supplements are assertedly subject to a "prior restraint" on their protected speech—the labeling of products. The Fifth, because the agency's approach is so vague as to deprive the producers of liberty (and property?) without due process.

Appellants do not challenge the concept of a prescreening system per se; their complaint is with the FDA's lack of guidance on which health claims will survive the prescreening process. But appellants never connected their vagueness concern with their oblique First Amendment prior restraint argument, and for that reason we need not decide whether prior restraint analysis applies to commercial speech. On the other hand, appellants' Fifth Amendment vagueness argument is squarely presented. Still, by prevailing on their APA claim, appellants would seem to gain the same relief—invalidation of the FDA's interpretation of the general standard and a remand for more guidance—as they would through a successful Fifth Amendment claim (or indeed a First Amendment prior restraint claim, if it had been properly presented and assuming arguendo that prior restraint analysis applies in the commercial speech context).

Consideration of this constitutional claim seems unnecessary because we agree with appellants that the APA requires the agency to explain why it rejects their proposed health claims—to do so adequately necessarily implies giving some definitional content to the phrase "significant scientific agreement." We think this proposition is squarely rooted in the prohibition under the APA that an agency not engage in arbitrary and capricious action. *See* 5 U.S.C. § 706(2)(A) (1994). It simply will not do for a government agency to declare—without explanation—that a proposed course of private action is not approved. To refuse to define the criteria it is applying is equivalent to simply saying no without explanation. Indeed, appellants' suspicions as to the agency's real reason for its volte-face on the general folate–neural tube defect claim highlight the importance of providing a governing rationale for approving or rejecting proposed health claims.

To be sure, Justice Stewart once said, in declining to define obscenity, "I know it when I see it," which is basically the approach the FDA takes to the term "significant scientific agreement." But the Supreme Court is not subject to the Administrative Procedure Act. Nor for that matter is the Congress. That is why we are quite unimpressed with the government's argument that the agency is justified in employing this standard without definition because Congress used the same standard in 21 U.S.C.A. § 343(r)(3)(B)(i). Presumably—we do not decide—the FDA in applying that statutory standard would similarly be obliged under the APA to give it content.

That is not to say that the agency was necessarily required to define the term in its initial general regulation—or indeed that it is obliged to issue a comprehensive definition all at once. The agency is entitled to proceed case by case or, more accurately, subregulation by subregulation, but it must be possible for the regulated class to perceive the principles which are guiding agency action. Accordingly, on remand, the FDA must explain what it means by significant scientific agreement or, at minimum, what it does not mean.

For the foregoing reasons, we hold invalid the four subregulations, 21 C.F.R. § 101.71(a), (c), (e); § 101.79(c)(2)(i)(G), and the FDA's interpretation of its general regulation, *id.* § 101.14. The decision of the district court is reversed, and the case is remanded to the district court with instructions to remand in turn to the FDA for reconsideration of appellants' health claims.

So ordered.

* * * * *

8.9 PROBLEM SUPPLEMENTS

8.9.1 Ephedra

Ephedra was widely promoted to aid weight loss, enhance sports performance, and increase energy. Ephedra, also called *Ma huang*, is a naturally occurring substance derived from a number of plants. The main active ingredient, ephedrine, is also chemically synthesized. When chemically synthesized, ephedrine is regulated as a drug.

In 1997 the FDA proposed a rule to establish a finding that a dietary supplement is adulterated if it contains 8 mg or more of ephedrine alkaloids. The rule would have also required certain warning statements.[49]

However, because of further information of adverse effects, on April 3, 2000, the FDA withdrew part of its proposed rule that had been published June 4, 1997.[50] The FDA noted that it was concerned about the safety of certain dietary ingredient level and the duration of use limit for these products. The FDA announced the availability of new adverse event reports and related information associated with these products, and also announced the FDA's plans to participate in a public forum to discuss this new information.[51]

On December 30, 2003, the FDA issued a consumer alert on the safety of dietary supplements containing ephedra. The FDA advised consumers to immediately stop using ephedra.[52] On February 6, 2004, the FDA issued a final rule prohibiting the sale of dietary supplements containing ephedrine alkaloids because it determined the supplements pose an unreasonable risk of illness or injury.

FDA's decision cited the RAND study, commissioned by the National Institutes of Health, which reviewed recent evidence on the risks and benefits of ephedra and ephedrine. The RAND study found some evidence of an effect of ephedra on short-term weight loss, and minimal evidence of an effect on performance enhancement in certain physical activities. However, the study also concluded that ephedra is associated with higher risks of mild to moderate side effects such as heart palpitations, psychiatric and upper gastrointestinal effects, and symptoms of autonomic hyperactivity, such as tremor and insomnia, especially when it is taken with other stimulants. The study reviewed over 16,000 adverse events reported after ephedra use and found about 20 "sentinel events," including heart attack, stroke, and death that occurred in the absence of other contributing factors.

After reviewing the RAND study and other recent studies of serious adverse events from ephedra, FDA concluded that ephedra as currently marketed had unreasonable safety risks. Under the DSHEA, the FDA may remove a dietary supplement from the market if it presents a significant or unreasonable risk of illness or injury when used according to its labeling or, in the absence of label directions, under ordinary conditions of use.[53] Alternatively the FDA may remove the dietary supplement if it poses an "imminent hazard to public health or safety."[54]

---

[49] 62 Fed. Reg. 30,677 to 30,724 (June 4, 1997).
[50] *Id.*
[51] 65 Fed. Reg. 17,474 to 17,477 (Apr. 3, 2000).
[52] Consumer Alert: FDA Plans Regulation Prohibiting Sale of Ephedra-Containing Dietary Supplements, http://www.fda.gov/oc/initiatives/ephedra/december2003/advisory.html (last visited Oct. 18, 2005).
[53] 21 U.S.C. § 342(f)(1)(A).
[54] *Id.* § 342(f)(1)(B).

Dietary supplement producers brought a lawsuit against FDA to challenge validity of the ban on ephedrine-alkaloid in dietary supplements. The United States District Court for the District of Utah overturned the ban in favor of the manufacturers.[55] FDA appealed, and the United States Court of Appeals reinstated the ban, finding that FDA properly conducted risk-benefit analysis, and that FDA's determination that no dosage level of ephedrine-alkaloid in dietary supplements was acceptable for market was not arbitrary and capricious.[56]

8.9.2   Androstenedione—Mark McGwire's Tonic

In 1998 homerun record holder Mark McGwire admitted that he used androstenedione as a performance-enhancing substance.[57] On March 11, 2004, the FDA warned manufacturers to stop distributing products containing androstenedione ("andro").[58] The FDA's enforcement theory is largely based on the health risks of andro. Androstenedione acts like a steroid once it is metabolized by the body, and therefore it can pose similar kinds of health risks as steroids. Andro products are generally advertised as dietary supplements that enhance athletic performance based on their claimed properties to stimulate muscle growth and increase production of testosterone.

A 2002 survey by Health and Human Service's National Institute on Drug Abuse found that about one out of forty high school seniors reported that they had used andro in the past year.[59] The survey also found that about one out of 50 tenth graders had taken andro in the previous year.

The National Collegiate Athletics Association, the National Football League, and the International Olympic Committee have banned androstenedione. The American Academy of Pediatrics, the American Medical Association, and other health professional groups have cautioned against the use of androstenedione because of its potential long-term adverse health consequences.[60]

8.9.3   FDA Warnings

FDA maintains a Web page devoted to warnings on problem supplements.[61] Supplements singled out as dangerous are as follows:

---

[55] Nutraceutical Corp. v. Crawford, 364 F. Supp. 2d 1310 (D. Utah 2005).
[56] Nutraceutical Corp. v. Von Eschenbach, 459 F.3d 1033 (10th Cir. (Utah) 2006).
[57] Charles E. Yesalis III, *Medical, Legal, and Societal Implications of Androstenedione Use*, 281 JAMA 2043–44 (1999).
[58] HHS Launches Crackdown on Products Containing Andro, *available at*: http://www.fda.gov/bbs/topics/news/2004/hhs_031104.html (last visited Oct. 18, 2005).
[59] *Id.*
[60] *Id.*
[61] FDA, Warnings and Safety Information, *available at*: http://www.cfsan.fda.gov/~dms/ds-warn.html (last accessed Mar. 1, 2008).

- Aristolochic acid—associated with nephropathy.
- Comfrey—contains pyrrolizidine alkaloids that are hepatotoxins in animals.
- Kava—potential risk of severe liver injury associated with the use.

8.9.4 Third-Party Certification

Some critics assert that the dietary supplement industry needs increased government regulation. Others argue that self-regulation is preferable.[62]

One option is third-party certification by organizations such as NSF International, which has a dietary supplement certification program. NSF certifies to NSF/ANSI Standard 173. The American National Standards Institute (ANSI) develops voluntary consensus standards in the United States and accredits organizations such as NSF. Standard 173 was issued as a draft in 2001 and approved as a full standard in 2005. NSF/ANSI Standard 173 certifies for formulation and production to comply with good manufacturing practices (GMPs).

Third-party certification may work so long as consumers make it profitable for firms to seek certification. That is, the increase in profit must exceed the extra cost of certification. An admitted shortcoming of third-party certification is that it only can address purity and quality of the ingredients; it does not address the accuracy of the claims made on labels or in advertising or the safety and effectiveness of the active ingredients.[63]

8.10 "STREET DRUG ALTERNATIVES"

"Street drug alternatives" are products promoted as alternatives to illicit street drugs that are intended to affect psychological states. Generally, these contain botanicals, but some contain vitamins and minerals. These products are not considered dietary supplements because they are intended for getting high (intended to affect the structure or function of the body), not as supplements to the diet. Marketing is what makes a product a drug (not the ingredients).

In *United States v. Undetermined Quantities of Articles of Drug*,[64] FDA seized products from three manufacturers of herbal mixtures marketed as street drug alternatives. FDA brought action for an order of condemnation and permanent injunction against manufacturers and their president. The court held, inter alia, that the products were not dietary supplements but unap-

---

[62] *See, e.g.*, Henry I. Miller and Peter VanDoren, *Food Risks and Labeling Controversies: Market-based alternatives to more government regulation of foods and dietary supplements,* 23:1 REGULATION 35–39 (Spring 2000).
[63] *Id.*
[64] 145 F. Supp. 2d 692 (D. Md. 2001).

proved new drugs within the meaning of the FD&C Act. The court refused "to carve out a statutory loophole for drug manufacturers attempting to profit from the illegal drug epidemic by masquerading potentially dangerous substances as legitimate dietary supplements."[65]

In addition, to the analysis under the FD&C Act definition of "drug," the court noted that the definition of "dietary supplement" requires that the product be labeled as a dietary supplement.[66] The defendants failed to label many of the products at issue as dietary supplements; therefore they could not be regarded as dietary supplements.

* * * * *

United States v. Undetermined Quantities of Articles of Drug

145 F. Supp. 2d 692 (D. Md. 2001)

WILLIAMS, District Judge:

Currently pending before the Court are Plaintiff's Motion for Summary Judgment and Defendant Perry Hitt's Motion for Partial Summary Judgment. The motions have been fully briefed by all parties. No hearing is deemed necessary. Upon consideration of the arguments made in support of, and opposition to, the respective motions, the Court makes the following determinations.

I. Factual Background

Hit Products, Inc., Riverdale Organics, and Dreamworlds (collectively "Defendants") manufacture, market, and distribute certain products that serve the basis for this controversy. Specifically, the products at issue are named as follows: "Herba Ghani," "Inda-Kind," "Hydro," "Sweet Green," "Chronix," "Rave X," "Rave Energy," "Utopia," "Shroomz," "Liquid X," "Liquid X Export," "Hashanna Oil," "Northern Heights," "Herbal Hash–Mean Green," "Herbal Hash–Honey-Blonde," and "Herbal Opium." The products are made from a mixture of herbs. Defendants specifically market the products via publications and the Internet to Generation Xers—the demographic of young adults aged 20 to 30 years old. The products promise users effects comparable to illegal street drugs that plague America's youth. The United States Food and Drug Administration ("FDA") categorizes such substances as "street drug alternatives" that qualify as misbranded and unapproved new drugs in violation of the Federal Food, Drug, and Cosmetic Act ("FD&C Act"). Under its statutory authority, the FDA seized Defendants' products. Thereafter the United States brought the instant action seeking an order of condemnation

---

[65] Undetermined Quantities, 145 F. Supp. 2d at 697.
[66] Id.

and permanent injunctive relief against the companies and their president, Perry Hitt.

II. Discussion

A. FDA's Guidance on Street Drugs Alternatives

The United States' arguments in this case essentially track the opinion delineated by the FDA in a policy statement issued in April of 2000. In the statement, the FDA defined street drug alternatives as herbal products that claim to mimic the euphoric effects of illegal street drugs. In the notice, the FDA announced its position that street drug alternatives constitute unapproved new drugs and misbranded drugs in violation of §§ 502 and 505 of the FD&C Act. The FDA also asserted that street drug alternatives did not fall within the definition of "dietary supplements" because such products are not "intended to supplement the diet," but rather to modify the psychological states of the user. Defendants' products fall within the scope of the FDA's definition of street drug alternatives.

. . . .

The Court believes Defendants' characterization of the seized products as "dietary supplements" constitutes a veiled attempt to circumvent federal anti-drug laws and the FD&C Act. This Court declines to carve out a statutory loophole for drug manufacturers attempting to profit from the illegal drug epidemic by masquerading potentially dangerous substances as legitimate dietary supplements. Such mischaracterizations are not only contrary to the language of the Dietary Supplement Health and Education Act of 1994 ("DSHEA") but also undermine the manifest purposes of the FD&C Act.

B. Dietary Supplements

. . . .

Few courts have interpreted this amendment to the FD&C Act by the DSHEA. Nonetheless, as a prerequisite to application of the DSHEA, the product must be labeled as a "dietary supplement." The vast majority of Defendants' products are not labeled in compliance with the Act and, therefore, cannot be classified as "dietary supplements" within the meaning of 21 U.S.C. § 321(ff).

Only the product "Utopia" is labeled as a "dietary supplement." However, this alone does not preclude FDA regulation as a "drug." The Second Circuit has ruled that a product may qualify as a "dietary supplement" under § 321(ff) for certain purposes under the FD&C Act and may also qualify as a "drug" under § 321(g)(1)(C). . . .

Here the FDA is not seeking to regulate "Utopia" based solely upon truthful and not misleading claims concerning its nutritive benefits. "Utopia's" labeling does not make such claims. Thus, even assuming that "Utopia" qualifies as a "dietary supplement" under § 321(ff), this status does not preclude the product from being classified as a "drug" under § 321(g)(1)(C) if it meets the statutory definition. Thus, the next issue is whether the products may be properly classified as drugs under the FD&C Act.

C. Drugs

The United States asserts that the Defendants' products are drugs because they are intended to affect the structure and function of the human body by altering the psychological states of users. Defendants counter that the products' appeal to "alternative" lifestyles does not elevate them to "drug" status. The FD&C Act defines "drug" as "articles (other than food) intended to affect the structure or any function of the body of man or other animals." 21 U.S.C. § 321(g)(1)(C). Thus "the definition of drug . . . require[s] not only that the article 'affect the structure or any function of the body,' but also that these effects be intended." *Brown & Williamson Tobacco Corp. v. FDA*, 153 F.3d 155, 163 (4th Cir.1998). Of primary significance in determining whether a product may be deemed a "drug" is its intended use or effect as gathered from the objective evidence disseminated by the vendor.

An examination of the labeling and promotional claims made by Defendants reflect an intent for the products to affect the function of the mind. The labeling for "Rave X" employs the phrases "ELECTRIFY YOUR SENSES," "BE HAPPY," and "RAGING POTENCY." The product promises "effects in 15–45 min[utes]" of use. "Rave Energy" is labeled as "the Strong Legal X alternative" and being "extremely potent." The label for "Utopia" states "Xperience the natural sensation" and promises that the product "lasts 4–5 hours." "Shroomz" includes a picture of a grinning figure with a mushroom atop his head suggesting to users "FEED YOUR HEAD." The label states that the product is of "Pharmaceutical quality & safety," "Full of shroomy goodness," and "lasts 3–5 hours with no unpleasant side effects." The labels for "Liquid X" and "Liquid X Export" explicitly state that they are "known in the club scene as Liquid Ecstasy." The labels described "Liquid X" and "Liquid X Export" as "unique & extremely powerful liquid infusion of enlivening organics," and "EXTREMELY POTENT" with "effects in 15–45 min." "Hashanna Oil" states that users can "obtain full hashanna effect with just a few drops. . . ." and "BE HAPPY." "Northern Heights" includes another smirking figure with the slogans "Intensify your Smoking adventures" and "Legal & Naturally Satisfying." The labels for "Herbal Hash–Mean Green" and "Herbal Hash–Honey-Blonde" describe the smokables as the "Ultra Chronic Blend" that comes "HIGHLY recommended." The labels instruct the user to "enjoy," "chill," and "repeat as needed every 2–4 hours." "Inda-Kind" is labeled as a "chronic smoke/incense blend" with the slogan "BE HAPPY." The label for "Hydro" describes the product as a "Stoney Hyrdophonic Smoke" with an "uncanny appearance and familiar effect." "Chronix" suggests the user "Get hooked" and "Step into the 60's." "Herbal Opium" is labeled as an "ultra-potent" herbal smoke "offered for those who seek enlightenment of the soul." The label for "Herba Ghani" states "high potency smoking herb" and "Ultra-Potent Smoke." The Court finds that the slogans and descriptions incorporated into the labels of the products evidence an intent for the products to have a mind-altering affect on the user.

Moreover, many of the products are marketed as substitutes for illegal drugs. For example, the name of the herbal smoking blend "Chronix" is strikingly similar to the word "Chronic." Chronic is commonly used by Generation Xers, Defendant's admitted target market, as slang for marijuana. The same is true for the other ultra potent smoking herbs marketed by Defendants as "Herbal Hash" (Hash) and "Herb Ghani" (Ghanja). "Shroomz" is used as slang for the psilocybin or psychotropic mushrooms. Defendants openly advertise that users identified "Liquid X" and "Liquid X Export" as Liquid Ecstasy. "X" and "E" are slang terms for Ecstasy. The catalog for Defendants' products explicitly states that the products are "for mood enhancement." Moreover, the labeling for several products indicates an intent that the user should feel the "effects" of the substances for a prescribed period of time.

The magazine advertisement for "Hydro," "Herbal Hash," "Inda-Kind," "Sweet Green," and "Herbal Opium" tout the products as "The Best Legal Buds in the Biz" and the "Strongest legal BUDS on the market." An advertisement for "Inda-Kind" includes testimonials stating that "Inda-Kind produced the strongest effects by far" and "did the trick, we're all stoked." Similarly, the advertisement for "Utopia" has a testimonial that states "6 hours feelin' fine." A testimonial for "Hyrdo" states "F_ing amazing Hydro totally works." Defendants advertised "Herbal Opium" as having a "very strong, long burning, stunning effect."

Here, there is no genuine dispute that the labels accompanied the products and the marketing materials were made to promote the products at issue. "[W]here ... there is no dispute concerning the wording and description of the labeling and other promotional materials or activities that relate to the products, the legal conclusion to be drawn therefrom can and should be resolved on summary judgment." After reviewing the labeling and promotional materials, the Court finds that the objective evidence demonstrates that the products at issue were intended to affect the function or structure of the mind by elevating the psychological condition of users. Accordingly, the Court holds that the seized products are drugs with the meaning for the FD&C Act.

D. New Drugs

The United States maintains that the products are unapproved new drugs. The FD&C Act defines a "new drug" as "[a]ny drug ... the composition of which ... is not generally recognized, among experts qualified by scientific training and experience to evaluate the safety and effectiveness of drugs, as safe and effective for use under the conditions prescribed, recommended, or suggested in the labeling thereof." 21 U.S.C. § 321(p)(1). Defendants assert that their products are generally recognized by qualified experts to be safe and effective and, therefore, cannot be categorized as "new drugs."

" '[G]eneral recognition' of effectiveness requires at least 'substantial evidence' of effectiveness for approval of [a new drug application]." Thus, " 'general recognition' requires a two-step showing: first, that there is general recognition in fact, i.e., that there is an expert consensus that the product is effective; and second, that the expert consensus is based upon 'substantial evidence' as defined in the Act and in FDA regulations." "General recognition of safety and effectiveness shall ordinarily be based upon published studies which may be corroborated by unpublished studies and other data and information." "Substantial evidence does not consist of the expressed opinions of experts hired to testify on behalf of one party or the other. Instead, it consists of adequate and well-controlled studies that must be generally available to the scientific community." The "FDA's 'judgments as to what is required to ascertain the safety and efficacy of drugs fall squarely within the ambit of the FDA's expertise and merit deference from [the courts].' "

The United States has submitted a declaration from Dr. Charles J. Ganley, a Drug Science Officer from the United States Drug Enforcement Administration (DEA). According to Dr. Ganley, the FDA conducted extensive searches of published and unpublished scientific literature on each product, by tradename and the combination of ingredients. The search did not reveal any studies or clinical data establishing the safety and effectiveness of any of the products in issue. The absence of literature establishing the safety and efficacy of a product is proof that the requisite general recognition does not exist. To rebut the government's case, Defendants rely on publications from THE HANDBOOK OF MEDICINAL HERBS and THE PHARMACOLOGY OF CHINESE HERBS as proof of the products' efficacy. These publications merely described the herbal ingredients contained in some of the products. The safety and efficacy of the complete product, not its components, is the central issue. As a matter of law, this evidence is insufficient to create a genuine dispute of material fact as to the lack of general recognition of the safety and efficacy of the Defendants' products. Therefore, the Court finds that the Defendants' products are "new drugs" under the FDA.

Before a new drug may be marketed in the United States, the FDA must approve a new drug application ("NDA") submitted by the manufacturer. 21 U.S.C. § 355(a). There is no dispute that the products at issue were marketed in the United States and did not go through the FDA approval process. Therefore the Court finds that there is no genuine dispute of material fact that the products in controversy are unapproved new drugs.

E. Misbranded

The United States maintains that the Defendants' products are misbranded because the directions on the labels are not based upon any clinical data that could substantiate the recommended levels for safe dosage. Defendants respond that the labeling is adequate for dietary supplements. Section 331 prohibits "[t]he introduction or delivery for introduction into interstate commerce of any food, drug, device, or cosmetic that is adulterated and misbranded." 21 U.S.C. § 331(a). A product is misbranded if it fails to include adequate directions for use. *See* 21 U.S.C. § 352(f)(1). " 'Adequate directions for use' means directions under which the layman can use a drug safely and for the purposes for which it is intended." 21 C.F.R. § 201.5. "Where nonprescription drugs are involved, the 'adequate directions for use' requirement insures full disclosure to the layman purchasing the drugs for self-treatment." "Courts must show substantial deference to the construction of a statute and regulations by an agency charged with their enforcement particularly where, as here, the agency is empowered not only to construe its governing statute, but additionally to make safety judgments delegated to it by Congress. . . ."

The labels for several disputed products generally instruct users to take a certain amount of the item. For example, "Rave X" instructs users to take 3 to 5 tablets. For "Rave Energy," the recommended dosage is 1 or 2 tablets. "Liquid X" and "Liquid X Export" suggest dosages from 1/4–1/2 tsp. "Herbal Hash–Mean Green" and "Herbal Hash–Honey-Blonde" generally instruct the user to place small pieces of the product on top of another herbal blend; smoke and enjoy; chill/engage in recreational activity; and repeat as need every 2 to 4 hours. "Northern Heights" directs users to add a few drops to smoking material; enjoy; and repeat as needed. Although laymen could understand the directions, there is no proof that the provided instructions for any of the Defendants' products are safe for the intended use of altering the psychological condition of the user. Essentially, in the absence of investigations or clinical data demonstrating the safety and efficacy of the drugs, there can be no adequate instruction for their safe use.

Furthermore, in the absence of "clinical proof, in the form of adequately controlled clinical studies, which establishes that . . . [the] product, is effective for any indicated use[,] [a]ny representation as to [its] proven efficacy is false and misleading, and therefore, [the product] is misbranded." Along similar lines, any representation as to the safety of a product in the absence of clinical proof renders the product misbranded under the FD&C Act. "Rave X," "Rave Energy," "Utopia," "Shroomz," "Hashanna Oil," "Inda-Kind," and "Herba Ghani" make representations that the products are safe or may be used to party safely. As discussed earlier, there are no clinical studies to substantiate the Defendants' claims of safety.

"Hydro," "Chronix," "Herbal Opium," and "Sweet Green" provide no instructions for their intended use. While there is no "Sweet Green" label for the Court to review, the determination of whether the product is misbranded may rest on its advertisements. Under the FD&C Act, "[a]ny printed material, including books and pamphlets, which refers to or explains the usefulness of a product and which is used, in any way, in its sale 'accompanies' the article in the statutory sense and constitutes 'labeling.' " The advertisements explain that "Sweet Green" is an "ultra-potent smoke," but provide no direction as to its safe use. Furthermore, "Utopia's" labeling as a dietary supplement is insufficient because it fails to list the quantity of each ingredient in the product.

There is no dispute that the products were introduced into interstate commerce as the products consisted of ingredients that crossed state lines and the products themselves were sold across jurisdictions. There is no evidence to suggest that the products fall within an exemption to the labeling requirements. Therefore, for the above discussed reasons, the Court finds the Defendants' products to be misbranded under § 352(f) of the FD&C Act.

F. Commercial Speech

Defendants assert that the policy of the FDA as set forth in its Guidance for Industry on Street Drug Alternatives, 65 Fed. Reg. 17,512 (Apr. 3, 2000), impermissibly limits commercial speech in violation of the First Amendment. Defendants also argue that the FDA cannot limit the marketing of otherwise legal products.

"[I]n order for commercial speech to be entitled to any First Amendment protection, the speech must first concern lawful activity and not be misleading." As explained in this opinion, the labeling and advertising for the unapproved new drugs at issue are misleading in that they claim to be safe and effective without any scientific support. Therefore, the products are entitled to no First Amendment protections. Secondly, the FDA's guidance does not prohibit the advertising of the Defendants' products. Rather, the advertisements and labels are merely used as proof of the Defendants' intended use of the products in establishing violations of the FD&C Act. The use of such materials to ascertain the intent of the manufacturer does not implicate the First Amendment. Accordingly, the Court finds that the FDA's Guidance for Industry on Street Drug Alternatives does not impinge upon any protected right to commercial speech under the First Amendment.

. . . .

I. Allegations against Perry L. Hitt as an Individual

Defendant, Perry Hitt, asserts that the action against him is improper because he is named as a defendant by virtue of his position as President of Hit Products, Inc., Riverdale Organics, and Dreamworlds. Mr. Hitt asserts that he did not personally participate in the manufacture, sale, or distribution of the products at issue. Thus, according to Mr. Hitt, he cannot be held liable for the corporate defendants' violations unless the Court pierces the corporate veil. The Court disagrees.

The FDA's statutory authority empowers the government to seek relief against corporate executives, as well as legal entities, in enforcement actions. "[C]orporate agents vested with the responsibility, and power commensurate with that responsibility, to devise whatever measures are necessary to ensure compliance with the Act bear a 'responsible relationship' to, or have a 'responsible share' in, violations." While the Supreme Court cases imposed criminal liability upon the individual corporate officers for violations of the FD&C Act, "the rationale for holding corporate officers criminally responsible for acts of the corporation, which could lead to incarceration, is even more persuasive where only civil liability is involved, which at most would result in a monetary penalty." The fact that a corporate officer could be subjected to criminal punishment upon a showing of a responsible relationship to the acts of a corporation that violate health and safety statutes renders civil liability appropriate as well."

Therefore, there is no need to "pierce the corporate veil" in order to hold Mr. Hitt responsible for the violations of the FD&C Act. Likewise, personal participation in "line" activity is not a prerequisite to the imposition of liability. Rather, the United States must show that Mr. Hitt had responsible relationship to, or a responsible share in the furtherance of, the transactions outlawed by the FD&C Act.

. . . .

III. Conclusion

For the reasons stated above, the Court will grant Plaintiff's Motion for Summary Judgment and deny Defendants' Motion for Partial Summary Judgment. The Court also finds that the circumstances warrant an order of condemnation as well as injunctive relief. An Order consistent with this Opinion will follow.

. . . .

* * * * *

8.11 ADVERTISING

The FTC is the federal agency that is generally responsible for regulating product advertising. The FTC and the FDA have worked closely on the advertising of dietary supplements. The FTC's truth-in-advertising law can be summed up in two rules:

1. Advertising must be truthful and not misleading.
2. Before disseminating an ad, advertisers must have adequate substantiation for all objective product claims.

8.11.1 Truthful and Not Misleading

FTC views a deceptive ad as one that contains a misrepresentation or omission that is likely to mislead consumers acting reasonably under the circumstances to their detriment. In general, FTC gives great deference to an FDA's determinations and standards regarding adequate support for a health claim.

Thus FTC and FDA generally arrive at the same conclusion when evaluating *unqualified* health claims.

Regarding *qualified* health claims, there are some limited instances when a carefully qualified health claim in advertising may be permissible under FTC law, where it would not be permitted in labeling. In part, this difference is inherent in the differences between labels and the broader medium of advertising, where there is more room and opportunity for qualifications that can prevent a claim from being misleading.

8.11.2 Claim Substantiation

In addition to conveying product ads truthfully, advertisers need to verify that there is adequate support for their claims. Before disseminating an ad, advertisers must have a reasonable basis for all product claims. What constitutes a reasonable basis depends on what claims are being made, how they are presented in the context of the entire ad, and how they are qualified. When evaluating claims, the FTC has typically applied a substantiation standard of "competent and reliable scientific evidence."

Substantiation applies to both express and implied claims. Therefore the first step in determining whether an ad complies with the law is to identify all express and implied claims that the ad conveys to consumers. Once the claims are identified, the scientific evidence is assessed to determine whether there is adequate support for those claims. The standard for substantiation for implied claims is the same as for express claims; that is, an advertiser cannot suggest a claim that they could not make directly.

* * * * *

FTC's Dietary Supplements Advertising Guide for Industry[67]

A. Identifying Claims and Interpreting Ad Meaning

1. Identifying Express and Implied Claims

The first step in evaluating the truthfulness and accuracy of advertising is to identify all express and implied claims an ad conveys to consumers. Advertisers must make sure that whatever they say expressly in an ad is accurate. Often, however, an ad conveys other claims beyond those expressly stated. Under FTC law, an advertiser is equally responsible for the accuracy of claims suggested or implied by the ad. Advertisers cannot suggest claims that they cannot make directly.

When identifying claims, advertisers should not focus just on individual phrases or statements, but rather should consider the ad as a whole, assessing the "net impression" conveyed by all elements of the ad, including the text,

---

[67] FTC, DIETARY SUPPLEMENTS: AN ADVERTISING GUIDE FOR INDUSTRY, *available at* http://www.ftc.gov/bcp/conline/pubs/bus-pubs/dietsupp.htm (last visited Nov. 28, 2005).

product name, and depictions. When an ad lends itself to more than one reasonable interpretation, the advertiser is responsible for substantiating each interpretation. Copy tests, or other evidence of how consumers actually interpret an ad, can be valuable. In many cases, however, the implications of the ad are clear enough to determine the existence of the claim by examining the ad alone, without extrinsic evidence. . . .

Depending on how it is phrased, or the context in which it is presented, a statement about a product's effect on a normal "structure or function" of the body may also convey to consumers an implied claim that the product is beneficial for the treatment of a disease. If elements of the ad imply that the product also provides a disease benefit, the advertiser must be able to substantiate the implied disease claim even if the ad contains no express reference to disease. . . .

2. When to Disclose Qualifying Information

An advertisement can also be deceptive because of what it fails to say. Section 15 of the FTC Act requires advertisers to disclose information if it is material in light of representations made or suggested by the ad, or material considering how consumers would customarily use the product. Thus, if an ad would be misleading without certain qualifying information, that information must be disclosed. For example, advertisers should disclose information relevant to the limited applicability of an advertised benefit. Similarly, advertising that makes either an express or implied safety representation should include information about any significant safety risks. Even in the absence of affirmative safety representations, advertisers may need to inform consumers of significant safety concerns relating to the use of their product. . . .

3. Clear and Prominent Disclosure

When the disclosure of qualifying information is necessary to prevent an ad from being deceptive, that information should be presented clearly and prominently so that it is actually noticed and understood by consumers. A fine-print disclosure at the bottom of a print ad, a disclaimer buried in a body of text, a brief video superscript in a television ad, or a disclaimer that is easily missed on an Internet web site, are not likely to be adequate. To ensure that disclosures are effective, marketers should use clear language, avoid small type, place any qualifying information close to the claim being qualified, and avoid making inconsistent statements or distracting elements that could undercut or contradict the disclosure. Because consumers are likely to be confused by ads that include inconsistent or contradictory information, disclosures need to be both direct and unambiguous to be effective. . . .

Qualifying information should be sufficiently simple and clear that consumers not only notice it, but also understand its significance. This can be a particular challenge when explaining complicated scientific concepts to a general audience, for example, if an advertiser wants to promote the effect of a supplement where there is an emerging body of science supporting that effect, but the evidence is insufficient to substantiate an unqualified claim. The advertiser should make sure consumers understand both the extent of scientific support and the existence of any significant contrary evidence. Vague qualifying terms—for example, that the product "may" have the claimed benefit or "helps" achieve the claimed benefit—are unlikely to be adequate. Furthermore, advertisers should not make qualified claims where the studies they rely on are contrary to a stronger body of evidence. In such instance even a qualified claim could mislead consumers. . . .

B. Substantiating Claims

In addition to conveying product claims clearly and accurately, marketers need to verify that there is adequate support for their claims. Under FTC law, before disseminating an ad, advertisers must have a reasonable basis for all express and implied product claims. What constitutes a reasonable basis depends greatly on what claims are being made, how they are presented in the context of the entire ad, and how they are qualified. The FTC's standard for evaluating substantiation is sufficiently flexible to ensure that consumers have access to information about emerging areas of science. At the same time, it is sufficiently rigorous to ensure that consumers can have confidence in the accuracy of information presented in advertising. A number of factors determine the appropriate amount and type of substantiation, including:

The Type of Product. Generally, products related to consumer health or safety require a relatively high level of substantiation.

The Type of Claim. Claims that are difficult for consumers to assess on their own are held to a more exacting standard. Examples include health claims that may be subject to a placebo effect or technical claims that consumers cannot readily verify for themselves.

The Benefits of a Truthful Claim, and The Cost/Feasibility of Developing Substantiation for the Claim. These factors are often weighed together to ensure that valuable product information is not withheld from consumers because the cost of developing substantiation is prohibitive. This does not mean, however, that an advertiser can make any claim it wishes without substantiation, simply because the cost of research is too high.

The Consequences of a False Claim. This includes physical injury, for example, if a consumer relies on an unsubstantiated claim about the therapeutic benefit of a product and foregoes a proven treatment. Economic injury is also considered.

The Amount of Substantiation that Experts in the Field Believe Is Reasonable. In making this determination, the FTC gives great weight to accepted norms in the relevant fields of research and consults with experts from a wide variety of disciplines, including those with experience in botanicals and traditional medicines. Where there is an existing standard for substantiation developed by a government agency or other authoritative body, the FTC accords great deference to that standard.

The FTC typically requires claims about the efficacy or safety of dietary supplements to be supported with "competent and reliable scientific evidence," defined in FTC cases as "tests, analyses, research, studies, or other evidence based on the expertise of professionals in the relevant area, that have been conducted and evaluated in an objective manner by persons qualified to do so, using procedures generally accepted in the profession to yield accurate and reliable results." This is the same standard the FTC applies to any industry making health-related claims. There is no fixed formula for the number or type of studies required or for more specific parameters like sample size and study duration. There are, however, a number of considerations to guide an advertiser in assessing the adequacy of the scientific support for a specific advertising claim.

1. Ads that Refer to a Specific Level of Support

If an advertiser asserts that it has a certain level of support for an advertised claim, it must be able to demonstrate that the assertion is accurate. Therefore, as a starting point, advertisers must have the level of support that they claim, expressly or by implication, to have. . . .

2. The Amount and Type of Evidence

When no specific claim about the level of support is made, the evidence needed depends on the nature of the claim. A guiding principle for determining the amount and type of evidence that will be sufficient is what experts in the relevant area of study would generally consider to be adequate. The FTC will consider all forms of competent and reliable scientific research when evaluating substantiation. As a general rule, well-controlled human clinical studies are the most reliable form of evidence. Results obtained in animal and in vitro studies will also be examined, particularly where they are widely considered to be acceptable substitutes for human research or where human research is infeasible. Although there is no requirement that a dietary supplement claim be supported by any specific number of studies, the replication of research results in an independently conducted study adds to the weight of the evidence. In most situations, the quality of studies will be more important than quantity. When a clinical trial is not possible (e.g., in the case of a relationship between a nutrient and a condition that may take decades to develop), epidemiologic evidence may be an acceptable substitute for clinical data, especially when supported by other evidence, such as research explaining the biological mechanism underlying the claimed effect.

Anecdotal evidence about the individual experience of consumers is not sufficient to substantiate claims about the effects of a supplement. Even if those experiences are genuine, they may be attributable to a placebo effect or other factors unrelated to the supplement. Individual experiences are not a substitute for scientific research. . . .

3. The Quality of the Evidence

In addition to the amount and type of evidence, the FTC will also examine the internal validity of each piece of evidence. Where the claim is one that would require scientific support, the research should be conducted in a competent and reliable manner to yield meaningful results. The design, implementation, and results of each piece of research are important to assessing the adequacy of the substantiation.

There is no set protocol for how to conduct research that will be acceptable under the FTC substantiation doctrine. There are, however, some principles generally accepted in the scientific community to enhance the validity of test results. For example, a study that is carefully controlled, with blinding of subjects and researchers, is likely to yield more reliable results. A study of longer duration can provide better evidence that the claimed effect will persist and resolve potential safety questions. Other aspects of the research results—such as evidence of a dose–response relationship (i.e., the larger the dose, the greater the effect) or a recognized biological or chemical mechanism to explain the effect—are examples of factors that add weight to the findings. Statistical significance of findings is also important. A study that fails to show a statistically significant difference between test and control group may indicate that the measured effects are merely the result of placebo effect or chance. The results should also translate into a meaningful benefit for consumers. Some results that are statistically significant may still be so small that they would mean only a trivial effect on consumer health.

The nature and quality of the written report of the research are also important. Research cannot be evaluated accurately on the basis of an abstract or an informal summary. In contrast, although the FTC does not require that studies be published and will consider unpublished, proprietary research, the publication of a peer-reviewed study in a reputable journal indicates that the research has received some measure of scrutiny. At the same time, advertisers should not rely simply on the fact that research is published as proof of the efficacy of a supplement. Research may yield results that are of sufficient interest to the scientific community to warrant publication, but publication does not necessarily mean that such research is conclusive evidence of a substance's effect. The FTC considers studies conducted in foreign countries as long as the design and implementation of the study are scientifically sound. . . .

4. The Totality of the Evidence

Studies cannot be evaluated in isolation. The surrounding context of the scientific evidence is just as important as the internal validity of individual studies. Advertisers should consider all relevant research relating to the claimed benefit of their supplement and should not focus only on research that supports the effect, while discounting research that does not. Ideally, the studies relied on by an advertiser would be largely consistent with the surrounding body of evidence. Wide variation in outcomes of studies and inconsistent or conflicting results will raise serious questions about the adequacy of an advertiser's substantiation. Where there are inconsistencies in the evidence, it is important to examine whether there is a plausible explanation for those inconsistencies. In some instances, for example, the differences in results are attributable to differences in dosage, the form of administration (e.g., oral or intravenous), the population tested, or other aspects of study methodology. Advertisers should assess how relevant each piece of research is to the specific claim they wish to make, and also consider the relative strengths and weaknesses of each. If a number of studies of different quality have been conducted on a specific topic, advertisers should look first to the results of the studies with more reliable methodologies.

The surrounding body of evidence will have a significant impact both on what type, amount and quality of evidence is required to substantiate a claim and on how that claim is presented—that is, how carefully the claim is qualified to reflect accurately the strength of the evidence. If a stronger body of surrounding evidence runs contrary to a claimed effect, even a qualified claim is likely to be deceptive. . . .

5. *The Relevance of the Evidence to the Specific Claim*

A common problem in substantiation of advertising claims is that an advertiser has valid studies, but the studies do not support the claim made in the ad. Advertisers should make sure that the research on which they rely is not just internally valid, but also relevant to the specific product being promoted and to the specific benefit being advertised. Therefore, advertisers should ask questions such as: How does the dosage and formulation of the advertised product compare to what was used in the study? Does the advertised product contain additional ingredients that might alter the effect of the ingredient in the study? Is the advertised product administered in the same manner as the ingredient used in the study? Does the study population reflect the characteristics and lifestyle of the population targeted by the ad? If there are significant discrepancies between the research conditions and the real life use being promoted, advertisers need to evaluate whether it is appropriate to extrapolate from the research to the claimed effect.

In drafting ad copy, the advertiser should take care to make sure that the claims match the underlying support. Claims that do not match the science, no matter how sound that science is, are likely to be unsubstantiated. Advertising should not exaggerate the extent, nature, or permanence of the effects achieved in a study, and should not suggest greater scientific certainty than actually exists. Although emerging science can sometimes be the basis for a carefully qualified claim, advertisers must make consumers aware of any significant limitations or inconsistencies in the scientific literature. . . .

C. Other Issues Relating to Dietary Supplement Advertising

In addition to the basic principles of ad meaning and substantiation discussed above, a number of other issues commonly arise in the context of dietary supplement advertising. The following sections provide guidance on some of these issues including: the use of consumer or expert endorsements in ads; advertising claims based on traditional uses of supplements; use of the DSHEA disclaimer in advertising; and the application to advertising of the DSHEA exemption for certain categories of publications, commonly referred to as "third party literature."

1. Claims Based on Consumer Experiences or Expert Endorsements

An overall principle is that advertisers should not make claims either through consumer or expert endorsements that would be deceptive or could not be substantiated if made directly.[68] It is not enough that a testimonial represents the honest opinion of the endorser. Under FTC law, advertisers must also have appropriate scientific evidence to back up the underlying claim.

Consumer testimonials raise additional concerns about which advertisers need to be aware. Ads that include consumer testimonials about the efficacy or safety of a supplement product should be backed by adequate substantiation that the testimonial experience is representative of what consumers will generally achieve when using the product. As discussed earlier, anecdotal evidence of a product's effect, based solely on the experiences of individual consumers, is generally insufficient to substantiate a claim. Further, if the advertiser's substantiation does not demonstrate that the results are representative, then a clear and conspicuous disclaimer is necessary. The advertiser should either state what the generally expected results would be or indicate that the consumer should not expect to experience the attested results. Vague disclaimers like "results may vary" are likely to be insufficient. . . .

When an advertiser uses an expert endorser, it should make sure that the endorser has appropriate qualifications to be represented as an expert and has conducted an examination or testing of the product that would be generally recognized in the field as sufficient to support the endorsement. In addition, whenever an expert or consumer endorser is used, the advertiser should disclose any material connection between the endorser and the advertiser of the

---

[68] The FTC has provided detailed guidance on this subject in its "Guides Concerning Use of Endorsements and Testimonials in Advertising" *available at*: http://www.ftc.gov/bcp/guides/endorse.htm (last accessed Sept. 4, 2008).

product. A material connection is one that would affect the weight or credibility of the endorsement, or put another way, a personal, financial, or similar connection that consumers would not reasonably expect. . . .

2. *Claims Based on Traditional Uses*

Claims based on historical or traditional use should be substantiated by confirming scientific evidence, or should be presented in such a way that consumers understand that the sole basis for the claim is a history of use of the product for a particular purpose. A number of supplements, particularly botanical products, have a long history of use as traditional medicines in the United States or in other countries to treat certain conditions or symptoms. Several European countries have a separate regulatory approach to these traditional medicines, allowing manufacturers to make certain limited claims about their traditional use for treating certain health conditions. Some countries also require accompanying disclosures about the fact that the product has not been scientifically established to be effective, as well as disclosures about potential adverse effects. At this time there is no separate regulatory process for approval of claims for these traditional medicine products under DSHEA and FDA labeling rules.

In assessing claims based on traditional use, the FTC will look closely at consumer perceptions and specifically at whether consumers expect such claims to be backed by supporting scientific evidence. Advertising claims based solely on traditional use should be presented carefully to avoid the implication that the product has been scientifically evaluated for efficacy. The degree of qualification necessary to communicate the absence of scientific substantiation for a traditional use claim will depend in large part on consumer understanding of this category of products. As consumer awareness of and experience with "traditional use" supplements evolve, the extent and type of qualification necessary is also likely to change.

There are some situations, however, where traditional use evidence alone will be inadequate to substantiate a claim, even if that claim is carefully qualified to convey the limited nature of the support. In determining the level of substantiation necessary to substantiate a claim, the FTC assesses, among other things, the consequences of a false claim. Claims that, if unfounded, could present a substantial risk of injury to consumer health or safety will be held to a higher level of scientific proof. For that reason, an advertiser should not suggest, either directly or indirectly, that a supplement product will provide a disease benefit unless there is competent and reliable scientific evidence to substantiate that benefit. The FTC will closely scrutinize the scientific support for such claims, particularly where the claim could lead consumers to forego other treatments that have been validated by scientific evidence, or to self-medicate for potentially serious conditions without medical supervision.

The advertiser should also make sure that it can document the extent and manner of historical use and be careful not to overstate such use. As part of this inquiry, the advertiser should make sure that the product it is marketing is consistent with the product as traditionally administered. If there are significant differences between the traditional use product and the marketed product, in the form of administration, the formulation of ingredients, or the dose, a "traditional use" claim may not be appropriate. . . .

3. Use of the DSHEA Disclaimer in Advertising

Under DSHEA, all statements of nutritional support for dietary supplements must be accompanied by a two-part disclaimer on the product label: that the statement has not been evaluated by FDA and that the product is not intended to "diagnose, treat, cure, or prevent any disease." Although DSHEA does not apply to advertising, there are situations where such a disclosure is desirable in advertising as well as in labeling to prevent consumers from being misled about the nature of the product and the extent to which its efficacy and safety have been reviewed by regulatory authorities. For example, a disclosure may be necessary if the text or images in the ad lead consumers to believe that the product has undergone the kind of review for safety and efficacy that the FDA conducts on new drugs and has been found to be beneficial for the treatment of disease. Failure to correct those misperceptions may render the advertising deceptive.

At the same time, the inclusion of a DSHEA disclaimer or similar disclosure will not cure an otherwise deceptive ad, particularly where the deception concerns claims about the disease benefits of a product. In making references to DSHEA and FDA review, advertisers should also be careful not to mischaracterize the extent to which a product or claim has been reviewed or approved by the FDA. Compliance with the notification and disclaimer provisions of DSHEA does not constitute authorization of a claim by FDA and advertisers should not imply that FDA has specifically approved any claim on that basis. . . .

4. Third-Party Literature

Dietary supplement advertisers should be aware that the use of newspaper articles, abstracts of scientific studies, or other "third-party literature" to promote a particular brand or product can have an impact on how consumers interpret an advertisement and on what claims the advertiser will be responsible for substantiating. For purposes of dietary supplement labeling, section 5 of DSHEA provides an exemption from labeling requirements for scientific journal articles, books, and other publications used in the sale of dietary supplements, provided these materials are reprinted in their entirety, are not false or misleading, do not promote a specific brand or manufacturer, are presented with other materials to create a balanced view of the scientific information, and are physically separate from the supplements being sold.

The FTC will generally follow an approach consistent with the labeling approach when evaluating the use of such publications in other contexts, such as advertising. Although the FTC does not regulate the content or accuracy of statements made in independently written and published books, articles, or other noncommercial literature, FTC law does prohibit the deceptive use of such materials in marketing products. The determination of whether the materials will be subject to FTC jurisdiction turns largely on whether the materials have been created or are being used by an advertiser specifically for the purpose of promoting its product. As a practical matter, publications and other materials that comply with the elements of the DSHEA provision, particularly with the requirement that such materials be truthful, not misleading and balanced, are also likely to comply with FTC advertising law.

III. Conclusion

Marketers of dietary supplements should be familiar with the requirements under both DSHEA and the FTC Act that labeling and advertising claims be truthful, not misleading and substantiated. The FTC approach generally requires that claims be backed by sound, scientific evidence, but also provides flexibility in the precise amount and type of support necessary. This flexibility allows advertisers to provide truthful information to consumers about the benefits of supplement products and, at the same time, preserves consumer confidence by curbing unsubstantiated, false, and misleading claims. To ensure compliance with FTC law, supplement advertisers should follow two important steps: (1) careful drafting of advertising claims with particular attention to how claims are qualified and what express and implied messages are actually conveyed to consumers; and (2) careful review of the support for a claim to make sure it is scientifically sound, adequate in the context of the surrounding body of evidence, and relevant to the specific product and claim advertised. . . .

* * * * *

NOTE

8.4. Further information. *See* FTC Policy Statement Regarding Advertising Substantiation, *available at*: http://www.ftc.gov/bcp/guides/ad3subst.htm and FTC's Enforcement Policy Statement on Food Advertising, *available at*: http://www.ftc.gov/bcp/policystmt/ad-food.htm.

PROBLEM EXERCISE

In this hypothetical exercise, you are counsel for the Pan Acea Corporation. Your company's product development staff bring you a new product concept. Their idea is sell food products containing the herb *Cattawumpus alba*.

C. alba has a thousand-year history of use as an herbal medicine in Asia for a variety of ailments, including stress relief. A number of journal articles mention *C. alba*, but one article questions whether the herb may cause cancer in laboratory mice. Pan Acea researchers have cultivated *C. alba*, and company employees tested it. These volunteers described a relaxing effect, and none experienced any negative symptoms.

The product development staff want to market two products with *Cattawumpus alba*: a capsule and a tea. The capsules would be called "Herbal Valium." The tea would be called "Stress Away Tea." Both products would bear the claim, "For the stress of daily living." Both product labels would include the Supplement Facts panel and the mandatory disclaimer, "This statement has not been evaluated by FDA."

Questions: What regulatory issues are raised with how Pan Acea wants to bring the products to market? How might the issues be resolved?

NOTES AND QUESTIONS

8.5.   What are dietary supplements? Before you read this chapter, what were thoughts about what dietary supplements are? How has your understanding changed?

8.6.   Public expectations. Do you think the system regulates dietary supplement to the extent the public believes?

8.7.   Clothing with dietary ingredients. Food additives are appearing in clothing with the claim and intent that the ingredients are absorbed through the skin. Companies have been marketing clothes impregnated with food ingredients, such as amino acids, vitamins, caffeine, and seaweed. How would these be regulated under the Food, Drug, and Cosmetic Act?

8.8.   The burden of proof on safety and manufacturer liability. How does the shift in the burden of proof on dietary supplement safety affect manufacturer liability?

8.9.   Incentives of the law. How does the current regulatory program encourage or discourage manufacturers from investing in research to show a dietary supplement's health value and safety?

8.10.   Self-regulation and market forces. Do you think market forces can regulate the dietary supplement industry? What factors work for or against self regulation? Miller and VanDoren propose third-party certification for dietary supplements as an alternative to government regulation. Henry I. Miller and Peter VanDoren, *Food Risks and Labeling Controversies: Market-based Alternatives to More Government Regulation of*

Foods and Dietary Supplements, 23:1 REGULATION 35–39 (Spring 2000). Do you think third-party certification would allow the public to distinguish the "good" dietary supplement firms from the "bad" ones? Will the market reward the conscientious firms? Miller and VanDoren note that third-party certification could assure of safety, but not effectiveness, of dietary supplements. Why? Does this mean that regulatory intervention is necessary regarding effectiveness of dietary supplement claims?

8.11. For more information. On dietary supplement labeling, see the FDA Office on Dietary Supplements at: http://www.cfsan.fda.gov/~dms/supplmnt.html. FDA's Ephedra Web page: http://www.cfsan.fda.gov/~dms/ds-ephed.html. FTC Advertising Guidelines for Dietary Supplements at: http://www.ftc.gov/bcp/conline/pubs/buspubs/dietsupp.htm. FDA—Office of Dietary Supplements at: http://www.cfsan.fda.gov/~dms/supplmnt.html. DSHEA also created the Office of Dietary Supplements with the National Institute of Health at: http://dietary-supplements.info.nih.gov/.

CHAPTER 9

Biotechnology and Genetically Engineered Organisms

9.1 INTRODUCTION

Biotechnology, in the broad sense, refers to all technological applications in biology. In common use, however, biotechnology refers to techniques used to modify the deoxyribonucleic acid (DNA) or the genetic material of an organism to achieve a desired trait. Biotechnology is also the shorthand term for an aggregation of scientific developments that include everything from stem cells applications and cloning to genetically engineered foods.

Genetically engineered (GE) refers to genetic modification through use of recombinant deoxyribonucleic acid (rDNA) techniques, or gene splicing, to give desired traits.

Genetically modified (GM) is commonly used as a synonym for genetically engineered. The term "genetically engineered" more precisely indicates that humans have directly engineered the DNA. In the broadest sense, all food crops have been genetically modified by humans using conventional cultivation and propagation techniques. Nonetheless, the term GM has become widely used and understood to mean genetic modification from use of rDNA techniques.

Genetically engineered organism (GEO) and **genetically modified organism (GMO)** refer to organisms that are genetically modified through use of rDNA techniques.

9.1.1 Background

James Watson and Francis Crick described the double-helix structure of DNA in 1953. Twenty years passed before scientists developed techniques for splicing DNA between organisms. Another two decades passed before the first food sold at retail contained such recombinant DNA (rDNA).

Today, by some estimates, nearly two-thirds of all foods on American grocery shelves contain genetically engineered ingredients. Many believe that biotechnology developers are poised to bring substantially more products to the market, and these products are likely to be more sophisticated and complicated modifications than in the past. The rapid growth and sophistication of the biotechnology fields is likely to bring challenges and novel issues to regulatory oversight.

* * * * *

Genetically Modified Foods: Experts View Regimen of Safety Tests as Adequate, but FDA's Evaluation Process Could Be Enhanced

General Accounting Office (May 2002)

. . .

Biotechnology offers a variety of potential benefits and risks. It has enhanced food production by making plants less vulnerable to drought, frost, insects, and viruses and by enabling plants to compete more effectively against weeds for soil nutrients. In a few cases it has also improved the quality and nutrition of foods by altering their composition. . . .

[T]he majority of modifications have been aimed at increasing crop yields for farmers by engineering a food plant to tolerate herbicides or attacks from pests such as insects and viruses (48 out of 62 modifications). Further, only two food plants have been altered to produce modified oil: the soybean and canola plants. According to industry officials, the modified soybean produces healthier oil. They also stated that the canola plant was modified to have a domestic source for laurate cooking oil. Because soybean oil is the most commonly consumed plant oil worldwide, scientists say that the new oil could significantly improve the health of millions of people.

For three key crops grown in the United States—corn, soybeans, and cotton—a large number of farmers have chosen to plant GM varieties. In 2001, GM varieties accounted for about 26 percent of the corn, 68 percent of the soybeans, and 69 percent of the cotton planted in the United States. These crops are the source of various ingredients used extensively in many processed foods, such as corn syrup, soybean oil, and cottonseed oil, and they are also major U.S. commodity exports. The United States accounts for about three-quarters of GM food crops planted globally.

However, the use of biotechnology has also raised concerns about its potential risks to the environment and people. For example, some people fear that common plant pests could develop resistance to the introduced pesticides in GM crops that were supposed to combat them. Further some fear that crops modified to be tolerant to herbicides could foster the evolution of "super weeds." Finally, some fear that scientists might unknowingly create or enhance a food allergen or toxin. Therefore, as biotechnology was being developed, U.S. scientists, regulators, and policy makers generally agreed that GM plants should be evaluated carefully before being put into widespread use. As a result the United States published a *Coordinated Framework for Regulation of Biotechnology* in 1986. This framework outlined the regulatory approach for reviewing GM plants, including relevant laws, regulations, and definitions of GM organisms.

Responsibility for implementing the coordinated framework fell primarily to three agencies: USDA, the Environmental Protection Agency (EPA), and FDA. Within USDA, the Animal and Plant Health Inspection Service (APHIS) bears the main responsibility for assessing the environmental safety of GM crops. The primary focus of APHIS' review is to determine whether or not a plant produced through biotechnology has the potential to harm natural habitats or agriculture. Developers can petition APHIS to exempt a GM plant from regulation once sufficient and appropriate data have been collected regarding the potential environmental impact of a GM plant.

To safeguard the environment and human health, EPA is responsible for regulating genetic modifications in plants that protect them from insects, bacteria, and viruses. These protectants are subject to the agency's regulations on the sale, distribution, and use of pesticides. EPA must review and grant a permit for field-testing plants with such protectants on more than 10 acres of land. Prior to commercialization of a GM plant with such a protectant, EPA reviews the application for approval of the protectant, solicits public comments, and may seek the counsel of external scientific experts.

FDA has primary authority for the safety of most of the food supply. The Federal Food, Drug, and Cosmetic Act establishes the standard for food safety as food being in an unadulterated condition. FDA established its basic policy regarding the review of GM foods in its 1992 *Policy on Foods Derived from New Plant Varieties*. According to this policy, FDA relies on companies developing GM foods to voluntarily notify the agency before marketing the foods. Notification leads to a two-part consultation process between the agency and the company that initially involves discussions of relevant safety issues and subsequently the company's submission of a safety assessment report containing test data on the food in question. At the end of the consultation, FDA evaluates the data and may send a letter to the company stating that the agency has no further questions, indicating in effect that it sees no reason to prevent the company from marketing the GM food. In 1997, FDA supplemented its 1992 Policy with the current *Guidance on Consultation Procedures*, clarifying procedures for the initial and final consultations. . . .

* * * * *

9.1.2 Regulatory Overview

Foods and pharmaceuticals produced by genetic engineering are not required to be labeled as such in the United States. The FDA does not consider the methods used to develop a new plant variety—whether hybridization, random mutation, or recombinant DNA methods—to be "material" within the meaning of "misleading" in section 201(n) as used in the Food, Drug, and Cosmetic Act (FD&C Act). For similar reasons GE products do not automatically require pre-market approval based on the development method. FDA does not consider foods—merely because they were derived by these new methods—to be different from other foods in any meaningful way or to present any different or greater safety concern than foods developed by traditional plant breeding. Therefore all new varieties are treated the same.[1]

This U.S. approach to the regulation of GMOs is often termed substantial equivalence.[2] This nomenclature, unfortunately, leads to misleading conclusions about how the US regulation of GMOs actually works. Both the European Union and the United States (and Codex Alimentarius) apply substantial equivalence as an analysis tool. Under substantial equivalence the attributes of new GM products are compared to conventional products that have been consumed for many years. New crop varieties produced through biotechnology—if they are "substantially equivalent"—are treated the same as their conventional counterparts.

Where the US regulatory approach differs from that of the European Union is that the US FDA does not consider the process used to developed new products. Only the new product's attributes are considered. In the European Union, the recombinant-DNA (rDNA) process itself is deemed material.

FDA presumes the biotechnology of the GMO *process* is irrelevant to the safety review. However, GM *products* are not presumed to be safe. Rather, substantial equivalence is a means of evaluation based on comparison to existing standard varieties. New varieties are examined based on their traits, which are compared to the most closely related conventional varieties. If the conventional version's traits are considered safe, then a GE variety's traits—that are substantially equivalent—would also be considered safe. For example, a new pea variety produced through rDNA techniques using only garden pea, *Pisum sativa*, genes would be substantially equivalent to the conventional pea varieties from which those genes were derived. On the other hand, the attributes of a GE food that lack substantial equivalence are subject to increased scrutiny. For instance, a gene from the bacteria, *Bacillus thuringiensis*, inserted into the garden pea would not be substantially equivalent to garden peas and would require evaluation of the environmental, health, and safety concerns.

Moreover each product must still meet existing food safety and environmental laws. Thus, if a GMO product has changed in some material way, the novel product must comply with applicable food safety and environmental laws. For example, if a peanut gene implanted in a tomato expressed a peanut protein, the product would require labeling of the allergenicity of the peanut protein. Other changes might trigger the food additive or new drug approval processes. Thus the FDA can require premarket approval when appropriate to evaluate products whose safety is not apparent. However, these decisions are based on the characteristics of final products, rather than on biotechnology process.

---

[1] *See* Foods Derived from New Plant Varieties, 57 Fed. Reg. 22,984, 22,990 (May 29, 1992).
[2] The term "substantial equivalence" is not used by FDA in its regulation of GMO foods and drugs, thereby avoiding confusion with the term's use in the regulation of medical devices.

The European Union, on the other hand, has established a more restrictive approach to regulating biotech products. From 1998 until 2004, the European Union effectively imposed a moratorium on GMO approvals and did not approve for marketing any new genetically modified organism. Since 2004, the European Union has begun slowly approving GMOs for marketing.

The EU approach is often termed the "precautionary principle." This terminology also creates unfortunate confusion. When one honestly examines the law in both the European Union and the United States, on paper there is little difference in the degree of precaution between the two approaches. The European Commission maintains that approval of new biotechnology products should not proceed if there is "insufficient, inconclusive, or uncertain" scientific data regarding potential risks. However, this is essentially the same approach as taken in the United States, with perhaps the insertion of one word, "plausible." In the United States, approval of new GE products should not proceed if there is insufficient, inconclusive, or uncertain scientific data regarding a plausible potential risk. Practical differences between the EU and the US approaches result more from social and political forces than incorporation of a precautionary principle in the law.

The European Union also requires labeling and traceability for all food and feed containing, consisting of, or produced from a GMO.[3] The label must indicate "[t]his product contains genetically modified organisms" or "produced from genetically modified [name of organism]." However, labeling is not based on safety precautions, but rather to inform consumers about the exact nature and characteristics of the food or feed so that they can make choices.

The FDA has been criticized (and praised) by various groups for its substantial equivalence policy. However, it is important to remember that the FDA is bound by the language of the FD&C Act. Following the decision in *Alliance for Bio-Integrity v. Shalala*,[4] mandatory labeling of GE foods is arguably not even within the FDA's power. U.S. regulators also stress that they consider the scientific evidence and that they exercise precaution in evaluating novel products.

9.2 FOOD SAFETY

*****

Genetically Modified Foods: Experts View Regimen of Safety Tests as Adequate, but FDA's Evaluation Process Could Be Enhanced

General Accounting Office, Rep. No. 02-566 (May 2002)

GM Foods Share the Same Types of Health Risks as Conventional Foods and Are Evaluated by Tests That Appear Adequate

---

[3] Council Regulation 1830/2003, 2003 O.J. (L 268) 24 (EC).
[4] Alliance for Bio-Integrity v. Shalala, 116 F.Supp. 2d 166 (2000).

All foods, including those from GM plants, pose the same types of inherent risks to human health: they can cause allergic or toxic reactions, or they can block the absorption of nutrients. Although some foods from GM plants have contained allergens, toxins, and antinutrients, scientists agree that the levels of these compounds have been comparable to those found in the foods' conventional counterparts. To reach such a finding, each GM food is evaluated using a regimen of tests. This regimen begins with tests on the source of the gene being transferred, proceeds to tests examining the similarity of the GM food to conventional varieties with known allergens, toxins, and antinutrients, and may include tests on the safety of the modified protein from the GM food in simulated digestive fluids. At every phase, test results are compared to the risk levels found in the food's conventional counterpart. If the risk levels are within the same range as those for the conventional food, the GM food is considered as safe as its conventional counterpart. Despite the limitations of individual tests, several experts agree that this regimen of tests has been adequate for ensuring the safety of GM foods.

All Foods Share the Same Three Risks, Which Are Evaluated in GM Foods

According to reports from the Organization for Economic Cooperation and Development, the Codex Alimentarius, and FDA, foods from GM plants pose three types of risk to human health: they can potentially contain allergens, toxins, or antinutrients. These risks are not unique to GM foods. People have consumed foods containing allergens, toxins, and antinutrients throughout human history. The small percentage of the population with food allergies (1–2 percent of adults and 6–8 percent of children) tries to prevent allergic reactions by avoiding offending foods. Additionally people commonly consume toxic substances in foods, but they usually do so at levels that are considered safe. People also frequently consume foods containing antinutrients, such as certain proteins that inhibit the digestion of nutrients in the intestinal tract, but common food preparation techniques, such as cooking, break down the antinutrients. Moreover consumption of a varied diet, in which a person is exposed to multiple nutrient sources, mitigates the risk of malnutrition from antinutrients, according to FDA officials and various academicians.

Because conventional foods contain allergens, toxins, and antinutrients, scientists recognize that food cannot be guaranteed to pose zero risk. The primary concern with the genetic modification of food with respect to human health, state industry officials, is the potential for unintentional introduction of a new allergen, an enhanced toxin, or an enhanced antinutrient in an otherwise safe food. For this reason, developers evaluate GM foods to determine if they are as safe as their conventional counterparts.

Allergic Reactions

An allergic reaction is an abnormal response of the body's immune system to an otherwise safe food. Some reactions are life threatening, such as anaphylactic shock. To avoid introducing or enhancing an allergen in an otherwise safe food, the biotech food industry evaluates GM foods to determine whether they are "as safe as" their natural counterparts. For example, in 1996 FDA reviewed the safety assessment for a GM soybean plant that can produce healthier soybean oil. As part of a standard safety assessment, the GM soybean was evaluated to see if it was as safe as a conventional soybean. Although soybeans are a common food allergen and the GM soybean remained allergenic, the results showed no significant difference between its allergenicity and that of conventional soybeans. Specifically, serums (blood) from individuals allergic to the GM soybean showed the same reactions to conventional soybeans.

Toxic Reactions

A toxic reaction in humans is a response to a poisonous substance. Unlike allergic reactions, all humans are subject to toxic reactions. Scientists involved in developing a GM food aim to ensure that the level of toxicity in the food does not exceed the level in the food's conventional counterpart. If a GM food has toxic components outside the natural range of its conventional counterpart, the GM food is not acceptable.

To date, GM foods have proven to be no different from their conventional counterparts with respect to toxicity. In fact, in some cases there is more confidence in the safety of GM foods because naturally occurring toxins that are disregarded in conventional foods are measured in the pre-market safety assessments of GM foods. For example, a naturally occurring toxin in tomatoes, known as tomatine, was largely ignored until a company in the early 1990s developed a GM tomato. FDA and the company considered it important to measure potential changes in tomatine. Through an analysis of conventional tomatoes, they showed that the levels of tomatine, as well as other similar toxins in the GM tomato, were within the range of its conventional counterpart.

Antinutrient Effects

Antinutrients are naturally occurring compounds that interfere with absorption of important nutrients in digestion. If a GM food contains antinutrients, scientists measure the levels and compare them to the range of levels in the food's conventional counterpart. If the levels are similar, scientists usually conclude that the GM food is as safe as its conventional counterpart. For example, in 1995 a company submitted to FDA a safety assessment for GM canola. The genetic modification altered the fatty acid composition of canola oil. To minimize the possibility that an unintended antinutrient effect had rendered the oil unsafe, the company compared the antinutrient composition of its product to that of conventional canola. The company found that the level of antinutrients in its canola did not exceed the levels in conventional canola.

To ensure that GM foods do not have decreased nutritional value, scientists also measure the nutrient composition, or "nutrition profile," of these foods. The nutrient profile depends on the food, but it often includes amino acids, oils, fatty acids, and vitamins. In the example previously discussed, the company also presented data on the nutrient profile of the GM canola and concluded that the significant nutrients were within the range of those in conventional canola. . . .

. . .

According to Experts, GM Food Safety Tests Have Been Adequate

Biotechnology experts whom we contacted from a consumer group, FDA, academic institutions, research institutions, the European Union, and biotechnology companies said that the current regimen of tests has been adequate for assessing the safety of GM foods. All but one expert considered the regimen of tests to be "good" or "very good" for ensuring the safety of GM foods for public consumption, and the remaining expert viewed the tests as "fair." While the experts noted that individual tests have limitations, most experts agreed that results from the regimen of tests provide the weight of evidence needed for scientists to make an accurate assessment of risk.

A distinction made by an academician and regulatory officials is that the available tests do not guarantee absolute safety of GM foods, but comparable safety. There is no assurance that even conventional foods are completely safe, since some people suffer from allergic reactions, and conventional foods can contain toxins and antinutrients. Because they have been consumed for many years, though, conventional foods are used as the standard for comparison in assessing the safety of GM foods, and experts note that the available tests are capable of making this comparison. . . .

. . .

Conclusions

Biotechnology experts believe that the current regimen of tests has been adequate for ensuring that GM foods marketed to consumers are as safe as conventional foods. However, some of these experts also believe that the agency's evaluation process could be enhanced. Specifically, FDA could verify companies' summary test data on GM foods, thus further ensuring the accuracy and completeness of this data. In addition, the agency could more clearly explain to the public the scientific rationale for its evaluation of these foods' safety, thereby increasing the transparency of, and public confidence in, FDA's evaluation process. By addressing these issues, FDA's assurance to consumers that GM foods are safe could be strengthened.

* * * * *

NOTES AND QUESTIONS

9.1. Other Safety Reviews. NATIONAL ACADEMY OF SCIENCES, INTRODUCTION OF RECOMBINANT DNA-ENGINEERED ORGANISMS INTO THE ENVIRONMENT: KEY ISSUES, National Academy Press (1987) (concluding that there was no evidence of the existence of unique hazards in the use of rDNA biotechnology and the risks associated with the introduction of rDNA biotechnology-derived organisms are the same in kind as those associated with the introduction of unmodified organisms and organisms modified by other methods). NATIONAL RESEARCH COUNCIL, FIELD TESTING GENETICALLY MODIFIED ORGANISMS: FRAMEWORK FOR DECISIONS, National Academy Press (1989) (concluding that "no conceptual distinction exists between genetic modification of plants and microorganisms by classical methods or by molecular techniques that modify DNA and transfer genes" and that the *product* of genetic modification or selection should be the primary focus for making decisions about the environmental introduction of a plant or microorganism and not the *process* by which the products were obtained). NATIONAL INSTITUTES OF HEALTH, NATIONAL BIOTECHNOLOGY POLICY BOARD REPORT (1992) (concluded that the risks associated with biotechnology are not unique, are associated with the products, not with the technology; biotechnology tends to reduce risks because the techniques are more precise and predictable; and risks of not pursuing biotechnology are likely to be greater than the risks of going forward). FAO/WHO, STRATEGIES FOR ASSESSING THE SAFETY OF FOODS PRODUCED BY BIOTECHNOLOGY, REPORT OF A JOINT FAO/WHO EXPERT CONSULTATION (1991) (concluding that GE techniques do not result in food which is inherently less safe than that produced by conventional ones). FAO/WHO, BIOTECHNOLOGY AND FOOD SAFETY, REPORT OF A JOINT FAO/WHO EXPERT CONSULTATION (1996) (reaffirming the conclusions and recommendations of the first FAO/WHO consultation). FAO/WHO, SAFETY ASPECTS OF GENETICALLY MODIFIED FOODS OF PLANT ORIGIN, REPORT OF A JOINT FAO/WHO EXPERT CONSULTATION ON FOODS DERIVED FROM BIOTECHNOLOGY (2000) (affirming that appropriate use of substantial equivalence, a comparative approach of similarities and differences between the GE food and its conventional counterpart, is the most appropriate strategy to assure the safety for GE foods). NATIONAL RESEARCH COUNCIL, GENETICALLY MODIFIED PEST-PROTECTED PLANTS: SCIENCE AND REGULATION, National Academy Press (2000) (reaffirming the principles of the 1987 NAS white paper; "properties of a genetically modified organism should be the focus of risk assessments, not the process by which it was produced.") THE ROYAL SOCIETY, GENETICALLY MODIFIED PLANTS FOR FOOD USE AND HUMAN HEALTH—AN UPDATE (2002), *available at*: http://image.guardian.co.uk/sys-files/Guardian/documents/2002/02/04/document-165.pdf (last visited Sept. 26, 2005) (found that the risks posed by GE plants are in principle no greater than those posed by conventionally derived crops or by plants introduced from other areas of the world).

9.3 FDA POLICY

The Food and Drug Administration (FDA) stated, "The statutory definition of 'food additive' makes clear that it is the intended or expected introduction of a substance into food that makes the substance potentially subject to food additive regulation. Thus, in the case of foods derived from new plant varieties, it is the transferred genetic material and the intended expression product or products that could be subject to food additive regulation, if such material or expression products are not GRAS."[5]

The FDA urges firms to consult with the FDA early in the development stages. The FDA has provided extensive guidance, including criteria and analytical steps that producers may follow to determine whether a food additive petition is appropriate.[6]

If a GE food is significantly different in function or structure, then it is treated as a food additive and must go through food additive review. However, even if the differences in a new GE food are not significant, the FDA recommends that firms communicate with the FDA early in the development process for a new plant variety. Although most of these products pose no risk, this approach addresses the possibility that material from a new plant variety might inadvertently enter the food supply before the developer has fully consulted with the FDA.[7] Early communication helps the FDA ensure that any potential food safety issues a new plant variety is resolved before any possible inadvertent introduction into the food supply.

The FDA reasons that the rapid acceleration of scientific advances expected over the next decade will lead to the development and commercialization of a greater number and diversity of bioengineered crops. "As the number and diversity of field tests for bioengineered plants increase, the likelihood that cross-pollination due to pollen drift from field tests to commercial fields and

---

[5] Foods Derived from New Plant Varieties, 57 Fed. Reg. 22,984, 22,990 (May 29, 1992); GRAS is the generally recognized as safe exception to the food additive requirements.
[6] For the FDA's latest guidance materials, visit CFSAN, Biotechnology, at http://www.cfsan.fda.gov/~lrd/biotechm.html#reg (last accessed Mar. 9, 2007).
[7] FDA Talk Paper, FDA Proposes Draft Guidance for Industry for New Plant Varieties Intended for Food Use (Nov. 19, 2004), *available at*: http://www.fda.gov/bbs/topics/ANSWERS/2004/ANS01327.html.

commingling of seeds produced during field tests with commercial seeds or grain may also increase. This could result in low-level presence in the food supply of material from new plant varieties that have not been evaluated through the FDA's voluntary consultation process for foods derived from new plant varieties (referred to as a "biotechnology consultation" in the case of bioengineered plants)."[8]

In addition to this early premarket communication, the FDA also expects developers to participate in the FDA's voluntary premarket consultation process when they commercialize a particular crop. At the date this was written, all new plant varieties developed through biotechnology that are intended for food and feed marketed in the United States completed the FDA consultation process before they entered the market.

9.3.1 Labeling

Basically FDA views the labeling of the method of variety development to be voluntary. FDA finds no reason to distinguish genetically engineered foods from foods developed through other methods of plant breeding. Therefore, GMOs are not required to be specially labeled to disclose the method of development, any more than other methods of breeding (e.g., somaclonal variation or cell culture). For example, sweet corn is not required to be labeled "hybrid sweet corn" because it was developed through cross-hybridization. Under the substantial equivalence approach, the FDA focuses on the final product—not the process used to develop the food—to determine how the food should be labeled.[9]

However, *if the composition of a food*—whether GMO or otherwise— *differs significantly from its conventional counterpart, that information would require labeling.* For example, if a food contained a major new sweetener, a new common or usual name or other labeling may be required. If a new food contains a protein derived from a food that commonly causes allergic reactions, labeling would be necessary to alert sensitive consumers because they would not expect to be allergic to that food. For example, peanut protein commonly produces very allergic reactions, and if transferred to another food, that food may require special labeling for protection of consumer health.

Numerous groups, such as the American Medical Association and the Institute of Food Technologists, support this approach to labeling. However, the European Union requires labeling of any product with GMO derived content of 0.9 percent or higher.[10] One argument advanced for the EU approach to labeling is to give consumers a choice when purchasing food. Others point out that USDA's National Organic Standards and its labeling system provides a choice to consumers who prefer non-GMO products.

---

[8] *Id.*
[9] CFSAN, FDA'S Policy for Foods Developed by Biotechnology (1995), *available at*: http://www. cfsan.fda.gov/~lrd/biopolcy.html (last accessed Mar. 8, 2007).
[10] Sec. 2, Art. 12, Council Regulation 1830/2003, 2003 O.J. (L 268) 24 (EC).

* * * * *

Voluntary Labeling Indicating Whether Foods Have or Have Not Been Developed Using Bioengineering

FDA, Guidance Document (2001)[11]

In determining whether a food is misbranded, FDA would review label statements about the use of bioengineering to develop a food or its ingredients under sections 403(a) and 201(n) of the act. Under section 403(a) of the act, a food is misbranded if statements on its label or in its labeling are false or misleading in any particular. Under section 201(n), both the presence and the absence of information are relevant to whether labeling is misleading. That is, labeling may be misleading if it fails to disclose facts that are material in light of representations made about a product or facts that are material with respect to the consequences that may result from use of the product. In determining whether a statement that a food is or is not genetically engineered is misleading under sections 201(n) and 403(a) of the act, the agency will take into account the entire label and labeling.

Statements about Foods Developed Using Bioengineering

FDA recognizes that some manufacturers may want to use informative statements on labels and in labeling of bioengineered foods or foods that contain ingredients produced from bioengineered foods. The following are examples of some statements that might be used. The discussion accompanying each example is intended to provide guidance as to how similar statements can be made without being misleading.

- "Genetically engineered" or "This product contains cornmeal that was produced using biotechnology."

The information that the food was bioengineered is optional and this kind of simple statement is not likely to be misleading. However, focus group data indicate that consumers would prefer label statements that disclose and explain the goal of the technology (why it was used or what it does for/to the food). Consumers also expressed some preference for the term "biotechnology" over such terms as "genetic modification" and "genetic engineering."

- "This product contains high oleic acid soybean oil from soybeans developed using biotechnology to decrease the amount of saturated fat."

This example includes both required and optional information. As discussed above in the background section, when a food differs from its tradi-

---

[11] *Available at*: http://www.cfsan.fda.gov/~dms/biolabgu.html (last visited Mar. 9, 2007).

tional counterpart such that the common or usual name no longer adequately describes the new food, the name must be changed to describe the difference. Because this soybean oil contains more oleic acid than traditional soybean oil, the term "soybean oil" no longer adequately describes the nature of the food. Under section 403(i) of the act, a phrase like "high oleic acid" would be required to appear as part of the name of the food to describe its basic nature. The statement that the soybeans were developed using biotechnology is optional. So is the statement that the reason for the change in the soybeans was to reduce saturated fat.

- "These tomatoes were genetically engineered to improve texture."

In this example, the change in texture is a difference that may have to be described on the label. If the texture improvement makes a significant difference in the finished product, sections 201(n) and 403(a)(1) of the act would require disclosure of the difference for the consumer. However, the statement must not be misleading. The phrase "to improve texture" could be misleading if the texture difference is not noticeable to the consumer. For example, if a manufacturer wanted to describe a difference in a food that the consumer would not notice when purchasing or consuming the product, the manufacturer should phrase the statements so that the consumer can understand the significance of the difference. If the change in the tomatoes was intended to facilitate processing but did not make a noticeable difference in the processed consumer product, a phrase like "to improve texture for processing" rather than "to improve texture" should be used to ensure that the consumer is not misled. The statement that the tomatoes were genetically engineered is optional.

- "Some of our growers plant tomato seeds that were developed through biotechnology to increase crop yield."

The entire statement in this example is optional information. The fact that there was increased yield does not affect the characteristics of the food and is therefore not necessary on the label to adequately describe the food for the consumer. A phrase like "to increase yield" should only be included where there is substantiation that there is in fact the stated difference.

Where a benefit from a bioengineered ingredient in a multi-ingredient food is described, the statement should be worded so that it addresses the ingredient and not the food as a whole; for example, "This product contains high oleic acid soybean oil from soybeans produced through biotechnology to decrease the level of saturated fat." In addition, the amount of the bioengineered ingredient in the food may be relevant to whether the statement is misleading. This would apply especially where the bioengineered difference is a nutritional improvement. For example, it would likely be misleading to make a statement about a nutritionally improved ingredient on a food that contains only a small amount of the ingredient, such that the food's overall nutritional quality would not be significantly improved.

FDA reminds manufacturers that the optional terms that describe an ingredient of a multi-ingredient food as bioengineered should not be used in the ingredient list of the multi-ingredient food. Section 403(i)(2) of the act requires each ingredient to be declared in the ingredient statement by its common or usual name. Thus, any terms not part of the name of the ingredient are not permitted in the ingredient statement. In addition, 21 C.F.R. § 101.2(e) requires that the ingredient list and certain other mandatory information appear in one place without other intervening material. FDA has long interpreted any optional description of ingredients in the ingredient statement to be intervening material that violates this regulation.

Statements about Foods That Are Not Bioengineered or That Do Not Contain Ingredients Produced from Bioengineered Foods

Terms that are frequently mentioned in discussions about labeling foods with respect to bioengineering include "GMO free" and "GM free." "GMO" is an acronym for "genetically modified organism" and "GM" means "genetically modified." Consumer focus group data indicate that consumers do not understand the acronyms "GMO" and "GM" and prefer label statements with spelled out words that mean bioengineering.

Terms like "not genetically modified" and "GMO free" that include the word "modified" are not technically accurate unless they are clearly in a context that refers to bioengineering technology. "Genetic modification" means the alteration of the genotype of a plant using any technique, new or traditional. "Modification" has a broad context that means the alteration in the composition of food that results from adding, deleting, or changing hereditary traits, irrespective of the method. Modifications may be minor, such as a single mutation that affects one gene, or major alterations of genetic material that affect many genes. Most, if not all, cultivated food crops have been genetically modified. Data indicate that consumers do not have a good understanding that essentially all food crops have been genetically modified and that bioengineering technology is only one of a number of technologies used to genetically modify crops. Thus, while it is accurate to say that a bioengineered food was "genetically modified," it likely would be inaccurate to state that a food that had not been produced using biotechnology was "not genetically modified" without clearly providing a context so that the consumer can understand that the statement applies to bioengineering.

The term "GMO free" may be misleading on most foods because most foods do not contain organisms (seeds and foods like yogurt that contain microorganisms are exceptions). It would likely be misleading to suggest that a food that ordinarily would not contain entire "organisms" is "organism free."

There is potential for the term "free" in a claim for absence of bioengineering to be inaccurate. Consumers assume that "free" of bioengineered material means that "zero" bioengineered material is present. Because of the potential for adventitious presence of bioengineered material, it may be necessary to conclude that the accuracy of the term "free" can only be ensured when there is a definition or threshold above which the term could not be used. FDA does not have information with which to establish a threshold level of bioengineered constituents or ingredients in foods for the statement "free of bioengineered material." FDA recognizes that there are analytical methods capable of detecting low levels of some bioengineered materials in some foods, but a threshold would require methods to test for a wide range of genetic changes at very low levels in a wide variety of foods. Such test methods are not available at this time. The agency suggests that the term "free" either not be used in bioengineering label statements or that it be in a context that makes clear that a zero level of bioengineered material is not implied. However, statements that the food or its ingredients, as appropriate, were not developed using bioengineering would avoid or minimize such implications. For example,

- "We do not use ingredients that were produced using biotechnology;"
- "This oil is made from soybeans that were not genetically engineered;" or
- "Our tomato growers do not plant seeds developed using biotechnology."

A statement that a food was not bioengineered or does not contain bioengineered ingredients may be misleading if it implies that the labeled food is superior to foods that are not so labeled. FDA has concluded that the use or absence of use of bioengineering in the production of a food or ingredient does not, in and of itself, mean that there is a material difference in the food. Therefore, a label statement that expresses or implies that a food is superior (e.g., safer or of higher quality) because it is not bioengineered would be misleading. The agency will evaluate the entire label and labeling in determining whether a label statement is in a context that implies that the food is superior.

In addition, a statement that an ingredient was not bioengineered could be misleading if there is another ingredient in the food that was bioengineered. The claim must not misrepresent the absence of bioengineered material. For example, on a product made largely of bioengineered corn flour and a small amount of soybean oil, a claim that the product "does not include genetically engineered soybean oil" could be misleading. Even if the statement is true, it is likely to be misleading if consumers believe that the entire product or a larger portion of it than is actually the case is free of bioengineered material. It may be necessary to carefully qualify the statement in order to ensure that consumers understand its significance.

Further, a statement may be misleading if it suggests that a food or ingredient itself is not bioengineered, when there are no marketed bioengineered varieties of that category of foods or ingredients. For example, it would be misleading to state "not produced through biotechnology" on the label of green beans, when there are no marketed bioengineered green beans. To not be misleading, the claim should be in a context that applies to the food type instead of the individual manufacturer's product. For example, the statement "green beans are not produced using biotechnology" would not imply that this manufacturer's product is different from other green beans.

Substantiation of Label Statements

A manufacturer who claims that a food or its ingredients, including foods such as raw agricultural commodities, is not bioengineered should be able to substantiate that the claim is truthful and not misleading. Validated testing, if available, is the most reliable way to identify bioengineered foods or food ingredients. For many foods, however, particularly for highly processed foods such as oils, it may be difficult to differentiate by validated analytical methods between bioengineered foods and food ingredients and those obtained using traditional breeding methods. . . . Because appropriately validated testing methods are not currently available for many foods, it is likely that it would be easier to document handling practices and procedures to substantiate a claim about how the food was processed than to substantiate a "free" claim.

. . . The national organic standards would provide for adequate segregation of the food throughout distribution to assure that non-organic foods do not become mixed with organic foods. The agency believes that the practices and record keeping that substantiate the "certified organic" statement would be sufficient to substantiate a claim that a food was not produced using bioengineering.

* * * * *

9.3.2   The Proposed New FDA Approach

In 2001 the FDA proposed rules that would require a manufacturer of a GMO food to provide the FDA with 120-day advance notice before the commercial distribution of such foods.[12] Basically the FDA would review the products as it has for voluntary consultations using scientific guidelines published by the FDA in 1992. The proposal would make the current practice of voluntary consultations mandatory.

As part of the notification, the manufacturer would provide information showing that the foods or feeds are as safe as their conventional counterparts. In those 120 days the FDA could veto the marketing under the general author-

---

[12]66 Fed. Reg. 4706–4738 (Jan. 18, 2001), *available at*: http://www.cfsan.fda.gov/~lrd/fr010118. html.

ity of the FD&C Act. The proposed rule would keep the substantial equivalence policy in place.

9.3.3 Enhancing the FDA's Oversight

Some people have questioned whether the current legal framework of government regulation provides the tools needed to effectively evaluate the future concerns, such as the environmental issues surrounding GE fish.[13] There are also questions whether the FDA has the expertise and resources necessary to conduct a comprehensive review of the more complex GMOs being produced, such as transgenic fish.[14]

Certainly, as these rapid scientific and technological changes hit the marketplace, there are new legal issues unforeseen when existing laws were enacted. FDA and USDA regulators will have to creatively stretch their authority to make old laws address the evolving next wave of GE products. For instance, FDA uses its authority over animal drugs to regulate GE fish. FDA believes that genetic engineering of animals falls under the FD&C Act definition of "drug" because rDNA techniques are "intended to affect the structure of function of the body of man or other animals."

One disadvantage of applying the approval requirements for drugs to GE products is that the drug approval process is closed to the public until after it is completed. This confidentiality protects trade secrets but cuts the public off from the agency's decision making. This lack of public participation and transparency may undermine the government's ability to gain public confidence in the decisions.[15]

Greater public involvement and greater transparency of the GMO evaluation process has also been recommended.[16] Without transparency it is hard for the public to trust that careful consideration of the risks was made before a GMO comes on the market.[17]

9.4 USDA'S ROLE

The USDA's Animal and Plant Health Inspection Service (APHIS) oversees the release of certain categories of plants and the field testing of GE crops.

---

[13] *See, e.g.,* Pew Initiative on Food and Biotechnology, Future Fish: Issues in Science and Regulation of Transgenic Fish (2003), which questions whether the FDA is legally empowered to address the environmental and ecological concerns some associate with genetically modified fish.

[14] *Id.*

[15] *Id.*

[16] *See, e.g.,* Genetically Modified Foods: Experts View Regimen of Safety Tests as Adequate, but FDA's Evaluation Process Could Be Enhanced, General Accounting Office, Rep. No. 02-566 (May 2002); *and* Organisation for Economic Co-operation and Development, Report of the Task Force for the Safety of Novel Foods and Feeds (May 17, 2000).

[17] *See* Pew Initiative on Food and Biotechnology, *supra* note 13.

APHIS has a mission of safeguarding the animal and plant resources of the United States from pests, noxious weeds, and disease. APHIS conducts agricultural quarantine inspections, monitors animal health, and carries out pest- and disease-eradication efforts, and enforces animal welfare laws.

The Plant Protection Act (PPA) of 2000[18] gives APHIS broad authority to regulate plant pests and noxious weeds in order to protect agriculture, public health, and the environment. APHIS implements its responsibilities for GE crops under part 340 regulation, "Introduction of Organisms and Products Altered or Produced through Genetic Engineering Which Are Plant Pests or Which There Is Reason to Believe Are Plant Pests."[19]

APHIS part 340 regulation defines "plant pest" broadly and includes known plant pests as well as virtually any organism from a listed genus known to contain a plant pest. APHIS has concluded that an organism from a genus known to contain a plant pest could "directly or indirectly" harm a plant.[20] Because most GE crops use a genus listed in part 340 as part of the modification process, most GE crops are considered to be regulated articles under part 340 regulation. However, certain GE crops might fall outside of this definition of authority.

Under part 340 regulation, GE crop developers are required to meet the APHIS regulatory requirements before releasing the crop into the environment.[21] Developers can gain authorization for field trials through a notification or by approval of a field trial permit. Notification for APHIS authorization for a field trial is available only for GE plants that meet certain eligibility criteria designed to exclude plants that are likely to pose risks to other plants or the environment.[22] When GE plants do not qualify for the notification procedure, or are denied field trial permission in response to a notification, the developer must submit a permit application.[23] A permit application requires more detailed information on the GE crop, the conduct of the field trial, and the procedures that will ensure containment during and after the trial.[24]

---

[18] Pub. L. Nos. 106–224, 114 Stat. 438 (codified in scattered sections of 7 U.S.C.).

[19] 7 C.F.R. §§ 340.0–340.9 (2005).

[20] *Id*. § 340.1 (definitions).

[21] *Id*. § 340.0.

[22] *Id*. § 340.3 ("Notification for the introduction of certain regulated articles"). The criteria are as follows: (1) the GM plant must be of a species that is not listed as a "noxious weed" or that otherwise has not been determined by APHIS to be a weed; (2) the introduced genetic material is "stably integrated" into the genome; (3) the function of the introduced genetic material is known, and its expression in the regulated article does not result in plant disease; (4) the introduced genetic material does not cause the production of an infectious entity, encode substances that are known or likely to be toxic to nontargeted organisms, or feed or live on the plant species or encode products intended for pharmaceutical use; (5) the introduced genetic sequences derived from plant viruses meet certain criteria to ensure that they do not pose a significant risk of the creation of any new plant virus; and (6) the plant has not been modified to contain certain specified genetic material from animal or human pathogens. *See id*. § 340.3(b) ("Regulated articles eligible for introduction under the notification procedure").

[23] *Id*. § 340(e)(5).

[24] *Id*. § 340.4(a).

9.5   EPA'S ROLE—THE SAFETY OF PESTICIDES IN BIOENGINEERED PLANTS

9.5.1   Pesticidal Substances in Food

The Environmental Protection Agency's (EPA) responsibility for GMOs derives from its regulatory jurisdiction over agricultural pesticides.[25] When a plant is genetically modified to contain a pesticidal trait, the EPA calls such trait a plant-incorporated protectant (PIP). The EPA regulates PIPs under the same statutes that apply to conventional chemical pesticides, which is to ensure that they are used in a manner that protects human health and the environment.

Under the Federal Insecticide, Fungicide, and Rodenticide Act (FIFRA),[26] the EPA decides whether and under what conditions a PIP may be used. Under the Food, Drug, and Cosmetic Act (FD&C Act),[27] the EPA decides whether and under what conditions the pesticidal substance may be present in food.

Under FIFRA, prior to commercial use, the EPA authorizes field testing by issuing experimental use permits (EUPs) that allow the use of the pesticide in the field to gather the data necessary to support an application for commercial use.[28] For PIPs that involve GE crops, the EPA cooperates with APHIS in the regulation of field trials. APHIS must also authorize such a trial under their notification and permit processes.

Under FIFRA no pesticide can be sold or used commercially until it has been approved (termed "registered") by the EPA for that use in response to an application submitted by the pesticide's developer (termed the "registrant"). To register a pesticide, the EPA must find that the pesticide will not cause "unreasonable adverse effects on the environment"[29] or "any unreasonable risk to man or the environment," including any dietary risk that is not allowable under the FD&C Act.[30] The EPA has broad authority to impose additional conditions and restrictions on use as needed to avoid unreasonable

---

[25] The term "pesticide" includes "any substance . . . intended for preventing, destroying, repelling, or mitigating any pest." 7 U.S.C. § 136(u).

[26] 7 U.S.C. §§ 136–136y.

[27] 21 U.S.C. §§ 321, 346, as amended by the Food Quality Protection Act, Pub. L. No. 104–170, 110 Stat. 1513.

[28] 40 C.F.R. §§ 172.1–172.59 (Experimental use permits). The EPA issues experimental use permits (EUPs) based on a showing by the field trial sponsor that limited planting of the crop will not lead to any unreasonable adverse effects. The EPA can impose various controls under EUPs that include data requirements for a notification, such as the identity of the microorganism constituting the microbial pesticide and a description of the proposed testing program, requirement of any information regarding potential adverse effects, and enforcement powers to seek penalties for violations.

[29] 7 U.S.C. § 136a(c)(5).

[30] *Id.* § 136a(bb).

adverse effects.[31] An example of such a restriction is that EPA restricted the use of StarLink corn to animal feed (discussed more below).

When a GE plant containing a PIP will be used in food, EPA may not register the PIP unless it has granted a tolerance or exempted the PIP from the tolerance requirement under section 408 of the FD&C Act. A tolerance limits the amount of a pesticide or PIP that can lawfully be present in food. Tolerances must be set at levels that ensure that the residues present a "reasonable certainty that no harm will result from aggregated exposure to the pesticide chemical residue, including all anticipated dietary exposures and all other exposures for which there is reliable information."[32] However, EPA may exempt a pesticide or PIP from the tolerance requirement when it does not consider a tolerance necessary for safety.[33]

9.5.2   StarLink Corn Investigation and Recall

In 2000, a consumer group reported that a bioengineered variety of corn not approved for human consumption had been found in taco shells. The corn, StarLink, was modified to contain a gene from the bacterium *Bacillus thuringiensis* that expresses a protein—Cry9C—toxic to certain insects that attack corn. The Cry9C corn was approved under the EPA responsibility for reviewing the safety of pesticide substances in bioengineered plants but was restricted for animal feed and industrial uses. The EPA did not approve the protein for human consumption because of lingering questions about Cry9C's potential to cause allergic reactions.

StarLink's developer, Aventis, was required to ensure that the bioengineered corn did not go into food. However, some became mingled with corn destined for human consumption. The presence of an unapproved pesticide in food means that the food is adulterated under the FD&C Act. The FDA began a full investigation. Kraft Foods, a producer of taco shells, initiated its own investigation and voluntarily recalled millions of taco shells as soon as an independent laboratory found that the shells contained the Cry9C gene. The FDA subsequently confirmed the presence of StarLink in the taco shells. Other recalls resulted.

Aventis agreed to buy back the 2000 StarLink crop. In a $110 million class action settlement, Aventis reimbursed farmers who grew StarLink corn. Farmers who had fields adjacent to StarLink corn fields or had StarLink corn commingled with theirs at grain elevators were also to be reimbursed for losses. Soon after the flurry of news surrounding StarLink, forty-four Americans complained that they became ill after eating foods containing StarLink corn.[34] A collaborative study of the health effects by CDC and FDA found no

---

[31] *Id*. § 136a(d).
[32] 21 U.S.C. § 346a(b)(2).
[33] *Id*. § 346a(c).
[34] *44 Claim Illness Was Caused by Biotech Corn in Food*, WASHINGTON POST, Nov. 29, 2000, at A10.

evidence that the allergic reactions experienced by the consumers were associated with hypersensitivity to Cry9C protein.[35] The difficulties evaluating the public health implications retrospectively led to CDC to highlight the importance of evaluating the allergic potential of genetically modified foods before they become available for human consumption.[36]

NOTES AND QUESTIONS

9.2. Religious and moral concerns. Some critics of biotechnology argue against GE crops for religious, moral, or ethical reasons. While to some people it just not right to meddle with nature in this way, such blanket objection to biotechnology neglects more ethical issues that it resolves. Should biotechnology that can save lives be prohibited? Should patients suffering from breast cancer be denied access to Herceptin[37] because it is from a GMO? Should the person suffering from diabetes be denied GMO insulin? Vitamin A deficiency is responsible for 500,000 cases of irreversible blindness annually, mostly in children.[38] Golden Rice is genetically engineered to contain beta-carotene in the grains and could greatly reduce the risk of vitamin A deficiency in developing countries where rice is a staple. Should children at risk for blindness and death be denied golden rice because it is a GMO?

9.3. Industrial farming. To date, commercial GMOs have overwhelmingly been products of large agricultural corporations. Some opponents of GMOs base their decision on opposition to large "factory farming." They wish to support small farmers, and see GMOs as tools of large agribusinesses squeezing out small farmers who cannot afford the technology. For example, rBST increases milk production per cow and thus could reduce the number of farms needed and squeezes out small dairies. On the other hand, millions of small farm growers willingly raise GE crops. *See* Gregory Conko and C. S. Prakash, *Can GM Crops Play a Role in Developing Countries?* PBI Bulletin, Biotechnology and Developing Countries: *The Potential and the Challenge* (2004 Issue 2), *available at*: http://www.pbi-ibp.nrc-cnrc.gc.ca/en/bulletin/2004issue2/page4.htm (last accessed Mar. 5, 2008).

---

[35] CDC, Investigation of Human Health Effects Associated with Potential Exposure to Genetically Modified Corn (June 11, 2001), *available at*: http://www.cdc.gov/nceh/ehhe/Cry9cReport/pdfs/cry9creport.pdf (last accessed Mar. 5, 2008).

[36] *Id.*

[37] Herceptin is a monoclonal antibody that is effective at treating certain tumors.

[38] J. H Humphrey et al., *Vitamin A Deficiency and Attributable Mortality in Under-5-Year-Olds*, 70 WHO BULLETIN 225–232 (1992), *available at*: http://whqlibdoc.who.int/bulletin/1992/Vol70-No2/bulletin_1992_70(2)_225-232.pdf.

PROBLEM EXERCISE

9.4. Hypothetical exercise. Novella Foods, Inc., has successfully created a higher protein corn by inserting genes from soybeans into the corn. Novella's testing has shown the product is as safe as any other corn or soybean. Novella has field tested its corn outside of the United States but now wants to import the product.

Questions. What contact, if any, does Novella need to have with FDA? When should they make that contact? What regulatory issues are likely to arise under existing statutes and existing and proposed regulations?

9.6   THE RIGHT TO KNOW

There have been calls for mandatory labeling requirement of GE food in the United States.[39] Proponents typically claim that the purpose of mandatory labeling is to inform consumers and farmers about the exact development method of the food or feed so that they can make informed choices.

FDA has taken a different approach, however, in considering the development method not to be relevant for regulatory status. The regulatory status of a food is based on characteristics of the food and not the nature of its development.

* * * * *

FDA, Statement of Policy: Foods Derived from New Plant Varieties

57 Fed. Reg. 22984 (May 29, 1992)

. . .

New methods of genetically modifying plants are being used to develop new varieties that will be sources of foods. These methods, including recombinant DNA techniques and cell fusion techniques, enable developers to make genetic modifications in plants, including some modifications that would not be possible with traditional plant breeding methods. This policy discusses the safety and regulatory status of foods derived from new plant varieties, including plants developed by the newer methods of genetic modification. . . .

Under this policy, foods, such as fruits, vegetables, grains, and their by-products, derived from plant varieties developed by the new methods of

---

[39] *See, e.g.*, Union of Concerned Scientists, http://www.ucsusa.org (last visited Sept. 26, 2005); Mothers for Natural Law, http://www.safe-food.org (last visited Sept. 26, 2005); Greenpeace International, http://www.greenpeace.org (last visited Sept. 26, 2005); *see also* Kirsten S. Beaudoin, *On Tonight's Menu: Toasted Cornbread with Firefly Genes?* 83 MARQUETTE LAW REVIEW 237, 278 (1999).

genetic modification are regulated within the existing framework of the act, FDA's implementing regulations, and current practice, utilizing an approach identical in principle to that applied to foods developed by traditional plant breeding. The regulatory status of a food, irrespective of the method by which it is developed, is dependent upon objective characteristics of the food and the intended use of the food (or its components). The method by which food is produced or developed may in some cases help to understand the safety or nutritional characteristics of the finished food. However, the key factors in reviewing safety concerns should be the characteristics of the food product, rather than the fact that the new methods are used.

The safety of a food is regulated primarily under FDA's postmarket authority of section 402(a)(1) of the act (21 U.S.C. § 342(a)(1)). Unintended occurrences of unsafe levels of toxicants in food are regulated under this section. Substances that are expected to become components of food as result of genetic modification of a plant and whose composition is such or has been altered such that the substance is not generally recognized as safe (GRAS) or otherwise exempt are subject to regulation as "food additives" under section 409 of the Act (21 U.S.C. § 348). Under the Act, substances that are food additives may be used in food only in accordance with an authorizing regulation.

In most cases the substances expected to become components of food as a result of genetic modification of a plant will be the same as or substantially similar to substances commonly found in food, such as proteins, fats and oils, and carbohydrates. As discussed in more detail in section V.C., FDA has determined that such substances should be subject to regulation under section 409 of the act in those cases when the objective characteristics of the substance raise questions of safety sufficient to warrant formal premarket review and approval by FDA. The objective characteristics that will trigger regulation of substances as food additives are described in the guidance section of this notice (section VII.). . . .

FDA has received several inquiries concerning labeling requirements for foods derived from new plant varieties developed by recombinant DNA techniques. Section 403(i) of the Act (21 U.S.C. § 343(i)) requires that a producer of a food product describe the product by its common or usual name, or in the absence thereof, an appropriately descriptive term (21 U.S.C. § part 101.3), and reveal all facts that are material in light of representations made or suggested by labeling or with respect to consequences which may result from use (21 U.S.C. § 343(a); 21 U.S.C. § 321(n)). Thus, consumers must be informed, by appropriate labeling, if a food derived from a new plant variety differs from its traditional counterpart such that the common or usual name no longer applies to the new food, or if a safety or usage issue exists to which consumers must be alerted.

For example, if a tomato has had a peanut protein introduced into it and there is insufficient information to demonstrate that the introduced protein could not cause an allergic reaction in a susceptible population, a label declaration would be required to alert consumers who are allergic to peanuts so they could avoid that tomato, even if its basic taste and texture remained unchanged. Such information would be a material fact whose omission may make the label of the tomato misleading under section 403(a) of the act (21 U.S.C. § 343(a)).

FDA has also been asked whether foods developed using techniques such as recombinant DNA techniques would be required to bear special labeling to reveal that fact to consumers. To date, FDA has not considered the methods used in the development of a new plant variety (e.g., hybridization, chemical or radiation-induced mutagenesis, protoplast fusion, embryo rescue, somaclonal variation, or any other method) to be material information within the meaning of section 201(n) of the act (21 U.S.C. 321(n)). As discussed above, FDA believes that the new techniques are extensions at the molecular level of traditional methods and will be used to achieve the same goals as pursued with traditional plant breeding. The agency is not aware of any information showing that foods derived by these new methods differ from other foods in any meaningful or uniform way, or that, as a class, foods developed by the new techniques present any different or greater safety concern than foods developed by traditional plant breeding. For this reason the agency does not believe that the method of development of a new plant variety (including the use of new techniques including recombinant DNA techniques) is normally material information within the meaning of 21 U.S.C. 321(n) and would not usually be required to be disclosed in labeling for the food.

. . .

* * * * *

FDA approved GE bovine growth hormone, recombinant bovine somatotropin (rBST), for use with dairy cows to increase milk production. FDA did not require labeling of the resulting milk because the rBST was safe and undetectable in the resulting milk. Vermont enacted a law requiring labeling of the use of rBST.

* * * * *

International Dairy Foods Ass'n v. Amestoy

92 F.3d 67 (2d Cir. 1996)

Before: ALTIMARI, MCLAUGHLIN, and LEVAL, Circuit Judges
Opinion: ALTIMARI, Circuit Judge

In 1993, the federal Food and Drug Administration ("FDA") approved the use of recombinant bovine somatotropin ("rBST") (also known as recombinant bovine growth hormone ("rGBH")), a synthetic growth hormone that increases milk production by cows. It is undisputed that the dairy products derived from herds treated with rBST are indistinguishable from products derived from untreated herds; consequently, the FDA declined to require the labeling of products derived from cows receiving the supplemental hormone.

In April 1994, defendant-appellee the State of Vermont ("Vermont") enacted a statute requiring that "[i]f rBST has been used in the production of milk or a milk product for retail sale in this state, the retail milk or milk product shall be labeled as such." The State of Vermont's Commissioner of Agriculture ("Commissioner") subsequently promulgated regulations giving those dairy manufacturers who use rBST four labeling options, among them the posting of a sign to the following effect in any store selling dairy products:

rBST Information
THE PRODUCTS IN THIS CASE THAT CONTAIN OR MAY CONTAIN MILK FROM rBST-TREATED COWS EITHER

(1) STATE ON THE PACKAGE THAT rBST HAS BEEN OR MAY HAVE BEEN USED, OR

(2) ARE IDENTIFIED BY A BLUE SHELF LABEL LIKE THIS [BLUE RECTANGLE], OR

(3) A BLUE STICKER ON THE PACKAGE LIKE THIS [BLUE DOT].

The United States Food and Drug Administration has determined that there is no significant difference between milk from treated and untreated cows. It is the law of Vermont that products made from the milk of rBST-treated cows be labeled to help consumers make informed shopping decisions.

It is not enough for appellants to show, as they have, that they were irreparably harmed by the statute; because the dairy manufacturers challenge government action taken in the public interest, they must also show a likelihood of success on the merits. We find that such success is likely.

In *Central Hudson*, the Supreme Court articulated a four-part analysis for determining whether a government restriction on commercial speech is permissible. We need not address the controversy concerning the nature of the speech in question—[whether] commercial or political—because we find that Vermont fails to meet the less stringent constitutional requirements applicable to compelled commercial speech.

Under *Central Hudson*, we must determine: (1) whether the expression concerns lawful activity and is not misleading; (2) whether the government's interest is substantial; (3) whether the labeling law directly serves the asserted interest; and (4) whether the labeling law is no more extensive than necessary. Furthermore, the State of Vermont bears the burden of justifying its labeling law. As the Supreme Court has made clear, "[t]his burden is not satisfied by mere speculation or conjecture; rather, a governmental body seeking to sustain a restriction on commercial speech must demonstrate that the harms it recites are real and that its restriction will in fact alleviate them to a material degree."

In our view, Vermont has failed to establish the second prong of the *Central Hudson* test, namely that its interest is substantial. In making this determination, we rely only upon those interests set forth by Vermont before the district court. ("[T]he *Central Hudson* standard does not permit us to supplant the precise interests put forward by the state with other suppositions.") As the district court made clear, Vermont "does not claim that health or safety concerns prompted the passage of the Vermont Labeling Law," but instead defends the statute on the basis of "strong consumer interest and the public's 'right to know'. . . ." These interests are insufficient to justify compromising protected constitutional rights.[40]

Vermont's failure to defend its constitutional intrusion on the ground that it negatively impacts public health is easily understood. After exhaustive studies, the FDA has "concluded that rBST has no appreciable effect on the composition of milk produced by treated cows, and that there are no human safety or health concerns associated with food products derived from cows treated with rBST." Because bovine somatotropin ("BST") appears naturally in cows, and because there are no BST receptors in a cow's mammary glands, only trace amounts of BST can be detected in milk, whether or not the cows received the supplement. Moreover, it is undisputed that neither consumers nor scientists can distinguish rBST-derived milk from milk produced by an untreated cow. Indeed, the already extensive record in this case contains no scientific evidence from which an objective observer could conclude that rBST has any impact at all on dairy products. It is thus plain that Vermont could not justify the statute on the basis of "real" harms.

We do not doubt that Vermont's asserted interest, the demand of its citizenry for such information, is genuine; reluctantly, however, we conclude that it is inadequate. We are aware of no case in which consumer interest alone was sufficient to justify requiring a product's manufacturers to publish the functional equivalent of a warning about a production method that has no discernable impact on a final product.

Although the Court is sympathetic to the Vermont consumers who wish to know which products may derive from rBST-treated herds, their desire is insuf-

---

[40] Although the dissent suggests several interests that if adopted by the state of Vermont *may* have been substantial, the district court opinion makes clear that Vermont adopted no such rationales for its statute. Rather, Vermont's sole expressed interest was, indeed, "consumer curiosity." The district court plainly stated that, "Vermont takes *no* position on whether rBST is beneficial or detrimental. However," the district court explained, "Vermont has determined that its consumers want to know whether rBST has been used in the production of their milk and milk products." 898 F.Supp. at 252 (emphasis added). It is clear from the opinion below that the state itself has not adopted the concerns of the consumers; it has only adopted that the consumers are concerned. Unfortunately, mere consumer concern is not, in itself, a substantial interest.

ficient to permit the State of Vermont to compel the dairy manufacturers to speak against their will. Were consumer interest alone sufficient, there is no end to the information that states could require manufacturers to disclose about their production methods. For instance, with respect to cattle, consumers might reasonably evince an interest in knowing which grains herds were fed, with which medicines they were treated, or the age at which they were slaughtered. Absent, however, some indication that this information bears on a reasonable concern for human health or safety or some other sufficiently substantial governmental concern, the manufacturers cannot be compelled to disclose it. Instead, those consumers interested in such information should exercise the power of their purses by buying products from manufacturers who voluntarily reveal it.

Accordingly, we hold that consumer curiosity alone is not a strong enough state interest to sustain the compulsion of even an accurate, factual statement, in a commercial context. Because Vermont has demonstrated no cognizable harms, its statute is likely to be held unconstitutional.

Conclusion

Because appellants have demonstrated both irreparable harm and a likelihood of success on the merits, the judgment of the district court is reversed, and the case is remanded for entry of an appropriate injunction.

LEVAL, Circuit Judge, dissenting:

I respectfully dissent. Vermont's regulation requiring disclosure of use of rBST in milk production was based on substantial state interests, including worries about rBST's impact on human and cow health, fears for the survival of small dairy farms, and concerns about the manipulation of nature through biotechnology. The objective of the plaintiff milk producers is to conceal their use of rBST from consumers. The policy of the First Amendment, in its application to commercial speech, is to favor the flow of accurate, relevant information. The majority's invocation of the First Amendment to invalidate a state law requiring disclosure of information consumers reasonably desire stands the Amendment on its ear. In my view, the district court correctly found that plaintiffs were unlikely to succeed in proving Vermont's law unconstitutional.

. . .

The interests which Vermont sought to advance by its statute and regulations were explained in the Agriculture Department's Economic Impact Statement accompanying its regulations. The Statement reported that consumer interest in disclosure of use of rBST was based on "concerns about FDA determinations about the product as regards health and safety or about recombinant gene technology"; concerns "about the effect of the product on bovine health"; and "concerns about the effect of the product on the existing surplus of milk and in the dairy farm industry's economic status and well-being." This finding was based on "consumer comments to Vermont legislative committees" and to the Department, as well as published reports and letters to the editors published in the press.

The state offered survey evidence which demonstrated similar public concern. Comments by Vermont citizens who had heard or read about rBST were overwhelmingly negative. The most prevalent responses to rBST use included: "Not natural," "More research needs to be done/Long-term effects not clear," "Against additives added to my milk," "Worried about adverse health effects," "Unhealthy for the cow," "Don't need more chemicals," "It's a hormone/Against hormones added to my milk," "Hurts the small dairy farmer," "Producing enough milk already."

On the basis of this evidence the district court found that a majority of Vermonters "do not want to purchase milk products derived from rBST-treated cows," and that the reasons included:

(1) They consider the use of a genetically engineered hormone in the production unnatural; (2) they believe that use of the hormone will result in increased milk production and lower milk prices, thereby hurting small dairy farmers; (3) they believe that the use of rBST is harmful to cows and potentially harmful to humans; and (4) they feel that there is a lack of knowledge regarding the long-term effects of rBST.

The court thus understandably concluded that "Vermont has a substantial interest in informing consumers of the use of rBST in the production of milk and dairy products sold in the state." . . .

Second, the majority distorts the meaning of the district court opinion. It relies substantially on Judge Murtha's statement that Vermont "does not claim that health or safety concerns prompted the passage of the Vermont Labeling Law," but "bases its justification . . . on strong consumer interest and the public's 'right to know'." . . . More likely, what Judge Murtha meant was that Vermont does not claim to know whether rBST is harmful. And when he asserted that Vermont's rule was passed to vindicate "strong consumer interest and the public's right to know," this could not mean that the public's interest was based on nothing but "curiosity," because the judge expressly found that the consumer interest was based on health, economic, and ethical concerns. . . .

To suggest that a government agency's failure to find a health risk in a short-term study of a new genetic technology should bar a state from requiring simple disclosure of the use of that technology where its citizens are concerned about such health risks would be unreasonable and dangerous. Although the FDA's conclusions may be reassuring, they do not guarantee the safety of rBST.

Forty years ago, when I (and nearly everyone) smoked, no one told us that we might be endangering our health. Tobacco is but one of many consumer products once considered safe, which were subsequently found to cause health hazards. The limitations of scientific information about new consumer products were well illustrated in a 1990 study produced at the request of Congress by the General Accounting Office. Looking at various prescription drugs available on the market, the study examined the risks associated with the drugs that became known only after they were approved by the FDA, and concluded:

> [E]ven after approval, many additional risks may surface when the general population is exposed to a drug. These risks, which range from relatively minor (such as nausea and headache) to serious (such as hospitalization and death) arise from the fact that preapproval drug testing is inherently limited.
>
> . . .
>
> In studying the frequency and seriousness of risks identified after approval, GAO found that of the 198 drugs approved by FDA between 1976 and 1985 for which data were available, 102 (or 51.5 percent) had serious post-approval risks, as evidenced by labeling changes or withdrawal from the market. All but six of these drugs . . . are deemed by FDA to have benefits that outweigh their risks. The serious post-approval risks are adverse reactions that could lead to hospitalization . . . severe or permanent disability, or death.

As startling as its results may seem, this study merely confirms a common-sense proposition: namely, that a government agency's conclusion regarding a product's safety, reached after limited study, is not a guarantee and does not invalidate public concern for unknown side effects.

In short, the majority has no valid basis for its conclusion that Vermont's regulation advances no interest other than the gratification of consumer curiosity, and involves neither health concerns nor other substantial interests. . . .

In my view, Vermont's multifaceted interest, outlined above, is altogether substantial. Consumer worries about possible adverse health effects from consumption of rBST, especially over a long term, is unquestionably a substantial interest. As to health risks to cows, the concern is supported by the warning label on Posilac, which states that cows injected with the product are at an increased risk for: various reproductive disorders, "clinical mastitis [udder infections] (visibly abnormal milk)," "digestive disorders such as indigestion, bloat, and diarrhea," "enlarged hocks and lesions," and "swellings" that may be permanent. As to the economic impact of increased milk production, caused by injection of rBST, upon small dairy farmers, the evidence included a U.S. Department of Agriculture economist's written claim that, "if rBST is heavily adopted and milk prices are reduced, at least some of the smaller farmers that do not use rBST might be forced out of the dairy business, because they would not be producing economically sufficient volumes of milk." Public philosophical objection to biotechnological mutation is familiar and widespread. . . .

Notwithstanding their self-righteous references to free expression, the true objective of the milk producers is concealment. They do not wish consumers to know that their milk products were produced by use of rBST because there are consumers who, for various reasons, prefer to avoid rBST. Vermont, on the other hand, has established a labeling requirement whose sole objective (and whose sole effect) is to inform Vermont consumers whether milk products offered for sale were produced with rBST. The dispute under the First Amendment is over whether the milk producers' interest in concealing their use of rBST from consumers will prevail over a state law designed to give consumers the information they desire. The question is simply whether the First Amendment prohibits government from requiring disclosure of truthful relevant information to consumers.

In my view, the interest of the milk producers has little entitlement to protection under the First Amendment. The case law that has developed under the doctrine of commercial speech has repeatedly emphasized that the primary function of the First Amendment in its application to commercial speech is to advance truthful disclosure—the very interest that the milk producers seek to undermine. . . .

The application of these principles to the case at bar yields a clear message. The benefit the First Amendment confers in the area of commercial speech is the provision of accurate, nonmisleading, relevant information to consumers. Thus, regulations designed to prevent the flow of such information are disfavored; regulations designed to provide such information are not.

The milk producers' invocation of the First Amendment for the purpose of concealing their use of rBST in milk production is entitled to scant recognition. They invoke the Amendment's protection to accomplish exactly what the Amendment opposes. And the majority's ruling deprives Vermont of the right to protect its consumers by requiring truthful disclosure on a subject of legitimate public concern.

I am comforted by two considerations: First, the precedential effect of the majority's ruling is quite limited. By its own terms, it applies only to cases where a state disclosure requirement is supported by no interest other than the gratification of consumer curiosity. In any case in which a state advanced something more, the majority's ruling would have no bearing.

Second, Vermont will have a further opportunity to defend its law. The majority's conclusion perhaps results from Vermont's failure to put forth sufficiently clear evidence of the interests it sought to advance. If so, the failure is remediable because it occurred only at the preliminary injunction stage. Trial on the merits has yet to be held. The majority has found on the basis of the evidence presented at the hearing that the plaintiffs are likely to succeed on the merits; it has, of course, not ruled on the ultimate issue. If Vermont succeeds at trial in putting forth clear evidence that its laws were in fact motivated by the concerns discussed above (and not merely by consumer curiosity), it will have shown a substantial interest sufficient to satisfy the requirements of the First Amendment.

* * * * *

In *Alliance for Bio-Integrity v. Shalala*,[41] the plaintiffs, a consortium of groups concerned about GE foods for environmental and religious reasons, brought an action to overturn the 1992 FDA's policy statement that GE food

---

[41] 116 F. Supp. 2d 166 (2000).

was not materially different from other food under the FD&C Act.[42] The FDA's policy statement indicates that rDNA was not a "material fact" under the FD&C Act that required labeling. The court held that the FDA lacks statutory authority to impose a label requirement based on consumer demands or religious concerns. Something more than a desire to know is needed to establish that information is material as defined by the FD&C Act and required to be labeled.

Several bills have been introduced in Congress that propose mandatory labeling of GE foods, such as "The Genetically Engineered Food Right to Know Act" and the "The Genetically Engineered Food Safety Act." However, none of these proposals has been successful.

* * * * *

Alliance for Bio-Integrity v. Shalala

116 F. Supp. 2d 166 (2000)

Technological advances have dramatically increased our ability to manipulate our environment, including the foods we consume. One of these advances, recombinant deoxyribonucleic acid (rDNA) technology, has enabled scientists to alter the genetic composition of organisms by mixing genes on the cellular and molecular level in order to create new breeds of plants for human and animal consumption. These new breeds may be designed to repel pests, retain their freshness for a longer period of time, or contain more intense flavor and/or nutritional value. Much controversy has attended such developments in biotechnology, and in particular the production, sale, and trade of genetically modified organisms and foods. The above-captioned lawsuit represents one articulation of this controversy.

Among Plaintiffs, some fear that these new breeds of genetically modified food could contain unexpected toxins or allergens, and others believe that their religion forbids consumption of foods produced through rDNA technology. Plaintiffs, a coalition of groups and individuals including scientists and religious leaders concerned about genetically altered foods, have brought this action to protest the Food and Drug Administration's ("FDA") policy on such foods in general, and in particular on various genetically modified foods that already have entered the marketplace. The parties have filed cross-motions for summary judgment on plaintiffs' multiple claims. Upon careful consideration of the parties' briefs and the entire record, the Court shall grant Defendants' motion as to all counts of Plaintiffs' Complaint.

I. Background

On May 29, 1992, the FDA published a "Statement of Policy: Foods Derived from New Plant Varieties" (Statement of Policy). In the Statement of Policy,

---

[42] FDA, Statement of Policy: Foods Derived from New Plant Varieties, 57 Fed. Reg. 22984 (May 29, 1992).

FDA announced that the agency would presume that foods produced through the rDNA process were "generally recognized as safe" (GRAS) under the Federal Food, Drug and Cosmetic Act ("FD&C Act"), 21 U.S.C. § 321(s), and therefore not subject to regulation as food additives. While FDA recommended that food producers consult with it before marketing rDNA-produced foods, the agency did not mandate such consultation. In addition, FDA reserved the right to regulate any particular rDNA-developed food that FDA believed was unsafe on a case-by-case basis, just as FDA would regulate unsafe foods produced through conventional means.

The Statement of Policy also indicated that rDNA modification was not a "material fact" under the FD&C Act, 21 U.S.C. § 321(n), and that therefore labeling of rDNA-produced foods was not necessarily required. FDA did not engage in a formal notice-and-comment process on the Statement of Policy, nor did it prepare an Environmental Impact Statement or Environmental Assessment. At least thirty-six foods, genetically altered through rDNA technology, have been marketed since the Statement of Policy was issued.

Plaintiffs filed a Complaint in this Court challenging the FDA's policy on six different grounds: (1) the Statement was not properly subjected to notice-and-comment procedures; (2) the FDA did not comply with the National Environmental Protection Act (NEPA) by compiling an Environmental Assessment or Environmental Impact Statement; (3) the FDA's presumption that rDNA-developed foods are GRAS and therefore do not require food additive petitions under 21 U.S.C. § 321(s) is arbitrary and capricious; (4) the FDA's decision not to require labeling for rDNA-developed foods is arbitrary and capricious; (5) the FDA's decision not to regulate or require labeling for rDNA-developed foods violates the Free Exercise Clause; and (6) the FDA's decision not to regulate or require labeling for rDNA-developed foods violates the Religious Freedom Restoration Act. Plaintiffs have also challenged on the third and fourth grounds each of FDA's specific decisions not to regulate 36 individual rDNA-produced products. The parties have filed cross-motions for summary judgment on all of Plaintiff's claims.

II. Discussion

. . .

A. *Subject Matter Jurisdiction*

. . .

The *Chaney* Court reasoned that courts reviewing agency action "need a meaningful standard against which to judge the agency's exercise of discretion." Individual agency decisions not to enforce a statute "involve a complicated balancing of a number of factors," and courts do not have a meaningful standard with which to evaluate the agency's balancing. Therefore, these decisions are "committed to agency discretion by law" and are not subject to judicial review. The Court noted that an agency's enforcement discretion may be limited when Congress has "set[] substantive priorities, or ... otherwise circumscribed an agency's power to discriminate among issues or cases it will pursue." When determining if an agency action is reviewable, courts looks to "whether the applicable statutes and regulations are drawn so that a court would have a meaningful standard against which to judge the agency's exercise of discretion."

This Circuit has recognized a distinction between agency decisions not to regulate an entire class of conduct, which are essentially policy choices, and individual nonenforcement decisions. When an agency has employed a formal procedure, such as notice and comment rulemaking, to announce a major policy decision not to regulate certain conduct, courts can use this procedure as "a focal point for judicial review." In the instant case, even without actual notice and comment procedures, the FDA's formal publication of the Statement of Policy provides a focal point for this Court's review of the agency's action. Moreover, this Court has a meaningful standard against which to judge the Statement of Policy. Congress's passage of the various statutes on which Plaintiffs rely here—the Administrative Procedure Act, the Federal Food Drug and Cosmetic Act, the National Environmental Protection Act, and the Religious Freedom Restoration Act—has limited the FDA's enforcement discretion. Although the Court may not review FDA's policy-laden individual enforcement decisions, the Court has jurisdiction to review whether or not FDA's Statement of Policy comports with congressional directives.

B. Notice and Comment

Plaintiffs argue that the Statement of Policy should be set aside because it was not subjected to notice and comment proceedings, as required under the Administrative Procedure Act ("APA"). While conceding that the Statement of Policy did not undergo a formal notice and comment process, Defendants maintain that the Statement of Policy is a policy statement or an interpretive rule not subject to notice and comment requirements. Plaintiffs contend instead that the Statement of Policy is a substantive rule, and that therefore it was improperly exempted from a formal notice and comment process.

A substantive rule, which must undergo a formal notice-and-comment process, is a rule that "implement[s]" a statute and has "the force and effect of law." Policy statements, on the other hand, are "statements issued by an agency to advise the public prospectively of the manner in which the agency proposes to exercise a discretionary power." Although the distinction between these categories is not entirely clear ... the Court of Appeals articulated a two-part test for determining when an agency action is a policy statement. Policy statements (1) must not impose any new rights or obligations, and (2) must "genuinely leave the agency and its decision-makers free to exercise discretion." In weighing these criteria, "the ultimate issue is the agency's intent to be bound." An agency's own characterization of its statement deserves some weight, but it is not dispositive. ...

By its very name, the Statement of Policy announces itself as a policy statement. More importantly, the plain language of the Statement suggests that it does not have a binding effect. For example, the Statement does not declare that transferred genetic material will be considered GRAS; rather, it announces that "such material is presumed to be GRAS." This presumption of safety is rebuttable, because FDA will "require food additive petitions in cases where safety questions exist sufficient to warrant formal premarket review by FDA to ensure public health protection." Rebuttable presumptions leave an agency free to exercise its discretion and may therefore properly be announced in policy statements. . . . *Mada-Luna v. Fitzpatrick* ("To the extent that the directive merely provides guidance to agency officials in exercising their discretionary powers while preserving their flexibility and their opportunity to make individualized determination[s], it constitutes a general statement of policy"); *accord Ryder Truck Lines, Inc. v. United States* ("As long as the agency remains free to consider the individual facts in the various cases that arise, then the agency action in question has not established a binding norm.").

In response to the argument that the Policy Statement vests broad discretion with the agency, Plaintiffs contend that the FDA's application of the Statement has given it a "practical effect" that has effectively bound the agency's discretion, as evidenced by the thirty-six genetically engineered foods that are currently on the market and not regulated by the FDA. Although courts will look to the "agency's actual applications" to determine the nature of an agency statement, such an inquiry occurs "where the language and context of a statement are inconclusive." Here, the plain language of the Statement clearly indicates that it is a policy statement that merely creates a presumption and does not ultimately bind the agency's discretion. Given this unambiguous language, this Court need not consider the agency's application of the Statement to determine the Statement's meaning.

Even if, as Plaintiffs argue, FDA has previously used notice-and-comment procedures to determine GRAS status, in the instant case FDA has not determined GRAS status but has rather announced a GRAS presumption. . . . *Panhandle Producers v. Econ. Regulatory Admin.* ("This court and others have consistently stated that an agency may announce presumptions through policy statements rather than notice-and-comment rulemaking."). The Statement of Policy creates a rebuttable presumption of GRAS that does not constrain the FDA's ability to exercise its discretion. *See id.* ("Presumptions, so long as rebuttable, leave such freedom [to exercise the agency's discretion]."). Because the Statement is a policy statement merely announcing a GRAS presumption, the omission of formal notice-and-comment procedures does not violate the Administrative Procedure Act.

C. NEPA

Plaintiffs have also alleged that FDA violated the National Environmental Protection Act (NEPA), by not performing an Environmental Assessment

(EA) or an Environmental Impact Statement (EIS) in conjunction with the Statement of Policy. NEPA requires "all agencies of the Federal Government ... [to] include in every recommendation or report on proposals for legislation and other major Federal actions significantly affecting the quality of the human environment, a detailed statement ... on the environmental impact of the proposed action."

"Major federal action," as defined in the Code of Federal Regulations, includes actions such as "adoption of official policy ... adoption of formal plans ... adoption of programs ... [and] approval of specific projects." For major federal actions, agencies must either prepare an EIS examining the environmental impact of the proposed action, prepare an EA determining whether or not to prepare an EIS, or claim that the action falls within a Categorical Exclusion, "a category of actions which do not individually or cumulatively have a significant effect on the human environment." If the agency is not engaging in a major federal action, NEPA requirements do not apply.

In the Statement of Policy, FDA announces that "the activities [FDA] may undertake with respect to foods from new plant varieties ... will [not] constitute agency action under NEPA." FDA's determination that the Statement is not a major federal action is essentially an interpretation of the meaning of "major federal action" in 42 U.S.C. § 4332(2)(c) and 40 C.F.R. § 1508.18. Agencies enjoy wide discretion in interpreting regulations, and the agency's interpretation will be upheld unless it is arbitrary and capricious.

The FDA's determination that the Statement was not a major federal action comports with the holdings of this Circuit, and is therefore neither arbitrary nor capricious. While declaring a rebuttable presumption that foods produced through rDNA technology are GRAS, the FDA has neither made a final determination that any particular food will be allowed into the environment, nor taken any particular regulatory actions that could affect the environment. In order to trigger the NEPA requirement of an EIS, the agency must be prepared to undertake an "'irreversible and irretrievable commitment of resources' to an action that will affect the environment." Because the FDA's presumption does not bind its decision-making authority, it has neither taken nor prepared to take the irreversible action that is necessary to require preparation of an EIS under Wyoming Outdoor Council. Evidencing this nonbinding effect is the FDA's 1993 decision to open the labeling issue for further discussion, requesting additional public comment on the possible implementation of a general labeling requirement.

Moreover agency decisions that maintain the substantive status quo do not constitute major federal actions under NEPA. Defendants maintain correctly that their actions have not altered the status quo because "rDNA-modified foods ... were regulated no differently before the publication of the Policy Statement than they are now." Because the announcement of a rebuttable presumption of GRAS does not affect the substantive regulatory status quo, it is not a major federal action.

The Statement of Policy is not only reversible and consistent with the status quo ante; it is also not properly an "agency action." The core of Plaintiff's NEPA claim is that FDA has failed to regulate rDNA-modified foods, and that this failure to act engenders environmental consequences. But NEPA applies only to agency actions, "even if inaction has environmental consequences. . . ."

In the instant case, FDA has not taken an overt action, but instead has merely announced a presumption that certain foods do not require special regulation. This presumption against regulation does not constitute an overt action, and is therefore not subject to NEPA requirements.

In sum, because FDA's Statement of Policy is reversible, maintains the substantive status quo, and takes no overt action, the Statement of Policy does not constitute a major federal action under NEPA. FDA was not required to compile an Environmental Assessment or an Environmental Impact Statement in conjunction with the Statement of Policy, and therefore its failure to do so does not violate NEPA.

D. GRAS Presumption

In their challenge to the FDA's Statement of Policy, Plaintiffs further claim that the Statement of Policy's presumption that rDNA-engineered foods are GRAS violates the GRAS requirements of the Federal Food, Drug, and Cosmetic Act ("FD&C Act"), 21 U.S.C. § 321(s), and is therefore arbitrary and capricious. The FD&C Act provides that any substance which may "become a component or otherwise affect[] the characteristics of any food" shall be deemed a food additive. A producer of a food additive must submit a food additive petition to FDA for approval unless FDA determines that the additive is "generally recognized [by qualified experts] . . . as having been adequately shown through scientific procedures . . . to be safe under the conditions of its intended use." *Id.*

In the Statement of Policy, FDA indicated that, under § 321(s),

> it is the intended or expected introduction of a substance into food that makes the substance potentially subject to food additive regulation. Thus, in the case of foods derived from new plant varieties, it is the transferred genetic material and the intended expression product or products that could be subject to food additive regulation, if such material or expression products are not GRAS.

57 Fed. Reg. at 22,990

Accordingly, FDA reasoned that the only substances added to rDNA engineered foods are nucleic acid proteins, generally recognized as not only safe but also necessary for survival. *See id.* ("Nucleic acids are present in the cells of every living organism, including every plant and animal used for food by humans or animals, and do not raise a safety concern as a component of food"). Therefore, FDA concluded that rDNA engineered foods should be presumed to be GRAS unless evidence arises to the contrary. *See id.* at 22,991 ("Ultimately, it is the food producer who is responsible for assuring safety."). The

Statement of Policy does acknowledge, however, that certain genetically modified substances might trigger application of the food additives petitioning process. In that vein, FDA recognized that "the intended expression product in a food could be a protein, carbohydrate, fat or oil, or other substance that differs significantly in structure, function, or composition from substances found currently in food. Such substances may not be GRAS and may require regulation as a food additive." . . .

This Court's evaluation of the FDA's interpretation of § 321(s) is framed by *Chevron, U.S.A. v. Natural Resources Defense Council*. . . . In other words, "a reviewing court's inquiry under *Chevron* is rooted in statutory analysis and is focused on discerning the boundaries of Congress' delegation of authority to the agency." To resolve the issue, "the question for the reviewing court is whether the agency's construction of the statute is faithful to its plain meaning, or, if the statute has no plain meaning, whether the agency's interpretation 'is based on a permissible construction of the statute.'" If this interpretation is "reasonable and consistent with the statutory scheme and legislative history." If this interpretation is "reasonable and consistent with the statutory scheme and legislative history," then the Court must defer to the agency. This inquiry into the agency's interpretation constitutes *Chevron* step two.

When Congress passed the Food Additives Amendment in 1958, it obviously could not account for the late twentieth-century technologies that would permit the genetic modification of food. The "object and policy" of the food additive amendments is to "require the processor who wants to add a new and unproven additive to accept the responsibility . . . of first proving it to be safe for ingestion by human beings." The plain language of § 321(s) fosters a broad reading of "food additive" and includes "any substance intended for use in producing, manufacturing, packing, processing, preparing, treating, packaging, transporting, or holding food; and . . . any source of radiation intended for any such use."

Nonetheless, the statute exempts from regulation as additives substances that are "generally recognized . . . to be safe under the conditions of its intended use. . . ." Plaintiffs have not disputed FDA's claim that nucleic acid proteins are generally recognized to be safe. Plaintiffs have argued, however, that significant disagreement exists among scientific experts as to whether or not nucleic acid proteins are generally recognized to be safe when they are used to alter organisms genetically. Having examined the record in this case, the Court cannot say that FDA's decision to accord genetically modified foods a presumption of GRAS status is arbitrary and capricious. "The rationale for deference is particularly strong when the [agency] is evaluating scientific data within its technical expertise." "In an area characterized by scientific and technological uncertainty[,] . . . this court must proceed with particular caution, avoiding all temptation to direct the agency in a choice between rational alternatives."

To be generally recognized as safe, a substance must meet two criteria: (1) it must have technical evidence of safety, usually in published scientific studies, and (2) this technical evidence must be generally known and accepted in the scientific community. Although unanimity among scientists is not required, "a severe conflict among experts . . . precludes a finding of general recognition." Plaintiffs have produced several documents showing significant disagreements among scientific experts. However, this Court's review is confined to the record before the agency at the time it made its decision. . . . *Walter O. Boswell Mem'l Hosp. v. Heckler* ("If a court is to review an agency's record fairly, it should have before it neither more nor less information than did the agency when it made its decision."). Therefore, the affidavits submitted by Plaintiffs that are not part of the administrative record will not be considered.

Nonetheless, Plaintiffs, pointing to the critical comments of lower level FDA officials insist that even the administrative record reveals a lack of general recognition of safety among qualified experts. However, lower level comments on a regulation "do[] not invalidate the agency's subsequent application and interpretation of its own regulation." Moreover, pointing to a 44,000 page record, the FDA notes that Plaintiffs have chosen to highlight a selected few comments of FDA employees, which were ultimately addressed in the agency's final Policy Statement. As a result, Plaintiffs have failed to convince the Court that the GRAS presumption is inconsistent with the statutory requirements.

E. Labeling

Plaintiffs have also challenged the Statement of Policy's failure to require labeling for genetically engineered foods, for which FDA relied on the presumption that most genetically modified food ingredients would be GRAS. Plaintiffs claim that FDA should have considered the widespread consumer interest in having genetically engineered foods labeled, as well as the special concerns of religious groups and persons with allergies in having these foods labeled.

The FD&C Act, 21 U.S.C. § 321(n), grants the FDA limited authority to require labeling. In general, foods shall be deemed misbranded if their labeling "fails to reveal facts . . . material with respect to consequences which may result from the use of the article to which the labeling . . . relates under the conditions of use prescribed in the labeling . . . or under such conditions of use as are customary or usual" 21 U.S.C. § 321(n). Plaintiffs challenge the FDA's interpretation of the term "material." Thus, the question is again one of statutory interpretation. As is apparent from the statutory language, Congress has not squarely addressed whether materiality pertains only to safety concerns or whether it also includes consumer interest. Accordingly, interpretation of the § 321(n)'s broad language is left to the agency. . . .

Because Congress has not spoken directly to the issue, this Court must determine whether the agency's interpretation of the statute is reasonable. Agency interpretations receive substantial deference, particularly when the agency is interpreting a statute that it is charged with administering. Even if the agency's interpretation is not "the best or most natural by grammatical or other standards," if the interpretation is reasonable, then it is entitled to deference.

The FDA takes the position that no "material change," under § 321(n), has occurred in the rDNA derived foods at issue here. Absent unique risks to consumer health[43] or uniform changes to food derived through rDNA technology, the FDA does not read § 321(n) to authorize an agency imposed food labeling requirement. More specifically irksome to the Plaintiffs, the FDA does not read § 321(n) to authorize labeling requirements solely because of consumer demand. The FDA's exclusion of consumer interest from the factors which determine whether a change is "material" constitutes a reasonable interpretation of the statute. Moreover, it is doubtful whether the FDA would even have the power under the FD&C Act to require labeling in a situation where the sole justification for such a requirement is consumer demand. . . .

Plaintiffs fail to understand the limitation on the FDA's power to consider consumer demand when making labeling decisions because they fail to recognize that the determination that a product differs materially from the type of product it purports to be is a factual predicate to the requirement of labeling. Only once materiality has been established may the FDA consider consumer opinion to determine whether a label is required to disclose a material fact. Thus, "if there is a [material] difference, and consumers would likely want to know about the difference, then labeling is appropriate. If, however, the product does not differ in any significant way from what it purports to be, then it would be misbranding to label the product as different, even if consumers misperceived the product as different." The FDA has already determined that, in general, rDNA modification does not "materially" alter foods, and as discussed in Section II.E, *supra*, this determination is entitled to deference. Given these facts, the FDA lacks a basis upon which it can legally mandate labeling, regardless of the level of consumer demand.

Plaintiffs also contend that the process[44] of genetic modification is a "material fact" under § 321(n) which mandates special labeling, implying that there

[43] In other contexts, the FDA has identified that the presence of an increased risk to consumer safety constitutes a "material change." *See, e.g.*, 49 Fed. Reg. 13,679 (pertaining to FDA requirement in 21 C.F.R. § 101.17(d)(1) that a special warning statement appear on the label of protein products intended for use in weight reduction due to health risks associated with very low calorie diets). Likewise, should a material consequence exist for a particular rDNA-derived food, the FDA has and will require special labeling. *See* Foods Derived from New Plant Varieties, 57 Fed. Reg. 22,984, 22, 991 (May 29, 1992) (discussing the requirement that laureate canola and high-oleic acid soybean oil have special labeling because they differ in composition and use from the traditional canola and soybean oil.). However, the Policy Statement at issue here provides only a very general rule regarding the entire class of rDNA derived foods. Thus, without a determination that, as a class, rDNA derived food pose inherent risks or safety consequences to consumers, or differ in some material way from their traditional counterparts, the FDA is without authority to mandate labeling.

[44] Disclosure of the conditions or methods of manufacture has long been deemed unnecessary under the law. The Supreme Court reasoned in 1924, "When considered independently of the product, the method of manufacture is not material. The act requires no disclosure concerning it." *U.S. v. Ninety-Five Barrels (More or Less) Alleged Apple Cider Vinegar*, 265 U.S. 438 (1924) (referring to the Food and Drug Act of June 30, 1906, 34 Stat. 768 (1906), precursor to the FDCA).

are new risks posed to the consumer. However, the FDA has determined that foods produced through rDNA techniques do not "present any different or greater safety concern than foods developed by traditional plant breeding," and concluded that labeling was not warranted. That determination, unless irrational, is entitled to deference. Accordingly, there is little basis upon which this Court could find that the FDA's interpretation of § 321(n) is arbitrary and capricious.

F. Free Exercise

Plaintiffs have argued that the Statement of Policy unconstitutionally violates their right to free exercise of religion by allowing unlabeled genetically engineered foods on the market. Under the Supreme Court's decision in *Employment Division v. Smith*, 494 U.S. 872 (1990), however, neutral laws of general applicability do not violate the Free Exercise Clause, even if the laws incidentally burden religion. Because it is not disputed that the Statement of Policy is neutral and generally applicable, Plaintiff's Free Exercise Claim must fail. . . .

G. Religious Freedom Restoration Act

Plaintiffs also claim that the Statement of Policy burdens their religion in violation of the Religious Freedom Restoration Act (RFRA). Congress enacted RFRA in reaction to the *Employment Division v. Smith* decision in order to "restore the compelling interest test" for Free Exercise issues. RFRA's definition of the compelling interest test provides that "government shall not substantially burden a person's exercise of religion even if the burden results from a rule of general applicability . . . [unless the rule is] (1) in furtherance of a compelling governmental interest; and (2) is the least restrictive means of furthering that compelling governmental interest." This test is not to be "construed more stringently or more leniently than it was prior to *Smith*."

. . .

Defendants concede that RFRA applies to the FDA. Assuming *arguendo* that Plaintiffs meet the RFRA requirement that their beliefs are sincerely held and can demonstrate an "honest conviction" desiring to avoid genetically engineered foods, Plaintiffs still must establish that Defendants have substantially burdened Plaintiffs' religion. A substantial burden does not arise merely because "the government refuses to conduct its own affairs in ways that comport with the religious beliefs of particular citizens." The Free Exercise Clause (as interpreted before *Smith* and incorporated into RFRA) does not require the government to take action to further the practice of individuals' religion. Indeed, were the government to take such action, it might bring itself precariously close to violating the First Amendment's Establishment Clause.

Arguing that the government does have some obligation to facilitate the practice of religion, Plaintiffs point to several cases involving prisoners, in which the government was required to provide nutritional information and alternative diets for inmates whose religious beliefs required dietary restrictions. . . . However, the prisoner cases cited by the Plaintiffs are inapposite to the issue before this Court. In this case, the Plaintiffs' liberty is not restricted and they are free to choose their food and may obtain their food from the source of their choosing.

Still, Plaintiffs argue that in the absence of labeling they are unable to know whether the foods they consume are genetically engineered or not. While the Court recognizes the potential inconvenience the lack of labeling presents for Plaintiffs, Defendant's decision to mandate labeling of genetically modified foods does not "substantially" burden Plaintiffs' religious beliefs. Furthermore, given that the FDA functions under statutory power granted by Congress and cannot exceed that power, Plaintiffs' argument on this point is probably better directed at Congress, than at the Defendant or this Court.[45] The Policy Statement does not place "substantial pressure" on any of the Plaintiffs, nor does it force them to abandon their religious beliefs or practices. Accordingly, Plaintiffs are not entitled to relief under RFRA.

Conclusion

For the foregoing reasons, the Court determines that Defendant's 1992 Policy Statement did not violate the Administrative Procedures Act, the National Environmental Policy Act, or the procedures mandated by the FD&C Act and FDA regulations. Furthermore, Defendant was not arbitrary and capricious in its finding that genetically modified foods need not be labeled because they do not differ "materially" from non-modified foods under 21 U.S.C. § 321(n). Finally, the Court finds that Defendant's Policy Statement does not violate the First Amendment Free Exercise Clause or RFRA, 42 U.S.C. § 2000bb-1(b). Hence, the Court denies Plaintiffs' motion for summary judgment and grants Defendant's motion for same. . . .

* * * * *

NOTES AND QUESTIONS

9.5. Benefits of GE crops. "Over the next half century genetic engineering could feed humanity and solve a raft of environmental ills—if only

---

[45] On November 16, 1999, the Genetically Engineered Food Right To Know Act was introduced in the House of Representatives. The bill proposed amending FDCA, the Federal Meat Inspection Act, and the Poultry Products Inspection Act to require that food that contains a genetically engineered material, or that is produced with a genetically engineered material, be labeled accordingly. *See* H.R. REP. No. 106-3377 (1999).

environmentalists would let it," Jonathan Rauch, *Will Frankenfood Save the Planet? Atlantic Monthly*, Oct. 2003. As the world population is growing, the demand for food is increasing. At the same time available arable land is decreasing. "We will not be able to feed the people of this millennium with the current agricultural techniques and practices. To insist that we can is a delusion that will condemn millions to hunger, malnutrition and starvation, as well as to social, economic and political chaos," Norman Norman, Op-Ed, *We Need Biotech to Feed the World*, WALL STREET JOURNAL, Dec. 6, 2000. Food quality is as much an issue as is food quantity. Energy from food may be available, but essential nutrients may be lacking. Biotechnology holds potential to help meet this important need by speeding the development of nutritionally improved cultivars. *See generally*, *Pew Initiative on Food and Biotechnology, Feeding the World: A Look at Biotechnology and World Hunger* (2004), *available at*: http://www.pewtrusts.org/uploadedFiles/wwwpewtrustsorg/Reports/Food_and_Biotechnology/pew_agbiotech_feed_world_030304.pdf (last accessed Mar. 4, 2008).

9.6.   **Flavr Savr deemed no different than conventional varieties**. The first genetically engineered food crop sold at retail in the United States was Calgene's Flavr Savr Tomato in 1994. The tomato was engineered by reinserting a tomato gene backward to create a longer shelf life. FDA did not require special labeling for the Flavr Savr tomato because the new tomato was deemed not significantly different from the range of commercial varieties. Do you agree or disagree with FDA's assessment? Why?

9.7.   **Voluntary disclosure of genetic engineering**. The FDA did not require special labeling for the Flavr Savr tomato. However, Calgene decided to provide special labeling, including point of sale information, to inform consumers that the new tomato has been developed through genetic engineering. Why do you think Calgene voluntarily disclosed their use of genetic engineering? Can you think of an advantage to Calgene of disclosing use of genetic engineering? For an insider's narrative of the development and failure of Flavr Savr, see Belinda Martineau, *First Fruit: The Creation of the Flavr Savr Tomato and the Birth of Biotech Foods* (2001).

9.8.   **The double standard for the risks of GE products**. Pioneer-Hybrid's Smart canola and other herbicide-resistant resistant varieties of canola have been produced through conventional breeding. These conventionally bred plants contain the same herbicide-resistant traits as genetically engineered canola. Thus the potential adverse environmental and health effects are the same (the risk for both is essentially nil). If the legitimate risks are the same, why is Smart canola allowed to be sold without special scrutiny or labeling but GE canola often prohibited?

9.9. Human genes inserted in other organisms. Since the late 1970s nearly all pharmaceutical insulin has been produced through biotechnology. The human gene for insulin was inserted into bacteria, and these bacteria produce human insulin. Before the 1970s, insulin was produced from cattle, pigs, and sheep. Thus GE insulin spares the sacrifice of untold numbers of animals. In addition the GE insulin is better for humans than cattle, pigs, and sheep insulin because the GE insulin is a better fit, being after all, human insulin. Do you find an ethical dilemma with inserting human genes in other organisms?

9.10. Homology of genes. All organisms share the same genetic language, the common language of DNA. This is why a human gene can be read and understood by a human cell, or even a plant cell. All organisms will read a DNA sequence, produce the same amino acid sequence, and produce in the same protein. Many of the biochemical processes, such as metabolism, energy use, and DNA manufacture, are shared among all animals, and many are shared among plants and animals. Accordingly, many of the genes controlling these processes are similar or identical. For example, the nematode *Caenorhabditis elegans* shares almost 7,000 genes with humans (out of 20,000 total). GE food producers clearly must be concerned about the religious and ethical preferences of consumers, when, for example, they transfer a pig gene into another food source. But what if a gene transferred from another species is homologous with a pig gene? For instance, what if a soybean gene that is homologous to a pig gene is transferred to wheat? For a detailed but readable explanation of the science of biotechnology, *see* Alan McHughen, *Pandora's Picnic Basket* (2000).

9.11. Should GE medicine be labeled. Nearly all diabetic persons who require insulin injects themselves with GE insulin. Should GE insulin be required to be labeled as such? Why or why not?

9.12. Should precaution apply equally to medicines and foods. Should the precautionary principle apply equally to medicines and foods? Why or why not? Are we more willing to accept a risk-benefit balance with medicine as opposed to foods? If yes, why?

9.13. If GE bananas cured baldness. One observation on consumer resistance on GMOs is that the first generation of GE products benefits only large corporations, and there is no benefit to consumers. If a banana were genetically engineered so that it cured male pattern baldness, would bald men be more accepting of GMOs?

CHAPTER 10

Food Terrorism

On September the 11th, the world learned how evil men can use airplanes as weapons of terror. Shortly thereafter, we learned how evil people can use microscopic spores as weapons of terror. Bioterrorism is a real threat to our country. It's a threat to every nation that loves freedom . . . Biological weapons are potentially the most dangerous weapons in the world. . . .

—President George W. Bush[1]

In response to the attacks of September 11, Congress sought to further safeguard the nation's food supply. In 2002, Congress passed a number of changes to the powers federal agencies, including the FDA and USDA under the Public Health Security and Bioterrorism Preparedness and Response Act of 2002, commonly called the Bioterrorism Act.[2] Under the Act, several new authorities and requirements were intended to protect the food supply.

10.1 THE U.S. FOOD SAFETY SYSTEM

Management of terroristic food contamination builds on the normal food safety system. Both unintentional and deliberate (criminal) foodborne diseases outbreaks can be managed by the same system. In particular, strong foodborne disease surveillance and response capacity is essential for responding to food terrorism—just as it is for an unintentional foodborne outbreak.[3]

Both FDA and USDA FSIS have important roles to play in ensuring a safe human food supply. They perform risk assessments on potential contamination, perfect testing procedures, and provide oversight at the borders. The USDA Animal and Plant Health Inspection Service (APHIS) has special on-farm responsibilities to analyze potential unusual incidences.

---

[1] Remarks by the President at Signing of H.R. 3448, the Public Health Security and Bioterrorism Response Act of 2002, in the Rose Garden (June 12, 2002), text and video *available at*: http://www. whitehouse.gov/news/releases/2002/06/20020612-1.html (last accessed Mar. 4, 2008).

[2] Pub. L. No. 107-188, 116 Stat. (2002).

[3] Food Safety Dep't., World Health Organization, TERRORIST THREATS TO FOOD: GUIDANCE FOR ESTABLISHING AND STRENGTHENING PREVENTION AND RESPONSE SYSTEMS 1 (2002), *available at*: http://www.who.int/foodsafety/publications/general/en/terrorist.pdf (last accessed Mar. 4, 2008).

Food Regulation: Law, Science, Policy, and Practice, by Neal D. Fortin
Copyright © 2009 Published by John Wiley & Sons, Inc.

The Centers for Disease Control and Prevention (CDC) works at scanning and analyzing outbreaks, works with state and local health departments, and would contribute epidemiological work should there be an outbreak. CDC also works with professionals in other countries and within the UN system when an outbreak occurs.

The Environmental Protection Agency (EPA) plays a role in certain aspects of food safety, especially with pesticide residues. The agency would also play a key role in any necessary cleanup of a contamination. In an actual event of bioterrorism, the President of the United States has stated that the controlling investigating authority is the Federal Bureau of Investigation (FBI) within the U.S. Department of Justice.

Other groups within the U.S. government conduct defensive research, including the National Institutes of Health (NIH) and the USDA Research, Education and Economics (REE) area. The U.S. Department of Defense, of course, must protect its own employees from food and water contamination, and it conducts research for that purpose. The Department of Homeland Security (DHS) has certain responsibilities at the border and in testing that relate to food and bioterrorism. To help on research, the system of land grant universities are funded through existing research authorities and research programs to protect the food supply of the United States by research activities to enhance response to emerging or existing terrorist threats to the food and agricultural system. These monies flow, for the most part, from REE at USDA.

In addition, states and local governments have responsibility in keeping the food supply safe and in identifying a disease outbreak. Responsibilities are distributed among various of departments in the states. Many states have resources and legal enforcement authorities that are not available to federal authorities. For instance, some states have recall authority to take control of food in the event of contamination. State veterinarians often have strong authorities and responsibilities to control and quarantine animal populations. Further, the state public health and epidemiology departments have communicable disease control authority.

The U.S. food safety system also is predicated on strong self-enforced compliance by food companies, distributors, wholesalers, and retailers. Food safety, whether routine or arising from terrorist activity, is best achieved through a cooperative effort between government and industry. This becomes more apparent when one realizes that the primary means for minimizing food risks lie with the food industry. A multi-stakeholder approach is essential for strong disease outbreak surveillance, investigation capacity, preparedness planning, effective communication, and response.[4]

In the case of terrorism the private sector is placed in a very difficult situation, not unlike the position in which they find themselves when subjected to criminal extortion threats. Their security personnel, human resources execu-

---

[4] *Id.*

tives, and scientific and quality assurance and food safety staffs may come under intense pressure and scrutiny as a situation escalates.

10.2 THE THREAT

> I, for the life of me, cannot understand why the terrorists have not . . . attacked our food supply because it is so easy to do.—Secretary of the Department of Health and Human Services, Tommy Thompson.[5]

The World Health Organization (WHO) has noted that "malicious contamination of food for terrorist purposes is a real and current threat" and such deliberate contamination could have global public health implications.[6] Unfortunately, there is abundant evidence that rogue states have conducted extensive research on terroristic methods, and there is evidence of the use of various chemical and pathogenic contaminants. Criminal events or extortion have been a periodic occurrence in the food arena for many years. Now the toxins or pathogens used in those events may be turned to destabilize confidence in the government, business, and in the food supply itself.

Among a long list of potential contaminants are ricin, anthrax, botulism toxin, and highly infectious diseases, such as glanders and tularemia.[7] Bioterrorism could also be targeted to do grave economic harm to crops or livestock. The former USSR bioweapons program did research on wheat rust and rice blast among other plant contaminants.

Some terrorist acts may be announced. Others may cause outbreaks of unknown origin. Some perpetrators may take credit in order to sow terror and sap confidence. Other perpetrators (especially those who may not want massive retaliation) may prefer to plan an event without credit, making it more difficult to diagnose, identify, and control the outbreak.

For instance, an outbreak of foot and mouth disease might be accidental or deliberate. In the immediate management of the outbreak, planned containment efforts would be implemented. However, it would be important to determine whether an outbreak of foot and mouth disease was naturally occurring or deliberate because a deliberate introduction would not necessarily follow a classic disease pattern and would not be as easily contained. If the perpetrator announces the event, responders may be helped to limit the outbreak and permit a more immediate response.

10.2.1 Illness and Death

The potential for foodborne pathogens to impact human health can be observed with unintended outbreaks. For instance, a 1994 outbreak of salmonellosis

[5] Mike Allen, *Rumsfeld to Remain at Pentagon; Thompson Quits HHS, Warns of Vulnerabilities*, Washington Post (Dec. 4, 2004) at A1.
[6] Food Safety Dep't., WHO, Terrorist Threats to Food, *supra* note 3.
[7] *See* CDC, Bioterrorism Agents/Diseases: http://www.bt.cdc.gov/agent/agentlist.asp (last accessed Mar. 5, 2008).

from *Salmonella enteritidis* contaminated ice cream caused 224,000 illnesses in 41 states.[8] An unintended 1991 outbreak of hepatitis A from consumption of clams in Shanghai, China, affected nearly 300,000 people.[9] The potential impact from deliberate food contamination clearly could be devastating. The potential effects of a terrorist act must be taken seriously.

10.2.2   Economic and Trade Effects

Food contamination can have enormous economic implications, and economic disruption may be a motive for a deliberate act. Contamination of Chilean grapes with cyanide in 1989 led to the recall of all Chilean fruit in Canada and the United States. Although no one was injured by the cyanide (no one was even know to have ingested any), the adverse publicity surrounding this incident resulted in damage amounted to several hundred million dollars and more than 100 growers and shippers going bankrupt.[10]

10.2.3   Social and Political Implications

Just as the diverse nature of the U.S. food supply makes it difficult to make it totally safe from terrorism, this diversity makes contamination of the entire food supply unlikely. However, the motive of terrorists might be political destabilization. The goal may be to create panic and destabilize the civil order.[11] Fear and anxiety may be caused by even a low number of illnesses. Mailings of envelopes containing anthrax in the United States resulted in considerable disruption and public anxiety with relatively limited dissemination of a biological agent.[12]

Even a relatively small outbreak may overwhelm local health resources and add to the public anxiety. When ten restaurant salad bars were intentionally contaminated in rural Oregon in 1984, the salmonellosis outbreak overwhelmed the local hospital. For the first time, all 125 hospital beds were filled, some patients had to be left in hallways, and many were angry and frightened, and some became violent, throwing urine and stool samples at hospital staff.[13]

---

[8]Thomas Hennessy et al., *A National Outbreak of* Salmonella enteritidis *Infections from Ice Cream,* 334 NEW ENGLAND JOURNAL OF MEDICINE 1281–6 (May 16, 1996).
[9]M. L. Halliday et al., *An Epidemic of Hepatitis A Attributable to the Ingestion of Raw Clams in Shanghai, China,* 164(5) JOURNAL OF INFECTIOUS DISEASES 852–9 (Nov. 1991).
[10]Robert S. Root-Bernstein, *Infectious Terrorism,* ATLANTIC MONTHLY (May 1991) (citing Raymond Zilinskas, of the Center for Public Issues in Biotechnology, in Perspectives in Biology and Medicine).
[11]Food Safety Dep't., WHO, TERRORIST THREATS TO FOOD, *supra* note 3 at 7.
[12]Jeremy Sobel, Ali S. Khan, and David L. Swerdlow, *Threat of Biological Terrorist Attack on the US Food Supply: the CDC Perspective,* 359 LANCET 874–880 (2002).
[13]JUDITH MILLER, STEPHEN ENGELBERG, AND WILLIAM BROAD, GERMS 19–20 (2001).

10.3   EXAMPLES OF CONTAMINATION

10.3.1   A *Salmonella* Tainted Election

Bhagwan Shree Rajneesh, a self-proclaimed guru, moved to rural Oregon in 1981. His followers took political control of the small nearby town of Antelope and renamed it Rajneesh. Then in 1984 his followers planned to run the whole county by stealing the local election.[14] Their plan was to make non-Rajneesh followers too sick to vote.[15]

The cult members spiked salad bars at ten restaurants in the nearby town of The Dalles with *Salmonella typhimurium*. Confirmed cases of salmonellosis numbered 751.[16] The salad bars were reportedly a test run, and the plan was to contaminate the local water supply at election time. Apparently the cult members did not expect such a large success because the outbreak brought in hordes of health officials and investigators from local, state, and national agencies.[17]

Within 48 hours of the outbreak, *Salmonella* was isolated from a patient's stool sample. Within another 48 hours, the germ was identified as *Salmonella typhimurium*. Although health officials were quick to pin down the bacterial agent and quick to shut down the salad bad, they put the blame on unintentional contamination by food handlers.[18]

In hindsight, there were clues that this was not a typical foodborne disease outbreak. For instance, it was an unusual strain of *Salmonella typhimurium*, and there was no common source for the food. However, unintentional contamination by food handlers could not be ruled out, and this was not a surprising conclusion for 1984. "It's rare that we have had to invoke evil intentions when bad luck and stupidity are usually to blame," Robert Tauxe, a CDC scientist, noted.[19]

It was more than a year before the truth came out. The Bagwan held a press conference and accused a cult member and her allies for the contamination of the salad bars (and other offenses).[20] The sabotage involved low-technology application of *Salmonella typhimurium*. Fortunately, no one died, and the cult's favored candidates did not win the election.

The lessons learned from this event were several. Attribution of unannounced sabotage can be difficult, perhaps particularly difficult for trained epidemiologists. The tools of epidemiology are not designed for criminal

---

[14] Lawrence K. Grossman, *The Story of a Truly Contaminated Election*, Columbia Journalism Review (Jan.–Feb. 2001)

[15] Miller et al., Germs (2001) (*see generally,* chapter 1, "The Attack").

[16] Thomas J. Torok, Robert V. Tauxe, et al., *A Large Community Outbreak of Salmonellosis Caused by Intentional Contamination of Restaurant Salad Bars,* 278 JAMA 389–395 (Aug. 6, 1997).

[17] Grossman, *supra* note 14.

[18] Miller, *supra* note 15.

[19] *Id.*

[20] *Id.* at 23.

investigations. Psychological denial among the scientific community and regulators delayed a full investigation. Response to criminal acts of food contamination brings together the criminal justice system and the public health system. Officials in the respective agencies may not be accustomed to working together, and approaches and techniques may be at odds with each other. For example, police agencies may wish to keep details of the incident secret to assist the investigation and to prevent copycat incidents. On the other hand, epidemiological and public health response requires public notification of details of the outbreak.

NOTES AND QUESTIONS

10.1. Tampering. The Federal Anti-Tampering Act of 1983 makes it a crime to tamper with any packaged consumer product. See the discussion on food safety in Chapter 6 for further information on the Federal Anti-tampering Act of 1983.

10.2. Additional reading. JUDITH MILLER et al., GERMS, chapter 1, "The Attack" (2001); *and* KEN ALIBEK WITH STEPHEN HANDELMAN, BIOHAZARD, Delta (2000).

10.3. Confidentiality. Access to classified or secret information may inform government decisions. This secrecy factor can make it more difficult for food regulators and company managers (without such access) to predict, prevent, and control potential terrorism in the food supply.

10.3.2   Pesticide Poisoning in Michigan

The criminal use of pesticide to contaminate and hurt consumers in Michigan is a ready example of the types of challenges facing response to terrorist criminals.

* * * * *

Nicotine Poisoning after Ingestion of Contaminated Ground Beef— Michigan, 2003

CDC, MMWR, MORBIDITY AND MORTALITY WEEKLY REPORT, Vol. 52, No. 18; 413; (May 9, 2003).

On January 3, 2003, the Michigan Department of Agriculture's (MDA) Food and Dairy Division and the U.S. Department of Agriculture (USDA) were notified by a supermarket of a planned recall of approximately 1,700 pounds of ground beef because of customer complaints of illness after eating the product. On January 10, the supermarket notified MDA that their laboratory had determined that the contaminant in the ground beef returned by customers with reported illness was nicotine. . . .

The recall was prompted by complaints from four families comprising 18 persons who became ill immediately after eating product sold on December 31 or January 1. Reported symptoms included burning of the mouth, nausea, vomiting, and dizziness. . . . MDA made routine notifications about the recall to local and state health departments. The product recall was issued on January 3 . . . followed by a press release. . . . After the initial recall notices, approximately 36 persons reported to the supermarket that they or their families had experienced illness after eating the product, and approximately 120 persons returned recalled product.

Company officials submitted samples of ground beef provided by the ill families to a private laboratory, where product testing for foodborne pathogens was negative. Additional testing for chemical contamination was conducted at a large regional medical center. On January 10, company officials notified MDA and USDA that nicotine had been presumptively identified in the ground beef samples tested by the second laboratory, which reported an assay result 1 week later of approximately 300 mg/kg nicotine in the submitted samples. The high nicotine concentrations found in the tested meat products prompted concerns of intentional contamination with a pesticide, which sometimes contains nicotine as an additive. USDA and the Federal Bureau of Investigation joined the investigation because interstate commerce could have been involved and intentional contamination was suspected. Because a legal investigation was initiated, federal authorities requested that information be released to the public only as necessary to avoid compromising any future criminal case. On January 17, the supermarket issued another press release and recall notice stating the implicated product contained an unspecified, nonbacterial contaminant that could not be made safe by cooking.

Contamination of the product was believed to have occurred at a single store rather than the meat-processing plant. The product was distributed directly from the plant to many other stores, including other stores in the supermarket chain; neither the processing plant nor any other store in the supermarket chain received complaints of illness. No nicotine-containing pesticides were reportedly used or sold in the store where the recalled product was sold.

On January 23, the local health department alerted hospital EDs and selected medical practices serving the area where the store was located. On January 24, after receiving confirmatory test results, the company issued another press release naming nicotine as the contaminant. This announcement was published and broadcast by local media.

The local health department conducted an epidemiologic investigation, including interviews of persons reporting illness, to assess the consistency of the clinical presentation and to establish a case definition. A case was defined as one or more symptoms (i.e., burning sensation to lips, mouth or throat, dizziness, nausea, vomiting, abdominal pain, diarrhea, sweating, blurred vision, headache, body numbness, unusual fatigue or anxiety, insomnia, tachypnea or dyspnea, and tachycardia or tachyarrythmias) in persons who ate ground beef product purchased from the supermarket on either December 31, 2002, or January 1, 2003, with symptom onset occurring within 2 hours of eating the product.

A total of 148 interviews were conducted with persons who reported they had experienced illness after eating the product and of family members and friends who also might have eaten the contaminated meat. Of those interviewed, 92 persons had illness consistent with the case definition. Patients had a median age of 31 years (range: 1–76 years), and 46 (50 percent) were female; 65 percent of the patients lived in the town where the implicated store was located. The majority of illness occurred during the time that the contaminated product was sold. Cases were identified as late as 49 days after the last date of potential sale, indicating that some persons froze and then ate the contaminated product after the first recall was issued. Of the 92 patients, four (3 percent) sought medical treatment, including two who reported to their personal physicians with complaints of vomiting and stomach pains and two who were evaluated in EDs. The two who were treated in the EDs included a man aged 39 years with atrial fibrillation and a woman aged 31 years who had nausea, vomiting, and complaint of rectal bleeding. Information is being collected on an additional 16 persons to assess whether their illnesses are consistent with the case definition, including a pregnant woman aged 24 years who was hospitalized for 1 day with episodic vomiting.

On February 12, a grand jury returned an indictment for arrest of a person accused of poisoning 200 pounds of meat at the supermarket with an insecticide called Black Leaf 40, which has a main ingredient of nicotine. The person was an employee of the supermarket at the time of the contamination. . . .

Editorial Note:

Deliberate contamination of food during its production and preparation has been reported infrequently. . . .

This investigation involved the private sector (i.e., the food retailer) and five government agencies, including local and state public health departments, the state agriculture department, and two federal agencies. Public health officials undertook an epidemiologic investigation that involved contacting affected persons and providing information to the public and clinicians about the health threat. It also was necessary to conduct a legal investigation in a rapid and relatively closed manner. Frequent contacts among the parties allowed for negotiation and consensus around most issues.

This incident underscores the importance of ensuring the safety and security of food supplies. Vigilance and heightened awareness for human poisonings caused by hazardous levels of chemical in the food supply are essential.

Clinicians should immediately report clusters of poisonings to public health officials, especially when presenting symptoms are unusual. Public health response capabilities addressing hazardous chemicals in food and other media need to be strengthened. Multiple agency coordination and cooperation of health, agriculture, and law enforcement officials at the local, state, and federal levels are critical for the detection and response to similar events, whether they are intentional or unintentional.

* * * * *

10.4 THE BIOTERRORISM ACT AND FDA'S NEW POWERS

The Bioterrorism Act[21] is a comprehensive law regarding public health, water, food, and some agriculture issues. It also contains requirements for interdepartmental coordination both in the development of countermeasures, response and training and research arenas and in the risk communications arena. Under the Act, FDA received several new authorities (some it had been seeking for many years):

Administrative detention. FD&C Act section 304 was amended to give FDA the authority to place articles of food under temporary detention if the agency believes the food "presents a threat of serious adverse health consequences or death to humans or animals."

Food facilities registration. FD&C Act section 415 was added to require registration of all food facilities, foreign and domestic.

Authority to require records. FD&C Act section 414 was added to authorize FDA to require establishment and maintenance of records relating to the manufacture and distribution of food.

Authority to inspect records. FD&C Act section 414 also gives FDA the authority to inspect and copy records relating to foods, if the agency has a reasonable belief that the food is adulterated and would have serious adverse health consequences to humans or animals.

Prior notice of imported food shipments. FD&C Act section 801 was amended to require that FDA be given prior notice of imported food shipments.

In these new powers, FDA's authority is not restricted to problems caused by bioterrorism. The new powers may be applied to more routine situations as well.

---

[21] A copy of the act and other information on its requirements is *available at*: http://www.fda.gov/ oc/bioterrorism/bioact.html (last accessed Mar. 4, 2008).

10.4.1 Food Facility Registration

Registration with FDA is required of all domestic and foreign food facilities that manufacture, process, pack, or hold food for human or animal consumption in the United States, unless the facility is exempt. Facilities exempt from registration include farms, retail food establishments whose principal business is selling food directly to consumers, restaurants; private residences, transport vehicles that hold food only in the usual course of business, nonprofit food establishments, fishing vessels not engaged in processing, and facilities regulated exclusively throughout the entire facility by the USDA. Foreign facilities are exempt from registration if food from the facilities undergoes further manufacturing, processing or packaging (of more than a de minimis nature) at another foreign facility before it is exported to the United States. Essentially, the last foreign facility that manufactures, processes, or packages the food, and any subsequent foreign facility that packs or holds the food, must register.

Foreign facilities required to register must also designate a U.S. agent who must live or maintain a place of business in the United States and be physically present in the United States. The registration requirement requires registration of covered facilities, not companies. A single company may be required to register multiple facilities.

If a facility is required to register, it is a violation of federal law to fail to register. If a foreign facility required to register fails to do so and food from that facility is offered for import into the United States, the Bioterrorism Act requires that the food be held at the port of entry.

10.4.2 Prior Notice of Food Imports

The prior notice regulations require advance notice to FDA of any shipment of human or animal food imported or offered for import into the United States. These notification requirements are in addition to, but to be incorporated with, existing notification procedures importers now follow when reporting their food imports—FDA's Operational and Administrative System for Import Support (OASIS) system and the Bureau of Customs and Border Protection's (Customs) Automated Broker Interface (ABI) of the Automated Commercial System (ACS). Prior notice must be received and confirmed electronically by FDA no more than five days before arrival. The prior notice must be submitted to FDA electronically.

10.4.3 New Penalties

New Animal Enterprise Terrorism Penalties USDA and its agencies already had extensive authorities for inspection and enforcement. However, the Bioterrorism Act added "Terrorism Penalties" to travel in interstate or foreign commerce or use or cause to be used the mail or any facility in interstate or foreign commerce for the purpose of causing physical disruption to the functioning of an animal enterprise. The penalties are graduated according to the seriousness of the economic damage and the bodily injury, and they range from fines to life imprisonment.

Debarment   The existing debarment provisions of FD&C Act section 306 were expanded by the Bioterrorism Act to included repeated or serious food import violations.

10.5   CONCLUSIONS

Potential threats must be understood as well as possible and accepted so that actions to diminish potential vulnerabilities move forward. Enhancement of existing food safety programs and implementation of reasonable security measures based on vulnerability assessments are key to prevention. Prevention is the first line of defense against of a food terrorism event, although prevention can never be completely effective. The widespread nature of the food sources and the global marketplace make complete prevention of deliberate sabotage impossible.

Preparing and coordinating the entire food chain for containment and diminished threat is important. Integrating the resources in both the public and private sectors from the senior management level, through the level of central scientific expertise and on to the field or line operators or inspection personnel, is critical. From farm to table, each segment must work to understand the role of the other in order to contain or manage an event.

NOTES AND QUESTIONS

10.4. Additional information. FDA's counterterrorism information Web page: http://www.fda.gov/oc/opacom/hottopics/bioterrorism.html (last accessed Mar. 4, 2008). Countering Bioterrorism and Other Threats to the Food Supply, *available at*: http://www.foodsafety.gov/~fsg/bioterr. html (last accessed Mar. 4, 2008). USDA FSIS, Food Defense and Emergency Response: http://www.fsis.usda.gov/Food Defense & Emergency Response/index.asp. Michelle Meadows, *The FDA and the Fight Against Terrorism*, FDA CONSUMER (Jan.–Feb. 2004), *available at*: http://www.fda.gov/fdac/features/2004/104 terror.html.

CHAPTER 11

Importation and Exportation

11.1 INTRODUCTION

United States law requires all imported foods meet the same food safety standards as foods produced in the United States. The Federal Food, Drug, and Cosmetic Act[1] (FD&C Act) and other laws that are designed to protect consumers' health and safety apply equally to domestic and imported products. For instance, imported foods must be pure, wholesome, safe to eat, and produced under sanitary conditions. Additionally all foods must bear the same informative and truthful labeling in English.

The Food and Drug Administration (FDA) and the United States Department of Agriculture (USDA) Food Safety Inspection Service (FSIS) share primary responsibility for ensuring the safety of food imported into the United States. FSIS has responsibility over meat, poultry, and some egg products. FDA regulates all other foods.

FSIS inspects each shipment of meat, poultry, and egg products imported to the United States. In addition FSIS is required to determine that the exporting country has a food safety inspection system for the products that is equivalent to the U.S. system.

In contrast, FDA lacks the statutory authority to impose an equivalency requirement for importation of FDA-regulated foods. The FDA generally must rely on inspections at the U.S. ports of entry to determine the safety of the imported foods.

This simple overview, however, masks a relatively complex set of interconnected regulations enforced by a number of different agencies. There are a number of redundancies in the import procedures because of overlapping requirements for specific food products. Moreover the basic structure of import regulation for most food was put in place in 1906 and the legal authorities and resources have not kept pace with the changes due to the globalization of the food supply and increased imports.

---

[1] Pub. L. No. 75-717, 52 Stat. 1040 (1938), as amended, 21 U.S.C. §§ 301–397 (2000).

Food Regulation: Law, Science, Policy, and Practice, by Neal D. Fortin
Copyright © 2009 Published by John Wiley & Sons, Inc.

11.2   THE MAJOR FEDERAL AGENCIES

To gain an understanding of the food safety regulatory system for imported foods, first one must sort through the alphabet soup of agencies that enforce the regulations. Eight federal agencies play a major role in the regulation of imported foods:

FDA—United States Food and Drug Administration
USDA—United States Department of Agriculture
  • APHIS—Animal and Plant Health Inspection Service
  • FSIS—Food Safety Inspection Service
  • NCIE—National Center for Import and Export (NCIE) Veterinary Services
CBP (Customs)—Bureau of Customs and Border Protection
EPA—Environmental Protection Agency
TTB—Alcohol and Tobacco Tax and Trade Bureau
NMFS—National Marine Fisheries Service

Both FSIS and FDA depend on working closely with the Bureau of Customs and Border Protection (CBP or "Customs").[2] Customs notifies FSIS and FDA of imported foods for the agencies' review. Customs holds imported food from commerce until the release by FSIS or FDA. Another major responsibility of Customs is to administer the Tariff Act of 1930, and assess and collect all duties, taxes, and fees on imported merchandise. The agency also administers and reviews import entry forms.

State agencies also play a role in import regulations. Imported product must conform to all the requirements in each of the 50 states where the product is sold—in addition to the federal laws of the United States. Fifty plus sets of differing regulations could be an immense burden to commerce, but generally, most state requirements are consistent with the federal requirements. One notable exception is California's Prop 65, which requires special warning statements on many products. Other exceptions exist but generally are smaller in scope and apply to a limited category of foods. For example, Michigan law requires a "last date of sale" on certain perishable foods, while the federal law is silent in this area.

11.3   THE FDA IMPORT PROCESS

FDA regulates the importation of most foods other than meat, poultry, and some egg products. FDA's authority derives from section 801 of the Food Drug

---

[2] FDA Memorandum of Understanding with Customs Service, 44 Fed. Reg. 53577 (Sept. 14, 1979).

and Cosmetic Act (FD&C Act), which authorizes FDA examination of foods, drugs, cosmetics, and medical devices offered for entry into the United States. This authority was largely put in place in 1906, when Congress passed the Pure Food and Drug Act of 1906. The provisions were carried over when the FD&C Act was enacted in 1938.

Foods may be imported into the United States if they meet the same standards as those foods that are produced domestically. However, imported food faces significant procedural and legal hurdles that are higher than domestic food products face. In particular, the standard for import denial is the product appears to be adulterated or misbranded, while domestic goods cannot be condemned unless they actually are shown to be adulterated or misbranded.[3] This creates a daunting standard of proof for an importer who wishes to challenge the FDA's determination.[4]

In addition importers face fewer constitutional protections than owners of food already in United States commerce.[5] For example, condemnation of a domestic food for adulteration deprives the owner of value, but import denial is not the taking of the importer's property.[6] There is no constitutional right to import goods into the United States, and due process protections apply only after the food enters United States commerce. Moreover the courts give FDA broad discretion in the measurement of defects in imported foods.[7]

11.3.1 Basic Import Procedure

Within five working days of the date of arrival of a shipment of food at a port of entry, the importer must file entry documents with U.S. Customs.[8] FDA is notified and reviews the importer's entry documents to determine whether a physical examination should be made or a sample taken for analysis. FDA's decision on whether to collect a sample is based on the nature of the product, FDA priorities, and the history of the commodity.

If the decision is made not to collect a sample, FDA sends a "May Proceed Notice" to Customs and the importer, and the shipment is released as far as FDA is concerned. If FDA sends a "Notice of Sampling" to Customs and the importer, the shipment must be held intact pending further notice, but the importer may move the shipment from the dock to another port or warehouse. New food products, food products placed on Import Alerts or Blocklisted

---

[3] FD&C Act § 801; 21 U.S.C. § 381.

[4] *See, e.g.,* Goodwin v. United States, 371 F. Supp. 433 (SD Cal 1972).

[5] Continental Seafoods, Inc. v. Schweiker, 674 F.2d 38 (D.C. Cir. 1982).

[6] Meserey v. United States, 447 F. Supp. 548 (D. Nev. 1977).

[7] Caribbean Produce Exchange, Inc. v. Secretary of Health and Human Services, 893 F.2d 3 (1st Cir. 1989).

[8] CFSAN, FDA, *FDA Import Procedures*, Industry Activities Staff Flyer, *available at:* http://www.cfsan.fda.gov/~lrd/import.html (1996) (last accessed Feb. 7, 2007).

foods (food products are those with negative import histories) are likely to be sampled.

This system—where the importer rather than FDA retains custody over shipments—has been criticized for allowing shipments that failed to meet U.S. safety standards to be distributed in domestic commerce.[9] Importers in some cases may have been able to provide substitutes for food targeted for inspection.

If the sample is found in compliance with requirements, FDA sends a Release Notice (FD-717) to Customs and the importer. If the sample appears to be in violation, FDA sends Customs and the importer a "Notice of Detention and Hearing" (FD-777). Nonconforming goods are kept in the possession of the owner if a bond covering potential liquidated damages is posted. If found to be nonconforming after sampling, a Notice of Refusal (FD-772) is issued, and the shipment will be ordered destroyed unless reconditioned or exported.

11.3.2 Prior Notice of Import

Prior notice is notification to FDA that an article of food or animal feed is being imported or offered for import into the United States in advance of the arrival of the article of food at the U.S. border.[10] The Bioterrorism Act of 2002[11] added a requirement to the FD&C Act that FDA receive prior notice of food imported into the United States.[12] On October 10, 2003, FDA published an interim final rule requiring submission to FDA of prior notice of food and animal feed that is imported or offered for import into the United States.[13]

Because of the broad definition of "food" under the FD&C Act, the FDA's prior notice requirement applies to some products also regulated by other

---

[9] *See, e.g.,* GAO, FOOD SAFETY: FEDERAL EFFORTS TO ENSURE THE SAFETY OF IMPORTED FOODS ARE INCONSISTENT AND UNRELIABLE, (GAO Report to the Chairman, Permanent Subcommittee on Investigations, Committee on Governmental Affairs, U.S. Senate) GAO/RCED-98-103 (Apr. 1998); GAO, FOOD SAFETY: FEDERAL EFFORTS TO ENSURE THE SAFETY OF IMPORTED FOODS ARE INCONSISTENT AND UNRELIABLE, GAO/RCED-98-103 (Apr. 30, 1998) and GAO/T-RCED-98-191 (May 14, 1998); *and* GAO, FOOD SAFETY: WEAK AND INCONSISTENTLY APPLIED CONTROLS ALLOW UNSAFE IMPORTED FOOD TO ENTER U.S. COMMERCE, Statement of Lawrence J. Dyckman, Director, Food and Agriculture Issues, Resources, Community, and Economic Development Division, Testimony before the Permanent Subcommittee on Investigations, Committee on Governmental Affairs, U.S. Senate, GAO/T-RCED-98-271 (Sept. 10, 1998).

[10] 68 Fed. Reg. 58974 (Oct. 10, 2003); *see also* FDA, Guidance for Industry: Prior Notice of Imported Food Questions and Answers, Edition 2 (May 2004), *available at*: http://www.cfsan.fda.gov/~pn/pnqagui2.html (last visited Dec. 27, 2006).

[11] Public Health Security and Bioterrorism Preparedness and Response Act of 2002 (the Bioterrorism Act) (Public Law 107-188).

[12] FD&C Act § 801(m) (21 U.S.C. § 381(m)), which was added by § 307 of the Bioterrorism Act.

[13] FDA, *supra* note 10.

agencies. For example, alcoholic beverages regulated by TTB must still comply with the FDA prior notice requirements. Live food animals that are subject to border inspections by APHIS are also subject to FDA's prior notice requirements (live food animals do not fall within the exclusive jurisdiction of USDA under the Federal Meat Inspection Act or Poultry Products Inspection Act, thus FDA and APHIS may both have jurisdiction over live animals).[14]

The requirement for prior notice to FDA does not alter the role of another agency, such as APHIS or TTB, or the requirements relating to that agency.[15] However, food under the exclusive jurisdiction of USDA at the time of importation is excluded from the prior notice requirement.

11.3.3 Import Food Facility Registration

The Bioterrorism Act also requires domestic and foreign facilities that manufacture, process, pack, or hold food for human or animal consumption in the United States to register with the FDA. Farms, fishing vessels not engaged in processing, and facilities regulated exclusively throughout the entire facility by the USDA are exempt from registration. Registration may be done by paper form, but FDA encourages electronic registration, which is available on FDA's Web site.[16]

11.3.4 Additional Forms for Certain Canned Foods, Milk, Cream, and Infant Formula

Besides the required entry forms, import registration, and prior notice of import, certain food products require specialized forms be submitted to the FDA. Firms must register and file processing information before shipping any low-acid canned food or acidified low-acid canned food into the United States.[17] This information must be provided to FDA for each applicable product at the time of importation. In addition the Federal Import Milk Act requires a permit for milk and cream imported into the United States.[18] These permits and registrations are in addition to the general registration and prior notice requirements.

Infant formula, in addition to meeting the laws and regulations governing foods generally, must meet additional statutory and regulatory requirements. The specific infant formula requirements are found in FD&C Act section 412 and 21 C.F.R. sections 106 and 107. In particular, all formulas marketed in the United States must meet federal nutrient requirements, and infant formula manufacturers must notify the FDA before marketing a new formula. This is

---

[14] 68 Fed. Reg. 58974 at 58991 (Oct. 10, 2003).
[15] *Id.*
[16] *See* http://www.fda.gov/furls.
[17] 21 C.F.R. §§ 108.25, 108.35 (2000).
[18] 21 U.S.C. §§ 141–149 (2000).

in addition to other notification requirements. If an infant formula manufacturer does not provide the information required in the notification for a new or reformulated infant formula, the formula is defined as adulterated under FD&C Act section 412(a)(1). These more stringent requirements were considered necessary because infant formula is often used as the sole source of nutrition by a vulnerable population during a critical period of growth and development.

Although the FDA's statutory authority is largely limited to inspections and tests of imported foods at the U.S. port of entry, with low-acid and acidified canned foods and infant formula, the FDA may request that foreign exporting firms grant FDA inspectors access to their plants. Nonetheless, the FDA conducts few foreign plant inspections. For example, there are almost 190,000 foreign food firms exporting food to the United States, yet FDA inspected fewer than 100 firms in fiscal year 2007.[19]

11.3.5   When a Violation Is Found

The FDA may refuse entry of an import shipment after a paperwork inspection and a physical examination. The FD&C Act authorizes the FDA to detain a regulated product that appears to be out of compliance with the FD&C Act. The FDA district office will then issue a "Notice of FDA Action," which identifies the nature of the violation to the owner or consignee of the goods.

11.3.6   When a Notice of Action Is Issued

Once a Notice of Action is issued, the importer has the following options:

- Request a hearing to defend the acceptability of the product
- Apply for permission to re-label or re-condition the product
- Re-exportation
- Judicial review of the Notice of Refusal

The importer has 10 days to request a hearing. The importer may then introduce evidence (written or oral) as to the acceptability of the shipment and of any analysis performed independently. FDA may review the submitted evidence or hold a hearing (known as a section 801 hearing). The Hearing Officer will review the submitted evidence and make a decision regarding acceptability.

The owner (or consignee) is entitled to an informal hearing regarding the admissibility of the goods; however, the hearing is less than FDA's full regula-

---

[19] GAO, FEDERAL OVERSIGHT OF FOOD SAFETY: FDA'S FOOD PROTECTION PLAN PROPOSES POSITIVE FIRST STEPS, BUT CAPACITY TO CARRY THEM OUT IS CRITICAL 6, GAO-08-435T (Jan. 29, 2008), *available at*: http://www.gao.gov/new.items/d08435t.pdf (last accessed Jan. 31, 2008).

tory hearing.[20] The hearing is the importer's only opportunity to present a defense of the importation or to present evidence as to how the shipment may be made eligible for entry.

The importer faces a steep burden of proof at such a hearing. In particular, the importer cannot demand that the FDA prove the source of contamination, and the FDA only has to prove that a product "appears adulterated."[21] This language in the FD&C Act indicates "Congress' intent to forego formal procedural requirements" for imports.[22] Typically the courts grant FDA broad deference and discretion in measuring and examining defects in imported foods.[23]

If the owner fails to submit evidence that the product is in compliance or fails to submit a plan to bring the product into compliance, FDA issues a second "Notice of FDA Action," which refuses admission to the goods. The goods must them be exported or destroyed within 90 days.[24] Re-exportation is within the discretion of FDA.[25]

11.3.7 Request for Authorization to Relabel or Perform Other Acts

If the importer loses at the hearing, section 801(b) permits the importer to request permission to "cure" or correct the nonconformity. An "Application for Authorization to Re-label or to Perform Other Acts" (FD-766) may be filed. If unsuccessful, the imported has 90 days to re-export the shipment.

The importer of detained goods may propose a manner in which detained food can be brought into compliance with the FD&C Act or be removed from coverage under the FD&C Act (i.e., rendered other than a food, drug, medical device, or cosmetic). The FDA may authorize relabeling or other action based on a timely submission of an appropriate completed request and the execution of sufficient bond.[26] The FDA notifies the importer of the approval or disapproval of the application to relabel or recondition. The FDA can charge the importer for the costs of supervision of the relabeling or reconditioning.[27] When approved, the FDA will state the conditions to be fulfilled, and the time limit within which to fulfill them.[28]

Judicial review is available to the appeal of the agency's decision to reject the shipment. The standard of review is whether the agency refused admission of the shipment in an arbitrary or capricious manner.

---

[20] 21 C.F.R. §§ 1.94 and 16.5(a)(2) (2000).

[21] 21 U.S.C. § 381 (2000).

[22] Seabrook Intl. Foods, Inc. v. Harris, 501 F. Supp. 1086 (DDC 1980).

[23] *See, e.g.,* Caribbean Produce Exchange, *supra* note 7.

[24] FDA, REGULATORY PROCEDURES MANUAL: CHAPTER 9 IMPORT OPERATIONS/ACTIONS (2002).

[25] 21 U.S.C. § 334(d)(1) (2000).

[26] 21 U.S.C. § 381(b) and 21 C.F.R. § 1.95 (2000).

[27] 21 C.F.R. § 1.99 (2000).

[28] 21 C.F.R. § 1.96 (2000).

11.3.8   Inspection after Reconditioning or Relabeling

After completion of relabeling or reconditioning, the importer provides FDA with notification of completion. At this point FDA may conduct a follow-up inspection, sampling, or both to determine compliance. FDA may also accept the statement from the importer and conduct no follow-up.

If the relabeling or reconditioning has been properly fulfilled, FDA will notify the owner or consignee that the admissible portion is no longer subject to detention or refusal of admission. This notice is usually identified as "Originally Detained and Now Released." Where a nonadmissible portion remains (rejects), that portion must be destroyed or re-exported under FDA or Customs supervision. A "Notice of Refusal of Admission" is issued for the rejected portion.

If the relabeling or reconditioning has not been successfully fulfilled, the FDA generally will not authorize a second relabeling or reconditioning unless the request includes an adjustment from the original method, and the applicant offers reasonable assurance that the second attempt will be successful. If an article is refused admission, such article must be destroyed or exported under Customs' supervision, generally within 90 days of receiving the Notice of Refusal.[29]

11.3.9   Good Agricultural Practices (GAPs)

In 1998 the FDA issued the GUIDE TO MINIMIZE MICROBIAL FOOD SAFETY HAZARDS FOR FRUITS AND VEGETABLES. This Guide recommends good agricultural practices (GAPs) and good manufacturing practices (GMPs) that growers, packers, and shippers should take to address common risk factors and reduce the food safety hazards potentially associated with fresh produce.[30]

Although the GAPs contain important guidelines for food safety practices, they are not binding on growers, packers, or shippers of food. Nonetheless, GAPs remain an important contribution to the control of food safety in imports. The FDA's actions can leverage its authority in ways that gain voluntary compliance with the GAPs. In particular, FDA can detain or ban shipments from growers or countries. For example, in 2002 the FDA banned all cantaloupe imports from Mexico.[31] This may explain some of the growth in the use of third-party audits and certification for GAPs. In addition market

---

[29] 21 U.S.C. § 801(a) (2000).

[30] FDA, GUIDANCE FOR INDUSTRY: GUIDE TO MINIMIZE MICROBIAL FOOD SAFETY HAZARDS FOR FRESH FRUITS AND VEGETABLES (1998), *available at*: http://www.foodsafety.gov/~dms/prodguid.html (last accessed Feb. 2, 2007).

[31] Linda Calvin, *Produce, Food Safety, and International Trade: Response to U.S. Foodborne Illness Outbreaks Associated with Imported Produce, in* ECON. RESEARCH SERV., USDA ECON. REPORT NO. 828, INTERNATIONAL TRADE AND FOOD SAFETY, 74, 78 (Jean C. Buzby, ed. 2003), *available at*: http://www.ers.usda.gov/Publications/AER828/ (last accessed Jan. 14, 2007).

forces can create a prophylactic factor that encourages adoption of better food safety measures, such as GAPs.

11.3.10 Enforcement

The FDA has been granted the power to obtain injunctive relief to prevent future violations or criminal penalties for repeat or egregious violators. In addition other penalties for adulteration and misbranding may apply.

Regarding imports, FDA may:

1. cause the destruction of the nonconforming goods if not re-exported (section 801(a));
2. require a bond payment for liquidated damages in the case of default, while the importer "cures" or relabels nonconforming goods (section 801(b)) (FDA may declare a bond violation in the case where the shipment has been distributed without a Notice to Proceed); or
3. order the seizure and condemnation of the nonconforming goods (section 304).

NOTES AND QUESTIONS

11.1. Additional information. For more information on the FDA Import Program, see: www.fda.gov/ora/import/ora_import_system.html (last accessed June 20, 2007).

11.4 USDA'S IMPORT SYSTEM

11.4.1 Food Safety Inspection Service (FSIS)

USDA-FSIS regulates the importation of meats and poultry and some egg products. The duty to inspect all commercial shipments of meat and poultry products entering the United States has been delegated to the USDA under the authority of the Federal Meat Inspection Act of 1958 (FMIA) and the Agricultural Marketing Agreement of 1937. The applicable USDA regulations appear generally at Title 9 of the Code of Federal Regulations (C.F.R.).

Before foreign firms can export meat or poultry into the United States, FSIS must have determined that the exporting country has a meat or poultry food safety system that is equivalent to that of the United States. When FSIS receives an application, the agency compares the foreign inspection system with the measures FSIS applies domestically. If FSIS determines that the foreign food regulatory system documentation meets all U.S. import requirements in the same or an equivalent manner, and provides the same level of public health protection attained in the United States, FSIS conducts an on-site audit of the entire foreign meat or poultry food regulatory system (or both). If a country completes these steps satisfactorily, FSIS publishes a proposed regulation that would add the country to FSIS' list of eligible import countries. FSIS must collect public comments on this proposed regulation and consider the comments before making a final decision as to whether the country will be eligible to import meat, poultry, or egg products into the United States.

The time from application to FSIS to the completion of an initial equivalence approval process normally requires three to five years. In October 2007, thirty-three countries were accepted by FSIS as eligible to import meat and poultry products to the United States (Table 11.1).

After initial approval, FSIS utilizes a three-part process to verify that a foreign meat or poultry food regulatory system continues to be equivalent to that of the United States. First, FSIS reviews documents, such as the laws, regulations, and implementing policies of a foreign food regulatory system, to ensure that the infrastructure remains in place. Next, FSIS conducts on-site food regulatory system audits at least annually in every country that exports meat or poultry products to the United States. Third, FSIS's continuous port-of-entry reinspection of products shipped from exporting countries provides evidence of how the foreign inspection systems are functioning. In contrast to FDA, FSIS inspectors visually check every imported shipment of foods under their jurisdiction at FSIS-approved import inspection stations. Most of these checks are for correct documentation and labeling. FSIS conducts more complete inspections and tests on a portion of the imported shipments to verify the effectiveness of the foreign food safety system. In 1997, 20 percent received more complete inspections,[32] whereas in the first three-quarters of FY 2007, only 11 percent received more complete inspections.[33] FSIS uses the term "reinspection" for its imported product inspections because the products have been previously inspected and passed by the importing country's inspection system.[34]

The same as importers of FDA-regulated products, importers of FSIS-regulated products must file an import notice and a bond with Customs within five days of the date that a shipment arrives at a port of entry. Unlike FDA law, which allows shipments to be moved out of FDA control, importers of FSIS-regulated food must hold their shipments at FSIS-registered warehouses for FSIS inspection until these shipments are released or refused entry.

---

[32] GAO, FOOD SAFETY: FEDERAL EFFORTS TO ENSURE THE SAFETY OF IMPORTED FOODS ARE INCONSISTENT AND UNRELIABLE, GAO/RCED-98-103 (Apr. 30, 1998) and GAO/T-RCED-98-191 (May 14, 1998).

[33] FSIS, U.S. Department of Agriculture, *Quarterly enforcement report, Table 3a, Imported meat and poultry products* 12 (2007) (Pounds of product presented, reinspected, and refused entry by fiscal year quarter), *available at:* http://www.fsis.usda.gov/PDF/QER_Q3_FY2007.pdf (accessed Sept. 17, 2007).

[34] Fed. Reg., *supra* note 14.

TABLE 11.1 Countries certified to export meat to the U.S.

Eligible Countries	Type of Product
Argentina	Meat
Australia	Meat; poultry (ratites only)
Belgium	Meat
Brazil	Meat
Canada	Meat; poultry; egg products
Chile	Meat
China	Poultry
Costa Rica	Meat
Croatia	Meat
Czech Republic	Meat
Denmark	Meat
Finland	Meat
France	Meat; poultry
Germany	Meat
Honduras	Meat
Hungary	Meat
Iceland	Meat
Ireland	Meat
Israel	Poultry
Italy	Meat
Japan	Meat
Mexico	Meat; poultry[a]
Netherlands	Meat
New Zealand	Meat; poultry (ratites only)
Nicaragua	Meat
Northern Ireland	Meat
Poland	Meat
Romania	Meat
San Marino	Meat
Spain	Meat
Sweden	Meat
United Kingdom	Meat; poultry
Uruguay	Meat

Note: Food Safety Inspection Service, U.S. Department of Agriculture, Eligible Foreign Establishments, updated June 5, 2007, *available at*: http://www.fsis.usda.gov/regulations_&_policies/Eligible_Foreign_Establishments/index.asp. (last accessed June 9, 2007).

[a]Mexico approved to export only processed poultry products slaughtered under federal inspection in the United States or in a country eligible to export slaughtered poultry to the United States.

However, before an importer may bring FSIS-regulated products into the United States, the importer must be certain its country has been accepted by FSIS to sell meat, poultry, or egg products in the United States. FSIS must determine that a country's federal inspection system is equivalent to that of the United States.

FSIS does not conduct food inspections in foreign countries, nor does it verify that individual foreign establishments are qualified to export to the United States. FSIS relies on its determination that a country has an equivalent food regulatory system that carries out appropriate inspection. A foreign establishment must obtain certification from its country's chief inspection official, who will certify to FSIS which establishments in the country meet the FSIS import requirements.

NOTES AND QUESTIONS

11.2. For more information, *see* USDA, *Importing Meat, Poultry and Egg Products to the United States*, Backgrounders/Key Facts (Dec. 2003), *available at*: http://www.fsis.usda.gov/oa/background/imports2003.htm (last accessed June 21, 2007).

11.4.2   Animal and Plant Health Inspection Service (APHIS)

Each meat, poultry, and egg product shipment enters the country under the authority of U.S. Customs and USDA's Animal and Plant Health Inspection Service (APHIS) before it is transferred to FSIS. APHIS is charged with protecting U.S. agricultural health, among many other responsibilities, such as regulating genetically engineered organisms and administering the Animal Welfare Act.

To accomplish its mission, one role of APHIS and the USDA Veterinary Services (VS) is to regulate the importation of animals and animal-derived materials to ensure that exotic animal and poultry diseases are not introduced into the United States. For example, APHIS works to prevent entry of foot-and-mouth disease or avian influenza. If no pest or disease of concern is detected or raised by the documentation, APHIS transfers control of the products to FSIS for visual inspections.

11.5   OTHER IMPORT CONTROLS

11.5.1   Customs (CBP)

All FDA and USDA regulated products imported into the United States must meet the Bureau of Customs and Border Protection (CBP or "Customs") requirements in addition to FDA and USDA requirements. The major responsibility of Customs is to administer the Tariff Act of 1930 as amended. Primary duties include assessment and collection of all duties, taxes, and fees on imported merchandise; administration and review of import entry forms; the enforcement of Customs' related laws; and administration of certain navigation laws and treaties.

There is a working agreement among the FDA, USDA, and Customs for cooperative enforcement. Products nonconforming with FDA or USDA requirements will be seized by Customs and released only after the agency receives written approval from the FDA or USDA, as applicable. Generally, FDA or USDA identify the violative food, but the refusal of admission, and subsequent re-exportation, or destruction of the food is carried out under the direction of Customs. In some cases actual supervision of destruction of violative food or the supervision of reconditioning may be conducted by FDA or FSIS personnel under a regional agreement. For example, where a port of entry is close to an FDA office, supervision is normally exercised by FDA. At remote ports, supervision is normally exercised by Customs.[35]

FDA has an electronic notification entry system, the Operational and Administrative System for Import Support (OASIS). When Customs receives electronic notifications of a food shipment entry, these are sent to FDA electronically via OASIS. FDA uses OASIS to electronically screen entries against criteria developed by FDA.

Articles offered for import into the United States (entries) that have a value greater than $1,250 are considered by Customs to be "formal" entries. One of the more important requirements for formal entries is the requirement for a bond. Under a formal entry bond, imported articles may be unconditionally released to importers, pending a determination of the admissibility (and amount of duty to be paid). The bond requires importers to redeliver the articles to Customs, upon demand of Customs, at any time. For example, Customs might demand redelivery of a food to allow FDA sampling or for re-exportation following refusal of admission. If the importer fails to redeliver the goods, Customs may institute proceedings to collect the liquidated damages provided for in the bond.[36]

Under FDA law importers generally maintain possession of the imported food before the FDA releases them. With perishable foods the shipment may begin entry into domestic commerce. This system has been criticized on a number of grounds.[37] Not all foods sampled, and later found violative, are returned by importers to Customs. In addition, even when food is returned, Customs does not always witness and verify that the violative food is properly disposed of; for instance, a landfill receipt may suffice. Customs also does not verify whether there has been a substitution when product is re-exported from the United States instead of destroyed. Finally, forfeiture of the bond is not always effective deterrence to ensure return of the food either because the value of the food may exceed the bond or because full damages often are not collected.[38]

---

[35] FDA, Regulatory Procedures Manual: Chapter 9 Import Operations/Actions (2002).

[36] See 19 C.F.R. § 113.62(k) and 21 C.F.R. § 1.97 (2000).

[37] GAO, Food Safety: Federal Efforts to Ensure the Safety of Imported Foods Are Inconsistent and Unreliable (GAO Report to the Chairman, Permanent Subcommittee on Investigations, Committee on Governmental Affairs, U.S. Senate) GAO/RCED-98-103 (Apr. 1998).

[38] Id.

11.5.2 Environmental Protection Agency

The Environmental Protection Agency (EPA) is not directly involved in regulation of imported food, but imported food must meet the same pesticide residue standards as domestic product. The law directs EPA to set limits on the pesticide residue remaining on food such that there is a reasonable certainty of no harm. These pesticide residue limits are called tolerances (some countries use the term "maximum residue limits" or MRLs). EPA sets the tolerances in the Code of Federal Regulations (C.F.R.) within title 40 C.F.R. part 180.[39]

The pesticide tolerances apply equally to imported and domestically produced food.[40] These tolerances are enforced by the USDA and FDA. The USDA enforces tolerances established for meat, poultry, and some egg products. The FDA enforces the tolerances established for other foods.

11.5.3 Alcohol and Tobacco Tax and Trade Bureau (TTB)

Importers seeking to import alcohol beverages into the United States must meet the requirements of the Federal Alcohol Administration Act enforced by the Alcohol and Tobacco Tax and Trade Bureau (TTB).[41] In particular, an importer must obtain the appropriate TTB-issued permit to import alcoholic products. Importers must maintain and staff a business office in the United States. In addition the importer must have a TTB-issued certificate of label approval (COLA). Finally, the importer must meet all state and local requirements, which may be in addition to federal requirements.

Alcoholic beverages are also defined as "food" under other statutes, so alcoholic beverages must also meet those additional general requirements. For example, the importer must ensure that the producer of the alcohol beverage is registered with the FDA and provide FDA with advance notification of an importation.

11.5.4 National Oceanic and Atmospheric Administration (NOAA)

More than 80 percent of the seafood that Americans consume is imported.[42] Seafood falls under the regulatory oversight of FDA. However, voluntary

---

[39] The tolerance information is codified within 40 C.F.R. § 180. However, for information on new tolerances or changes to tolerances not yet codified, search the EPA's Web site http://www.epa. gov/pesticides/search.htm or the Federal Register (*available at:* http://www.gpoaccess.gov/fr/index. html).

[40] For more information on pesticide tolerances, *see* U.S. Environmental Protection Agency, Pesticides and Food: What the Pesticide Residue Limits are on Food, *available at:* http://www.epa.gov/ pesticides/food/viewtols.htm (last accessed Jan. 3, 2007).

[41] For more information, *see* http://www.ttb.gov/index.shtml (last accessed Feb. 7, 2007).

[42] GAO, David M. Walker, Comptroller General of the United States, Testimony before the Subcommittee on Agriculture, Rural Development, FDA, and Related Agencies, Committee on Appropriations, House of Representatives, *Federal Oversight of Food Safety: High-Risk Designation Can Bring Needed Attention to Fragmented System*, GAO-07-449T (2007).

inspection programs within the National Oceanic and Atmospheric Administration (NOAA) of the U.S. Department of Commerce provide important support for FDA's regulatory role.

Administered through the 1946 Agricultural Marketing Act, NOAA also offers a voluntary inspection service to the industry. The NOAA Seafood Inspection Program offers establishment sanitation inspection, process and product inspection, product grading, product lot inspection, laboratory analyses, training, and consultation.[43] While this is not strictly speaking an import program, the service is provided in foreign countries as well as the United States.

Products that are inspected and meet the requirements under the program can bear one of the agency's official marks, such as U.S. Grade A, Processed under Federal Inspection (PUFI), and lot inspection marks. The program is available for all edible products, ranging from whole fish to formulated products, as well as fishmeal products for animal foods.

NOAA also plays a role in seafood imports through its division of the National Marine Fisheries Service (NMFS). NMFS is responsible for the management, conservation, and protection of living marine resources. The agency assesses and predicts the status of fish stocks, ensures compliance with fisheries regulations, and works to reduce wasteful fishing practices and to prevent lost economic potential associated with over fishing, declining species, and degraded habitats.

In these roles NMFS may put restrictions on the import of certain marine species. For instance, to implement recommendations of the International Commission for the Conservation of Atlantic Tunas, NMFS has banned the import of undersized Atlantic swordfish and extended dealer permitting and reporting requirements to swordfish importers.[44]

11.6 COUNTRY-OF-ORIGIN LABELING

Under the Tariff Act of 1930, as amended,[45] most imported product, including foods, are required to be marked with the country of origin for the ultimate purchaser. However, the Tariff Act allows retailers to display loose produce without country of origin labeling. In addition foods that are "substantially transformed" in the United States are not subject to this requirement. Under the Federal Meat Inspection Act,[46] all meat products imported into the United States must bear country of origin labeling on the containers in which the products are shipped.

---

[43] U.S. Department of Commerce, *USDC Seafood Inspection Program*, http://seafood.nmfs.noaa.gov/publications.htm (last accessed Mar. 25, 2007).
[44] 50 C.F.R. § 635.46 (2000).
[45] 19 U.S.C. § 1304 (2000).
[46] 21 U.S.C. § 601 *et seq.*(2000).

The Farm Security and Rural Investment Act of 2002 (more commonly known as the 2002 Farm Bill) contained new country-of-origin labeling (COOL) requirements to be effective September 30, 2004, for beef, lamb, pork, fish, perishable agricultural commodities, and peanuts. In 2004, implementation of COOL was postponed for all covered commodities except wild and farm-raised fish and shellfish until September 30, 2006. On November 10, 2005, implementation was again postponed for all covered commodities except wild and farm-raised and shellfish until September 30, 2008.

COOL requires that retailers notify their customers of the country of origin for beef, pork, and lamb, fish, shellfish, peanuts, and perishable agricultural commodities. Seafood labeling must also distinguish between "wild fish" and "farm-raised fish." COOL also imposes recordkeeping requirements for any person supplying a covered commodity to a retailer. All points of the supply chain before the retailer—including growers, importers, distributors, handlers, packers, and processors—must make available to the next purchaser in the supply chain information about the country of origin and, if applicable, wild or farm-raised claims, and this information must be maintained for two years.[47]

11.7   CHALLENGES FACING IMPORT REGULATION

Ensuring the safety of imported food is a daunting task as Americans consume a continually increasing amount of imported food. The United States is moving from a nation self-sufficient in its food supply to one that is increasingly dependent on other countries.[48] Imports in 2006 accounted for about 16 percent of the total vegetable supply and about 44 percent of the total United States fruit supply. The quantity of imported food is escalating (Table 11.2 and Figure 11.1) while FDA's resources to inspect them are not keeping up.[49]

U.S. regulation of imported food has been criticized for many years with—the U.S. Government Accountability Office (GAO—formerly the General Accounting Office) being a frequent critic.[50] This criticism crescendos as consumption of imported food rises. A number of recent foodborne illness outbreaks illustrate that imported foods can challenge the United States regulatory system:

---

[47] 7 C.F.R. § 60.400(b)(1) and (3).

[48] Richard Gilmore, *US Food Safety under Siege?* 22 NATURE BIOTECHNOLOGY 1503–1505 (2004).

[49] FDA, FDA SCIENCE AND MISSION AT RISK: REPORT OF THE SUBCOMMITTEE ON SCIENCE AND TECHNOLOGY (2007), *available at*: http://www.fda.gov/ohrms/dockets/ac/07/briefing/2007-4329b_02_01_FDA%20Report%20on%20science%20and%20Technology.pdf (last accessed Jan. 31, 2008).

[50] *See, e.g.,* GAO *supra* note 9.

TABLE 11.2   Food entries (imports) under the jurisdiction of FDA (millions)

Year	
1985	0.95
1989	1.1
1990	1.1
1991	1.15
1992	1.15
1993	1.3
1994	1.8
1995	1.9
1996	2.4
1997	2.7
1998	3
1999	3.7
2000	4.3
2003	6
2004	7
2005	9
2006	10 (projected)
2007	11.7 (projected)

Source: FDA, Performance Plan 2002, *available at:* http://www.fda.gov/ope/fy02plan/food.html (2002) (last accessed June 11, 2007); and FDA, Statement of Robert E. Brackett, Director, Center for Food Safety and Applied Nutrition, before the Committee on Agriculture, Nutrition and Forestry, United States Senate, *available at:* http://www.fda.gov/ola/2005/counterterrorism0720.html (July 20, 2005) (last accessed June 11, 2007).

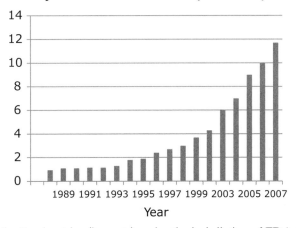

Figure 11.1   Food entries (imports) under the jurisdiction of FDA (millions)

- In 1996, more than 1,465 cases of *Cyclospora* foodborne illness were reported in the United States and Canada with Guatemalan raspberries identified as the most likely source.[51]
- In 1997, more than 200 schoolchildren and teachers in Michigan contracted hepatitis A from strawberries shipped from Mexico.[52]
- In 2000, some 47 people became ill from *Salmonella poona* with cantaloupe from Mexico implicated as the source.[53]
- In 2002, some 58 became ill from *Salmonella poona* in the United States and Canada with Mexican cantaloupe implicated.[54]
- In 2003, around 555 people fell ill and 3 people died from hepatitis A associated with green onions imported from Mexico.[55]

Currently, FDA inspects approximately 1 percent of the food imported under its jurisdiction. In 2003, Congress increased funding for import inspections, and in that year, FDA hit a high of 1.3 percent of food imports inspected.[56] Nonetheless, in 2004, the acting FDA Commissioner told Congress that the agency's border inspectors were being swamped by the increasing number of imports and predicted that the Agency would inspect less than 1 percent of food imports in 2007.[57]

While import regulation by USDA relies primarily on its determination that a country has an equivalent food safety regulatory system, nonetheless, USDA checks every import shipment of meat, poultry, and egg products (at least for paper compliance). Therefore concern arises over whether FDA's inspection resources are adequate. FDA, however, lacks any statutory authority that would allow the agency to require exporting countries to have inspection systems equivalent to the United States. GAO has criticized this lack of authority.[58]

---

[51] Linda Calvin et al., *Response to a Food Safety Problem in Produce: A Case Study of a Cyclosporiasis Outbreak in* GLOBAL FOOD TRADE AND CONSUMER DEMAND FOR QUALITY (Barry Krissoff, Mary Bohman, Julie Caswell eds. 2002).

[52] Yvan Hutin et al., *A Multistate, Foodborne Outbreak of Hepatitis A*, 340 NEW ENG. J. MED. 595–602 (1999) (cases were reported in other states as well).

[53] J. Anderson et al., *Multistate Outbreaks of* Salmonella *Serotype Poona Infections Associated with Eating Cantaloupe from Mexico—United States and Canada, 2000–2002,* 51(46) MORBIDITY MORTALITY WEEKLY REPORT. 1044–1047 (Nov. 22, 2002).

[54] *Id.*

[55] Centers for Disease Control, *Hepatitis A Outbreak Associated with Green Onions at a Restaurant—Monaca, Pennsylvania,* MORBIDITY MORTALITY WEEKLY REPORT, at 1155–1157 (Nov. 28, 2003).

[56] FDA Week, Vol. 12 No. 33, Inside Washington Publishers (Aug. 18, 2006).

[57] *Id.*

[58] GAO, FOOD SAFETY: WEAK AND INCONSISTENTLY APPLIED CONTROLS ALLOW UNSAFE IMPORTED FOOD TO ENTER U.S. COMMERCE, Statement of Lawrence J. Dyckman, Director, Food and Agriculture Issues, Resources, Community, and Economic Development Division, Testimony Before the Permanent Subcommittee on Investigations, Committee on Governmental Affairs, U.S. Senate, GAO/T-RCED-98-271 (Sept. 10, 1998).

FDA has been criticized for the small percentage (roughly 1 percent) of imported items inspected.[59] Critics of the FDA system point out the much higher percentage (8 to 15 percent) of meat and poultry products that is reinspected by USDA field inspectors. However, the different systems employed by these agencies to track their work load (food entries versus pounds of product) makes comparison difficult.

Moreover, the FDA system of food safety regulation—for both domestic and imported food—relies on a small number of inspections. What is most important is not review of the numbers or percentages of *product* inspections by FDA—a method of review that can never provide more than a weak and limited opportunity to ensure safety—but rather a review is needed of whether FDA has adequate regulatory authority over the *processing* of food abroad, where there is greater opportunity to ensure safety. A spotlight on the small number of FDA import inspections draws attention away from the important point that port-of-entry inspections alone may never provide acceptable protection.

A 1991 report by the Advisory Committee on the Food and Drug Administration called point-of-entry inspections an "anachronism" and considered the process of inspecting a final product to ensure conformity to standards "totally discredited" as a means of ensuring regulatory compliance.[60] It is widely accepted that a prevention-based system, such as HACCP, is more effective and efficient at ensuring food safety. Simply put, prevention better ensures safety than end product testing alone.[61]

A prevention-based inspection system would require inspection of hazard analysis and risk controls in the country of origin. Unfortunately, FDA lacks the statutory authority to impose an equivalency requirement for importation of foods, generally has no review authority in or over foreign countries, and must rely on inspections at the U.S. ports of entry to determine the safety. Although FDA applies a risk-based approach for targeting inspections of import shipments, the use of the term "risk based" is somewhat misleading because FDA lacks the statutory authority to apply a scientific risk-based approach to imported food safety, as this would require authority to reach back to the country of origin.

11.8 EXPORT

11.8.1 Export Exemption

Foods that are intended for export from this country are *not* required to comply with the requirements of the FD&C Act, so long as the food product:

---

[59] *The Dangers of Imported food*, 7 THE WEEK 315 (2007), at 15.
[60] U.S. Department of Health and Human Services, Advisory Committee on the Food and Drug Administration, FINAL REPORT OF THE ADVISORY COMMITTEE ON THE FOOD AND DRUG ADMINISTRATION (May 1991).
[61] *Id.*

- meets the specifications of the importing country,
- is not in conflict with the laws of the importing country, and
- outside shipping container clearly indicates that the product is for export.

If the food product meets the listed criteria above, it will be exempted from the adulteration and misbranding provisions of the FD&C Act unless re-introduced into domestic commerce. Distribution in interstate state commerce nullifies the exemption. This is known as the "export exemption" of section 801(e) of the Act. This section has been criticized by some as the "commodity dumping" provision. Note that adulterated or misbranded foods may not enter the United States but may be exported after seizure and condemnation by FDA.

Of course, to claim the exemption, the exporter must demonstrate to FDA (if contested), with good evidence, that the product conforms to the laws of the importing country. This is typically shown by an "export certificate" issued by the importing country.

11.8.2 Import for Export

Among other changes, the Bioterrorism Act of 2002 introduced new import for export provisions. These provisions increase the FDA's authority regarding the oversight and monitoring of certain foods, food additives, color additives, dietary supplements, drugs, and devices that are imported into the United States for further processing and export. The Act requires the importer to submit to FDA the following information at the time of importation of food additives, color additives, or dietary supplements:

1. A statement that the article is intended to be further processed and exported.
2. The identity of the manufacturer of the article, and each processor, packer, distributor, or other entity that had possession of the article in the chain of possession from the manufacturer to the importer of the article.
3. Such certificates of analysis as are necessary to identify the article.

Under the Act, the initial owner or consignee of the product must execute a bond providing for the payment of liquidated damages in the event of default. The bond is issued by U.S. Customs. In addition the Act requires the initial owner or consignee of an article to maintain records on the use or destruction of the article or portions of it, and to provide these records to FDA upon request. The initial owner or consignee is also required to submit a report to FDA, upon request, that provides an accounting of the export or destruction of the imported article or portions, and the manner in which such owner or consignee complied with the requirements of the Act.

Finally, under the Act, FDA may refuse admission of any article if there is credible evidence or information indicating that the article is not intended to be further processed or incorporated into a product to be exported.

In the case of the *United States v. 76,552 Pounds of Frog Legs*, the government brought action to condemn adulterated frog legs allegedly brought into the United States in violation of customs statutes and of the Federal Food and Drug and Cosmetic Act. Claimants moved for release of the article or, alternatively, for release of the article for reconditioning. The District Court held that the claimants were not entitled to the benefit of a statutory exemption that would allow the frog legs to be exported, that the United States was entitled to a judgment of condemnation, but that the claimants were entitled to conditional repossession of the frog legs in order that they might be brought into compliance by reconditioning.

* * * * *

U.S. v. 76,552 Pounds of Frog Legs

423 F.Supp. 329 (1976)

GARZA, Chief Judge

This is an action for condemnation of certain adulterated frog legs that were brought into the United States in violation of 21 U.S.C.A. § 301, § 331, and § 334(d)(1) and 19 U.S.C.A. § 1592. . . . The original claimants, Manuel Sanchez and Progressive Sea Products, Inc. (hereinafter Progressive), filed Motions for Release of Article contending this Court has authority, prior to condemnation of the food, to grant its release for export if the provisions of 21 U.S.C.A. § 381 are met. . . . there is no genuine issue of material fact regarding whether the food involved herein should be condemned. . . .

Agent Gene Nicko, of the United States Customs Service, received information from a confidential informer that certain frog leg shippers were engaged in illegal activity. He was told that frog leg importers are forced to sell, at depressed prices, all contaminated frog legs that had been refused importation entry into the United States. Salvage buyers, who bought this food product at reduced prices, would subsequently illegally reintroduce this food into the United States at regular prices. His information was that Manuel Sanchez, Jr., owner of Progressive Sea Products, Inc., was buying large quantities of rejected frog legs, and consigning them to Jose Mendoza, a buyer in Mexico, for export. Once out of the country, the contaminated frog legs would be packed under other sea products and brought back into the United States and sold as wholesome food products. Pursuant to this information, Agent Nicko began his investigation. . . .

Nicko decided to check out the local cold storage facilities in Brownsville. When he interviewed the manager, Walter Brimmer, he was told that some of Progressive's employees were in the back of the facility repackaging frog legs at that very moment. The agent investigated and discovered that Penninsular

Brand Frog Legs, Products of India, were being repackaged into boxes marked "Products of Mexico, packed by Industrias G.M.E., Mexico, D.F.," but this was not being done under the supervision of a Customs Officer. He was told the boxes from India were in such poor condition they necessitated repackaging. . . .

The Customs Agent felt this indicated the frog legs were not intended for export and that the repacking had been done to conceal the true nature of the frog legs. When it was discovered that Progressive had in fact sold contaminated frog legs in domestic commerce when they should have been under bond but were not, both the Secretary of the Treasury and the Secretary of Health, Education and Welfare decided to seek forfeiture of the food for various violations of the Customs statutes and the Food and Drug Act. . . .

The statutory import–export exemption in 21 U.S.C.A. § 381 does not apply to claimant's, Progressive Sea Products, Inc., frog legs. The clear wording of the statute provides that articles intended for export under the provisions of the statute are not exempt if such article is sold or offered for sale in domestic commerce. The record establishes that some of the frog legs in the shipments under seizure were sold and offered for sale in domestic commerce. The statute further provides that the Secretary shall seek condemnation if the article is not exported within 90 days and an extension is not granted under the appropriate regulations. The record establishes that the frog legs, if ever intended for export, were not exported within 90 days and an extension was not granted. . . .

This Court holds that the claimant is not entitled to the benefits of the statutory exemption, 21 U.S.C.A. § 381, allowing the exportation of adulterated frog legs which have been refused entry. This Court further holds the Secretary of Health, Education and Welfare, through the FDA, did not abuse its discretionary power to revoke claimant's import–export privilege under § 381 and seek condemnation. After considering the threshold question of the applicability of a statutory exemption, this Court must next consider the further question, whether the United States has proved that the food should be condemned and if condemned what disposition should be made. . . .

The frog legs are a food within the meaning of 21 U.S.C.A. § 321(f)(1). It is undisputed they are adulterated within the meaning of 21 U.S.C.A. § 342(a)(1) with pathogenic *Salmonella*. They were introduced into interstate commerce within the meaning of 21 U.S.C.A. § 334(a), when they were shipped by truck, without bond, from New York, New York, to Brownsville, Texas, and not exported before expiration of the 90 day grace period. Once part of the shipment was sold in interstate commerce, the entire shipment can be deemed to have entered interstate commerce. This Court holds the frog legs are subject to condemnation pursuant to 21 U.S.C.A. § 334(a), (d), and the United States is entitled to a judgment of condemnation. A Summary Judgment of condemnation is this day being entered.

Having determined that the frog legs should be condemned, this Court must next determine whether to order the frog legs destroyed or to allow the claimant conditional repossession. This Court has discretionary power to permit the claimant's attempt to salvage a potentially valuable food regardless of the claimant's mala fides. Federal Courts must, however, protect the public interest in keeping from the channels of commerce food products so adulterated as to injure or endanger health, and to ensure that food products are properly branded so the consumer can know there is no misrepresentation as to substance, and that the food purchased is what it purports to be.

This Court holds the claimant is not entitled to the benefit of the import–export provisions found in 21 U.S.C.A. § 334(d)(1). As previously discussed, the claimant cannot comply with all the requirements found in 21 U.S.C.A. § 381 and this is a condition precedent to invocation of the import–export provisions found in 21 U.S.C.A. § 334(d)(1). Since the claimant offered part of the frog legs for sale in domestic commerce, he is not entitled to the benefits of its import–export provisions. Additionally, Section 334(d)(1) is not available where food is condemned because it is injurious to health. This is necessary to prevent adulterated food from being commingled with good lots of the same food and again offered for import under conditions that would make the adulteration difficult to detect. The evidence presented at the trial, and the guilty plea entered in Criminal No. 76-B-112, prove that the claimant had cause for believing and knew that the frog legs were adulterated and were sold in domestic commerce. Claimant's motion to be allowed to export the frog legs is denied. . . .

* * * * *

11.8.3 Export Certificates

Firms exporting products from the United States are often asked by foreign customers or foreign governments to supply certification that the food is subject to the Federal Food, Drug, and Cosmetic Act and other acts the FDA administers. Under the FDA Export Reform and Enhancement Act of 1996 FDA is authorized to issue certificates for drugs, animal drugs, and devices within 20 days of receipt of a request for such a certificate. In addition to issuing export certificates for approved or licensed products, the FDA also issues export certificates for unapproved products that meet the requirements of sections 801(e) or 802 of the FD&C Act.

For more information and the FDA Procedure for Obtaining Certificates for Export are *available at*: http://www.cfsan.fda.gov/~lrd/certific.html.

NOTES AND QUESTIONS

11.3. **Import regulations of other countries.** One of the best resources in English for the information on the food importation regulations of various countries is the Foreign Agricultural Service (FAS) of the U.S.

Department of Agriculture (USDA) http://www.fas.usda.gov/. They service publish a synopsis of various countries' food regulations through the GAINS Reports, Attaché Reports, Import and Export Guides, FAIRS report, etc. Food and Agricultural Import Regulations and Standards (FAIRS) http://www.fas.usda.gov/itp/ofsts/fairs_by_country.asp.

To obtain information by subject, Attaché Reports—USDA—Foreign Agriculture Service: http://www.fas.usda.gov/scriptsw/attacherep/default.asp. *Note*: Find the abbreviation for the country (i.e., Brazil = BR), and then try Option 3 from this page to bring up more for that country than just a search by country.

11.4. **State meat inspection programs limited to intrastate.** Foreign countries with inspection program equivalent to FSIS are allowed to sell meat interstate in the United States. However, state meat inspection programs must be certified by USDA as at least equal to the FSIS inspection program. Nonetheless, meat inspected under a state inspection program is limited to intrastate commerce.

11.5. **Compare and contrast.** Compare and contrast the different approaches to import inspection given to FDA and FSIS.

11.6. **One percent inspections.** FDA inspects less than 1 percent of the food imported under its jurisdiction. How does this reflect on the import inspections system? Discuss how this compares to FDA inspection of domestic goods.

11.7. **What is an "inspection?"** Note the various uses of the term "inspection" in the context of imported goods. How does an FDA "inspection" differ from a USDA "re-inspection"?

11.8. **Condemned.** What does "condemned" mean in the context of imported goods? What are the options for handling condemned imported goods?

11.9. **Export exemption example.** An exporter wants to export dried figs to an Australian purchaser, who uses them in coffee flavors. FDA seizes the figs prior to export because the product is adulterated with insect larvae. The exporter shows that the figs are properly labeled and are intended for export not for domestic sale, and thus fall under the export exemption. What if the impure materials are not permitted under Australian law—should the figs be allowed for export? What options are available to the exporter? *See United States v. Catz American Co.*, 53 F.2d 425 (9th Cir. 1931).

11.10. **Export exemption criticism.** The United States has been criticized for its export exemption. Can you think of a reason why?

PART V

INSPECTION AND ENFORCEMENT

CHAPTER 12

Federal Enforcement

12.1 INTRODUCTION

This chapter provides an overview of the regulatory enforcement powers of the federal food safety agencies, with the emphasis on the Food and Drug Administration (FDA). A summary of some aspects of the state authority and the U.S. Department of Agriculture's (USDA's) Food Safety and Inspection Service (FSIS) powers are also included.

12.1.1 A Note on Materials

Statutes Because the food regulatory enforcement authority is largely statutory, the statutes are the most important reference source. When examining an enforcement issue, review of the statute—particularly, the prohibited acts section and the definitions—usually provides the most important references.

Regulations Both FDA and FSIS have promulgated extensive regulations to implement the enforcement provisions of their statutory frameworks. Regulations have essentially the same force and effect as the laws they implement, but regulations generally are more detailed than statutory language. FDA's regulations are codified in Title 21 C.F.R. USDA's are codified in Title 9 C.F.R.

When new regulations are promulgated, the agency will publish in the Federal Register a "preamble" that contains the explanations, analyses, and comments on proposed and final regulations. Preamble language provides insight into understanding the regulations. In addition, the preambles are important because they represent the agency's contemporaneous reasoning for whatever action was being proposed or taken. Because public comments are required to be considered by administrative agencies when they promulgate rules, a large part of many lengthy preambles consists of public comment analysis. For anyone desiring a thorough understanding of the regulations, the preambles are valuable reading.

Food Regulation: Law, Science, Policy, and Practice, by Neal D. Fortin
Copyright © 2009 Published by John Wiley & Sons, Inc.

Agency Policies and Guidance Documents A variety of written materials describe the agencies' enforcement philosophy, procedures, and practices. These materials include policies, guidance documents, directives, and operating manuals. In most cases these materials are readily accessible on the FDA and FSIS Web sites. Some examples are as follows:

- FDA's <u>Compliance Program Guidance Manual</u> (CPGM)—compliance programs and program plans and instructions directed to field personnel for project implementation.[1]
- FDA's <u>Compliance Policy Guides</u> (CPG)—compliance policy and regulatory action guidance for FDA staff.[2]
- FDA's <u>Regulatory Procedures Manual</u> (RPM)—FDA regulatory procedures and practices for use by FDA personnel.[3]
- FDA's <u>Action Levels for Poisonous or Deleterious Substances in Human Food and Animal Feed</u>—lists action levels for unavoidable poisonous or deleterious substances, which are established by the FDA to control levels of contaminants in human food and animal feed.[4]

12.1.2 The Role of States

In carrying out their food enforcement responsibilities, both FDA and FSIS depend on state and local government agencies to assist them. There is a considerable amount of overlap and sharing of food enforcement authority between the states and FDA. Most states have adopted food and drug laws with nearly identical provision as the FD&C Act. With the similarities in definitions of adulteration and misbranding, and similarity in enforcement authorities, FD&C Act violations generally also result in violations that state authorities can pursue.

By comparison, the scope of FSIS' enforcement responsibility for food is more narrowly defined, encompassing primarily meat and poultry. However, FSIS also shares its regulatory responsibility with state officials.[5]

State authorities often have broader power or additional enforcement tools that are unavailable to FDA. For example, most states have licensing requirements, so state officials can usually revoke or suspend food establishment operating licenses. In addition, most state officials have the authority to place

---

[1] *Available at*: http://www.fda.gov/ora/cpgm/default.htm (last accessed Mar. 10, 2008).
[2] *Available at*: http://www.fda.gov/ora/compliance_ref/cpg/default.htm (last accessed Mar. 10, 2008).
[3] *Available at*: http://www.fda.gov/ora/compliance_ref/rpm/default.htm (last accessed Mar. 10, 2008).
[4] *Available at*: http://vm.cfsan.fda.gov/~lrd/fdaact.html (last accessed Mar. 10, 2008).
[5] *See, e.g.*, FSIS Directive 5720.2 Rev. 2, Cooperative Inspection Programs.

embargoes on the spot without the need to first go to court to seize violative products.

Finally, FDA often commissions state officials to conduct inspections or gather evidence for the federal government. FD&C Act section 702(a) expressly authorizes FDA to act through state officials.

12.1.3 Public Records

The activities of the government agencies largely are matters of public record, including enforcement activities. Increasingly this information is becoming available over the Internet. For example, FDA places its warning letters are placed on its Web site as part of FDA's Electronic Freedom of Information Reading room at http://www.fda.gov/foi/warning.htm.

The FDA Enforcement Report is published weekly by the Food and Drug Administration. It contains information on enforcement actions taken in connection with agency regulatory activities. The FDA Enforcement Report is available at http://www.fda.gov/opacom/Enforce.html.

Many companies check periodically to see whether their competitors have received warning letters from FDA. Other companies check enforcement reports to track problems with suppliers and ingredients.

12.2 STATUTORY AUTHORITIES

FDA's food enforcement powers are based primarily in the Federal Food, Drug, and Cosmetic Act (FD&C Act). FSIS enforcement powers derive from the Federal Meat Inspection Act (FMIA), the Poultry Products Inspection Act (PPIA), and the Egg Products Inspection Act.

A number of Supreme Court rulings have noted that the FD&C Act and other public safety laws should be construed broadly to achieve their intended purposes. Put in simplest words, the courts tend to favor public health over the commercial rights of affected companies.[6]

12.2.1 Prohibited Acts

FD&C Act section 301 enumerates the acts prohibited by the statute. Section 301 is involved in every enforcement case initiated by FDA. We have already covered many of the substantive requirements of the act and will not repeat those here. The prohibitions are enforceable by a number of remedies provided by the FD&C Act and elsewhere.

---

[6] *See, e.g.,* United States v. Lexington Mill & Elevator Co., 232 U.S. 399 (1914); United States v. Dotterweich, 320 U.S. 277 (1943); United States v. Sullivan, 332 U.S. 689 (1948); *and* United States v. Park, 421 U.S. 658, 672 (1975).

12.2.2   FDA's Enforcement Discretion

Congress intentionally wrote the FD&C Act with broad discretionary powers for FDA. FDA's powers and responsibilities, however, have never been matched with enough resources to enforce all issues within its oversight. Nonetheless, FDA's resources generally have been sufficient to bring enforcement action against flagrant violation and violations that threaten to become widespread.[7]

Necessarily, FDA must also decline to take regulatory action against some violations. FDA authority to do so has been upheld in court actions. For example, the National Milk Producers Federation brought suit to compel the FDA to take action against two substitute cheese products. The court held that FDA's enforcement proceedings were discretionary.[8]

12.2.3   Role of the Justice Department

Enforcement discretion is not the exclusive choice of FDA. The decision on court action must be shared by the U.S. Department of Justice (DOJ) and the United States Attorney for the judicial district in which FDA seeks judicial remedy. The FDA and USDA, like most federal agencies, lack statutory power to make independent court appearances. This system is designed for efficiency. With dozens of federal agencies and sub-agencies, if all were given power to go to court independently, the duplication of effort would be inefficient and the fragmentation would weaken the federal government's overall enforcement efforts.

The U.S. Congress gave both FDA and FSIS statutory authority to initiate court actions, such as seizure actions and injunctions. The regulatory agency attorneys, however, do not litigate these cases. The agencies make requests to DOJ attorneys, who have final discretion on whether or not to litigate.

Court actions are resource-intensive for the both the regulatory agencies and the DOJ. Therefore, stringent criteria must be met before prosecution. There are several layers of review within FDA before a case is referred to DOJ. This intense use of limited resources is one of the major reasons that FDA seeks remedies through regulatory actions short of court, the primary one being warning letters.

---

[7] *See* Peter Barton Hutt, *FDA Reduces Economic Regulation of Food Industry*, LEGAL TIMES OF WASHINGTON 31 (Aug. 30, 1981). (Also noting that one method of reducing demands on scarce regulatory resources employed by FDA was to write more specific and detailed, and less judgmental, regulations so that enforcement could be more cut and dried with fewer disputes.)

[8] NMPF v. Harris, 653 F.2d 339 (8th Cir. 1981); *See also,* Heckler v. Chaney, 470 U.S. 821 (1985). (FDA declined to prosecute state officials who administered a drug for a purpose not on the label—lethal injection for execution of a convicted murderer. The court held that the FDA's decisions not to take certain enforcement actions are not subject to judicial review under the APA.)

12.3 ENFORCEMENT JURISDICTION

12.3.1 Introduction into Interstate Commerce

One defense to enforcement of the FD&C Act is that the act generally requires the product being introduced into interstate commerce. In particular, a number of challenges have been made to FDA's authority to seize food based on the wording in the prohibited acts of the FD&C Act.

* * * * *

Sec. 331 [331].—Prohibited acts

The following acts and the causing thereof are prohibited:

(a) The introduction or delivery for *introduction into interstate commerce* of any food, drug, device, or cosmetic that is adulterated or misbranded.

(b) The adulteration or misbranding of any food, drug, device, or cosmetic *in interstate commerce.*

(c) The receipt *in interstate* commerce of any food, drug, device, or cosmetic that is adulterated or misbranded, and the delivery or proffered delivery thereof for pay or otherwise.

(d) The introduction or delivery for introduction *into interstate commerce* of any article in violation of Section 344 or 355 of this title. . . .

* * * * *

However, an important principal of statutory interpretation is that a statute is given reasonable interpretation (construction) to affect the purpose of the act. Applying this principal and others to the interpretation to the FD&C Act, the courts have generally upheld FDA's authority and prevented the creation of loopholes. For example, "introduction into interstate commerce" has been interpreted broadly; the sale or contract for sale can be considered the introduction into interstate commerce, although the product may not have actually been transported over interstate lines. *United States v. 7 Barrels . . . Spray Dried Whole Eggs*, 141 F.2d 767 (7th Cir. 1944). Further, introducing into interstate commerce includes sale or delivery to another person who will introduce it into interstate commerce. *United States v. Sanders*, 196 F.2d 895 (10th Cir. 1952).

The FD&C Act penalty provisions apply regardless of the status of the person holding the food. For example, a public warehouse company may be liable for the adulteration of food, although another company owns the food and the warehouse had nothing to do with transport in interstate commerce. Mere holding of the food after shipment in interstate commerce is sufficient to bring the product under the FD&C Act.[9]

---

[9] United States v. Wiesenfeld Warehouse Co., 376 U.S. 86 (1964).

12.3.2 Held for Sale after Shipment in Interstate Commerce

To ensure the protection of the FD&C Act over food that was being held for sale after shipment in interstate commerce, Congress added section 301(k). FD&C Act section 301(k) expressly extends the FDA's authority over products that are held for sale *after* shipment in interstate commerce. This provision ensures FDA has authority to respond to adulteration and misbranding that occurs after shipment. *United States v. Sullivan*, 332 U.S. 689 (1948).

Use of an ingredient that was shipped in interstate commerce has been held to be sufficient to confer federal jurisdiction on a food and fall within the scope of the FD&C Act. In *United States v. 40 Cases . . . "Pinocchio Brand . . . Blended Oil,"* the packing house asserted that its Pinocchio Brand Blended Oil fell outside regulation under the FD&C Act because the blended oil was made within state boundaries, although the firm admitted that the individual oils used in the blending had been shipped in interstate commerce. The court reasoned that it would undermine the purpose of the FD&C Act if blended oil constituted a "different product" from the individual oils, when the blend contains ingredients that were shipped interstate.

* * * * *

United States v. 40 Cases . . . "Pinocchio Brand . . . Blended . . . Oil"

289 F.2d 343 (2nd Cir. 1961)

LUMBARD, Chief Judge

The single question before us on this appeal is whether § 304(a) of the Federal Food, Drug, and Cosmetic Act, authorizes the United States to proceed against and seize mislabeled or adulterated cans of blended vegetable oils mixed entirely within the State of New York from various oils shipped under proper labels from other states and foreign countries. Section 304(a) permits seizure of food which is "adulterated or misbranded when introduced into or while in interstate commerce or while held for sale (whether or not the first sale) after shipment in interstate commerce. . . ." The district judge held that the blended oil was a "new product" and therefore not the same as those shipped in interstate commerce. He dismissed the libel.

In its libel, the allegations of which must be taken as true on review of the decision to dismiss, the United States charged that forty cases of six one-gallon cans of "Pinocchio" brand oil were delivered to the La Gondola Food Corporation in Syracuse; that they were being held for sale after interstate shipment; that the cans were labeled "25 per cent pure olive oil"; that examination showed that the cans contained little or no olive oil; and that the oil was therefore "adulterated" within the meaning of § 402(b)(2) or "misbranded" within the meaning of § 403(a) of the Federal Food, Drug, and Cosmetic Act, 21 U.S.C.A. §§ 342(b)(2), 343(a).

An answer to the libel was filed by the A.M.S. Packing Company, which alleged that it had blended and packed the oil attached under the libel; that the blend was as represented on the labels of the cans; and that all steps in the manufacturing and/or blending of the oils had taken place within the State of New York. On the basis of its allegation that the blending had been done in New York, the company put in issue the jurisdiction of the federal court and moved to dismiss the libel.

The United States did not in the district court or here challenge the truth of the company's assertion that the blending process was done entirely within the State of New York, nor did it claim that the blended oil was carried across any state line. It is also undisputed that the various oils from which the blend was made had been shipped under proper labels from New Jersey, Illinois, and Georgia, and that olive oil had been transported to the company's plant in Ozone Park, New York, from Spain, Italy, and Tunisia. The United States contends that although the component oils were correctly labeled when shipped interstate, the misbranding or adulteration which occurred during or after the blending of the oils brought them within the compass of the federal act as articles of food held for sale after interstate shipment.

From 1938 to 1948, the Food, Drug, and Cosmetic Act provided for condemnation of articles of food that were adulterated or misbranded only "when introduced into or while in interstate commerce." This language was held not to authorize seizure of food that was pure and properly labeled while in interstate commerce and became adulterated or misbranded only after the interstate voyage had been completed. In 1948, however, Congress amended 304(a) so as also to permit seizure of food that is adulterated or misbranded "while held for sale (whether or not the first sale) after shipment in interstate commerce." 62 Stat. 582. Had the company in this case not mixed the oils it received from various sources but instead pasted new misleading labels on the containers in which they were shipped in interstate commerce or otherwise adulterated the oils, seizure would have been authorized. The appellee would have us hold here that the blending of the oils which had been transported in interstate commerce took the final product out from under federal regulation although each of its separate components was being held for sale after shipment in interstate commerce. We do not agree.

The original Food and Drugs Act of 1906 authorized seizure only if the adulterated or misbranded food or drug was being transported in interstate commerce, "or, having been transported, remains unloaded, unsold, or in original unbroken packages." The "original package" limitation was omitted when the Food, Drug, and Cosmetic Act of 1938 supplanted the prior law. The appellee cannot therefore prevail here on the theory that the oil seized was not in the same container as was used for its shipment in interstate commerce.

The appellee contends that the packing company here did more than merely break open the original package. It argues that the processing of the oils in

Ozone Park, New York, created a new product—a blend of many different vegetable oils—which was not the same as the food transported in interstate commerce and therefore could not be seized as food held for sale after shipment in interstate commerce. We disagree.

In enacting the 1948 amendment to the Pure Food, Drug, and Cosmetic Act, Congress sought to fill the gap in the regulatory scheme pointed out in *United States v. Phelps Dodge Mercantile Co., supra,* by subjecting to condemnation food which had been adulterated or misbranded after coming to rest within a state but before being sold to a consumer. The interest of the federal government in ensuring that such food meets minimum standards of purity and is not misbranded arises out of its supervisory function over interstate commerce. The House and Senate reports both referred expressly to the congressional desire to protect the integrity of interstate products so as not to depress the demand for goods that must travel across state lines. This interest surely extends to products such as olive oil, which a New York consumer would probably recognize as out-of-state of foreign in origin.

Moreover in this case all the components of the oil blend had been transported in interstate commerce, and the completed mixture was being held for sale as "oil"—the very same type of food which had traveled across the state line. This is not a case in which oil which was transported interstate was used as one of many ingredients in a finished product which in no way resembled the food which had crossed state lines. Oil may come in many varieties, but to the unsophisticated consumer one oil blend is much like another. We would be undermining the remedial legislative purpose of consumer protection were we to deny the power to seize misbranded articles on the ground that such foods as corn oil, peanut oil, soya bean oil, and olive oil when mixed constitute a "different product" from a blend of less than all or from a pure measure of any one of them.

The appellee relies heavily, as did Judge Foley in dismissing the libel, on *United States v. An Article or Device Consisting of 31 Units.* That case has been limited considerably by a more recent decision by the same district judge. In any event, the case is distinguishable, since the misleading aspect of the pamphlets which accompanied the condemned device related there to remedial qualities purportedly possessed by one entire apparatus, which had been assembled wholly within the State of Michigan. None of the separate components which had been transported in interstate commerce was misbranded; it was the labor of assembling the components, which when assembled purportedly constituted an electrotherapy device, which was misrepresented. Here, however, the misbranding related directly to the percentage content of the olive oil shipped in interstate commerce. Congress' policy may not extend to assuring a consumer that all representations regarding any device or food made up of components transported in interstate commerce are true, but it at least goes so far as to assure him that the interstate elements themselves are not misrepresented.

Congress surely intended the provisions of the Food, Drug, and Cosmetic Act to apply to foods processed within a state, after shipment in interstate commerce, as was the case here. The statute must be read and applied broadly in order to effectuate its remedial purpose. We have no doubt of the power of Congress so to protect the public with respect to foodstuffs which have been shipped in interstate and foreign commerce, and we reverse the order of the district court dismissing the libel.

* * * * *

12.3.3 FD&C Act Interstate Commerce Presumption

Although the FD&C Act contains an interstate commerce requirement with respect to products over which FDA may exercise its enforcement authority, 21 U.S.C. § 379a creates a statutory presumption that all FDA-regulated products have moved in interstate commerce. This presumption relieves FDA of the burden of coming forward with proof of actual movement of the products in interstate commerce before initiating enforcement action. Of course, a party defending against the FDA action still has the option to prove that movement in interstate commerce was lacking. This is accomplished by the introduction of evidence in court sufficient to overcome the presumption.

12.3.4 FDA Jurisdiction over Restaurants

In 1975, the General Accounting Office (GAO) issued a report, Federal Support for Restaurant Sanitation Found Largely Ineffective. The GAO report criticized restaurant sanitation and estimated "that about 90 percent of the 14,736 restaurants [in the study] were insanitary." The report went on to note that FDA had jurisdiction over restaurant food that had been shipped in interstate commerce, but that FDA relied on state and local governments to regulate restaurants. "However, local governments generally have been ineffective in regulating restaurant sanitation and, generally, the States' monitoring of these programs has been minimal. . . ."

In anticipation of the GAO report, FDA proposed federal regulations on the sanitation of food service establishments.[10] After opposition from state officials, FDA re-evaluated its own priorities and abandoned the proposal.[11] FDA recognized—with hundreds of thousands of restaurants and millions of meals served daily—the agency could not inspect or regulate more than an insignificant percentage of these establishments, and the primary responsibility had to remain with the state and local governments.

More on the important role that federal–state cooperation plays in food regulation is presented in a separate chapter on state laws.

[10] 39 Fed. Reg. 35438 (Oct. 1, 1974).
[11] 42 Fed. Reg. 15428 (March 22, 1977).

12.4 ADMINISTRATIVE ENFORCEMENT

12.4.1 List of Inspectional Observations

While it often is not considered an enforcement action, FDA's issuance of a List of Inspectional Observations (FD Form 483) is the most common first step before application of FDA's enforcement powers. Form 483 is provided to a firm after a facility inspection.

FDA expects corrective action by a company that addresses the inspector's observations specifically and systemically to prevent recurring violations. Because failure to correct the inspectional observations may result in enforcement action, and because the observations may be a first step toward enforcement action, Form 483 is best viewed as an enforcement action.

12.4.2 Warning Letters

FDA Warning Letters Warning letters are the FDA's most commonly used formal means of notifying a regulated firm that FDA believes the firm is in violation of the FD&C Act. Warning letters may be issued by one of FDA's District Offices or by one of the agency's centers (e.g., the Center for Food Safety and Applied Nutrition, or CFSAN). In March 2002 the FDA's Office of Chief Counsel began reviewing all warning letters. This policy was adopted to ensure that the letters going out from various offices are consistent with FDA's centralized policies and to ensure that they are legally sufficient under the FD&C Act. Warning letters are essentially FDA telling a violator to "knock it off."[12]

Warning letters explain what FDA alleges the law violations to be. They give the recipient, usually the president or CEO of the company, a short, specified period of time to respond, often 15 working days. The letter will include a warning that failure to respond may result in formal law enforcement action, without further notice. The two most common precipitating events for a warning letter are as a follow-up to an FDA inspection and notification of adverse result from an FDA laboratory test results of a product sample.

Every warning letter should be taken seriously by the recipient. A prompt and thoughtful written response may prevent possible additional enforcement action. FDA can and will take formal enforcement action—including seizure or prosecution—against nonresponsive companies. FDA's specific course of action, however, will depend on a number of factors, such as the seriousness and number of prior violations.

The most important purpose of a warning letter is to put the company receiving it on notice that it has received FDA's official position with respect

---

[12] FDA's Dr. John Jennings wrote in the margin of a report from subordinates about a violation "that summed up the very heart and soul of the regulatory letter. These words were, 'Tell them to knock it off.'" James O'Reilly, FOOD AND DRUG ADMINISTRATION, 6–5 (quoting Pines, *Regulatory Letters, Publicity and Recalls*, 31 FOOD, DRUG COSM. L.J. 352 (1976).

to allegations of law violation. The fact that a company received and ignored a warning letter can become persuasive evidence in court that a firm demonstrates a history of past violations or a pattern of violative behavior.

When writing a warning letter response, a business should realize that their letter may be made public through a Freedom of Information Act (FOIA) request. Response letters therefore should be written with awareness that competitors and the general public may read the letter.

In some situations a firm may wish to request that FDA post the firm's response letter along with the FDA warning letter. As a pilot program, FDA agreed to grant some of these requests (FDA reserves the discretion to not grant the request). It is not unusual for a company to want to tell its side of the story alongside FDA's allegations. However, it is not necessarily in a company's best interests to have its disagreement with an FDA warning letter response so publicly displayed. A firm may also wish their response letter to be posted when the response indicates that all violations have been promptly correctly. Of course, each situation is different and should be carefully evaluated before requesting public display of a response letter.

FDA warning letters may be viewed on FDA's Web site at <u>http://www.fda.gov/foi/warning.htm</u>.

Two examples of warning letters that are typical in style and format follow below. The names of the companies and individuals to whom they were addressed are redacted. However, remember that this information is available to the public.

* * * * *

Food and Drug Administration
New Orleans District
Southeast Region
6600 Plaza Drive Suite 400
New Orleans, Louisiana 70127
Telephone: (504) 253–4519
Facsimile: (504) 253–4520

March 18, 2003
WARNING LETTER NO. 2003-NOL-12
FEDERAL EXPRESS
Overnight Delivery

Mr. ****, President and Owner
A *** Company

Dear Mr. **:

The U.S. Food and Drug Administration (FDA) inspected your spice processing and associated food storage warehouse facility, *** during January 21, 22, and 28, 2003. The inspection was conducted to determine compliance with FDA's Current Good Manufacturing Practice requirements in Manufacturing,

Packing, or Holding Human Food, Title 21, Code of Federal Regulations (C.F.R.), Part 110. During the inspection, our investigators documented numerous insanitary conditions, which caused the ingredients and finished food products manufactured, packed, and/or held at your facility to become adulterated. The adulterated ingredients and finished food products are in violation of Sections 402(a)(3) and 402(a)(4) of the Federal Food, Drug, and Cosmetic Act, in that they consist in whole or in part of filthy substances, including rodent excreta pellets, and had been held under insanitary conditions whereby they may have become contaminated with filth.

Evidence of rodent activity was observed in, on, and near foods stored in your spice processing and food storage facility. This evidence included live and dead rodents, rodent excreta pellets, rodent urine stains, and gnawed food packaging. Evidence of rodent gnawing and general rodent activity was observed on several different food product packaging material including, but not limited to, poppy seeds, black pepper, and meat tenderizer. Our FDA laboratory confirmed the findings of rodent excreta pellets, rodent urine stains, rodent hair, and gnawed packaging based on samples taken from your facility during the inspection.

Our investigation of the general conditions in the spice processing and storage facility revealed: approximate $1'' \times 10''$ and $2'' \times 4''$ openings to the outdoors at the bottom of the northeast comer door and approximate $1'' \times 3''$ and $2'' \times 3''$ openings to the outdoors at the bottom of the southwest comer door. In addition, our investigators documented a can of insecticide ([redacted]) stored adjacent to finished product in your spice processing an

The above listed violations are not intended to be an all-inclusive list of deficiencies at your facility. It is your responsibility to ensure that your facility is operated in a sanitary manner.

We are aware that on January 21, 2003, you voluntarily destroyed a 50 pound sack of blue poppy seeds, a 50 pound sack of black pepper, a 50 pound sack of meat tenderizer, and two, 50 pound sacks of whole ground mustard. We are also aware that you made promises to our investigators during the inspection to correct some of the observed deficiencies.

At the conclusion of the inspection, our investigators presented to you a list of deficiencies on a Form FDA 483, Inspection Observations. You should take prompt action to correct these violations. Failure to promptly correct these deviations may result in regulatory action being initiated by FDA without further informal notice. Such actions may include the initiation of seizure, injunction, or prosecution actions in federal court.

You should notify this office, in writing, within 15 days of receipt of this letter, of the steps you have taken to correct the noted violations, including an explanation of each step being taken to prevent the recurrence of similar violations. If corrective action cannot be completed within 15 days, please state the reason for the delay and the time by which the corrections will be completed.

Your response should be directed to Rebecca A. Asente, Compliance Officer, U.S. Food and Drug Administration, 6600 Plaza Drive, Suite 400, New

Orleans, Louisiana 70127. Should you have any questions concerning the contents of this letter, you may contact Ms. Asente at (504) 253–4519.

Sincerely,

/s/

F. Dwight Herd
Acting District Director

New Orleans District
Food and Drug Administration
New England District
One Montvale Avenue
Stoneham, Massachusetts 02180
(781) 596–7700
FAX: (781) 596–7896

July 2, 2002
WARNING LETTER
NWE-21-02W
VIA FEDEX

****, Owner
**** Candy Company, Inc.
*** Main Street, Winter Hill
Somerville, Massachusetts 02145

Dear Mr. xxx:

The Food and Drug Administration (FDA) conducted an inspection of your facility located at *** Main Street, Somerville, Massachusetts, on February 5, 6, 8, 21 and March 1, 2002. Based on our review of product labels collected during the inspection, we have determined that your Pistachio Cream, Maple Walnut, Swiss Fudge, Peanut Butter Chips, and Peanut Butter Melts are misbranded within the meaning of Section 403(i)(2) of the Federal Food, Drug, and Cosmetic Act (the act) as follows:

- The products Pistachio Cream, Maple Walnut, and Swiss Fudge are fabricated from two or more ingredients, but the labels fail to bear the common or usual name of each ingredient in the product, as required by Section 403(i)(2) of the Act and by Title 21 of the Code of Federal Regulations Section 101.4(a)(1) [21 C.F.R. 101.4(a)(1)]. Specifically, the labels for your Pistachio Cream and Maple Walnut candies list the ingredient convertit and your Swiss Fudge label lists the ingredient white couventure. These do not appear to be the common or usual names for these ingredients.
- Your products Peanut Butter Chips and Peanut Butter Melts list the standardized foods peanut butter and chocolate in the ingredient listing. Furthermore, the standardized food milk chocolate is identified in the

ingredient lists for Swiss Fudge, and Butter Crunch. However, the ingredient statements for these products fail to bear the common or usual name of each ingredient in the standardized foods, as required by 21 C.F.R. 101.4(b)(2). This requirement may be met either by parenthetically listing the component ingredients contained in each of the standardized foods after the name of the standardized food, or by listing the component ingredients without listing the standardized food itself. Under the first alternative, the component ingredients must be listed in descending order of predominance in the standardized food; under the second, the component ingredients must be listed in descending order of predominance in the finished food.

The above violations are not meant to be an all-inclusive list of deficiencies in your labeling or at your facility. You should take prompt action to correct these violations, to establish procedures whereby such violations do not recur, and to review your operations and your product labels to ensure compliance with all applicable laws and regulations. Ensuring compliance with the laws and regulations is your responsibility.

Failure to do so may result in regulatory action without further notice. These actions include, but are not limited to, seizure and/or injunction. Information related to FDA laws and regulations, including the act and 21 C.F.R., may be obtained through links at www.fda.gov.

Please notify this office in writing within fifteen (15) working days from the date you receive this letter of the steps you have taken to correct the violations. For corrections that you cannot complete within the fifteen (15) working days, state the justification for the delay and your time frame for completion. Please provide documentation of the corrections as they are made, including copies of any revised labels, and explain your plan for preventing such violations in the future.

Please send your reply to the Food and Drug Administration, Attention: Bruce R. Ota, Compliance Officer, One Montvale Avenue, Fourth Floor, Stoneham, Massachusetts 02180.

Sincerely,

/s/

Gail T. Costello
District Director
New England District Office

* * * * *

FDA Cyber Letters   In early 2000, FDA created a new form of regulatory correspondence called "cyber letters." FDA issues these letters via e-mail to firms that market products on the Internet. Like warning letters, cyber letters describe alleged violations of the FD&C Act, and they should be taken seriously by their recipients. If the agency asks for a response to the letter, one should be provided promptly.

"Cyber letters" show that FDA, like a number of federal and state regulatory agencies, is monitoring commercial Internet marketing practices and following up with enforcement actions. Internet commercial activity in the United States is subject to a significant amount of regulation and enforcement. In addition, Internet marketers may receive similar enforcement correspondence from investigators at the FTC.

FDA Untitled Correspondence   FDA also sends out "untitled" correspondence. These letters are termed "untitled" because they lack a heading or title (e.g., "Warning Letter"). These untitled letters typically are used for regulatory issues that FDA does not consider serious enough to warrant a warning letter.

The letters often give the recipient a chance to respond or react in a way that will prevent further problems with the agency. The letters also serve as official notice from the agency that the company has been informed about FDA's concerns. Therefore, because they create a record, these kinds of letters also should be given careful thought. Written responses should be sent if appropriate.

Untitled correspondence typically ends with something like the following:

> At the end of the inspection, the FDA investigator left a list of inspectional observations at your firm. We have received your firm's written response, dated to the FDA483. Copies of this response and the FDA483 are enclosed.
>
> While this inspection found deficiencies of your quality system that would warrant a warning letter if not corrected, your written response has satisfied us that you either have taken or are taking appropriate corrective actions. At this time, the FDA does not intend to take further action based on these inspectional findings. The agency is relying on your commitment regarding corrective actions and, should we later observe that the deviations form the quality system regulation have not been remedied, future regulatory action (e.g., seizure, injunction and civil penalties) may be take without further notice.

FSIS Letters of Warning and Notices of Intended Enforcement Action (NOIE)   FSIS enforcement officials are also authorized to issue correspondence that puts regulated establishments on notice of alleged violations for which the Agency wants corrective action, specifically, Letters of Warning and Notices of Intended Enforcement Action (NOIE). All correspondence should be treated with the highest level of importance if it mentions any compliance or enforcement problem with the company's products. Responses to letters and notices should be timely and appropriate.

FSIS "letters of warning" are sent out by the district offices or headquarters for relatively minor violations of the law, namely those for which the agency does not intend to pursue formal law enforcement action. These letters put the establishments receiving them on notice that continued violations may result in formal action, including criminal prosecution.

"Notices of Intended Enforcement Action" (NOIE) are more serious written communications from FSIS. Unlike FDA warning letters, they do not afford the recipient 15 business days to respond before possible enforcement action. These notices are sent when the "Inspector in Charge" of an establishment has made a determination that repeated violations of one type or another warrant withholding the marks of inspection or suspending inspection. The Notice will explain the reasons for the intended enforcement action and give the recipient establishment three business days to contest the basis for the action or to show how it has come into compliance or will do so. Receipt of an NOIE requires immediate and serious attention.

12.4.3 Recalls

Recalls, where a manufacturer or importer retrieves violative product, are a common remedy for violative product. Firms are not required to report voluntary recalls, but FDA requests that they be advised of recalls involving product noncompliant with the FD&C Act.[13] Reporting recalls to FDA is a good policy because it maintains a good relationship with the FDA, demonstrates that the firm is responsible and capable of handling their own noncompliance, and lets FDA know that an enforcement action is unnecessary.

FDA lacks the general authority to order a recall. However, most manufacturers cooperate with FDA's requests for voluntary recall as a way to head off FDA enforcement action. If a firm does not comply with FDA's request to recall a product, FDA may take legal action under the FD&C Act, typically seizure of available product. FDA may also seek an injunction of the firm, including a court demand for recall of the product.

FDA guidelines for companies to follow when recalling FD&C Act violative products are published in 21 C.F.R. part 7.[14] These guidelines make clear that FDA expects these firms to take full responsibility for product recalls, including follow-up checks to ensure that recalls were successful. The guidelines also call on manufacturers and distributors to develop contingency plans for product recalls that can be put into effect if and when needed. The guidelines categorize all recalls into one of three classes according to the level of hazard involved.

- Class I recalls are for dangerous or defective products that predictably could cause serious health problems or death. Examples of products that

---

[13] Some voluntary recalls, of course, are conducted for reasons other than violations of the FD&C Act.

[14] *See, also*, FDA's GUIDANCE FOR INDUSTRY: PRODUCT RECALLS, INCLUDING REMOVALS AND CORRECTIONS, *available at*: http://www.fda.gov/ora/compliance_ref/recalls/ggp_recall.htm (last visited Sept. 10, 2008).

could fall into this category are a food containing botulinum toxin or food with undeclared allergens.

- Class II recalls are for products that might cause a temporary health problem or pose only a slight threat of a serious nature.
- Class III recalls are for products that are unlikely to cause any adverse health reaction but violate FDA labeling or manufacturing regulations. Examples are off-taste, color, or leaks in a bottled drink, and lack of English-language labeling on a retail food.

While the vast majority of recalls are voluntary, FDA does have authority to order a recall in specified circumstances, which are:

- medical devices when there is a reasonable possibility that a device could cause "adverse health consequences or death";[15]
- licensed biological products when the product presents an "imminent or substantial hazard to the public health";[16]
- infant formula when the product lacks the required nutrients or is other-wise adulterated or misbranded[17]

12.4.4 Debarment

A company or person who is convicted of certain FD&C Act violations can be "debarred" from future FDA-regulated activities. The disbarment penalty was enacted in 1992 in response to a generic drug scandal of the late 1980s as a means of preventing fraud and misconduct, such as bribery.

The Bioterrorism Act added a provision to FD&C Act section 306 to provide the penalty of debarment for repeated or serious food import violations. Specifically, debarment may be imposed when a "person has been convicted of a felony for conduct relating to the importation into the United States of any food; or . . . has engaged in a pattern of importing . . . adulterated food that presents a threat of serious adverse health consequences or death to humans or animals."[18] Debarment is a complete prohibition "from importing . . . food . . . or . . . offering . . . [food] for import into the United States."[19]

The Bioterrorism Act also makes it a prohibited act under the FD&C Act to import or offer "for import into the United States . . . an article of food by, with the assistance of, or at the direction of, a person [who has been] debarred."[20]

---

[15] FD&C Act § 518(e), 21 U.S.C. 360h(e).
[16] Public Health Services Act § 351(d)(2), 42 U.S.C. § 262(d)(2).
[17] FD&C Act § 412, 21 U.S.C. § 350a.
[18] 21 U.S.C. § 335a(b)(3).
[19] 21 U.S.C. § 335a(b)(1)(C).
[20] 21 U.S.C. § 331(cc): "The importing or offering for import into the United States of an article of food by, with the assistance of, or at the direction of, a person debarred under section 306(b)(3)."

12.4.5 Import Detentions

Import products do not have the same legal protection as product already in interstate commerce. Federal officials can bar incoming shipments and order their re-export or destruction with less Due Process concern than for comparable domestic products. Import detentions were covered in more detail in Chapter 11.

12.4.6 Civil Penalties

Civil Money Penalty Authority Congress authorized FDA to obtain civil money penalties for certain product violations. These consist of seven areas:

- Medical devices
- Prescription drug marketing
- Radiation-emitting products
- Biological products
- Generic drugs
- Mammography quality
- Childhood vaccination

Civil money penalties are fines FDA may assess administratively or through complaint in a federal court.

Consent Decrees Although FDA lacks statutory authority to seek civil money penalties for food products, consent decree settlements may result in the equivalent of a civil fine. Negotiated settlements with FDA have the virtue of solving problems with conservation of expenditure of effort in court for both sides.

In a recent case the agency recouped profits of $100 million from a company, and successes such as this may lead to this type of enforcement settlement used more frequently.[21] One theory for seeking such large settlements is "disgorgement of profits" wrongfully obtained from the sale of violative products. Another theory is restitution, where the money obtained from a company charged with wrongdoing is used to compensate victims of the wrongdoing. These are "equitable remedies" that do not require statutory language in the FD&C Act.

Recently FDA's policy has been to name individuals, including the company CEO in most consent decrees. FDA's principal reasons for this policy is that the Agency wants to be able to hold an individual or individuals personally accountable for ensuring that corrections are made, and because named individual defendants serve as public examples for a deterrent effect.

---

[21] See Eric Bloomberg, *Abbott Laboratories Consent Decree and Individual Responsibility under the Federal Food Drug and Cosmetic Act*, 55 FOOD & DRUG L.J. 145 (2000).

12.4.7 Withdrawal of Product Approvals

Some FDA-regulated products require FDA to approve an application before they can be marketed. The FD&C Act provides that such approvals may be withdrawn under certain circumstances, such as for misbranding or when new information indicates that a product is no longer safe.

Generally, before withdrawal of product approval, the FDA will publish a notice in the Federal Register of their intent to withdraw and provide the opportunity for comment. However, when a public health hazard exists, FDA may temporarily suspend approval without prior notice or opportunity for a hearing.

12.4.8 Inspection-Related Enforcement Powers of FSIS

Related to FSIS's inspection authority are a number of unique enforcement options provided to the agency. Under the <u>Federal Meat Inspection Act</u> (FMIA) and the <u>Poultry Products Inspection Act</u> (PPIA), meat and poultry establishments operate under a "continuous inspection" system, where an inspector is required to be present when the establishment is in operation (although, in practice, not necessarily every minute).

Some of these inspections will be discussed in more detail in the following chapter on inspections. Below is a summary of FSIS' major inspection-related enforcement powers.[22]

Regulatory control action. FSIS inspectors may invoke a number of different enforcement powers, including retention of the product, rejection of equipment, slowing or stopping of lines, or refusal to allow the processing of specifically-identified product.[23]

Withholding action. Action that refuses to allow the marks of inspection to be applied to products[24] may be taken against all products in an establishment or just product from a particular process. Withholding actions may be taken with or without prior notification to the establishment.

Suspension. The interruption in the assignment of FSIS inspection employees to all or part of an establishment, which is a suspension of inspection, has the practical effect of shutting down the operations of the facility. It therefore has a more impact than either a withholding or regulatory control action. As with withholding actions, suspensions may be imposed without prior notification.[25]

Withdrawal of inspection. The total withdrawal of the grant of federal inspection puts an establishment out of business. Withdrawal of inspection

---

[22] Rules of Practice, Food Safety and Inspection Service (FSIS), Final Rule, 64 Fed. Reg. 66,541 (Nov. 29, 1999) codified in 9 C.F.R. part 500.
[23] 9 C.F.R. § 500.1(a).
[24] 9 C.F.R. § 500.1(b).
[25] 9 C.F.R. § 500.6.

comes only after an administrative hearing. The conditions that warrant grounds for withdrawal of inspection usually are serious violations plus being "unfit to engage in any business requiring inspection."[26]

Withdrawal of inspection is sought infrequently, usually only following criminal convictions. The "prior criminal convictions" become evidence that a person is "unfit to engage in business" under the inspection regulations. Other grounds for being found unfit to engage in business is when establishment personnel assault, intimidate, or interfere with federal inspection service.[27]

12.5   FDA CIVIL COURT ACTIONS

12.5.1   Seizure

Seizures are civil judicial actions that have traditionally been one of FDA's primary enforcement tools because seizures provide an expeditious means of removing violative products from the marketplace. Section 304 of the FD&C Act provides FDA's seizure authority.[28] Seizures are less resource intensive for the FDA than many other actions, such as injunctions and criminal prosecution. In addition seizures can be processed through FDA very quickly, if necessary. Therefore seizures are considered an important and effective enforcement tool relative to the resources expended and compliance achieved.

Generally, before a seizure is initiated, an FDA district will recommend a seizure action to FDA headquarters. The appropriate center at FDA headquarter considers issues such as prior warning, the significance of the violation, current status of the firm, pending and adjudicated seizure actions, the public health risk, and the amount of the product to be seized. After review, by the center and a final legal review by the Office of General Counsel, the seizure action is transmitted to the appropriate FDA district office. This central review is designed to maintain uniformity, equity, and credibility in FDA's decision-making process for seizures (the agency uses the same process of analysis for injunctions).

FDA brings its recommendation for seizure actions to the U.S. attorney in the state where the product is located. If the Department of Justice (DOJ) accepts the FDA recommendation, the U.S. attorney files a complaint in federal district court on behalf of the FDA. A seizure is an *ex parte* order obtained from the federal court under rules dating back to early admiralty law through an *in rem* [29] action against the named goods.

After the complaint is filed, the federal district court issues a warrant for the arrest (seizure) of the product. Seizures run against the goods, not the company that owns them.[30] Thus, seizure cases have unusual names, such as "*United States v. 10,000 prophylactic devices with holes.*"

---

[26] 9 C.F.R. § 500.1(c).
[27] Rules of Practice, *supra* note 22.
[28] 21 U.S.C. § 334.
[29] *In rem* means the action is against the things, rather than against a person.
[30] 21 U.S.C. § 334.

FDA must act with the Due Process restrictions. Nonetheless, the restrictions are viewed narrowly because of the public health and safety concern underlying FDA powers.[31] The due process rights are preserved by allowing an immediate seizure with a post-seizure hearing.[32] At the hearing the court will decide whether the allegations in the complaint have been proven, and therefore to condemn the product, or if the court finds the complaint has not been proven, to release the goods from seizure. Condemnation only sustains the government's position, and the firm may be allowed to recondition the product (correct the defects)—such as relabeling, sorting good product from bad, or cleaning.[33]

If a seizure is contested, the FDA has the burden of proving its case by a preponderance of the evidence.[34] However, nearly all seizures result in condemnation, often by default or consent. Partly, this is due to the extensive review involved, which results in strong cases. However, the other practical reason is that the owner of the seized goods generally finds it in his or her best interest to quickly resolve the seizure. For example, reconditioning—if it will be permitted—cannot begin until the product is first condemned. In addition, while the contested case proceeds, perishable commodities will lose value.[35]

12.5.2 USDA

FSIS's seizure authority is also statutory, granted expressly under the FMIA and PPIA. 21 U.S.C. § 673 (FMIA) and 21 U.S.C. § 467b (PPIA). In FSIS plants that are subject to continuous inspection, FSIS inspectors may administratively detain product they consider to be adulterated. This prevents the product from leaving the plant.

Expanded Administrative Detention Authority The FDA previously was without authority to immediately embargo or detain food that the Agency believed to be in violation of the FD&C Act. However, section 303 of the Bioterrorism Act granted FDA limited detention authority. FDA can order food to be detained up to 20 days (or 30 days if needed to pursue seizure or injunction) if the official "has credible evidence or information indicating that [the food] presents a threat of serious adverse health consequences or death to humans or animals."[36]

---

[31] *See* United States v. Article of Device . . . Theramatic, 715 F2d 1339 (9th Cir. 1983).

[32] Juici-Rich Prods, Inc. v. Lowe, 735 F. Supp. 1387 (CD Ill. 1990).

[33] *See, e.g.*, United States v. 43½ Gross Rubber Prophylactics Labeled in Part "Xcello's Prophylactics," 65 F. Supp. 534 (1946).

[34] *See, e.g.*, United States v. 60 28-Capsule Bottles . . . "Unitrol," 325 F.2d 513 (3d Cir. 1963).

[35] As long as the government's action was reasonable, the claimant who wins a contested seizure case is not likely to recover the lost value of perishable goods. United States v. 2,116 Boxes Boned Beef, 516 F. Supp. 321 (1981) *aff'd* 726 F.2d. 1481 (10th Cir. 1984).

[36] 15 21 U.S.C. § 334(h)(1)(A).

* * * * *

Bioterrorism Act sec. 303 (Section 304 of the Federal Food, Drug, and Cosmetic Act (21 U.S.C. 334))

(h) **Administrative Detention of Foods**.—

(1) Detention authority.—

(A) In general.—An officer or qualified employee of the Food and Drug Administration may order the detention, in accordance with this subsection, of any article of food that is found during an inspection, examination, or investigation under this Act conducted by such officer or qualified employee, if the officer or qualified employee has credible evidence or information indicating that such article presents a threat of serious adverse health consequences or death to humans or animals.

(B) Secretary's approval.—An article of food may be ordered detained under subparagraph (A) only if the Secretary or an official designated by the Secretary approves the order. An official may not be so designated unless the official is the director of the district under this Act in which the article involved is located, or is an official senior to such director.

(2) Period of detention.—An article of food may be detained under paragraph (1) for a reasonable period, not to exceed 20 days, unless a greater period, not to exceed 30 days, is necessary, to enable the Secretary to institute an action under subsection (a) or section 302. The Secretary shall by regulation provide for procedures for instituting such action on an expedited basis with respect to perishable foods.

(3) Security of detained article.—An order under paragraph (1) with respect to an article of food may require that such article be labeled or marked as detained, and shall require that the article be removed to a secure facility, as appropriate. An article subject to such an order shall not be transferred by any person from the place at which the article is ordered detained, or from the place to which the article is so removed, as the case may be, until released by the Secretary or until the expiration of the detention period applicable under such order, whichever occurs first. This subsection may not be construed as authorizing the delivery of the article pursuant to the execution of a bond while the article is subject to the order, and section 801(b) does not authorize the delivery of the article pursuant to the execution of a bond while the article is subject to the order.

* * * * *

Remedies and Consequences   Once a federal court orders goods to be seized, the company may attempt to reclaim them for purposes of reconditioning; otherwise, they may be destroyed. If FDA approves a reconditioning plan, the result usually will be a consent decree for reconditioning. FDA inspectors then oversee the relabeling or reprocessing by the company, and FDA sends the court a report of the results of the recondition. The court then can dismiss the seizure order and the goods are released to the claimant.

If the reconditioning fails, the goods are destroyed under FDA's supervision. If the court agrees that destruction must be ordered because the goods were unsafe or were unlikely to meet FDA standards even after reconditioning, then FDA prepares for the court an order of forfeiture. Federal marshals either destroy the forfeited products or otherwise dispose of them in accordance with FDA instructions.

Once a seizure motion has been served, the seized goods cannot be moved without permission of the court. The goods are legally considered to be under the control of the court regardless of their physical location. Violation of a seizure order is contempt of the federal court.

Role of the States in Seizure   FDA has a longstanding, effective system of working with state agencies on seizures because must state possess authority under state laws to embargo goods immediately.

12.5.3 Injunction

An injunction is the command of a federal court that imposes an enforceable order against named persons. Because injunctions are resource intensive for the FDA (e.g., the injunctions must be monitored), injunctions are rarely sought by the FDA and generally are only used when all other enforcement actions have been exhausted without success. Recurrent violations are generally the cause for seeking an injunction.

Injunctions can be prohibitions, such as an order to cease violative behavior, or mandatory commands, such as an order to clean up a facility. Injunctions can shut down a company's operation until compliance with the requirements is achieved; or even put a person or company permanently out of business.[37] One of the few limits on injunction may be the authority to order a recall. The courts have split on the question of whether the FD&C Act authorizes recalls as part of the injunctive relief.

FDA's injunction authority derives from FD&C Act section 302, which permits the agency to stop conduct alleged to violate the Act when there is a likelihood that violations will continue.[38] FSIS' authority to enjoin violative practices derives from both the FMIA and the PPIA.[39]

---

[37] On very rare occasions, persons have been barred from ever engaging in a certain business again.
[38] 21 U.S.C. § 332.
[39] 21 U.S.C. § 674 (FMIA) and 21 U.S.C. § 467c (PPIA).

12.5.4   Contempt Action

Contempt is the judicial power to enforce court orders. The FDA may bring a contempt motion when a person refuses a warrant for inspection or when an injunction or seizure has been violated. Criminal contempt can result in imprisonment.

12.5.5   Destruction of Products without a Hearing

The FD&C Act provides that FDA must proceed under section 334 with an action to seize and condemn products. This necessarily provides judicial oversight of the issuance of a seizure and the opportunity for a hearing before food or other product may be destroyed.

However, in many cases the FDA will refer a case to state authorities, whose powers of seizure may be greater than FDA. States may have the power to seize and even destroy food or other products without a hearing. This power raises Fourth Amendment implications, which are addressed in *North American Cold Storage Co. v. City of Chicago*, 211 U.S. 306 (1908).

Under its police power, the State has the inherent power to seize and destroy—without a hearing—food that is unwholesome or unfit. Although the Fourteenth Amendment requires notice and opportunity to be heard, the hearing may be provided after the destruction of the property in circumstances where public nuisance and importance of protecting the public health are involved. Due Process is satisfied by the right of the party whose property was destroyed to have a right of action after the seizure.

In addition, the power of the State to destroy food that is unfit for human consumption is not taken away because some value may remain in the food for other uses (e.g., animal feed) when the product has been kept to be sold at some time as food. See *Reduction Company v. Sanitary Works*, 199 U.S. 306, and *Gardner v. Michigan*, 199 U.S. 325.

In *North American Cold Storage Co.* the complainant challenged the constitutionality of an ordinance allowing summary seizure and destruction of food because the ordinance did not provide for notice and opportunity to be heard before such destruction. The Court held that the ordinances are not unconstitutional as depriving persons of property without due process of law. The reasoning is explained in the case that follows.

* * * * *

North American Cold Storage Co. v. City of Chicago et al.
211 U.S. 306 (1908)

Statement by Mr. Justice PECKHAM:

The bill of complaint in this case . . . was filed against the city of Chicago and the various individual defendants in their official capacities—commissioner of health of the city of Chicago, secretary of the department of health, chief food inspector of the department of health, and inspectors of that department, and policemen of the city—for the purpose of obtaining an injunction under the circumstances set forth in the bill. It was therein alleged that the complainant was a cold storage company, having a cold storage plant in the city of Chicago, and that it received, for the purpose of keeping in cold storage, food products and goods as bailee for hire; . . . that it received some 47 barrels of poultry on or about October 2, 1906, from a wholesale dealer, in due course of business, to be kept by it and returned to such dealer on demand; that the poultry was, when received, in good condition and wholesome for human food, and had been so maintained by it in cold storage from that time, and it would remain so, if undisturbed, for three months; that on the 2d of October, 1906, the individual defendants appeared at complainant's place of business and demanded of it that it forthwith deliver the 47 barrels of poultry for the purpose of being by them destroyed, the defendants alleging that the poultry had become putrid, decayed, poisonous, or infected in such a manner as to render it unsafe or unwholesome for human food. The demand was made under 1161 of the Revised Municipal Code of the City of Chicago for 1905, which reads as follows:

> Every person being the owner, lessee, or occupant of any room, stall, freight house, cold storage house, or other place, other than a private dwelling, where any meat, fish, poultry, game, vegetables, fruit, or other perishable article adapted or designed to be used for human food shall be stored or kept, whether temporarily or otherwise, and every person having charge of, or being interested or engaged, whether as principal or agent, in the care of or in respect to the custody or sale of any such article of food supply, shall put, preserve, and keep such article of food supply in a clean and wholesome condition, and shall not allow the same, nor any part thereof, to become putrid, decayed, poisoned, infected, or in any other manner rendered or made unsafe or unwholesome for human food; and it shall be the duty of the meat and food inspectors and other duly authorized employees of the health department of the city to enter any and all such premises above specified at any time of any day, and to forthwith seize, condemn, and destroy any such putrid, decayed, poisoned, and infected food, which any such inspector may find in and upon said premises.

The complainant refused to deliver up the poultry, on the ground that the section above quoted of the Municipal Code of Chicago, in so far as it allows the city or its agents to seize, condemn, or destroy food or other food products, was in conflict with that portion of the 14th Amendment which provides that no state shall deprive any person of life, liberty, or property without due process of law; nor deny to any person within its jurisdiction the equal protection of the laws.

After the refusal of the complainant to deliver the poultry the defendants stated that they would not permit the complainant's business to be further conducted until it complied with the demand of the defendants and delivered up the poultry, nor would they permit any more goods to be received into the warehouse or taken from the same, and that they would arrest and imprison any person who attempted to do so, until complainant complied with their demand and delivered up the poultry. Since that time the complainant's business has been stopped and the complainant has been unable to deliver any goods from its plant or receive the same.

The bill averred that the attempt to seize, condemn, and destroy the poultry, without a judicial determination of the fact that the same was putrid, decayed, poisonous, or infected, was illegal; and it asked that the defendants, and each of them, might be enjoined from taking or removing the poultry from the warehouse, or from destroying the same, and that they also be enjoined from preventing complainant delivering its goods and receiving from its customers, in due course of business, the goods committed to its care for storage.

In an amendment to the bill the complainant further stated that the defendants are now threatening to summarily destroy, from time to time, pursuant to the provisions of the above-mentioned section, any and all food products which may be deemed by them, or either of them, as being putrid, decayed, poisonous, or infected in such manner as to be unfit for human food, without any judicial determination of the fact that such food products are in such condition. . . .

Mr. Justice Peckham, after making the foregoing statement, delivered the opinion of the court:

In this case the ordinance in question is to be regarded as in effect a statute of the state, adopted under a power granted it by the state legislature, and hence it is an act of the state within the 14th Amendment. . . .

We think there was jurisdiction, and that it was error for the court to dismiss the bill on that ground. . . . The bill contained a plain averment that the ordinance in question violated the 14th Amendment, because it provided for no notice to the complainant or opportunity for a hearing before the seizure and destruction of the food. A constitutional question was thus presented to the court, over which it had jurisdiction, and it was bound to decide the same on its merits. . . . A constitutional question being involved, an appeal may be taken directly to this court from the circuit court.

Holding there was jurisdiction in the court below, we come to the merits of the case. The action of the defendants, which is admitted by the demurrer, in refusing to permit the complainant to carry on its ordinary business until it delivered the poultry, would seem to have been arbitrary and wholly indefensible. Counsel for the complainant, however, for the purpose of obtaining a decision in regard to the constitutional question as to the right to seize and destroy property without a prior hearing, states that he will lay no stress here upon that portion of the bill which alleges the unlawful and forcible taking possession of complainant's business by the defendants. He states in his brief as follows:

> There is but one question in this case, and that question is, Is section 1161 of the Revised Municipal Code of Chicago in conflict with the due process of law provi-

sion of the 14th Amendment, is this: that it does not provide for notice and an opportunity to be heard before the destruction of the food products therein referred to? . . .

The general power of the state to legislate upon the subject embraced in the above ordinance of the city of Chicago, counsel does not deny. Nor does he deny the right to seize and destroy unwholesome or putrid food, provided that notice and opportunity to be heard be given the owner or custodian of the property before it is destroyed. We are of opinion, however, that provision for a hearing before seizure and condemnation and destruction of food which is unwholesome and unfit for use is not necessary. The right to so seize is based upon the right and duty of the state to protect and guard, as far as possible, the lives and health of its inhabitants, and that it is proper to provide that food which is unfit for human consumption should be summarily seized and destroyed to prevent the danger which would arise from eating it. The right to so seize and destroy is, of course, based upon the fact that the food is not fit to be eaten. Food that is in such a condition, if kept for sale or in danger of being sold, is in itself a nuisance, and a nuisance of the most dangerous kind, involving, as it does, the health, if not the lives, of persons who may eat it. A determination on the part of the seizing officers that food is in an unfit condition to be eaten is not a decision which concludes the owner. The ex parte finding of the health officers as to the fact is not in any way binding upon those who own or claim the right to sell the food. It a party cannot get his hearing in advance of the seizure and destruction, he has the right to have it afterward, which right may be claimed upon the trial in an action brought for the destruction of his property; and in that action those who destroyed it can only successfully defend if the jury shall find the fact of unwholesomeness, as claimed by them. The often-cited case of *Lawton v. Steele*, substantially holds this. By the 2d section of an act of the legislature of the state of New York of 1800 it was provided that any "net . . . for capturing fish which was floated upon the water or found or maintained in any of the waters of the state," in violation of the statutes of the state for the protection of fish, was a public nuisance, and could be abated and summarily destroyed, and that no action for damages should lie or be maintained against any person for or on account of seizing or destroying such nets. Nets of the kind mentioned in that section were taken and destroyed by the defendant, and the owner commenced action against him to recover damages for such destruction. That portion of the section which provided that no action for damages should lie was applicable only to a case where the seizure or destruction had been of a nature amounting to a violation of the statute, and of course did not preclude an action against the person making a seizure if not made of a net which was illegally maintained. The seizure and destruction were justified by the defendant in the action, and such justification was allowed in the state courts and in this court. Mr. Justice Brown, in delivering the opinion of this court, said:

Nor is a person whose property is seized under the act in question without his legal remedy. If in fact his property has been used in violation of the act, he has no just reason to complain; if not, he may replevy his nets from the officer seizing them, or, if they have been destroyed, may have his action for their value. In such cases the burden would be upon the defendant to prove a justification under the statute. As was said by the supreme court of New Jersey, in a similar case: "The party is not, in point of fact, deprived of a trial by jury. . . ." Indeed, it is scarcely possible that any actual injustice could be done in the practical administration of the act.

The statute in the above case had not provided for any hearing of the question of violation of its provisions, and this court held that the owner of the nets would not be bound by the determination of the officers who destroyed them, but might question the fact by an action in a judicial proceeding in a court of justice. The statute was held valid, although it did not provide for notice or hearing. And so in *People ex rel. Copcutt v. Board of Health*, the question arose in a proceeding by certiorari, affirming the proceedings of the board of health of the city of Yonkers, by which certain dams upon the Nepperhan River were determined to be nuisances and ordered to be removed. The court held that the acts under which the dams were removed did not give a hearing in express terms nor could the right to a hearing be implied from any language used in them, but that they were valid without such provision, because they did not make the determination of the board of health final and conclusive on the owners of the premises wherein the nuisances were allowed to exist; that before such a final and conclusive determination could be made, resulting in the destruction of property, the imposition of penalties and criminal punishments, the parties proceeded against must have a hearing, not as a matter of favor, but as a matter of right, and the right to a hearing must be found in the acts; that if the decisions of these boards were final and conclusive, even after a hearing, the citizen would, in many cases, hold his property subject to the judgments of men holding ephemeral positions in municipal bodies and boards of health, frequently uneducated, and generally unfitted to discharge grave judicial functions. It was said that boards of health under the acts referred to could not, as to any existing state of facts, by their determination make that a nuisance which was not in fact a nuisance; that they had no jurisdiction to make any order or ordinance abating an alleged nuisance unless there were in fact a nuisance; that it was the actual existence of a nuisance which gave them jurisdiction to act. There being no provision for a hearing, the acts were not void nevertheless, but the owner had the right to bring his action at common law against all the persons engaged in the abatement of the nuisance to recover his damages, and thus he would have due process of law; and if he could show that the alleged nuisance did not in fact exist, he will recover judgment, notwithstanding the ordinance of the board of health under which the destruction took place.

The same principle has been decided by the supreme judicial court of Massachusetts. The case of *Salem v. Eastern R. Co.* was an action brought to recover moneys spent by the city to drain certain dams and ponds declared by the board of health to be a nuisance. The court held that, in a suit to recover such expenses incurred in removing a nuisance, when prosecuted against a party on the ground that he caused the same, but who was not heard, and had no opportunity to be heard upon the questions before the board of health, such party is not concluded in the findings or adjudications of that board, and may contest all the facts upon which his liability is sought to be established.

Miller v. Horton is in principle like the case before us. It was an action brought for killing the plaintiff's horse. The defendants admitted the killing, but justified the act under an order of the board of health, which declared that the horse had the glanders, and directed it to be killed. The court held that the decision of the board of health was not conclusive as to whether or not the horse was diseased, and said that: "Of course there cannot be a trial by jury before killing an animal supposed to have a contagious disease, and we assume that the legislature may authorize its destruction in such emergencies without a hearing beforehand. But it does not follow that it can throw the loss on the owner without a hearing. If he cannot be heard beforehand he may be heard afterward. The statute may provide for paying him in case it should appear that his property was not what the legislature had declared to be a nuisance, and may give him his hearing in that way. If it does not do so, the statute may leave those who act under it to proceed at their peril, and the owner gets his hearing in an action against them."

And in *Stone v. Heath* the court held that, under the statute, it had no power to restrain the board of health from abating nuisances and from instituting proceedings against plaintiff on account of his failure to abate them, as provided for in the statute, because the board of health had adjudged that a nuisance existed and had ordered it to be abated by the plaintiff, yet still the question "whether there was a nuisance, or whether, if there was one, it was caused or maintained by the parties charged therewith, may be litigated by such parties in proceedings instituted against them to recover the expenses of the abatement, or may be litigated by the parties whose property has been injured or destroyed in proceedings instituted by them to recover for such loss or damage, and may also be litigated by parties charged with causing or maintaining the nuisance in proceedings instituted against them for neglect or failure to comply with the orders of the board of health directing them to abate the same." In that way they had a hearing and could recover or defend in case there was no nuisance.

Complainant, however, contends that there was no emergency requiring speedy action for the destruction of the poultry in order to protect the public health from danger resulting from consumption of such poultry. It is said that the food was in cold storage, and that it would continue in the same condition it then was for three months, if properly stored, and that therefore the defendants had ample time in which to give notice to complainant or the owner and have a hearing of the question as to the condition of the poultry; and, as the ordinance provided for no hearing, it was void. But we think this is not required. The power of the legislature to enact laws in relation to the public health being conceded, as it must be, it is to a great extent within legislative discretion as to whether any hearing need be given before the destruction of unwholesome food which is unfit for human consumption. If a hearing were to be always necessary, even under the circumstances of this case, the question at once arises as to what is to be done with the food in the meantime. Is it to remain with the cold storage company, and, if so, under what security that it will not be removed? To be sure that it will not be removed during the time necessary for the hearing, which might frequently be indefinitely prolonged, some guard would probably have to be placed over the subject-matter of investigation, which would involve expense, and might not even then prove effectual. What is the emergency which would render a hearing unnecessary? We think when the question is one regarding the destruction of food which is not fit for human use, the emergency must be one which would fairly appeal to the reasonable discretion of the legislature as to the necessity for a prior hearing, and in that case its decision would not be a subject for review by the courts. As the owner of the food or its custodian is amply protected against the party seizing the food, who must, in a subsequent action against him, show as a fact that it was within the statute, we think that due process of law is not denied the owner or custodian by the destruction of the food alleged to be unwholesome and unfit for human food without a preliminary hearing. The cases cited by the complainant do not run counter to those we have above referred to.

Even if it be a fact that some value may remain for certain purposes in food that is unfit for human consumption, the right to destroy it is not, on that account, taken away. The small value that might remain in said food is a mere incident, and furnishes no defense to its destruction when it is plainly kept to be sold at some time as food.

The decree of the court below is modified by striking out the ground for dismissal of the bill as being for want of jurisdiction, and, as modified, is affirmed.

Mr. Justice Brewer dissents.

* * * * *

12.6 CRIMINAL ACTIONS

12.6.1 Strict Liability

Criminal law generally requires *mens rea*, criminal or specific intent, to find criminal culpability. However, apart from constitutional due process safeguards, few of the criminal law standards apply to the prosecution of violators under the Food, Drug, and Cosmetic Act. The reason is that the controlling standard in food law prosecutions is strict liability.

Strict liability means that guilt applies regardless of intent or conventional fault. Ignorance of the violation, lack of intent to commit a violative act, and lack of personal involvement are not defenses against FD&C Act violations. Liability for the crime arises out of one's authority and power to avoid the violation. Strict liability makes the defense of a criminal charge much more difficult.

The rationale for this stringent level of liability is society's need for a very tough deterrent where irreparable harm may be done to the public health by violative products. Violations—whether by error or intent—can have such public health consequences that it justifies a harsh standard. Put another way, the FD&C Act imposes a heavy burden of responsibility in return for the privilege of doing business in an important area of public welfare.[40] Without deterrence through personal liability of company officials, corporations might otherwise view the penalties (fines) as the cost of doing business.[41]

A further advantage of such a strict liability scheme is that increased legal liability exposure not only creates deterrence but also fosters an environment where voluntary cooperation is far more likely. FDA prosecutes very few criminal cases, but their existence helps ensure cooperation from other regulated firms.

The legal standard for individual criminal liability under the FD&C Act is laid out in the landmark case, *United States v. Park*, 421 U.S. 658 (1975). *Park* held a corporate CEO responsible regardless of whether he had actual knowledge of the insanitary conditions in the company's food storage warehouse:

> the act imposes not only a positive duty to seek out and remedy violations when they occur but also, and primarily, a duty to implement measures that will insure that violations will not occur. The requirements of foresight and vigilance imposed on responsible corporate agents are beyond question demanding, and perhaps onerous, but are no more stringent than the public has a right to expect of those who voluntarily assume positions of authority in business enterprises whose services and products affect the health and well-being of the public that supports them.
>
> —*United States v. Park*, 421 U.S. 658, 672 (1975).

In *United States v. Park*, the defendant, the president of a grocery store chain, was individually convicted of causing adulteration of food, and he appealed. The grocery store chain pled guilty and did not appeal. The Supreme Court held, inter alia, that the trial court's instructions—which instructed the jury that the defendant need *not* have personally participated in the situation that caused the alleged adulteration—adequately focused on the issue of defendant's authority respecting the conditions that formed the basis of the alleged violations. Thus, to find guilt, the jury must only find that

---

[40] Smith v. California, 361 U.S. 147 (1959).
[41] United States v. Dotterweich, 320 U.S. 277, 282–83 (1943).

defendant "had a responsible relation to the situation" and that by virtue of his position defendant had authority and responsibility to deal with such conditions.

* * * * *

United States v. Park

421 U.S. 658 (1975)

Mr. Chief Justice BURGER delivered the opinion of the Court.

We granted certiorari to consider whether the jury instructions in the prosecution of a corporate officer under § 301(k) of the Federal Food, Drug, and Cosmetic Act, 21 U.S.C. § 331(k), were appropriate under *United States v. Dotterweich.*

Acme Markets, Inc., is a national retail food chain with approximately 36,000 employees, 874 retail outlets, 12 general warehouses, and four special warehouses. Its headquarters, including the office of the president, respondent Park, who is chief executive officer of the corporation, are located in Philadelphia, Pa. In a five-count information filed in the United States District Court for the District of Maryland, the government charged Acme and respondent with violations of the Federal Food, Drug and Cosmetic Act. Each count of the information alleged that the defendants had received food that had been shipped in interstate commerce and that, while the food was being held for sale in Acme's Baltimore warehouse following shipment in interstate commerce, they caused it to be held in a building accessible to rodents and to be exposed to contamination by rodents. These acts were alleged to have resulted in the food's being adulterated within the meaning of 21 U.S.C. §§ 342(a)(3) and (4), in violation of 21 U.S.C. § 331(k).

Acme pleaded guilty to each count of the information. Respondent pleaded not guilty. The evidence at trial demonstrated that in April 1970 the Food and Drug Administration (FDA) advised respondent by letter of insanitary conditions in Acme's Philadelphia warehouse. In 1971 the FDA found that similar conditions existed in the firm's Baltimore warehouse. An FDA consumer safety officer testified concerning evidence of rodent infestation and other insanitary conditions discovered during a 12-day inspection of the Baltimore warehouse in November and December 1971. He also related that a second inspection of the warehouse had been conducted in March 1972. On that occasion the inspectors found that there had been improvement in the sanitary conditions, but that "there was still evidence of rodent activity in the building and in the warehouses and we found some rodent- contaminated lots of food items."

The government also presented testimony by the Chief of Compliance of the FDA's Baltimore office, who informed respondent by letter of the conditions at the Baltimore warehouse after the first inspection. There was testimony by Acme's Baltimore division vice president, who had responded to the letter on behalf of Acme and respondent and who described the steps taken to remedy the insanitary conditions discovered by both inspections. The government's final witness, Acme's vice president for legal affairs and assistant secretary, identified respondent as the president and chief executive officer of the company and read a bylaw prescribing the duties of the chief executive officer. He testified that respondent functioned by delegating "normal operating duties," including sanitation, but that he retained "certain things, which are the big, broad, principles of the operation of the company," and had "the responsibility of seeing that they all work together."

At the close of the government's case in chief, respondent moved for a judgment of acquittal on the ground that "the evidence in chief has shown that Mr. Park is not personally concerned in this Food and Drug violation." The trial judge denied the motion, stating that *United States v. Dotterweich* was controlling.

Respondent was the only defense witness. He testified that, although all of Acme's employees were in a sense under his general direction, the company had an "organizational structure for responsibilities for certain functions" according to which different phases of its operation were "assigned to individuals who, in turn, have staff and departments under them." He identified those individuals responsible for sanitation, and related that upon receipt of the January 1972 FDA letter, he had conferred with the vice president for legal affairs, who informed him that the Baltimore division vice president "was investigating the situation immediately and would be taking corrective action and would be preparing a summary of the corrective action to reply to the letter." Respondent stated that he did not "believe there was anything (he) could have done more constructively than what (he) found was being done."

On cross-examination, respondent conceded that providing sanitary conditions for food offered for sale to the public was something that he was "responsible for in the entire operation of the company," and he stated that it was one of many phases of the company that he assigned to "dependable subordinates." Respondent was asked about and, over the objections of his counsel, admitted receiving, the April 1970 letter addressed to him from the FDA regarding insanitary conditions at Acme's Philadelphia warehouse. He acknowledged that, with the exception of the division vice president, the same individuals had responsibility for sanitation in both Baltimore and Philadelphia. Finally, in response to questions concerning the Philadelphia and Baltimore incidents, respondent admitted that the Baltimore problem indicated the system for handling sanitation "wasn't working perfectly" and that as Acme's chief executive officer he was responsible for "any result which occurs in our company."

At the close of the evidence, respondent's renewed motion for a judgment of acquittal was denied. The relevant portion of the trial judge's instructions to the jury challenged by respondent is set out in the margin.[42] Respondent's counsel objected to the instructions on the ground that they failed fairly to reflect our decision in *United States v. Dotterweich, supra*, and to define "responsible relationship." The trial judge overruled the objection. The jury found respondent guilty on all counts of the information, and he was subsequently sentenced to pay a fine of $50 on each count.

The Court of Appeals reversed the conviction and remanded for a new trial. That court . . . stated that as "a general proposition, some act of commission or omission is an essential element of every crime." It reasoned that, although our decision in *United States v. Dotterweich*, had construed the statutory provisions under which respondent was tried to dispense with the traditional element of "awareness of some wrongdoing," the Court had not construed them as dispensing with the element of "wrongful action." The Court of Appeals concluded that the trial judge's instructions "might well have left the jury with the erroneous impression that Park could be found guilty in the absence of 'wrongful action' on his part," and that proof of this element was required by due process. It held, with one dissent, that the instructions did not "correctly state the law of the case," and directed that on retrial the jury be instructed as to "wrongful action," which might be "gross negligence and inattention in discharging . . . corporate duties and obligations or any of a host of other acts of commission or omission which would 'cause' the contamination of food."

The Court of Appeals also held that the admission in evidence of the April 1970 FDA warning to respondent was error warranting reversal, based on its conclusion that, "as this case was submitted to the jury and in light of the sole issue presented," there was no need for the evidence and thus that its prejudicial effect outweighed its relevancy. . . .

---

[42] "In order to find the Defendant guilty on any count of the Information, you must find beyond a reasonable doubt on each count. . . .

"Thirdly, that John R. Park held a position of authority in the operation of the business of Acme Markets, Incorporated.

"However, you need not concern yourselves with the first two elements of the case. The main issue for your determination is only with the third element, whether the Defendant held a position of authority and responsibility in the business of Acme Markets.

"The statute makes individuals, as well as corporations, liable for violations. An individual is liable if it is clear, beyond a reasonable doubt, that the elements of the adulteration of the food as to travel in interstate commerce are present. As I have instructed you in this case, they are, and that the individual had a responsible relation to the situation, even though he may not have participated personally.

"The individual is or could be liable under the statute, even if he did not consciously do wrong. However, the fact that the Defendant is pres(id)ent and is a chief executive officer of the Acme Markets does not require a finding of guilt. Though, he need not have personally participated in the situation, he must have had a responsible relationship to the issue. The issue is, in this case, whether the Defendant, John R. Park, by virtue of his position in the company, had a position of authority and responsibility in the situation out of which these charges arose." *Id.*, at 61–62.

We granted certiorari because of an apparent conflict among the Courts of Appeals with respect to the standard of liability of corporate officers under the Federal Food, Drug, and Cosmetic Act as construed in *United States v. Dotterweich, supra*, and because of the importance of the question to the Government's enforcement program. We reverse.

I

The question presented by the Government's petition for certiorari in *United States v. Dotterweich, supra*, and the focus of this Court's opinion, was whether "the manager of a corporation, as well as the corporation itself, may be prosecuted under the Federal Food, Drug, and Cosmetic Act of 1938 for the introduction of misbranded and adulterated articles into interstate commerce." In *Dotterweich*, a jury had . . . convicted *Dotterweich*, the corporation's president and general manager. The Court of Appeals reversed the conviction on the ground that only the drug dealer, whether corporation or individual, was subject to the criminal provisions of the Act, and that where the dealer was a corporation, an individual connected therewith might be held personally only if he was operating the corporation "as his alter ego."

In reversing the judgment of the Court of Appeals and reinstating Dotterweich's conviction, this Court looked to the purposes of the Act and noted that they "touch phases of the lives and health of the people which, in the circumstances of modern industrialism, are largely beyond self-protection." It observed that the Act is of "a now familiar type" which "dispenses with the conventional requirement for criminal conduct—awareness of some wrongdoing. In the interest of the larger good it puts the burden of acting at hazard upon a person otherwise innocent but standing in responsible relation to a public danger."

Central to the Court's conclusion that individuals other than proprietors are subject to the criminal provisions of the Act was the reality that "the only way in which a corporation can act is through the individuals who act on its behalf." The Court also noted that corporate officers had been subject to criminal liability under the Federal Food and Drugs Act of 1906, and it observed that a contrary result under the 1938 legislation would be incompatible with the expressed intent of Congress to "enlarge and stiffen the penal net" and to discourage a view of the Act's criminal penalties as a "license fee for the conduct of an illegitimate business." . . .

II

The rule that corporate employees who have "a responsible share in the furtherance of the transaction which the statute outlaws" are subject to the criminal provisions of the Act was not formulated in a vacuum. Cases under the Federal Food and Drugs Act of 1906 reflected the view both that knowledge or intent were not required to be proved in prosecutions under its criminal provisions, and that responsible corporate agents could be subjected to the liability thereby imposed. Moreover, the principle had been recognized that a corporate agent, through whose act, default, or omission the corporation committed a crime, was himself guilty individually of that crime. The principle had been applied whether or not the crime required "consciousness of wrongdoing," and it had been applied not only to those corporate agents who themselves committed the criminal act but also to those who by virtue of their managerial positions or other similar relation to the actor could be deemed responsible for its commission.

In the latter class of cases, the liability of managerial officers did not depend on their knowledge of, or personal participation in, the act made criminal by the statute. Rather, where the statute under which they were prosecuted dispensed with "consciousness of wrongdoing," an omission or failure to act was deemed a sufficient basis for a responsible corporate agent's liability. It was enough in such cases that, by virtue of the relationship he bore to the corporation, the agent had the power to prevent the act complained of.

The rationale of the interpretation given the Act in *Dotterweich*, as holding criminally accountable the persons whose failure to exercise the authority and supervisory responsibility reposed in them by the business organization resulted in the violation complained of, has been confirmed in our subsequent cases. Thus, the Court has reaffirmed the proposition that "the public interest in the purity of its food is so great as to warrant the imposition of the highest standard of care on distributors. In order to make 'distributors of food the strictest censors of their merchandise,' the Act punishes 'neglect where the law requires care, or inaction where it imposes a duty.' 'The accused, if he does not will the violation, usually is in a position to prevent it with no more care than society might reasonably expect and no more exertion than it might reasonably exact from one who assumed his responsibilities." Similarly, in cases decided after *Dotterweich*, the Courts of Appeals have recognized that those corporate agents vested with the responsibility, and power commensurate with that responsibility, to devise whatever measures are necessary to ensure compliance with the Act bear a "responsible relationship" to, or have a "responsible share" in, violations.

Thus *Dotterweich* and the cases which have followed reveal that in providing sanctions which reach and touch the individuals who execute the corporate mission—and this is by no means necessarily confined to a single corporate agent or employee—the Act imposes not only a positive duty to seek out and remedy violations when they occur but also, and primarily, a duty to implement measures that will insure that violations will not occur. The requirements of foresight and vigilance imposed on responsible corporate agents are beyond question demanding, and perhaps onerous, but they are no more stringent than the public has a right to expect of those who voluntarily assume positions of authority in business enterprises whose services and products affect the health and well-being of the public that supports them.

The Act does not, as we observed in *Dotterweich*, make criminal liability turn on "awareness of some wrongdoing" or "conscious fraud." The duty imposed by Congress on responsible corporate agents is, we emphasize, one that requires the highest standard of foresight and vigilance, but the Act, in its criminal aspect, does not require that which is objectively impossible. The theory upon which responsible corporate agents are held criminally accountable for "causing" violations of the Act permits a claim that a defendant was "powerless" to prevent or correct the violation to "be raised defensively at a trial on the merits." If such a claim is made, the defendant has the burden of coming forward with evidence, but this does not alter the government's ultimate burden of proving beyond a reasonable doubt the defendant's guilt, including his power, in light of the duty imposed by the Act, to prevent or correct the prohibited condition. Congress has seen fit to enforce the accountability of responsible corporate agents dealing with products which may affect the health of consumers by penal sanctions cast in rigorous terms, and the obligation of the courts is to give them effect so long as they do not violate the Constitution.

III

We cannot agree with the Court of Appeals that it was incumbent upon the District Court to instruct the jury that the government had the burden of establishing "wrongful action" in the sense in which the Court of Appeals used that phrase. The concept of a "responsible relationship" to, or a "responsible share" in, a violation of the Act indeed imports some measure of blameworthiness; but it is equally clear that the government establishes a prima facie case when it introduces evidence sufficient to warrant a finding by the trier of the facts that the defendant had, by reason of his position in the corporation, responsibility and authority either to prevent in the first instance, or promptly to correct, the violation complained of, and that he failed to do so. The failure thus to fulfill the duty imposed by the interaction of the corporate agent's authority and the statute furnishes a sufficient causal link. The considerations which prompted the imposition of this duty, and the scope of the duty, provide the measure of culpability. . . .

Viewed as a whole, the charge did not permit the jury to find guilt solely on the basis of respondent's position in the corporation; rather, it fairly advised the jury that, to find guilt, it must find respondent "had a responsible relation to the situation," and "by virtue of his position . . . had . . . authority and responsibility" to deal with the situation. . . .

IV

Our conclusion . . . suggests as well our disagreement with that court concerning the admissibility of evidence demonstrating that respondent was advised by the FDA in 1970 of insanitary conditions in Acme's Philadelphia warehouse. We are satisfied that the Act imposes the highest standard of care and permits conviction of responsible corporate officials who, in light of this standard of care, have the power to prevent or correct violations of its provisions. . . .

Respondent testified in his defense that he had employed a system in which he relied upon his subordinates, and that he was ultimately responsible for this system. He testified further that he had found these subordinates to be "dependable" and had "great confidence" in them. By this and other testimony respondent evidently sought to persuade the jury that, as the president of a large corporation, he had no choice but to delegate duties to those in whom he reposed confidence, that he had no reason to suspect his subordinates were failing to insure compliance with the Act, and that, once violations were unearthed, acting through those subordinates he did everything possible to correct them.

Although we need not decide whether this testimony would have entitled respondent to an instruction as to his lack of power, had he requested it, the testimony clearly created the "need" for rebuttal evidence. That evidence was not offered to show that respondent had a propensity to commit criminal acts, or that the crime charged had been committed; its purpose was to demonstrate that respondent was on notice that he could not rely on his system of delegation to subordinates to prevent or correct insanitary conditions at Acme's warehouses, and that he must have been aware of the deficiencies of this system before the Baltimore violations were discovered. The evidence was therefore relevant since it served to rebut respondent's defense that he had justifiably relied upon subordinates to handle sanitation matters. And, particularly in light of the difficult task of juries in prosecutions under the Act, we conclude that its relevance and persuasiveness outweighed any prejudicial effect.

Reversed.

[Dissent by Mr. Justice Stewart, with whom Mr. Justice Marshall and Mr. Justice Powell join, omitted.]

* * * * *

12.6.2 Fines, Prison

Section 303 of the FD&C Act provides jail and fines for conviction. The generic violation provision is imprisonment for not more than a year and/or a fine of $1,000.

Often the amount of the fine is dwarfed by other payments agreed to in settlements and consent decrees. For example, in 2002, a pharmacist charged with selling diluted drugs, and he pleaded guilty to 8 counts of product tampering and 6 counts each of misbranding and adulterating drugs, in violation of the Federal Food, Drug, and Cosmetic Act (FD&C Act). He was sentenced to 30 years in prison, and required to pay a $25,000 fine. In addition he had to pay $10.4 million in restitution.

12.7 OTHER REMEDIES AND CONCERNS

12.7.1 Publicity

Section 705 of the FD&C Act provides the FDA with the authority to seek publicity to warn the public of violative products and possible adverse consequences associated with the use of the regulated products. While FDA does not generally issue press releases of its enforcement actions, it can occur.

Adverse publicity of this nature can be devastating to a business or a product. Such publicity can cause more damage than the cost of the action proposed by the agency. Adverse publicity affects not only the short-term problem a company is facing; it also may extend to future product liability claims and sometimes to shareholder lawsuits.

This is one tool that FDA has used to compensate for lack of general power to recall violative products—an FDA press release about adverse health effects of a product can result in an immediate end to a product's marketability. This tool also provides FDA power to persuade companies to cooperate in a voluntary recall rather than face adverse publicity.

12.7.2 Referral to State Agencies

Most states have adopted laws similar to the Federal Food, Drug, and Cosmetic Act (FD&C Act). However, in some cases the states may have greater power than the FDA. In addition to summary seizure and condemnation authority, states require almost all food establishments to obtain a license to operate within their jurisdictions. Therefore states can suspend a firm's food establishment license.

12.7.3 Postenforcement Compliance Monitoring

A food establishment faces another kind of enforcement experience in the aftermath of a formal agency enforcement action. The agency will conduct intensified follow-up monitoring activities of facilities after significant violations are discovered. Return visits after problem inspections or other types of enforcement issues should be expected. Attentive responses to warning letters, FDA Form 483 observations, and careful follow-through in making the changes and corrections promised in responses to federal (or state) agencies, can reduce the number and intensity of monitoring visits by government officials after an enforcement event.

12.7.4 Criminal Code Charges

The Justice Department (DOJ), when it brings FD&C Act complaints forward on behalf of the FDA, is not limited to the charges specified in the FD&C Act. The DOJ may also bring forward charges based on the provisions of the federal criminal code based on willful misconduct involving FDA-regulated product. For example, a firm that shipped adulterated product and then conceals that shipment from the FDA inspectors and investigators may be prosecuted both for adulteration under the FD&C Act and for obstruction of justice.

For example, in 2002, during a routine FDA inspection of the Jeppi Nut Company, located in Baltimore, Maryland, FDA inspectors uncovered evidence of unsanitary conditions and extensive rodent infestation. Sometime prior to that inspection, the firm had expanded operations to a second building located nearby. The firm's owner did not post any signs or indicate that his company was operating from the second building. He also denied having any operations in that building when asked by the FDA inspectors. Following the FDA inspection, surveillance conducted in the evening revealed Jeppi Nut employees moving items out of the second building. A subsequent inspection of the building determined that those food products were contaminated by rodents and insects. The firm's owner was convicted of a felony count of violating title 18 U.S.C. § 1505—Obstruction of Justice.

Knowingly submitting false information to the FDA to obtain a product approval may be prosecuted under both the FD&C Act and for the criminal charge of submitting false information to a federal agency. In some cases of disseminating false information, violations of federal mail fraud and wire fraud statutes may result. When two or more company employees work together to mislead the government or consumers, criminal conspiracy charges may result.

Some criminal cases are developed by DOJ based on information brought to them by FDA. However, the DOJ may also conduct criminal investigations with the support of the FBI. Moreover, some cases are developed from information supplied by other businesses. For example, FDA received information in 1997 from an Illinois business that advised FDA that Nutritional Source, Inc., of Paducah, Kentucky, was mislabeling rolls, donuts, and cookies. Nutritional Source purchased high caloric rolls, donuts, and cookies at wholesale prices and labeled them as low caloric. He then sold the products to the health food industry at an inflated price. These rolls and donuts had a fat content of 33 grams of fat per roll/donut but were labeled them as a fat content of 3 to 5 grams of fat per roll/donut. The person running the company was convicted of title 18 U.S.C. § 1341—Mail Fraud; and title 18 U.S.C. § 2—Aiding and Abetting, and sentenced to 15 months incarceration, and 36 months supervised release after incarceration. In addition, he is not allowed to purchase, distribute, or sell any food product regulated by any government agency.

NOTES AND QUESTIONS

12.1. Voluntary recalls. Why would a firm want to report its voluntary recalls?

12.2. Condemnation. In discussion of seizures, the term "condemnation" has a special and narrow meaning than used in everyday speech. What is this?

12.3. Due process and seizures without hearings. Why can the agency seize and destroy products without a hearing? Doesn't this violate due process?

12.4. Strict liability. What do you think of the strict liability standard for violations of the FD&C Act? Is it fair? Why does this standard seem to offend some sense of fairness?

CHAPTER 13

Inspections

13.1 INTRODUCTION

An inspection is the official examination of property, persons, or documents by a government representative for a regulatory purpose. FDA inspections involve intrusion into premises and thus raise the possibility of running afoul of the Fourth Amendment's prohibition on unreasonable governmental search and seizure.

This chapter explores the constitutional definition of a search and the circumstances when an inspection constitutes a search. This chapter also provides understanding of what procedures regulators must followed before, during, and after an inspection to assure compliance with Fourth Amendment and FD&C Act protections. Finally, this chapter addresses measures that regulated firms may take to ensure a successful inspection.

Inspections are the primary source of information for FDA's enforcement actions. Moreover, the statutory power to conduct inspections is designed to supports the enforcement provisions of the food safety laws.

There are two objectives of an inspection: to determine compliance with the law and to gather evidence for enforcement if there is noncompliance. This first purpose—determining whether there is compliance—separates health and safety inspections from police searches. Police searches are conducted with the specific aim of obtaining evidence for use in criminal prosecution. Inspections are designed to assure the public that applicable safety standards are being met. This assurance serves both the public and the regulated industry, which depends on public confidence for marketability of their product.

The optimist focuses on the compliance objective and looks at the inspection as an opportunity to demonstrate the strength of the firm's controls over the purity of the food and their conformance to applicable requirements. Most inspections do not find serious violations, and no enforcement action is initiated. To the optimist, each inspection is a learning experience.[1]

---

[1] *See* JAMES T. O'REILLY, FOOD AND DRUG ADMINISTRATION § 20:11 (2d ed. 2004).

Food Regulation: Law, Science, Policy, and Practice, by Neal D. Fortin
Copyright © 2009 Published by John Wiley & Sons, Inc.

The pessimist, on the other hand, looks at each inspection as a potential disaster at his or her doorstep.[2] Evidence gathered during inspections is the key to most FDA prosecutions and seizures. Each inspection could be a potential prelude to enforcement action.

Whether you are an optimist or a pessimist or somewhere in between, training and preparation are essential to a successful inspection. The latter part of this chapter discusses how to plan and prepare for a successful inspection.

13.2   CONSTITUTIONAL LIMITS

Constitutional safeguards restrict certain aspects of FDA inspections. The Fourth Amendment to the U.S. Constitution provides that:

> The right of the people to be secure in their persons, houses, papers, and effects against unreasonable searches and seizures shall not be violated, and no Warrants shall issue, but upon probable cause, supported by oath or affirmation, and particularly describing the place to be searched, and the persons or things to be seized.

The Fourth Amendment protects against official harassment and arbitrary or improper intrusion by the government into the lives of its citizens. In part, this protection derives from the requirement that law enforcement officials not conduct a search until they have first persuaded a judge or magistrate that there is a specific reason to suspect that a search of a particular place will disclose a specified violation of the law.[3] In addition, the "probable cause" requirement is to ensure that searches are not arbitrary or capricious.

If health and safety inspections were constitutionally equated with police searches, inspectors would need to go before a judge or magistrate prior to each inspection. Inspectors would have to describe the premises to be searched, the purpose of the search, and the specific violations likely to be discovered. In addition, the inspectors would have to provide reasonable grounds for suspecting that these specific violations would be found. Obviously such requirements would be impractical with the type of random, unannounced inspections that we consider basic to health and safety regulation.

The Supreme Court has balanced the special needs of health and safety inspections with the requirements of the Fourth Amendment. The Court developed a special approach for "administrative warrants." Two hallmarks of this approach are:

- no requirement of specific probable cause for routine health and safety inspections authorized by statute, and
- less stringent administrative warrant requirements for such inspections.

---

[2] *Id.*

[3] There are a number of exceptions to the warrant rule, such as searches incident to a lawful arrest.

An administrative warrant need only establish that an inspection is to be conducted pursuant to a preexisting neutral administrative plan or procedure authorized by law. For example, if an owner refuses to grant an inspector voluntary access to inspect the facility, the inspector would apply for an administrative search warrant. In application for the warrant, the inspector would only be required to describe the inspection schedule or policy and attest that the inspection has been routinely scheduled as part of the administrative plan for regulation under the law.[4]

It should be noted that the courts have recognized that citizens expect greater privacy within their homes than within a business. The right to be left alone in our own homes and private affairs (free from governmental intrusion) is a fundamental right in our society, one that is implicitly recognized by the Bill of Rights.

Thus the courts will require an agency to prove a more compelling need before authorizing inspection of a private home than will be required to obtain a warrant to search a business. The FDA defers to this distinction in its Investigation Operations Manual, which states:

> All inspections where the premises are also used for living quarters must be conducted with a warrant for inspection unless:
>
> Owner Agreeable—The owner or operator is fully agreeable and offers no resistance or objection whatsoever or;
>
> Physically Separated—The actual business operations to be inspected are physically separated from the living quarters by doors or other building construction. These would provide a distinct division of the premises into two physical areas, one for living quarters and the other for business operations, and you do not enter the living area.[5]

On occasion, invocation of constitutional rights can limit the scope of inspections. However, in practice most firms voluntarily concede expanded scope of inspection.

13.2.1 Statutory Power for Inspections

The statutory power to inspect is not unlimited.[6] Administrative searches must be conducted under the authorization of a valid statute. The law authorizing the inspection must be constitutional. The inspection must further a public interest advanced by the statute. In addition, the person who conducts the search or investigation must have authority to do so.

---

[4] Camara v. Municipal Court, 387 U.S. 523 (1967); *see also,* Marshall v. Barlow's, Inc., 346 U.S. 307 (1978).
[5] FDA, *Premises Used for Living Quarters*, INVESTIGATIONS OPERATIONS MANUAL, 501.02 (2004), *available at*: http://www.fda.gov/ora/inspect_ref/iom/ChapterText/500.html#501.02 (last visited Aug. 4, 2004).
[6] 21 U.S.C. § 374.

Moreover, the scope of the inspection must conform to the limits of the authorizing statute. For example, the FDA's inspection powers are not uniform across all product categories. An inspector of medical devices holds more authority than one who inspects food.

FDA gained inspection authority in 1938 with section 704 of the FD&C Act. Section 704 applies a general standard to all FDA-regulated products. Section 704 states in part that the agency's "duly designated" officers or employees, upon presenting appropriate credentials and a written notice, are "authorized (A) to enter, at reasonable times, any factory, warehouse, or establishment in which food, drugs, devices or cosmetics are manufactured, processed, packed or held . . . or to enter any vehicle being used to transport or hold such food . . . , and (B) to inspect, at reasonable times and within reasonable limits and in a reasonable manner, such factory, warehouse, establishment or vehicle and all pertinent equipment, finished and unfinished materials, containers, and labeling therein."

Section 704 permits physical inspection of the facility, equipment, labeling, and products, but it is conspicuously silent about records, reports, and files. In addition to the broad authority of section 704, FDA is granted specific authority to inspect the records for prescription drugs and restricted medical devices.[7] Because Congress was selective in granting access to records, this implies that Congress intended records of other products not to be subjected to inspection under section 704.[8] FDA may attempt to review records, and many firms cooperate, but there is a gap between statutory authority and the degree of regulatory right to review a firm's records.

There is nothing improper with, and the law does not prevent an FDA investigator from asking a regulated firm to provide access to records despite the investigator's lack of statutory grounds to demand them. In fact, many times firms will want to voluntarily share a document with the FDA. Under some circumstances such openness is prudent.

Nonetheless, regulated firms should be aware that the law imposes no constitutional duty upon FDA to warn the inspected persons of their rights to avoid self-incrimination. When an FDA investigator asks for records, and the firm voluntarily provides them, these records are admissible as evidence against the firm. By providing the records—often merely in response to a request or question—the firm will be considered to have waived their right to object.

The Trade-Offs   The decision on when a firm should waive some legal rights is one of the toughest aspects of planning for inspection. There are trade-offs involved. Foremost, establishing a professional and pleasant relationship with

---

[7] 21 U.S.C. § 374.

[8] The legal maxim *inclusio unius eat exclusio alterius* would interpret the listing by Congress of two items as meaning exclusion of all others. However, the issue is not closed until Congress or the Supreme Court speaks to it.

the inspector can minimize disruption to the firm and expedite the inspection. A cooperative person's demeanor communicates that there is nothing to hide. In addition, a cooperative person is more likely to be trusted by the inspector to correct marginal deficiencies.

On the other hand, refusals of requests may be met with inspector suspicion and increased zeal for discovering violations. This trade-off involves not just increased scrutiny but real costs. An uncooperative attitude may provide grounds for a finding that a firm deserves court action, seizure, or increased inspection frequency.

It also bears mentioning that the firm must be correct in their decision to refuse access. Denial of access to areas where the agency is authorized to inspect is ground for prosecution for obstruction or refusal of inspection. Such a refusal is a separate violation of the FD&C Act.[9]

Generally, when denied access to records or areas that the agency believes are under its authority, the agency will seek an administrative search warrant. If the inspector obtains a search warrant and is still denied access, the federal marshals (or state police in the case of state agency) can arrest the person for refusing. In addition to violation of the FD&C Act, refusal can be punished as contempt of court.[10]

Decisions on such trade-offs should be made by a firm's management well in advance of inspection. Those decisions should be memorialized in written policy. In deciding on firm policy and training for employees, review of FDA's Investigation Operations Manual, the FDA Compliance Policy Guides, and other compliance materials can help make the firm aware of the types of questions and records that may be raised.

13.2.2   Record Access under the Bioterrorism Act

Section 306 of the Bioterrorism Act sets out a set of conditions when FDA has authority to access certain records for any food (and packaging) that is sold or offered for sale in the United States. This provision requires maintenance of records that identify the immediately prior source and the immediately subsequent recipient for all food (including packaging) so that FDA can access relevant information if the agency "has a reasonable belief that an article of food is adulterated and presents a threat of serious adverse health consequences or death to humans or animals."

13.2.3   The Warrantless Inspection Exception

In *United States v. Jamieson-McKames Pharmaceuticals, Inc.*,[11] pharmaceutical manufacturers and two of their corporate officers were convicted of conspiracy

---

[9] 21 U.S.C. § 331(e)-(f).

[10] The general view is that there are two violations when access is refused when there is a warrant: under the FD&C Act and for contempt. *Becton, Dickinson & Co. v. FDA*, 448 F. Supp. 776 (N.D. N.Y. 1978), *aff'd.* 598 F.2d 1175 (2d Cir. 1978).

[11] 651 F2d 532 (8th Cir. 1981).

to violate the FD&C Act, and convicted of counterfeiting, misbranding, and adulterating drugs. The defendants appealed. The Court of Appeals held that the drug-manufacturing industry fell within an exception to search warrant requirement.

* * * * *

United States v. Jamieson-McKames Pharmaceuticals, Inc.

651 F2d 532 (8th Cir. 1981)

ARNOLD, Circuit Judge

. . . Jamieson-McKames Pharmaceuticals, Inc. (Jamieson-McKames) is a Missouri corporation with its principal place of business in St. Louis, Missouri. The company manufactured, purchased, packaged, labeled, distributed, and sold drugs . . .

On October 29, 30, and 31, and November 3, 1975, federal and state agents entered and searched the premises of Jamieson-McKames Pharmaceuticals, Inc., and Pharmacare, Inc., . . . Samples of drugs were taken, documents were taken, quantities of drugs were embargoed, the premises and contents photographed, and machinery seized. . . .

Thereafter, on May 12, 1977, defendants were charged in an 11-count indictment with counterfeiting, adulterating, and misbranding drugs and conspiracy to counterfeit, adulterate, and misbrand drugs. The indictment also charged that the defendants committed all of these acts with the intent to defraud and mislead, rendering such felonies punishable under 21 U.S.C. § 333(b).

II. The Fourth Amendment

The appellants contend that their Fourth Amendment rights were violated by the failure of the court to suppress evidence seized by government agents from the defendants' business premises . . . The seizures at the Wentzville pharmacy were conducted on the authority of a notice to inspect authorized by 21 U.S.C. § 374(a). The employee in charge was given a copy of the notice to inspect, but no warrant to inspect was obtained.

The Supreme Court has held that warrantless searches are generally unreasonable, and that commercial premises as well as homes are within the Fourth Amendment's protection. An exception from the search-warrant requirement has, however, been delineated for industries "long subject to close supervision and inspection," and "pervasively regulated business(es)." *Colonnade* involved the liquor industry, and *Biswell* the interstate sale of firearms. The threshold question therefore is whether the drug-manufacturing industry should be included within this class of closely regulated businesses.

The appellants argue that the drug-manufacturing industry is no more closely regulated than any number of industries involved in interstate commerce, and that therefore the rule of *Marshall v. Barlow's, Inc., supra*, requiring a warrant in the absence of consent before an administrative search can take place, should apply. In *Barlow's*, the Supreme Court held that warrantless searches authorized by § 8(a) of the Occupational Safety and Health Act, 29 U.S.C. § 657(a), violated the Fourth Amendment. There, however, the government sought to inspect work areas not open to the public on the premises of an electrical and plumbing contractor. In *Barlow's* the argument that all businesses involved in interstate commerce had "long been subject to close supervision" of working conditions was urged by the Secretary of Labor but explicitly rejected by the Court. In rejecting this argument and others, the Court specifically preserved the *Colonnade-Biswell* exception to the warrant requirement. The Court indicated that there were other industries, covered by regulatory schemes applicable only to them, where regulation might be so pervasive that a *Colonnade-Biswell* exception to the warrant requirement could apply.

Such warrantless searches are upheld because "when an entrepreneur embarks on such a business, he has chosen to subject himself to a full arsenal of governmental regulation," and "in effect consents to the restrictions placed on him" Further, in the face of a long history of government scrutiny, such a proprietor has no "reasonable expectation of privacy."

We think the drug-manufacturing industry is properly within the *Colonnade-Biswell* exception to the warrant requirement. The drug-manufacturing industry has a long history of supervision and inspection. The present Food, Drug, and Cosmetic Act has its origins in the Food and Drug Act of 1906. That Act was an attempt by Congress "to exclude from interstate commerce impure and adulterated food and drugs . . ." and to prevent the transport of such articles "from their place of manufacture."

The *Biswell* Court acknowledged that the history of regulation of interstate firearms traffic was "not as deeply rooted" as the history of liquor regulation, but included firearms within the warrant exception because their regulation was of "central importance to federal efforts to prevent violent crime and to assist the states in regulating the firearms traffic within their borders." This passage teaches that the nature of the federal or public interest sought to be furthered by the regulatory scheme is important to our analysis. It is difficult to overstate the urgent nature of the public-health interests served by effective regulation of our nation's drug-manufacturing industry. Furthermore, virtually every phase of the drug industry is heavily regulated, from packaging, labeling, and certification of expiration dates, to prior FDA approval before new drugs can be marketed. The regulatory burdens on the drug-manufacturing industry are weighty, and that weight indicates that the drug manufacturer accepts the burdens as well as the benefits of the business and "consents to the regulations placed on him."

The final lesson of *Barlow's* is that the reasonableness of warrantless searches is dependent on the "specific enforcement needs and privacy guarantees of each statute." In Barlow's the Court was unconvinced that requiring

OSHA officials to obtain administrative warrants when consent to inspect was withheld would cripple the effectiveness of the enforcement scheme. . . .

Regulation of the drug industry differs from the OSHA situation in another significant way. The class sought to be protected by OSHA regulation of safety of work areas is made up of employees who are in the workplace itself and free to report violations at any time. The protected class in the area of drug-manufacturing is the consuming public, which has no way of learning of violations short of illness resulting from the consumption of defective drug products. In this sense the enforcement needs of drug-industry regulation are considerably more critical than those before the Court in Barlow's.

As for privacy guarantees, the Supreme Court points out that a warrant provides assurances that the proposed "inspection is reasonable under the Constitution, is authorized by statute, and is pursuant to an administrative plan containing specific neutral criteria." The notice of inspection used in this case satisfies at least some of these criteria. It informs the "owner or agent in charge" (s 374(a)) of the "scope and objects of the search." Although the notice of inspection makes no express reference to reasonableness under the Constitution, it clearly states that notice is given pursuant to 21 U.S.C. § 374, which is enacted by the Congress. The name of the firm and address is also prominently listed. Further, the notice of inspection reproduces large portions of § 374(a), stating the areas and objects to be searched; that the inspection is to take place at reasonable times; that certain records are to be made available to the inspector; that each inspection must be made with reasonable promptness; that each inspection must be accompanied by a separate notice; and that the purpose of any inspection of a prescription-drug operation is discovery of information "bearing on whether prescription drugs (are being) adulterated or misbranded within the meaning of" the Act or on other violations of the Act. . . .

In sum, the authorizing statute now before the Court was not painted with so broad a brush as the one rejected in Barlow's, the enforcement needs are more critical in the drug-manufacturing field, and the interests of the general public are more urgent. We hold that inspections authorized by § 374 are "reasonable" and therefore not inconsistent with the Fourth Amendment. Thus this case falls within the "carefully defined classes of cases" which are an exception to the search-warrant requirement.[12] We share, to a degree,

---

[12] Several courts have addressed this or similar issues. In accord with our conclusion is *United States v. New England Grocers Supply Co.*, 488 F.Supp. 230 (D.Mass.1980). This case involved evidence seized during inspection of a food-supply warehouse pursuant to § 374. *United States v. Acri Wholesale Grocery Co.*, 409 F.Supp. 529 (S. Iowa 1976), decided prior to *Barlow's*, found warrantless searches authorized by § 374 to be consistent with those allowed in *Biswell. United States v. Business Builders, Inc.*, 354 F.Supp. 141 (N.Okla.1973), was another pre-*Barlow's* case involving food. Relying on *Biswell*, the court said: "It would be an affront to common sense to say that the public interest is not as deeply involved in the regulation of the food industry as it is in the liquor and firearms industries. (Footnote omitted). One need only call to mind recent cases of deaths occurring from botulism. Modern commerce has devised such an efficient and rapid means of

the fears expressed by appellants that many businesses are thoroughly regulated by the United States, and that an undue extension of our rationale might obliterate much of the Fourth Amendment's protection. On balance, however, we are persuaded that the capacity for good or ill of the manufacture of drugs for human consumption is so great that Congress had power to enact § 374(a).

Having concluded that drug manufacturing is a "pervasively regulated" industry does not end our inquiry, but establishes only that Congress has broad authority to place restrictions on that industry that might otherwise violate the Fourth Amendment. A question remains as to whether the conduct of the government in this case conforms with the statutory scheme provided by the Congress. . . .

After emphasizing the long history of regulation of the sale of liquor, the Court acknowledged the broad authority of Congress to provide for inspections under the liquor laws. Under the existing statutory scheme, however, the Court concluded that Congress had not provided for "forcible entries without a warrant." This conclusion was based on Congress's provision of a separate penalty for a dealer's refusal to permit an inspection. Compare *Biswell, supra.* Consent being absent, the search was held invalid, and the evidence suppressed.

The Federal Food, Drug, and Cosmetic Act contains provisions, similar to those addressed in *Colonnade*, which punish refusals to permit inspections by imprisonment up to one year, or a fine of not more than $1,000, or both. It follows therefore, as in *Colonnade*, that an inspection pursuant to a § 374 notice to inspect is authorized only when there is a valid consent. If consent is withheld, a separate violation of the Act occurs, and the FDA inspectors are required to obtain a warrant before the inspection can proceed. . . .

We add a word of clarification as to the meaning of the term "consent" as we intend it in this context. We do not mean, by imposing a requirement of "consent," to require a factual determination as to whether appellants, with respect to the Wentzville site, knowingly and understandingly relinquished a known right. The question is whether appellants refused to permit entry or

distribution of food products to the consumer that a batch of contaminated food may cause widespread illness and death before the public can be warned and the contaminated products removed from the market." 354 F.Supp. at 143. In *United States v. Del Campo Baking Mfg. Co.*, 345 F.Supp. 1371 (D.el.1972), still another food case, the court considered the regulatory scheme under the Federal Food, Drug, and Cosmetic Act to be as pervasive as the licensing scheme in *Biswell*. Post-*Biswell* cases holding a search warrant required in the absence of consent include *United States v. Roux Laboratories*, 456 F.Supp. 973 (M.Fla.1978) (where consent refused, a warrant is necessary). See also *United States v. Litvin*, 353 F.Supp. 1333 (D.C.1973), considering a motion to suppress evidence seized during a routine inspection of a food warehouse. The court treats the food industry as a closely regulated one but requires a warrant when consent to inspect is withheld. Compare the discussion of *Colonnade* in text *infra*. Cf. *Donovan v. Dewey*, 452 U.S. 594, 101 S.Ct. 2534, 69 L.Ed.2d 262 (1981) (mining industry held within the *Colonnade-Biswell* exception).

inspection, thereby violating 21 U.S.C. § 331 (f). If they did so refuse, then FDA was obliged to obtain an administrative warrant in order to effect the inspection, and could also seek a separate criminal prosecution for the refusal itself. If appellants did not refuse to permit entry or inspection, then they "consented" to the search and seizure, as we use that term here. This formulation, while it may not answer every question that may arise with respect to searches and seizures pursuant to § 374 notices of inspection, seems to us to be the most logical way to harmonize Biswell and Colonnade. As the court said in *United States v. Litvin*:

> Therefore, as in *Colonnade*, if defendant refused entry to the Food and Drug Administration inspectors, he was in violation of § 331(f) and subject to one-year imprisonment and/or $1,000 fine. But if defendant refused entry, the inspectors had no right to enter the storeroom, and following *Colonnade*, any evidenc seized as a result of the search must be suppressed.

... Appellants next argue that the inspections were part of an ongoing criminal investigation, and that therefore a warrant issued on less than criminal probable cause was not sufficient to authorize a search. It is our view that a warrant based on an administrative showing of probable cause is valid in this pervasively regulated industry. To hold otherwise would be inconsistent with our conclusion, already expressed, that warrantless entry under a notice of inspection does not violate the Fourth Amendment in the drug-manufacturing field. Probable-cause standards are relaxed because the business person engaged in this industry has a lesser expectation of privacy. . . .

Appellants next argue that certain statements made by the defendants to FDA agents during the searches were inadmissible at trial because *Miranda* warnings were not given. *See Miranda v. Arizona*. The district court held that Miranda was not applicable because "the evidence failed to establish that defendants . . . were in a custodial situation, subject to arrest."

Evidence presented at trial showed that FDA agents are without authority to make arrests, that the defendants' movements were not restricted during the time of the search, and that there were no threats or coercion. Evidence also indicated that appellants' employees were free to go about their business, and that consultation with attorneys was not limited. There is ample evidence to support the district court's finding, and the statements were therefore properly admitted at trial. . . .

* * * * *

13.2.4   Consent to Inspect

As a rule, administrative inspections must be conducted pursuant to an administrative search warrant unless consent has been given to conduct the search. In some circumstances consent is imposed as a matter of law, and there is no obligation to obtain a search warrant. For example, consent to reasonable inspections may be considered a condition of obtaining a state license.

An exception to the requirement of a warrant for administrative search warrant has been delineated by the Supreme Court for industries "long subject to close supervision and inspection," *Colonade Catering Corp. v. U.S.*, 397 U.S. 72 (1970)[13] and "pervasively regulated business[es]," *U.S. v. Biswell*, 406 U.S. 311 (1972).

As a practical matter, the issue is rarely tested because if a firm refuses FDA entry for an inspection, it is FDA's policy to leave and return with a search warrant and U.S. marshals. If a firm refuses to accept the search warrant, the firm and its officials face contempt charges along with any violations of the FD&C Act.

In the following case, *United States v. Thriftimart, Inc.*,[14] the defendants argued that fact that inspectors for the FDA had not warned the defendant's warehouse managers of their rights to insist upon a warrant before allowing a food inspection. The court held that the possibility that managers were unaware of precise nature of their rights under Fourth Amendment did not render the managers' consent unknowing or involuntary, as managers were presented with a clear opportunity to object to inspection and their manifestation of assent, no matter how casual, could reasonably be accepted as waiver of warrant.

* * * * *

United States v. Thriftimart, Inc.
429 F.2d 1006 (1970)

MERRILL, Circuit Judge:

Appellants have been convicted of violations of the Federal Food, Drug & Cosmetic Act. Upon inspection, food in four company warehouses had been found to be infested with insects. . . .

The inspectors did not have search warrants nor did they advise the warehouse managers that they had a right to insist upon a search warrant. . . . The precise issue raised is whether the informal and casual consent to search given by the warehouse managers made it unnecessary to secure a search warrant. Appellants argue that a waiver of search warrant "cannot be conclusively presumed from a verbal expression of assent. The court must determine from all the circumstances whether the verbal assent reflected an understanding,

---

[13] *Colonnade Catering Corp. v. United States, supra,* involved Internal Revenue Service agents, suspecting a violation of the federal excise tax law, sought to inspect an establishment where liquor was being served. The owner refused to open a locked storeroom and asked if the agents had a search warrant. The agents answered that they didn't need a warrant, broke the lock, entered, and found evidence of regulatory violations.

[14] 429 F.2d 1006 (1970).

uncoerced, and unequivocal election to grant officers a license which the person knows may be freely and effectively withheld." Since the managers were not warned that they had a right to refuse entry and since there was no proof that they knew they had such a right, appellants argue that the consent was not effective to remove the need for a search warrant. . . .

It is clear, therefore, that the administrative search is to be treated differently than the criminal search. The issue in this case is whether the body of law that has grown up around the definition of consent to a search in the criminal area should mechanically be applied to the inspection of a warehouse.

In a criminal search the inherent coercion of the badge and the presence of armed police make it likely that the consent to a criminal search is not voluntary. Further, there is likelihood that confrontation comes as a surprise for which the citizen is unprepared and the subject of a criminal search will probably be uninformed as to his rights and the consequences of denial of entry. . . . These circumstances are not present in the administrative inspection. The citizen is not likely to be uninformed or surprised. Food inspections occur with regularity. As here, the judgment as to consent to access is often a matter of company policy rather than of local managerial decision. FDA inspectors are unarmed and make their inspections during business hours. Also, the consent to an inspection is not only not suspect but is to be expected. The inspection itself is inevitable. Nothing is to be gained by demanding a warrant except that the inspectors have been put to trouble—an unlikely aim for the businessman anxious for administrative goodwill.

We hold that the absence of coercive circumstances and the credibility of a consent given to an inspection justify a departure from the Schoepflin rule in cases of administrative inspection. Here, the managers were asked for permission to inspect; the request implied an option to refuse and presented an opportunity to object to the inspection in an atmosphere uncharged with coercive elements. The fact that the inspectors did not warn the managers of their right to insist upon a warrant and the possibility that the managers were not aware of the precise nature of their rights under the Fourth Amendment did not render their consent unknowing or involuntary. They, as representatives of Thriftimart, Inc., were presented with a clear opportunity to object to the inspection and were asked if they had any objection. Their manifestation of assent, no matter how casual, can reasonably be accepted as waiver of warrant.

In conclusion, we hold that in the context of the exclusionary rule a warrantless inspectorial search of business premises is reasonable when entry is gained not by force or misrepresentation, but is, with knowledge of its purpose, afforded by manifestation of assent. Lack of warrant under these circumstances did not render the inspections unreasonable under the Fourth Amendment. . . .

Judgment affirmed.

* * * * *

13.2.5 Statements by Firm Representatives

The language of section 704 authorizes FDA to "inspect" and does not expressly authorized FDA inspectors to interview company employees. As a practical matter, inspectors often engage in discussions with a variety of corporate representatives, not only the person(s) assigned to accompany them.

A firm should establish, in advance, a policy on statements by company employees. Decide which employees are authorized to talk with the FDA inspectors. Keep in mind that any admission by an employee may be used against the firm (or the employee) in law enforcement proceedings, including criminal prosecution.[15]

Again, FDA inspectors are not required to give Miranda warnings before interviewing company representatives during inspections.[16] On the other hand, the silence of an employee may also be used against the firm as an implied admission, particularly when it would be reasonable to disagree with a statement.[17] Therefore a policy that balances between the two extremes is usually best—for example, specifying a few individuals who may answer the inspector's questions, and training them when to answer and when to defer to the firm's counsel for response.

Affidavits In addition to interviews, FDA inspectors may ask company representatives to sign affidavits. Affidavits are used to record and document facts about products that are known by a person. The main uses are to establish the movement of food in interstate commerce, identify the product and lot for a sample, and to identify specific responsibility for a violation.

Company representatives are not required to sign affidavits during inspections. Firms are advised to establish and communicate to their employees a clear policy on this subject. The balancing act can be similar for affidavits as it is for oral communication, but written documents should generally be reviewed by a firm's counsel before signing.

---

[15] Hearsay is generally inadmissible in court. FRE 802. However, an admission is not hearsay. In short, an admission is anything that would haunt you if repeated in court. More specifically, an admission is an admission by a party-opponent when the statement is offered against the party and is the party's own statement (words or act). FRE 801(d)(2). Hearsay is defined as "a statement [words or acts], other than one made by the declarant while testifying at the trial or hearing, offered in evidence to prove the truth of the matter asserted." FRE 801(c). N.B.: There can be implied admissions, such as admission by silence, when a reasonable person would deny a statement, a person's non-denial may be introduced to show that he agreed or accepted it as true. *US v. Hoosier*, 542 F.2d 687, 688 (6th Cir. 1976). However, silence when under arrest can never be an admission when exercising Fifth Amendment rights. *Id.*

[16] Of course, this means government inspectors are not required to tell you that anything you say may be used against you. The inspector's lack of arrest power also means that silence can be used against a party or a firm when the silence is an implied admission. An admission by silence occurs when a reasonable person would deny a statement; a person's non-denial may be introduced to show that he agreed or accepted it as true. *US v. Hoosier* 542 F.2d 687, 688 (6th Cir. 1976).

[17] *Id.*

When FDA encounters a refusal to sign, FDA's policy is to prepare the document and ask the employee to read it. If he or she declines, FDA will read it to them. FDA will ask him or her to correct and initial any errors in the person's own handwriting.

If the firm still does not sign the affidavit, the inspector will write a statement noting the refusal, the reason, and the situation, near the bottom of the affidavit. For example: "Refused to sign on advice of company counsel. Affiant refused to read the statement, but on hearing it read aloud avowed the statement to be true."[18]

13.2.6   Scope of Inspections

Section 704 of the FD&C Act does not provide FDA with a general power to inspect establishment records. Certain categories of processing and operations, however, are subject to greater oversight authority. For example, FDA has the authority to inspect low-acid and acidified canned food processing records.[19] Medical devices and drugs are also subject to FDA inspection of "records, files, papers, processes, controls, and facilities" with exceptions for financial, research, and personnel data.

The absence of express authority to inspect certain records does not prevent FDA inspectors from asking to see such records. If the records are voluntarily disclosed, FDA may examine the records. While FDA cannot use fraudulent methods to obtain access to records, FDA is under no obligation to educate a business in the extent of their Fourth Amendment rights or the limits of FDA's authority under the FD&C Act.

Records of Sources and Recipients of Food   As part of FDA new powers under the Bioterrorism Act, all persons who manufacture, process, pack, transport, distribute, receive, hold, or import food (excluding farms and restaurants) must maintain records that identify the immediately prior source and the subsequent recipient for all food (including packaging). FDA holds authority to access relevant information if the agency "has a reasonable belief that an article of food is adulterated and presents a threat of serious adverse health consequences or death to humans or animals."[20]

---

[18] *See* FDA, INVESTIGATIONS OPERATIONS MANUAL 2007, sec. 4.4.8.2—*Refusal to Sign the Affidavit* (2007), *available at*: http://www.fda.gov/ora/inspect_ref/iom/ChapterText/4_4.html#4.4.8.2 (last accessed Mar. 31, 2007).
[19] 21 C.F.R. § 108.35.
[20] 21 U.S.C. § 350c(a).

13.2.7   Photographs

* * * * *

Is a Picture Worth More Than 1,000 Words?

Neal D. Fortin, 1 JOURNAL OF FOOD LAW AND POLICY 239-268 (Fall 2005)[21]

. . . .

Section 704 of FD&C Act[22] empowers FDA to enter and inspect any establishment in which food, drugs, devices, or cosmetics are manufactured, processed, packed, or held, for introduction into interstate commerce or after such introduction.[23] FD&C Act specifies that this inspection authority covers all pertinent equipment, finished and unfinished materials, containers, and labeling.[24] However, the Act is silent on photography during inspections.

In addition, section 704 provides that, with certain limitations, the inspection authority extends to all food records and other related information when FDA has a reasonable belief that an article of food is adulterated and presents a threat of serious, adverse health consequences or death to humans or animals.[25] When the inspection pertains to prescription drugs, nonprescription drugs intended for human use, or restricted medical devices, the FDA's inspection authority is broader yet and extends to "all things therein (including records, files, papers, processes, controls, and facilities)."[26]

---

[21] Also available at: http://papers.ssrn.com/sol3/papers.cfm?abstract_id=910581 (last accessed Mar. 31, 2007).

[22] 21 U.S.C. § 374.

[23] 21 U.S.C. § 374(a)(1) reads in pertinent part:

    (a)(1) For purposes of enforcement of this chapter, officers or employees duly designated by the Secretary, upon presenting appropriate credentials and a written notice to the owner, operator, or agent in charge, are authorized:

        (A) to enter, at reasonable times, any factory, warehouse, or establishment in which food, drugs, devices, or cosmetics are manufactured, processed, packed, or held, for introduction into interstate commerce or after such introduction, or to enter any vehicle being used to transport or hold such food, drugs, devices, or cosmetics in interstate commerce; and

        (B) to inspect, at reasonable times and within reasonable limits and in a reasonable manner, such factory, warehouse, establishment, or vehicle and all pertinent equipment, finished and unfinished materials, containers, and labeling therein. . . .

[24] *Id.*

[25] 21 U.S.C. § 374(a)(1)(B).

[26] 21 U.S.C. § 374(a)(1)(B) reads in pertinent part:

In the case of any factory, warehouse, establishment, or consulting laboratory in which prescription drugs, nonprescription drugs intended for human use, or restricted devices are manufactured, processed, packed, or held, the inspection shall extend to all things therein (including records, files, papers, processes, controls, and facilities) bearing on whether prescription drugs, nonprescription drugs intended for human use, or restricted

The FDA's Position on Its Authority to Photograph

The FDA policy on photography during establishment inspections[27] is published in the agency's Investigations Operations Manual (IOM).[28] IOM, chapter 5, subchapter 523, "Photographs–Photocopies," discusses the taking of photographs during inspections.[29] IOM cites examples of conditions or practices that may be "effectively documented by photographs," such as evidence of rodent or insect infestation, contamination of raw materials or finished products, and employee practices contributing to contamination or to violative conditions.[30] IOM states, "[s]ince photographs are one of the most effective and useful forms of evidence, everyone should be taken with a purpose. Photographs should be related to insanitary conditions contributing or likely to contribute filth to the finished product, or to practices likely to render it injurious or otherwise violative."[31]

FDA directs its inspectors:[32]

Do not request permission from management to take photographs during an inspection. Take your camera into the firm and use it as necessary just as you use other inspectional equipment. If management objects to taking photographs, explain that photos are an integral part of an inspection and present an accurate picture of plant conditions. Advise management the United States [c]ourts have held that photographs may lawfully be taken as part of an inspection.[33]

FDA's operational policy not to request permission to take photographs often raises the ire at regulated firms for its seeming rudeness. The rationality

---

devices which are adulterated or misbranded within the meaning of this chapter, or which may not be manufactured, introduced into interstate commerce, or sold, or offered for sale by reason of any provision of this chapter, have been or are being manufactured, processed, packed, transported, or held in any such place, or otherwise bearing on violation of this chapter.

[27] FDA also provides policy guidance with its COMPLIANCE POLICY GUIDES, COMPLIANCE PROGRAM GUIDANCE MANUAL, and its REGULATORY PROCEDURES MANUAL, *available at*: http://www.fda.gov/ ora/compliance_ref/cpg/default.htm; http://www.fda.gov/ora/cpgm/default.htm; and http://www. fda.gov/ora/compliance_ref/rpm/default.htm (last accessed Mar. 31, 2007).

[28] OFFICE OF REGULATORY AFFAIRS, FDA, INVESTIGATIONS OPERATIONS MANUAL (IOM) 2005, *available at*: http://www.fda.gov/ora/inspect_ref/iom/ [hereinafter IOM] (last accessed Mar. 31, 2007).

[29] *Id.* at chapter 5, subchapter 523.

[30] *Id.*

[31] *Id.*

[32] "Inspector" and "field investigator" are terms often used interchangeably for field agents of FDA. While both are general terms and can apply to a variety of activities, the term inspector is used throughout this article to distinguish inspections (where a Form FDA 482, Notice of Inspection, is issued) from various investigations, particularly criminal investigations. In 1992–93, FDA added armed criminal investigators, and FDA's criminal investigations raise other constitutional issues, such as *Miranda* warnings, which are not required during administrative inspections. *See, e.g., United States v. Gel Spice Co., Inc.*, 773 F.2d 427 (2d Cir. 1985).

[33] IOM, *supra* note 28, at 523.01.

of FDA's policy, however, must be determined with the context of FDA's section 704 inspection authority and relevant case law.

The Scope of Section 704 Inspection Authority

The scope of the FDA's authority for inspections under section 704 is general with few specific constraints. The most specific constraint is a limit on the FDA's access to financial data, sales data other than shipment data, pricing data, personnel data (other than data as to qualification of technical and professional personnel), and research data (other than data relating to new drugs, antibiotic drugs, and devices and subject to reporting requirements).[34]

FD&C Act also sets a few procedural requirements. Before entering an establishment or inspecting, the FDA inspector must present appropriate credentials and a written notice to the owner, operator, or agent in charge.[35] The FDA inspector may inform a firm of the purpose of the inspection (routine, complaint investigation, pre-approval, etc.). However, the FDA's Notice of Inspection form[36] does not specifically supply the reason for the inspection.[37] In addition, the notice of inspection is not required to include the reasons for the inspection or what the inspector expects to find.[38]

The major constraint on FDA is a rule of reasonableness. Inspections must be "at reasonable times and within reasonable limits and in a reasonable manner."[39] The reasonableness of the time, limits, and manner of inspections has only occasionally been litigated; but when an inspection's reasonableness has been challenged, courts largely determine reasonableness based on whether FDA met the procedural requirements of Section 704.[40] Reasonableness will also be determined from the facts of each situation, such as the enforcement needs under the statute and whether an unnecessary burden is placed on a firm.[41]

---

[34] 21 U.S.C. § 374(a)(1).

[35] 21 U.S.C. § 374(a)(1).

[36] FDA, Notice of Inspection Form FDA-482, *available at*: http://www.fda.gov/ora/inspect_ref/iom/exhibits/x510a.html.

[37] *Id.*

[38] Daley v. Weinberger, 400 F. Supp. 1288 (E.D.N.Y. 1975), *aff'd.* 536 F.2d 519 (2d Cir. 1976) and *cert. denied* 430 U.S. 930 (1977); *see also* United States v. Jamieson-McKames Pharm., Inc., 651 F. 2d at 538 (8th Cir. 1981) ("The notice of inspection used in this case satisfies at least some of these criteria. It informs the 'owner or agent in charge' of the 'scope and objects of the search.'").

[39] 21 U.S.C. § 374(a)(1)(B).

[40] *See, e.g.,* Gel Spice, 601 F. Supp. at 1228 (holding that photographing was not unreasonable where "the agents were in the warehouse pursuant to lawful authority and followed all procedural requirements mandated under 21 U.S.C. § 374").

[41] *See, e.g., Jamieson-McKames,* 651 F.2d at 537 (noting that the reasonableness of the warrantless search is dependent on the "specific enforcement needs and privacy guarantees of each statute").

Refusal to Permit Inspection

Refusal to permit an FDA inspector to duly[42] enter and inspect a regulated facility is a violation of section 301(f) of FD&C Act.[43] FDA considers a section 301(f) refusal to be a refusal to permit an inspection or prohibiting an inspector from obtaining information to which FDA is entitled by law.[44] A refusal may be a partial refusal, for example, a refusal to permit access to some records or some parts of a facility to which FDA is authorized to inspect.

Whether a refusal to allow photographs is a refusal (or partial refusal) of inspection under section 301(f) remains an issue of debate.[45] In the absence of explicit language in the statute, it has been contended that refusal to permit photography should not be considered a section 301(f) refusal of inspection.[46]

As a matter of legal interpretation, if photography is a reasonable part of a section 704 inspection, then refusal to permit photography would be a section 301(f) violation, "The refusal to permit entry or inspection as authorized by section 374 (i.e., 704) of this title."[47] Nonetheless, it remains arguable that a court would not find a 301(f) violation, a refusal to permit inspection, when a firm courteously refused to consent to photography, but otherwise allowed the inspection. Particularly when the immediate issue will have been resolved by a search warrant, a court may be reluctant to mete out punishment.

The controversy is unlikely to be resolved by the courts because the circumstances foreclose the two basic occasions for litigation. The first occasion is the pursuit of a complaint for refusal to permit photography. The second is the FDA's use of search warrants, which preclude the need for other judicial action.

FDA has not yet pursued a complaint for the refusal to permit photography and is unlikely to do so in the future.[48] In part, this is because the issue is arguable, but the likely reason for such reluctance is arguably due to pragmatism in marshalling limited resources. The FDA's powers and responsibilities have never been matched with enough resources to enforce all issues within its oversight. Therefore, the agency must decline to take action against some vio-

---

[42] The inspector presents proper identification and a valid inspection notice during a reasonable time as required by FD&C Act Section 704, 21 U.S.C. § 374 (2000).

[43] 21 U.S.C. § 331(f) (stating that "[t]he refusal to permit entry or inspection as authorized by section 374 of this title is a prohibited act").

[44] IOM, *supra* note 28, at § 514.

[45] Branding & Ellis, *Underdeveloped, supra* note 8 at 12:

> Whether a refusal to allow photographs is an actual refusal of the inspection under section 704 is not settled. . . . An investigator may characterize a firm's nonconsent to the taking of photographs as a refusal of the inspection or of information. In the absence of explicit legal authority in the statute, however, such nonconsent should not, as a matter of legal interpretation, be referred to as a refusal of the inspection.

[46] *Id.*

[47] 21 U.S.C. § 331(f).

[48] FDA has never prosecuted a firm for failure to permit photography. E-mail from Evelyn DeNike, Consumer Affairs Officer, FDA (Aug. 29, 2005) (on file with the author).

lations, and the FDA's authority to do so has been upheld in court actions.[49] In addition, enforcement discretion is not the exclusive choice of FDA. The United States Department of Justice (DOJ) and the United States Attorney for the judicial district in which FDA seeks judicial remedy also share discretion in filing court actions. Court actions are resource-intensive for both FDA and DOJ, and the agencies perform several layers of review before a case can proceed. All of these factors combine to make the FDA's pursuit of a complaint for failure to permit photography unlikely.

The lack of a case on point also exists because, if a firm refuses to permit photography, and FDA determines photography is necessary, FDA will seek an administrative search warrant.[50] The FDA's boilerplate language for administrative search warrants includes authorization of photography. Once the search warrant is issued, refusal to permit inspection photography in the face of search warrant authority mutes the issue of authority under FD&C Act. After an FDA inspector obtains a search warrant, federal marshals will execute it. At that point, refusal to permit inspection can result in arrest by the federal marshals. Refusal in the face of a search warrant is punishable by judicial contempt of court sanctions[51] in addition to separate criminal violations under FD&C Act.[52] Additionally, refusal to permit inspection in such circumstances might result in seizures and injunctive actions.

Photographic evidence can be very damaging.[53] Because the issue of the legality of a firm refusing to permit photography absent a warrant is unlikely to be settled by the courts, and because the risk of prosecution is remote, many firms are likely to continue to refuse to consent to photography.[54] Thus, the

---

[49] *See, e.g.*, National Milk Producers Fed'n v. Harris, 653 F.2d 339 (8th Cir. 1981) (holding that FDA's enforcement proceedings were discretionary); Heckler v. Chaney, 470 U.S. 821 (1985) (holding that FDA's decisions not to take certain enforcement actions are not subject to judicial review under the APA).

[50] IOM, *supra* note 28, at 523.01 ("If management refuses, advise your superior so legal remedies may be sought to allow you to take photographs, if appropriate.").

[51] *See, e.g.*, Becton, Dickinson & Co. v. FDA, 448 F. Supp. 776, 780 n.6 (N.N.Y. 1978) *aff'd.* 589 F2d. 1175 (2d Cir. 1978) ("This [c]ourt cannot, however, condone the actions of the defendants in refusing to abide by a Writ lawfully issued by this [c]ourt. . . . This cuts against all notions of law and order, and sets the stage for an obviously intolerable confrontation in every case in which a search warrant is issued.").

[52] 21 U.S.C. § 331(e) and (f) (Supp. 2005), amended by Pub. L. 109-59, Tit. VII, § 7202 (d), (e), 119 Stat 1913 (2005) (amended Aug. 10, 2005).

[53] *See* IOM, *supra* note 28, at 523 ("Since photographs are one of the most effective and useful forms of evidence, everyone should be taken with a purpose. Photographs should be related to insanitary conditions contributing or likely to contribute filth to the finished product, or to practices likely to render it injurious or otherwise violative.").

[54] Firms should be aware that there might be repercussions for refusing to permit photography beyond FDA returning with a search warrant. For example, such an action may make the inspector suspicious, more vigilant, and increase the frequency and duration of inspections. Inspectors may increase scrutiny when the actions or attitude of a firm appear suspicious. In addition, the inspectors always retain a degree of discretion. An uncooperative attitude on the part of firm management may well result in an uncooperative attitude by the inspector.

status quo is likely to continue where some firms refuse consent, and FDA seeks an administrative warrant when the agency considers photography necessary to complete their inspection.

In summary, FD&C Act provides FDA with the power to enter and inspect regulated establishments. The statute applies a general rule of reasonableness. The FDA's policy is not to request permission to photograph during inspections, but to proceed taking photographs unless stopped. Refusal to permit an FDA inspector to enter and inspect is a violation of FD&C Act, but it is unclear whether a firm would be prosecuted for refusing permission to take photographs absent a warrant.

. . . .

Fourth Amendment Constraints

Government inspections are a form of search and thus are constrained by the Fourth Amendment.[55] Except in carefully defined circumstances, the Fourth Amendment requires government agents to obtain a search warrant before inspecting private premises.[56]

Inspections under FD&C Act are within one of those exceptions. FDA is not required to obtain a search warrant to inspect an establishment regulated under Section 704, so long as the inspection is conducted reasonably as to time, place, and method.[57] An individual search warrant is not necessary because FD&C Act serves as a substitute for a search warrant.[58]

Such warrantless inspections have been held to be fully consistent with the Fourth Amendment.[59] The Supreme Court has upheld warrantless inspections for industries "long subject to close supervision and inspection"[60] and for

---

[55] The Fourth Amendment provides: "The right of the people to be secure in their persons, houses, papers and effects, against unreasonable searches and seizures, shall not be violated, and no Warrants shall issue, but upon probable cause, supported by Oath or affirmation, and particularly describing the place to be searched, and the persons or things to be seized." U.S. CONST. amend. IV.

[56] Under the Fourth Amendment, "except in certain carefully defined classes of cases, a search of private property without proper consent is 'unreasonable' unless it has been authorized by a valid search warrant," *Camara v. Municipal Ct.*, 387 U.S. 523, 528–29 (1967). *See also See v. City of Seattle*, 387 U.S. 541, 543: "The businessman, like the occupant of a residence, has a constitutional right to go about his business free from unreasonable official entries upon his private commercial property. The businessman, too, has that right placed in jeopardy if the decision to enter and inspect for violation of regulatory laws can be made and enforced by the inspector in the field without official authority evidenced by warrant."

[57] United States v. New England Grocers Supply Co., 488 F. Supp. 230, 238–39 (D. Mass. 1980); *see also* United States v. Business Builders, Inc., 354 F. Supp. 141, 143 (N. Okla. 1973); United States v. Del Campo Baking Mfg. Co., 345 F. Supp. 1371, 1376–77 (D. Del. 1972).

[58] *Id.*

[59] *See, e.g.,* New England Grocers Supply Co., 488 F. Supp. 238.

[60] Colonnade Catering Corp. v. United States, 397 U.S. 72, 77 (1970) (addressing the Bureau of Alcohol, Tobacco, and Firearms' inspectional authority over liquor).

"pervasively regulated business[es]."[61] This search warrant exception is often called the *Colonnade-Biswell* exception, so named for the paired rulings that delineate the exception.[62]

Under the *Colonnade-Biswell* exception, the government may conduct a search of a "closely regulated" commercial business without a warrant if three criteria are met.[63] First, the regulatory inspection scheme must be supported by a "substantial" government interest.[64] Second, warrantless inspections must be "necessary to further [the] regulatory scheme."[65] Third, "the statute's inspection program, in terms of the certainty and regularity of its application, [must] provid[e] a constitutionally adequate substitute for a warrant."[66] In other words, the statute must be "sufficiently comprehensive and defined that the owner of commercial property cannot help but be aware that his property will be subject to periodic inspections undertaken for specific purposes," and the inspection program must be "carefully limited in time, place and scope."[67]

Application of the Colonnade-Biswell Exception to Photography

Numerous court decisions support the application of the *Colonnade-Biswell* exception to inspections authorized under FD&C Act.[68] Businesses regulated under FD&C Act and subject to Section 704 inspections would have a difficult battle convincing a court that *Colonnade-Biswell* does not apply. As the court noted in *United States v. Business Builders, Inc.:*[69]

> It would be an affront to common sense to say that the public interest is not as deeply involved in the regulation of the food industry as it is in the liquor and firearms industries.[70] One needs only to call to mind recent cases

---

[61] United States v. Biswell, 406 U.S. 311, 316 (1972) (regarding a warrantless inspection of a pawnshop, which was federally licensed to sell guns pursuant to the Gun Control Act of 1968). A system of warrantless inspections was deemed necessary "if the law is to be properly enforced and inspection made effective." *Id.*

[62] *See, e.g.*, Donovan v. Dewey, 452 U.S. 594, 602 (1981).

[63] New York v. Burger, 482 U.S. 691, 702 (1987).

[64] *Id.*

[65] *Id.*

[66] *Id.* at 703.

[67] *Id.*

[68] *See generally* Daniel H. White, Annotation, *Validity of Inspection Conducted under Provisions of Federal Food, Drug, and Cosmetic Act (21 U.S.C.A. § 374(9)) Authorizing FDA Inspectors to Enter and Inspect Food, Drug, or Cosmetic Factory, Warehouse, or Other Establishment*, 18 A.L.R. Fed. 734 (2004); *see also Jamieson-McKames Pharm.*, 651 F.2d 532 (regarding a drug manufacturing industry); *New England Grocers Supply Co.*, 488 F. Supp. 230 (involving a food-supply warehouse); United States v. Acri Wholesale Grocery Co., 409 F. Supp. 529 (S. Iowa 1976) (food); United States v. Business Builders, Inc., 354 F. Supp. 141 (N. Okla. 1973) (food); and United States v. Del Campo Baking Mfg. Co., 345 F. Supp. 1371 (D. Del. 1972) (food).

[69] Business Builders, 354 F. Supp. at 143.

[70] Presumably, federal interest in liquor is pecuniary, due to the great amount of taxes collected from that industry. Likewise, federal interests in firearms are the prevention of violent crime.

of deaths occurring from botulism. Modern commerce has devised such an efficient and rapid means of distribution of food products to the consumer that a batch of contaminated food may cause widespread illness and death before the public can be warned and the contaminated products removed from the market.[71]

The *Colonnade-Biswell* exception to the Fourth Amendment's warrant requirement is considered constitutionally acceptable largely because businesses that are subject to comprehensive, government regulatory supervision have a "reduced expectation of privacy."[72] The Supreme Court discussed this reduced expectation in *New York v. Burger*: This expectation is particularly attenuated in commercial property employed in "closely regulated" industries. The Court observed in *Marshall v. Barlow's, Inc.*: "Certain industries have such a history of government oversight that no reasonable expectation of privacy, could exist for a proprietor over the stock of such an enterprise."[73]

The *Colonnade-Biswell* exception permits warrantless inspections of a closely supervised and pervasively regulated industry because "when an entrepreneur embarks on such a business, he has chosen to subject himself to a full arsenal of governmental regulation,"[74] and "'in effect consents to the restrictions placed on him.'"[75] In light of such a history of government scrutiny, such a business has no "reasonable expectation of privacy."[76]

Applying this reduced expectation of privacy to the FD&C Act inspections begs the rhetorical question: Is there any expectation of privacy from photography in areas where FDA has the authority to inspect? Common sense dictates that where FDA has the statutory authority to inspect—to observe, document, and sample—there is no Fourth Amendment expectation of privacy.[77] Thus, there would be no Fourth Amendment protection against FDA photographing areas where FD&C Act authorizes FDA to inspect.[78]

Under Fourth Amendment jurisprudence, the test is whether "the government's intrusion infringes on the personal and societal values protected by the Fourth Amendment."[79] Thus, photography by the FDA inspectors

---

However, it would seem to this Court that the public health and welfare under any system of values would be more important than revenue and suppression of criminal activity. *Id.* at n.1.

[71] *Id.* at 143.

[72] New York v. Burger, 482 U.S. 691, 702 (1987).

[73] *Id.* at 700.

[74] Marshall v. Barlow's, Inc., 436 U.S. 307, 313 (1978).

[75] *Id.* (citing Almeida-Sanchez v. United States, 413 U.S. 266, 271 (1973)).

[76] *Id.*

[77] While photography may be deemed more intrusive into privacy that mere visual observation in some circumstances, it seems unlikely that this would be the case in the context of a regulatory inspection where the statute gives the authority to inspect, document conditions, and sample.

[78] Again, attorneys must be careful when speaking with their clients. The author's experience is that some clients easily believe they have an inherent or constitutional right not be photographed.

[79] Oliver v. United States, 466 U.S. 170, 171 (1984).

during a duly authorized inspection would violate the Fourth Amendment only if the business manifested a subjective expectation of privacy of the area photographed that society accepts as objectively reasonable.[80]

However, businesses regulated by FDA are well aware that during inspections the FDA inspectors will view and document observations in the establishment and take samples. Accordingly, an FDA-regulated firm (a business subjected to close supervision and pervasive regulation) would have no reasonable expectation of privacy in the areas under an inspection.[81] Moreover, the photography would be merely cumulative or duplicative of the inspector testimony, reports, and samples, which mitigates the intrusiveness of an inspection.[82] In the face of such government scrutiny, a business has no reasonable expectation of privacy against photography. Accordingly, the Fourth Amendment would not protect against photography of areas and items legitimately subject to FDA inspection. Nonetheless, the Fourth Amendment provides other protections, such as restraint against breaking and entering without a warrant.[83]

No Authorization for Forced Entry without a Warrant

If a business denies FDA entry to inspect, no language in FD&C Act authorizes FDA to force entry or inspection. Absent express statutory authority to force entry or inspection without a warrant, the Fourth Amendment prevents authorization by implication.[84] The Supreme Court in *Colonnade Catering* sets out the reasoning behind this protection:

> Where Congress has authorized inspection but made no rules governing the procedure that inspectors must follow, the Fourth Amendment and its various restrictive rules apply. . . . [T]his Nation's traditions . . . are strongly opposed to using force without definite authority to break down doors. . . . Congress has broad authority to fashion standards of reasonableness for searches and seizures. Under the existing statutes, Congress selected a standard that does not include forcible entries without a warrant. It resolved the issue, not by authorizing forcible, warrantless entries, but by making it an offense for a licensee to refuse admission to the inspector.[85]

---

[80] For application of this standard, *see California v. Greenwood*, 486 U.S. 35 (1988) (finding no reasonable expectation of privacy in garbage bags left at the curb).

[81] *See* Marshall, 436 U.S. at 313.

[82] *See, e.g.*, Acri Wholesale Grocery Co., 409 F. Supp. 529 (finding that there was no unlawful or unwarranted intrusion by photography). The court noted, "Moreover, in this case the photographs introduced into evidence at trial were merely cumulative of the inspectors' testimony regarding the insanitary conditions in the warehouse." *Id.* at 533.

[83] *See, e.g.*, King v. City of Ft. Wayne, Ind., 590 F. Supp. 414, 428 (N. Ind. 1984).

[84] *See* Colonnade Catering Corp., 397 U.S. at 77 ("[T]his Nation's traditions that are strongly opposed to using force without definite authority to break down doors.").

[85] *Id.* at 77.

While *Colonnade* involved the federal liquor law, the provisions in FD&C Act are similar to those addressed in *Colonnade*.[86] Congress provided no authority in FD&C Act for FDA to force entries without a warrant, but Congress did make it an offense to refuse permission to enter or inspect.[87] This issue, at least with respect to FD&C Act, was addressed in *United States v. Jamieson-McKames Pharmaceuticals, Inc.*[88] The court found that the *Colonnade-Biswell* exception applied to inspections under FD&C Act, but that the Act did not authorize FDA to force entry or inspection where consent was withheld.[89] The court found that if consent were withheld, a separate violation of FD&C Act would occur; but the FDA inspectors are required to obtain a warrant before the inspection can proceed.[90]

Although the *Jamieson-McKames* court stated, "that an inspection pursuant to a [section] 374 [i.e., 704] notice to inspect is authorized only when there is a valid consent," this ruling followed the *Colonnade* decision.[91] In *Colonnade*, the Court found that, in the absence of statutory authorization by Congress to force entry, the Fourth Amendment restricted the government from *forcible* entry for inspection.[92]

Thus, out of context, the statement that a section 704 inspection "is only authorized where there is valid consent," would be misleading.[93] More precisely, consent is not necessary for a valid FDA inspection under FD&C Act, but FD&C Act does not authorize FDA, absent consent, to force an entry or inspection without a warrant.[94]

---

[86] *See, e.g.,* Jamieson-McKames, 651 F.2d at 539.

[87] 21 U.S.C. § 331(f) (Supp. 2005) (stating that it is a prohibited act to refuse "to permit entry or inspection as authorized by section 374 of this title").

[88] Jamieson-McKames, 651 F.2d 532.

[89] *Id.* at 539–40 ("It follows, therefore, as in *Colonnade*, that an inspection pursuant to a [section] 374 notice to inspect is authorized only when there is a valid consent. If consent is withheld, a separate violation of the Act occurs, and the FDA inspectors are required to obtain a warrant before the inspection can proceed.").

[90] *Id.* Other cases that hold that a search warrant is required in the absence of consent include *United States v. Roux Lab., Inc.*, 456 F. Supp. 973 (M. Fla. 1978) and United States v. Litvin, 353 F. Supp. 1333 (D.C. 1973).

[91] Jamieson-McKames, 651 F.2d at 539–40.

[92] Colonnade, 397 U.S. at 77: "Congress has broad authority to fashion standards of reasonableness for searches and seizures. Under the existing statutes, Congress selected a standard that does not include forcible entries without a warrant. It resolved the issue, not by authorizing forcible, warrantless entries, but by making it an offense for a licensee to refuse admission to the inspector."

[93] For example, under *Colonnade*, 397 U.S. at 77, clients may well hear that consent to inspection is required for a valid inspection and fail to hear that it is a violation of the FD&C Act to deny consent to FDA for an authorized inspection. This has been the author's experience in practice. Some need to be told bluntly, "FDA can't break down your door if you don't let them in, but FDA has the authority to inspect, and refusing their entry to inspect violates the law."

[94] Business Builders, 354 F. Supp. at 143 ("In effect, the statute takes the place of a valid search warrant. Thus, consent is immaterial and Defendants do not contend that the inspection was conducted unreasonably as to time, place, or method.").

This precision in important because the circumstances of a valid inspection without consent exist where a firm gave consent but the consent was invalid—for example, where consent to inspection was involuntary because consent was only given under threat[95] of prosecution.[96] Hypothetically, invalid consent might also occur where a firm's employee lacks the authority to grant consent to FDA but nonetheless permitted entry. Foremost, the Fourth Amendment restriction on forcible entry would not extend protection to situations where there was no force used but where was consent was vague or ambiguous. Consent in such situations is moot because the regulated firms are required to comply with a warrantless regulatory inspection.[97]

This matter relates to the reason why the issue of the FDA's authority to take photographs remains largely unexplored by the courts. When a firm refuses to permit FDA to take photographs during an inspection, FDA lacks the authority to take photographs forcibly.[98] Therefore, faced with a refusal to permit photography, FDA must obtain an administrative search warrant if they consider photographs necessary for their inspection.[99] Thus, the issue of the FDA's authority to take photographs never comes to a head. Once a search warrant is obtained, the issue of the reasonableness of photography under FD&C Act becomes moot.[100]

Therein rests the heart of the issue: whether, absent the specific mention of photography in section 704, FDA is nevertheless empowered to take photographs. As discussed above, an FDA-regulated industry would have no reasonable expectation of privacy against photography in the areas and items under inspection.[101] Therefore, the legitimate areas and items of the FDA inspection would not be legitimate areas for Fourth Amendment protection against photography.[102] Absent Fourth Amendment protection, this issue will revolve around interpretation of FD&C Act.

. . . .

[95] A person who believes that their consent is required for an FDA inspection will naturally construe FDA's explanation of the penalties for failure to permit inspection as a threat.

[96] *See, e.g.,* Business Builders, 354 F. Supp. 142 (explaining defendants who argued there was no valid consent when they allowed inspection because they were threatened with criminal prosecution if they refused to permit an inspection). "[I]t is this [c]ourt's conclusion that in the circumstances of this case, neither consent nor a search warrant is necessary." *Id.*

[97] *See* United States v. Articles of Drug, 568 F. Supp. 1182, 1185 (N. Cal. 1983) ("To the extent [the defendant] thought he was cooperating with a regular FDA inspection, consent is not an issue because [he] was required to comply with a warrantless regulatory inspection. . . .").

[98] *See, e.g.,* Jamieson-McKames, 651 F.2d at 539–40.

[99] This is essentially FDA's policy. *See* IOM, *supra* note 28, at § 523.

[100] FDA, given prioritization of limited resources, is also unlikely to bring forward a complaint for refusal to permit inspection solely for a refusal to permit photography, although this might be considered a partial refusal. Section 331(f) makes refusal to permit entry or inspection as authorized by section 374 of FD&C Act a prohibited act. 21 U.S.C. § 331 (f). Refusal to permit inspection is discussed further below.

[101] *See supra* Section II.A.

[102] *Id.*

Conclusion

The language of FD&C Act section 704 provides FDA with broad inspectional authority based on a flexible standard of reasonableness. The statutory language and the case law support the conclusion that FDA may lawfully take photographs during a section 704 inspection so long as the inspection is otherwise lawfully conducted, the FDA's procedural requirements for inspection are met, and the photography is within the normal course of the inspection.

In narrow circumstances there may be a viable issue of whether the inspection itself (including the taking of the photographs) was reasonable. However, such determinations are likely to be ruled in the FDA's favor. The outcome of these decisions, of course, will be heavily dependent upon the particular facts and circumstances of the inspection and a firm's regulatory history.

Therefore, the common statement that the language of FD&C Act section 704 does not expressly authorize FDA to take photographs during inspections presents a misleading perspective. A fair summary of FD&C Act and the case law is that FD&C Act authorizes FDA to take photographs within the contours of a lawful inspection to advance the purposes of the Act.

Refusal to consent to photography as part of a lawful section 704 inspection would be a partial refusal to permit the inspection, but it is arguable whether such a refusal would result in conviction for violation of section 301(f).[103] Nevertheless, FDA is unlikely to pursue such a complaint—in part, because the issue remains unsettled—largely because of the pragmatic use of limited resources. Because this issue is unlikely to be settled by the courts and because of the damaging nature of photographic evidence, many firms are likely to continue to refuse to consent to photography, and FDA will be forced to seek administrative search warrants.

Considering the time and expense to the government of suspending an inspection, requesting a search warrant, and returning with federal marshals, efficiency calls for instructional language to be added to FD&C Act to make explicit the FDA's authority to take photographs during an inspection. Congress could accomplish this simply by placing language in section 704 that states an inspection "includes, but is not limited to, photography." Alternatively, Congress could amend the Act and place the cost of refusing permission to photograph on the firm refusing such an inspection. This alternative could be accomplished by providing FDA with the authority to issue an administrative fine for refusal to permit photography. Such an administrative fine provision would also clarify the FDA's authority to take photographs, and refusals to permit photography would decline. An additional benefit of such amendments to FD&C Act would be to eliminate an area in the law that encourages conflict between the FDA and regulated firms.

However, other than the controversy over the legality of refusing to consent to photography during an FDA inspection, the law on the FDA's authority to

---

[103] 21 U.S.C. § 331(f).

take photographs is clear. In essence, the FDA's authority to take photographs is coextensive with the agency's authority to enter and inspect.

As the saying goes, "A picture is worth 1,000 words,"—*not* that a picture is more than 1,000 words. Under the law, inspection photography is no more intrusive than other documentation. Where FDA has the authority to enter, inspect, and document the conditions in an establishment, the agency holds the authority to take photographs. In the twenty-first century photographs are a reasonable way for FDA to document conditions in regulated establishments.

* * * * *

Because FDA inspectors are instructed not to seek consent for photographs, a firm must decide in advance what position to take on this issue. A firm that refuses to allow photographs must explicitly state this before the inspection begins. Post-inspection efforts to block admission of photographs taken with consent, or without express refusal, will likely fail.

Firms that inform FDA inspectors of a no-photos policy should do so in a calm and professional manner. FDA inspectors will usually comply with the company policy (at least temporarily). If the FDA inspectors decide they need photographs, they will seek a search warrant. Arguably a firm that refuses to allow the FDA inspector to take photographs could be prosecuted for a partial refusal to permit inspection; however, this is unlikely.

A firm can strengthen its no-photos position by putting its policy in writing after careful consideration of legitimate concern for protection of trade secrets and to exclude glass from certain manufacturing areas. These concerns apply to all outsiders, so that should be reflected in the policy. "No Cameras" signage may be in order.

Firms with a no-photo policy should be aware that state inspectors may have different statutory authority than the FDA. In addition, government inspectors monitoring a firm's supply of government contracts (i.e., supplying food, medical devices, or drugs to the government) have greater authority to examine certain records. If the firm refuses information, its status as a qualified supplier may be jeopardized.

An inspector's purpose in taking photographs will be to capture on film perceived law violations. Some firms may be concerned that an inspector's photos will be unrepresentative of the company's total operation. If a company chooses to allow photographs, a firm may choose to take duplicate pictures of what the inspector photographs, as well as additional photos to show a broader spectrum of the operation, and to show correcting of any violation captured on film by the inspector.

However, there is a trade-off with a firm taking their own photographs. These photos can and have been used as additional evidence against a firm. Some firms may wish to refrain from photographing. In such cases the firm should request copies of the FDA photographs under the Freedom Information Act.

13.3   PLANNING FOR THE INSPECTION

> Financial preparation should include an awareness that capital investment in compliance, when the firm learns of a problem needing correction, may save the managers a jail term in egregious cases. Fixing up after the inspection may be too little, too late to avoid a recall or prosecution recommendation.
>
> —JAMES T. O'REILLY, FOOD AND DRUG ADMINISTRATION
> § 20:11 (2d ed. 2004)

The primary purpose of FDA inspections is to determine whether products are being handled and manufactured in compliance with the applicable law. When the FDA finds violative conditions or practices during an inspection, FDA will collect evidence to document the violation. Evidence includes the inspector's observations and reports, but may also include copies of company records and reports, photographs, and samples.

You may expect the inspector to be professional and polite, but they have a job to do. Their job does not include informing a firm about the limits of FDA's authority or whether items asked for by the inspector are required to be provided.

13.3.1   Policies

For this reason a regulated firm is well advised to have legal counsel provide defensive education to company managers. This can save a firm a great deal of money, not to mention untold headaches and trouble. For instance, firm management should determine with counsel which documents must be provided, and which documents or information will be voluntarily provided if requested. Other company policies include handling photography, who is granted authority to sign affidavits, and so forth. In some cases it may be advisable for a firm to prepared legal papers in advance to support a partial refusal of inspection, such as the firm's refusal to allow photography.

Of course, maintaining compliance is the surest means for having a successful inspection. Management must effectively communicate its expectation that all laws and regulations are to be met. Too often this is left unspoken or there is a misperception about priorities. All sorts of pressures can force a company out of compliance. Company policy and procedures must be designed to ensure compliance under the worst of conditions. Financial hardship often is associated with noncompliance. Company managers should remember that financial hardship is no defense under the law, and FDA has a policy of charging company leaders for violations.

13.3.2   Training

Training is essential to successful inspections. Staff must be trained in policy, of course, but they must also know the technical aspects of their jobs and the law well enough so that they can do their jobs well enough to ensure compliance with the applicable requirements. "Ignorance of the law is no excuse," is not just a platitude; in food and drug law, it is the rule. Expect FDA to treat noncompliance out of ignorance the same as an intentional violation.

Policy should designate who will accompany FDA during the inspection and provide them with appropriate training. For large companies, more than one person should be designated this role. For complex operations, it is also good to designate someone as a dedicated scribe. The scribe can keep notes on what was asked and answered, concern raised, and so forth. This unburdens the escort so they can focus on the FDA inspector.

13.3.3   FDA's Notice of Inspection

FDA inspectors are required to produce credentials and a written notice of inspection (FDA Form 482) to the "owner, operator or agent in charge" of a facility before beginning an inspection.[104] However, they are not required to give advance notice of the inspection. A firm may ask and be told the purpose of the inspection (e.g., routine random or investigation of a complaint), but FDA is not required to explain the reason for the inspection or what they expect to find.

Form-482 lists the name and address of the facility being inspected, and bears the signature(s) and typed or printed name(s) of the FDA employee(s) conducting the inspection. The form also cites the sections of the law that grant inspection authority.

13.3.4   Inspectional Observations (FDA Form 483)

Section 704(b) of the FD&C Act requires FDA inspectors to provide company representatives, upon completion of inspections and before leaving the premises, with a "report in writing setting forth any conditions or practices observed by him which, in his judgment, indicate that any food in such establishment" is contaminated with filth or has been "prepared, packed, or held under insanitary conditions" and thus may have become contaminated or injurious to health. Note that this provision of the act does not cover all violations of the FD&C Act, but FDA uses Form 483 to record most violations.

The FDA inspector will review (or request to review) the inspection listing with a company official before leaving. This is an opportunity for the company to immediately correct any mistakes or misunderstandings on the inspection. Although this does not mean one should debate with the inspector (which is generally unwise). This is also any opportunity for the company representative to obtain clarification for anything that is not understood. However, occasionally a firm will have a policy to withhold comment during the exit discussion, and comment later in writing.

---

[104] FD&C Act § 704(a).

If possible, any deficiencies noted should corrected during the inspection or before the inspector leaves. A firm may wish to respond to FDA after the inspection with a follow-up letter explaining how and when any deficiencies were corrected. However, both the Form 483 and responses to them are subject to public disclosure, and this should be kept in mind when preparing a response. If mention of confidential information is unavoidable, it must be designated as such.

Being in violation of the Act does not mean FDA will take enforcement action. Most FDA inspections and 483s do *not* result in enforcement action by FDA. Sometimes FDA issues warning letters following after the inspection. These letters may restate much of the Form 483 listings. If a 483 consists of observations the agency considers violative but not warranting further or immediate regulatory action, the firm will not receive a follow-up letter.

13.3.5 Times of Inspection

Section 704 authorizes accredited FDA personnel to enter and inspect certain establishments at "reasonable times and within reasonable limits and in a reasonable manner." What is a "reasonable time" is not defined in the statute, but has generally been held to be any time that regulated activities are being conducted at a facility.

Inspection authority generally encompasses facilities and products but may cover records, files, and other documents. The Act specifically exempts from FDA inspection certain types of information including financial, sales, and pricing (other than shipment data), research data, and certain personnel data unless such information directly has a bearing on whether a product may be adulterated, misbranded, or otherwise in violation of applicable laws or regulations.

13.3.6 Samples

When FDA collects a physical sample during an inspection, a written receipt for the sample (form FDA-484) must be provided to the owner, operator, or agent in charge. Receipts apply to not only product food samples but also any physical evidence, such as insects, rodent filth, foreign material, residues, or bacterial swabs. Documentary evidence such as labels, copies of production records, invoices, and shipping records are not listed on the FDA-484.

The receipt must be issued upon completion of the inspection before leaving the premises. In the case of food samples collected during an inspection, FDA is required to provide a written report on any sample analyzed for evidence of being filthy, putrid, or decomposed.[105] Sample analysis is generally done in FDA laboratories soon after the inspection.

---

[105] FD&C Act § 704(d). This provision applies only to samples foods and does not extend to samples of drugs, devices, or cosmetics that are collected during FDA inspections. This provision also does not apply to analysis for factors other than filth or decomposition.

The firm should take, label, and store duplicates of everything the inspector samples. If the firm asks FDA to give back part of its sample for the firm's lab to examine, FDA is required by the statute to do so.[106] This allows the firm to duplicate or challenge FDA's results.

Representative samples are usually sought. If the sample collection method was biased or the FDA procedures for sample collection were not followed, the agency's evidence might be challenged when it is presented to the court. The specific procedures for particular products can be found in the FDA's Investigation Operations Manual or the FDA Compliance Policy Guides.

13.3.7 Follow-up Information

As an alternative to an outright refusal to provide information to an inspector, it may be prudent to offer to provide the information later. Such an offer should always be made subject to the firm's management approval. Supplying additional information gives the firm an opportunity to go on the record with a clarification of anything listed in the inspection report that the firm believes is incorrect.

13.3.8 Etiquette for Dealing with FDA Inspectors

Interacting with inspectors requires a special combination of diplomacy and technical competency. Poor interactions with the inspector can trigger hostility. Being inspected by the government is stressful, and personal conflict can erupt over a minor issue. One hopes and should expect the inspector to remain professional and objective, but never forget that an uncooperative attitude from firm management will be factored in when assessing regulatory compliance and the need for enforcement action.

On the other hand, quality interactions with the inspector can defuse conflict and foster a productive, professional interaction. For example, the person interacting with the inspector should understand how the firm operates and be able to explain how procedures affect the food products. A cooperative attitude and reasonable explanations can sometimes head off citation.

13.4 FSIS INSPECTION AUTHORITY

The Federal Meat Inspection Act (FMIA) and the Poultry Products Inspection Act (PPIA) are the principal statutes providing USDA-FSIS with inspection authority. FSIS inspects meat and poultry under a "continuous inspection" system, which means that an inspector is assigned to every FSIS-regulated establishment and is required to be present when the establishment is in operation. However, as a practical matter, this does not necessarily mean every

[106] 21 U.S.C. § 374(d).

minute throughout the operation. In contrast, FDA-regulated food establishments typically will be visited by an FDA inspector less than once a year.

The agency inspects two different types of facilities—slaughtering plants and processing plants. Historically, the inspector's primary role in slaughtering plants has been to visually inspect animals before and after slaughter. This visual inspection system is still used. However, with the growing recognition that carcass-by-carcass inspection is a less than optimal way to detect food safety problems, the agency has been moving toward inspection methods based on HACCP.

In 1996, FSIS issued regulations requiring the implementation of HACCP systems for meat and poultry processing plants. The effective date of the regulations was phased in on different dates for different size establishments beginning in 1997. As a consequence of HACCP implementation, the role of inspectors in meat and poultry processing plants has shifted from primarily sensory inspection of the operation for evidence of sanitation problems to tasks involving the monitoring of compliance with regulatory performance standards.

13.4.1   Major Enforcement Powers

FSIS continuous inspection authority carries with it major enforcement powers. These powers were summarized in the preceding chapter on enforcement powers, but are explained again in the context of inspections in this chapter.[107]

Regulatory control actions is defined to encompass a number of different enforcement powers inspectors may invoke:

1. Retention of product
2. Rejection of equipment or facilities
3. Slowing or stopping of lines
4. Refusal to allow the processing of specifically-identified product.[108]

Among the circumstances possibly precipitating regulatory control actions are the following:

- Insanitary conditions or practices
- Product adulteration or misbranding
- Conditions that prevent the inspector from determining that the product being packed or processed is not adulterated or misbranded
- Inhumane handling or slaughtering of livestock[109]

---

[107] Rules of Practice, Food Safety and Inspection Service (FSIS), Final Rule, 64 Fed. Reg. 66,541 (Nov. 29, 1999) (codified in 9 C.F.R. Part 500).
[108] 9 C.F.R. § 500.1(a).
[109] 9 C.F.R. § 500.2(a).

Regulatory control actions typically are taken in response to relatively nonserious situations involving "specific amounts of product or generally well-defined deficiencies such as crushed and open cartons or malfunctioning equipment."[110]

Withholding action is the refusal to allow the marks of inspection to be applied to products, which effectively stops the sale of the products. A withhold action may be taken against all products in an establishment or just product from a particular process, and may be taken with or without prior notification to the establishment. Withholding actions are generally more significant than regulatory control actions and affect a larger part of an establishment. In most cases withholding actions are taken because of systemic problems, such as HACCP plan inadequacies.

Consequently, the steps necessary to correct a problem that resulted in a withholding action are more complex than those necessary to resolve a problem that resulted in a regulatory control action. Correction for a withholding action is likely to require a HACCP plan reassessment and modification, or revision of sanitation Standard Operating Procedures (SOPs).

Conditions that may trigger a withholding action without prior notification to the establishment include the following:[111]

- Production and shipment of adulterated or misbranded product
- Lack of a HACCP plan in accordance with regulatory requirements
- Lack of Sanitation SOP's in accordance with regulatory requirements
- Sanitary conditions such that products in the establishment are or would be rendered adulterated
- Violating the terms of a regulatory control action
- Assault, threat to assault, intimidation of, or interference with FSIS official
- Failure to destroy condemned product in accordance with regulatory requirements

USDA may take a withholding action "after an establishment is provided prior notification and the opportunity to demonstrate or achieve compliance" for a number of reasons:[112]

- Inadequacies in HACCP system; multiple or recurring incidents of noncompliance
- Improper implementation of or failure to maintain Sanitation SOP's
- Failure to maintain sanitary conditions as required

---

[110] FSIS Rules of Practice, 1999 Final Rule, 64 Fed. Reg. at 66,543.
[111] 9 C.F.R. § 500.3.
[112] 9 C.F.R. § 500.4.

- Multiple or recurring incidents of noncompliance
- Failure to collect and analyzed for pathogens performance standards

Suspension is an interruption in the assignment of USDA program employees to all or part of an establishment.[113] Suspension of inspection has the practical effect of shutting down the operations of the facility; it therefore has a more substantial impact than either a withholding or regulatory control action. FSIS stated,

> When public health is a concern, FSIS immediately suspends inspection until the problem is corrected. FSIS refuses to mark product as "inspected and passed" or retains an establishment's meat or poultry products if the Agency determines that meat or poultry products are adulterated. . . . Such actions typically are discontinued when the adulterated products have been destroyed or properly controlled, or when the deficiencies or noncompliances are corrected satisfactorily.[114]

Suspension typically is used as an enforcement tool after an establishment fails to correct a situation involving a withholding action, or when the nature of the noncompliances are such that the corrective action, such as HACCP plan reassessment or changes in the establishment's operation, may take a significant amount of time to implement.[115] Decisions to suspend inspection generally are taken at the FSIS District Manager level or higher.[116]

Suspensions may be imposed without prior notification. The conditions for imposing suspension without prior notification are the same as those for taking withholding action without prior notification, with one exception. One additional reason is listed for imposing suspension without prior notification: "because the establishment is handling or slaughtering animals inhumanely."[117]

Withdrawal of inspection, or total loss of inspection, puts an establishment out of business. FSIS inspectors do not themselves have the authority to permanently withdraw inspection. The agency seeks withdrawal of the grant of federal inspection through an administrative hearing. The grounds for withdrawal of inspection include serious violations plus being "unfit to engage in any business requiring inspection."[118]

13.4.2 Records Access

FSIS inspectors may obtain access to a wide variety of records in meat and poultry packing and processing establishments.

---

[113] 9 C.F.R. § 500.1(c).
[114] FSIS Rules of Practice, 1999 Final Rule at 66,541.
[115] Id. at 66,543.
[116] Id.
[117] 9 C.F.R. § 500.3(b).
[118] 9 C.F.R. § 500.6.

13.4.3   FSIS' International Inspection Activities

In addition to its domestic inspection responsibilities, FSIS is responsible for reviewing the inspection systems of all foreign countries that have been found eligible to export meat and poultry to the United States By law, those systems must be equivalent to the U.S. inspection system.

When imported meat and poultry are offered for entry into the United States, they are inspected again by FSIS officials. As noted above, FDA-regulated food is now subject to certain provisions of the Bioterrorism Act.

13.4.4   Supreme Beef v. USDA

The *Supreme Beef v. USDA* decision was discussed earlier regarding HACCP and food safety. *Supreme Beef* is often touted as standing for a rebuke of the USDA's application of HACCP principals and transition to HACCP inspections.[119] HACCP remains in use by USDA FSIS:

> The Pathogen Reduction/HACCP rule continues to be the United States Department of Agriculture's most important tool for ensuring the safety of meat and poultry. USDA's Food Safety Inspection Service' means of ensuring that there is process control under Pathogen Reduction/HACCP—the use of generic *E. coli* data—is undisturbed by the *Supreme* decision and will be enhanced by the introduction of consumer safety officers into the Agency's inspection force. FSIS intends to continue to use *Salmonella* testing as one means of determining whether a plant is controlling pathogens. In addition FSIS has requirements that address specific pathogens, including zero tolerance for *E. coli O157:H7* in raw ground beef and zero tolerance for any pathogen in cooked products. The Pathogen Reduction/HACCP rule requires plants to determine points in their processes where contamination can occur and develop methods to control it. If a plant does not have an adequate Pathogen Reduction/HACCP plan or does not have an adequate sanitation program in place to produce safe products, FSIS can withhold the mark of inspection or suspend inspection at a plant.
> . . .
> The *Salmonella* performance standard continues to be a part of USDA's Pathogen Reduction/HACCP inspection system. *Salmonella* testing in grinding plants will be used in conjunction with other information to verify that Pathogen Reduction/HACCP systems and sanitation systems are under control. Grinding plants that fail two *Salmonella* sample sets will be subject to an in-depth review of the plant's food safety systems. If deficiencies are identified in these systems, FSIS may initiate enforcement action. In addition USDA will continue to test for other pathogens such as *E. coli* O157:H7 and *Listeria monocytogenes*.[120]

---

[119] *See, e.g.*, Food Safety and Inspection Service, United States Department of Agriculture, Myths & Facts: Inaccuracies in News Articles Concerning the decision by the U.S. Court of Appeals for the Fifth Circuit in *Supreme Beef Processors, Inc. v. USDA* (Dec. 2001), *available at*: http://www.fsis.usda.gov/OA/news/2001/supremem&f.htm.

[120] *Id.*

Nonetheless, many believe the decision dealt a serious blow to USDA's food safety and inspection reforms.[121] Basically the appeals court upheld a lower court ruling that the USDA lacks the authority to shut down a meat-processing plant based on failure to meet *Salmonella* performance standards for incoming raw ingredients, when those standards were used as a proxy for insanitary conditions in the plant.

* * * * *

Supreme Beef Processors, Inc. v. USDA

275 F.3d 432 (5th Cir. 2001)

Certain meat inspection regulations promulgated by the Secretary of Agriculture, which deal with the levels of *Salmonella* in raw meat product, were challenged as beyond the statutory authority granted to the Secretary by the Federal Meat Inspection Act. The district court struck down the regulations. We hold that the regulations fall outside of the statutory grant of rulemaking authority and affirm.

. . . .

The *Salmonella* performance standards set out a regime under which inspection services will be denied to an establishment if it fails to meet the standard on three consecutive series of tests. The regulations declare that the third failure of the performance standard "constitutes failure to maintain sanitary conditions and failure to maintain an adequate HACCP plan . . . for that product, and will cause FSIS to suspend inspection services." The performance standard, or "passing mark," is determined based on FSIS's "calculation of the national prevalence of *Salmonella* on the indicated raw product."

. . . .

The difficulty in this case arises, in part, because *Salmonella*, present in a substantial proportion of meat and poultry products, is not an adulterant per se, meaning its presence does not require the USDA to refuse to stamp such meat "inspected and passed." This is because normal cooking practices for meat and poultry destroy the *Salmonella* organism, and therefore the presence of *Salmonella* in meat products does not render them "injurious to health" for purposes of § 601(m)(1). *Salmonella*-infected beef is thus routinely labeled "inspected and passed" by USDA inspectors and is legal to sell to the consumer.

. . . .

Not once does the USDA assert that *Salmonella* infection indicates infection with § 601(m)(1) adulterant pathogens. Instead, the USDA argues that the *Salmonella* infection rate of meat product correlates with the use of patho-

---

[121] *PBS Frontline: Modern Meat, available at*: http://www.pbs.org/wgbh/pages/frontline/shows/meat/evaluating/supremebeef.html (last accessed Apr. 23, 2004).

gen control mechanisms and the quality of the incoming raw materials. The former is within the reach of § 601(m)(4), the latter is not.

IV

Because we find that the *Salmonella* performance standard conflicts with the plain language of 21 U.S.C. § 601(m)(4), we need not reach *Supreme's* numerous alternative arguments for invalidating the standard, which were not addressed by the district court.

V

We AFFIRM and REMAND with instructions that the final judgment of the district court be amended to include the National Meat Association.

* * * * *

NOTES AND QUESTIONS

13.1. Inspections and the Fourth Amendment. Warrantless searches are generally unreasonable violations of the Fourth Amendment. How can the FDA conduct inspections without warrants?

13.2. Pervasively regulated industry exception. Why do you think the Supreme Court accepted the "pervasively regulated business" exception?

13.3. FDA's photography policy. What do you think of FDA's policy of not informing firms that the agency plans to take photographs and not asking permission? Would you suggest an alternative policy? What would be the consequences of an alternative policy?

CHAPTER 14

State Laws and Their Relationship to Federal Laws

14.1 INTRODUCTION

This chapter provides an overview of the regulatory role the states provide in the regulation of food. The relationship between the federal and state laws is examined with a particular eye toward the scope of state authority and the doctrines of implicit and explicit federal preemption.

14.1.1 Overview of the Role of States

In carrying out their food enforcement responsibilities, both FDA and FSIS depend on state and local government agencies to assist them. There is a considerable amount of overlap and sharing of food enforcement authority between the states and FDA.

Most states have adopted food and drug laws with nearly identical provision as the Food, Drug, and Cosmetic Act (FD&C Act). With the similarities in definitions of adulteration and misbranding, and similarity in enforcement authorities, FD&C Act violations generally also result in violations that state authorities can pursue. In addition FDA often commissions state officials to conduct inspections or gather evidence for the federal government. FD&C Act section 702(a) expressly authorizes FDA to act through state officials.

This is more than a passing fact of interest. A practical effect on food establishments is that a visit of an inspector may be based on both state and federal authority. When determining enforcement options, government regulators may cherry pick from the various state and federal authorities, prohibited acts, and penalties.

In 2006, for instance, over 80 percent of FDA inspections were conducted by state officials on contract to FDA.[1] FDA is responsible for about 210,000

---

[1] FDA Week., *FDA Inspects Significantly Less Than 1 Percent of Food Imports*, 12 FDA WEEK No. 33, Inside Washington Publishers (Aug. 18, 2006).

Food Regulation: Law, Science, Policy, and Practice, by Neal D. Fortin
Copyright © 2009 Published by John Wiley & Sons, Inc.

establishments. In fiscal year 2005, FDA conducted 4,573 food safety program inspections, but only 3,400 in 2006. FDA contracts out nearly 17,000 inspections to the states, but this still totals only 20,000 inspections and less than ten percent of the establishments under FDA's oversight.[2]

14.1.2 FDA Jurisdiction over Restaurants

The FDA reliance on state and local governments is perhaps most apparent in the regulation of restaurants. In 1975, the General Accounting Office (GAO) issued a report, Federal Support for Restaurant Sanitation Found Largely Ineffective. The GAO report criticized restaurant sanitation and estimated "that about 90 percent of the 14,736 restaurants [in the study] were insanitary." The report went on to note that FDA had jurisdiction over restaurant food that had been shipped in interstate commerce, but that FDA relied on state and local governments to regulate restaurants. "However, local governments generally have been ineffective in regulating restaurant sanitation and, generally, the States' monitoring of these programs has been minimal. . . ."

In anticipation of the GAO report, FDA proposed federal regulations on the sanitation of food service establishments.[3] After opposition from state officials, FDA re-evaluated its own priorities and abandoned the proposal.[4] FDA recognized—with hundreds of thousands of restaurants and millions of meals served daily—their agents could not inspect or regulate more than an insignificant percentage of these establishments, and the primary responsibility had to remain with the state and local governments.

14.1.3 FSIS

By comparison, the scope of FSIS' enforcement responsibility for food is more narrowly defined, encompassing primarily meat and poultry. FSIS also shares its regulatory responsibility with state officials through cooperative inspection programs that are deemed equal to FSIS's program.[5] However, the role that states may play is more constricted by the FMIA and PPIA.

14.1.4 Related Authorities

State food regulatory authority is generally delegated to either the state department of agriculture or the department of public health. In most, the authority over restaurants is divided from the authority over food manufactur-

---

[2] *Id.*
[3] 39 Fed. Reg. 35438 (Oct. 1, 1974).
[4] 42 Fed. Reg. 15428 (Mar. 22, 1977).
[5] *See, e.g.*, FSIS Directive 5720.2 Rev. 2, Cooperative Inspection Programs.

ing and other retail sales, such as grocery stores. Michigan is a notable exception, where authority over all food establishments is delegated to the state department of agriculture. Often regulatory authority over restaurants is further delegated to local public health agencies.

Only a few states have a separate FDA-like agency with a broad range of oversight. For example, in many states there is no regulatory authority established over pharmaceuticals or medical devices. In other states, the state board of pharmacy or the state board of health may be given responsibility over these areas, but nonetheless, have ill-defined enforcement authority. For these reasons the state Attorney Generals usually are the lead investigators of drug and medical device related issues, and even on food-related issues that pertain to broader matters such as advertising and use of the Internet.

14.1.5 National and State Cooperation

FD&C Act section 301(k) extends FDA's jurisdiction over the adulteration of food after shipment interstate, and thus FDA has jurisdiction over virtually all restaurants, food vending machines, and grocery stores. As a practical matter, FDA recognized that it lacked the resources to police the hundreds of thousands of establishments across the nation.

One tool on which FDA relies to achieve it mission is cooperative development of model codes. There are a number of federal–state model food sanitation programs. The oldest cooperative sanitation code is the Grade "A" Pasteurized Milk Ordinance (PMO).[6] Begun in 1924, the PMO has been adopted in most states and is periodically updated. Originally part of the Public Health Service (PHS), administration of this program was transferred to FDA in 1968. The National Conference on Interstate Milk Shipments was established to complement the PMO.

The National Shellfish Sanitation Program (NSSP) is another federal–state model code. NSSP was established in 1925 to assure sanitary shellfish products. The FDA and the shellfish industry developed an Interstate Shellfish Sanitation Conference (ISSC) patterned after the National Conference on Interstate Milk Shipments.

14.1.6 Organizations Fostering Uniformity

Under the United States system of states, uniformity of regulatory law is a major issue for national and international food companies. In many cases the differences are small and do not create a burden on interstate commerce. For example, Michigan's temperature requirement for smoked fish is that they are be held "at or below 38 degrees Fahrenheit (3.3 degrees Celsius),"[7] which is colder than the federal requirement.

---

[6] Originally the code was called the *Standard Milk Ordinance* in 1924, then the *Pasteurized Milk Ordinance*.
[7] MICH. ADMIN. CODE r 285.569.4 (2000).

DISCUSSION QUESTION

14.1. Until 1994, 43 Pennsylvania Statute section 405 required every bakery product sold in Pennsylvania to bare the legend, "Registered with Pennsylvania Department of Agriculture," in full text, or an abbreviated form "Reg. Penna. Dept. Agric." This statutory requirement was enacted in 1933 and, to my knowledge, was never legally challenged. What would happen if all 50 states enacted a requirement similar to Pennsylvania's?

The Association of Food and Drug Officials (AFDO) was established in 1896 to foster uniformity in the adoption and enforcement of food, drug, medical devices, cosmetics, and product safety laws, rules, and regulations. Associations, such as AFDO, demonstrate the option of achieving uniformity through cooperation and education.

AFDO provides a mechanism and a forum where regional, national, and international issues are deliberated and resolved uniformly to provide the best public health and consumer protection in the most expeditious and cost-effective manner. In addition, AFDO provides a number of model uniform laws, such as their Model Consumer Commodity Salvage Code, Model Veterinary Drug Code, and Model Water Vending Machine Regulation. These models are often adopted by states and other units of government and achieve a level of uniformity. For example, the AFDO Model Food Law helped foster adoption of relatively uniform state food law.

In 1968, AFDO in cooperation with the FDA prepared a model Retail Food Store Sanitation Code. This model served to foster uniformity in state food law regarding the inspection of grocery stores. Similar model codes had been published for other areas of retail food sales. In 1935, the Conference of State and Territorial Officials and the Nation Restaurant Code Authority in cooperation with the federal Public Health Service (PHS) prepared the model Food Service Sanitation Ordinance for restaurants. In 1957, the PHS prepared a Vending of Food and Beverages Ordinance. (This function of the PHS was transferred to FDA in 1968.)

Over time the distinctions between grocery stores, vending machine operations, and restaurants became blurred. Grocery stores include food service operations, and restaurant often have packaged retail foods. All three functions may occur at a single location. Therefore, the concept for a Unicode evolved.

FDA announced a Unicode project in 1987 and made a first draft available in 1988. After a name change, the Food Code was first published in 1993. With the support of the Conference for Food Protection, FDA revises the Food Code on four-year intervals. The Conference for Food Protection (CFP) began in 1971, sponsored jointly by FDA and the American Public Health Association (APHA). CFP makes recommendations to FDA on the FDA Model Food Code.

The FDA publishes the Food Code as a model document and reference for regulatory agencies. In legal effect, it is neither federal law nor federal regulation and is not preemptive, but may be adopted and used by agencies at all levels of government that have responsibility for managing food safety risks at retail. States are left free to adopt the National Code in its entirety or to fashion their own versions of the Code.[8]

NOTES

14.2. Prospective adoption of model codes and federal regulations. One practical difficulty in achieving uniformity is that some states may adopt prospectively adopt updates in model codes and federal regulations. Nondelegation clauses in state constitutions may require state legislative enactment for each change.

14.2 STATE INSPECTION AND ENFORCEMENT POWERS

Most states have adopted laws similar to the Federal Food, Drug, and Cosmetic Act (FD&C Act). The state authority, however, varies considerably based on the class of products. Most states regulate food products, for example. Many states require drug manufacturers to register with the state, and about a dozen require registration of drug distributors operating with the state. Almost all food establishments are required to obtain a license for the states where they operate.

14.2.1 Summary Seizure and Condemnation

Licensing and registration provide an additional power to the state authorities. Because the states can suspend a firm's license or registration, they have the power to shut down an operation.

In other ways most states have greater power than the FDA. For instance, most state officials have the authority to place embargoes on the spot without the need to first go to court to seize violative products. In addition, to summary seizure power, many states also provide summary condemnation authority to their inspectors.

14.2.2 Destruction of Products without a Hearing

The FD&C Act provides that FDA must proceed under section 334 with an action to seize and condemn products. This necessarily provides judicial

[8]For adoption information, *see* FDA, *Real Progress in Food Code Adoptions, available at:* www.cfsan.fda.gov/~ear/fcadopt.html.

oversight of the issuance of a seizure and the opportunity for a hearing before food or other product may be destroyed.

However, in many cases the FDA will refer a case to state authorities, whose powers of seizure usually are greater than FDA. States generally have the power to seize adulterated or misbranded food without a warrant, for example, whereas FDA must usually seek judicial approval.

States also often have the power to seize and destroy food or other products without a hearing. This power raises Fourth Amendment implications. Nonetheless, where public nuisance and importance of protecting the public health are involved, the Fourteenth Amendment requirement for an opportunity to be heard may be fulfilled by an opportunity for hearing after the destruction of the property. This is discussed in greater detail in chapter 13 and in *North American Cold Storage Co. v. City of Chicago*, 211 U.S. 306 (1908).

14.3   STATE WARNING REQUIREMENTS

California's Proposition 65 requires a warning statement on any product that exposes a person to a chemical in more than "no significant risk" amounts of chemicals (natural or synthetic) "known to the state to cause cancer."[9] This requirement is perhaps the most notorious example of non-uniformity between state and federal requirements.

Congress considered preempting food warning labels many times since the 1990 amendments to the FD&C Act, but so far has not done so. California courts have upheld Proposition 65 against arguments that the FDA labeling system for new drugs is preemptive of state labeling requirements.[10] On the other hand, the San Francisco Superior Court held that Proposition 65 is preempted by the FD&C Act with respect to warnings for mercury in canned tuna. The decision was based on multiple reasons, but the judge gave "substantial deference" to FDA's position that the proposed warnings were preempted by the FD&C Act due to conflict with the federal policy to provide information about the risks and benefits of eating canned tuna through advisories.[11]

14.4   FEDERAL PREEMPTION OF STATES

The trend for the last hundred years has been increasing federal authority with the role of the federal government encroaching on traditional areas of state

---

[9] Cal, Health & Safety Code § 25249 (1986).
[10] Dowhal v. SmithKline Beecham Consumer Healthcare, 100 Cal App. 4th 8, 122 Cal Rptr. 2d 246 (2002).
[11] People v. Tri-Union Seafoods, LLC, et al., S.F. County Sup. Ct. Consolidated Case Nos. CGC-01-402975 and CGC-04-432394 (May 11, 2006).

powers. In part, this is due to changes in the economy, such as increasing nationalization and internationalization.

The U.S. Constitution gives Congress plenary power over interstate commerce. The courts have liberally interpreted these powers to include not only items in interstate commerce but also items and actions affecting interstate commerce. Congress's power to legislate under the Commerce Clause is broad and yet limited to matters that truly affect interstate commerce.[12] Nonetheless, the relevant and enduring precedent of *Wickard v. Filburn*[13] indicate that even a trivial impact on interstate commerce falls with the scope of federal regulatory authority.[14]

This broad application of the meaning of "interstate" seems to go against common sense because it is different from our everyday use of the word. After all, a retail establishment operating only within a single state is not directly involved in interstate commerce. Similarly the regulation of an in-state retail establishment would fall within traditional state powers. However, due to the national nature of the commerce, in-state commercial activities can clearly affect interstate commerce. Therefore "interstate commerce" includes activities that only very indirectly affect interstate commerce. For example, if an unapproved food additive were both produced and sold only in Michigan, that food additive business could be considered to indirectly affect interstate commerce because out-of-state firms compete against the in-state firm.

The end result is that today most commerce is interstate or has an interstate impact, and thus is under federal purview. The courts have interpreted interstate commerce so widely that nearly all food businesses fall under federal jurisdiction.

The Supremacy Clause of the United States Constitution means that when a federal law (within constitutional limits) imposes substantive requirements that conflict with state law, the federal requirement will prevail.[15] Federal law can preempt the state law under the Supremacy Clause of the Constitution in four different ways:

- Express preemption of inconsistent state law
- A comprehensive federal scheme occupies the field
- Direct conflict between federal and state law
- State law stands as an obstacle to the purposes and objectives of Congress

---

[12] *See* United States v. Lopez, 514 U.S. 549, 567 (1995); U.S. v. Morrison, 529 U.S. 598, 618 (2000).

[13] 317 U.S. 111 (1942).

[14] In particular, in *United States v. Lopez*, 514 U.S. 549, 567 (1995) the Supreme Court acknowledged the continuing vitality of *Wickard v. Filburn*, 317 U.S. 111 (1942) (noting that although Filburn's own contribution to the demand for wheat could have been trivial by itself, that was not "enough to remove him from the scope of federal regulation where, as here, his contribution, taken together with that of many others similarly situated, is far from trivial.")

[15] U.S. CONST. Art. VI.

14.4.1   Express Preemption of Inconsistent State Law

Under the Supremacy Clause, when acting within constitutional limits, Congress is empowered to preempt state law by so stating in express terms.[16] Express preemption of an inconsistent state law occurs when a federal statute prohibits a state from enacting conflicting state law or additional requirements in state law.[17]

Meat Inspection Programs   Express federal preemption occurs under the Federal Meat Inspection Act (FMIA), the Poultry Products Inspection Act (PPIA), and the Egg Products Inspection Act (EPIA). Congress included prohibitions within those acts against any "additional" or different" state requirements.

Thus, under these acts the USDA has complete regulatory oversight over meat slaughter, processing, and distribution. Soon after enactment of the preemption provision in the FMIA in 1967[18] the meat industry sought to invalidate Michigan's Comminuted Meat Law. The law established ingredient standards for sausage, ground beef, and other comminuted meats that were more stringent than those of USDA. The industry won the battle, but it took 14 years and came at a high cost. In *Armour & Co. v. Ball*, 438 F.2d 76 (6th Cir. 1972), the court concluded that the federal law preempted the state law.

State Meat Inspection Programs   The Federal Meat Inspection Act (FMIA) and the Poultry Products Inspection Act (PPIA) authorizes states with inspection programs certified by the USDA as "at least equal to" the federal program to inspect meat and/or poultry products that will be distributed intrastate. With little interest in direct marketing of meat and poultry in the 1970s and 1980s, there was not much effort to support the state programs. This combined with an adversarial relationship between state programs and the USDA caused many state programs to be dropped.

Today, interest in direct market and niche markets has grown, and so has interest in state meat inspection programs. Some states have reinstated meat inspection programs and others are considering doing so. State programs have the reputation of providing more technical support and guidance than the USDA.

Although state meat inspection program must be certified as at least equal to the USDA inspection program, meat inspected under a state inspection program is limited to intrastate commerce. Ironically, foreign countries with inspection program not equal but equivalent to FSIS are allowed to sell meat interstate in the United States. A bill introduced by Senator Thomas Daschle,

---

[16] U.S. Const. art. 6, cl. 2.
[17] *See, e.g.*, Hillsborough Co. v. Automated Medical Labs, Inc., 471 U.S. 707, 712 (1985).
[18] 21 U.S.C. § 467e.

"New Markets for State Inspected Meat Act of 1999," S.1988 would have allowed interstate marketing.[19]

FD&C Act   The FD&C Act generally has no explicit preemption of the states. There are, however, specific preemptive provisions for special categories of medical devices, vaccines, infant formula, and nutritional labeling of food.[20] Thus federal preemption under the FD&C Act is the exception rather than the rule.

14.4.2   Comprehensive Federal Scheme That Occupies the Field

Under the Supremacy Clause, the intent of Congress to preempt all state law in a particular area may be inferred where the scheme of federal regulation is sufficiently comprehensive that it is a reasonable inference that Congress left no room for supplementary state regulation.[21] On the other hand, when the federal law lacks regulation over a particular aspect of regulation, then the states may be free to write their own requirements. For example, a state law may require a "last date of sale" because the FD&C Act has no provision related to open date labeling.[22]

The case of *Florida Lime and Avocado Growers, Inc. v. Paul*[23] demonstrates the reluctance of the courts to invalidate a state law without express preemption by Congress. The case also demonstrates how the burden on interstate commerce plays an important role in these cases.

In the *Florida Lime and Avocado Growers* case the Florida avocado growers sought to enjoin enforcement of California statute, which gauged maturity of avocados by oil content, when applied against Florida avocados that were certified as mature under federal regulations. The U.S. Supreme Court held, inter alia, that the statute does not offend the supremacy clause of the federal Constitution.

* * * * *

Florida Lime and Avocado Growers, Inc. v. Paul

373 U.S. 132 (1963)

Mr. Justice BRENNAN delivered the opinion of the Court.

Section 792 of California's Agricultural Code, which gauges the maturity of avocados by oil content, prohibits the transportation or sale in California of

---

[19] *Available at:* http://thomas.loc.gov/cgi-bin/query/z?c106:s.1988: (last accessed Mar. 25, 2007).
[20] 21 U.S.C. §§ 343-1, 360k.
[21] Hillsborough Co. v. Automated Medical Labs, Inc., 471 U.S. 707 (1985).
[22] *See, e.g.*, Grocery Manufacturers of America v. Dept of Public Health, 379 Mass. 70, 393 N.E.2d 881 (1979).
[23] 373 U.S. 132 (1963).

avocados which contain "less than 8 percent of oil, by weight . . . excluding the skin and seed."[24] In contrast, federal marketing orders approved by the Secretary of Agriculture gauge the maturity of avocados grown in Florida by standards which attribute no significance to oil content. This case presents the question of the constitutionality of the California statute insofar as it may be applied to exclude from California markets certain Florida avocados which, although certified to be mature under the federal regulations, do not uniformly meet the California requirement of 8 percent of oil.

Appellants in No. 45, growers and handlers of avocados in Florida, brought this action in the District Court for the Northern District of California to enjoin the enforcement of § 792 against Florida avocados certified as mature under the federal regulations. Appellants challenged the constitutionality of the statute on three grounds: that under the Supremacy Clause, Art. VI, the California standard must be deemed displaced by the federal standard for determining the maturity of avocados grown in Florida. . . .

Almost all avocados commercially grown in the United States come either from Southern California or South Florida. The California-grown varieties are chiefly of Mexican ancestry, and in most years contain at least 8 percent oil content when mature. The several Florida species, by contrast, are of West Indian and Guatemalan ancestry. West Indian avocados, which constitute some 12 percent of the total Florida production, may contain somewhat less than 8 percent oil when mature and ready for market. They do not, the District Court found, attain that percentage of oil "until they are past their prime." . . . The experts who testified at the trial disputed whether California's percentage-of-oil test or the federal marketing orders' test of picking dates and minimum sizes and weights was the more accurate gauge of the maturity of avocados. In adopting his calendar test of maturity for the varieties grown in South Florida, the Secretary [of the U.S. Department of Agriculture] expressly rejected physical and chemical tests as insufficiently reliable guides for gauging the maturity of the Florida fruit.

. . . Whether a State may constitutionally reject commodities which a federal authority has certified to be marketable depends upon whether the state regulation "stands as an obstacle to the accomplishment and execution of the full purposes and objectives of Congress." By that test, we hold that § 792 is not such an obstacle; there is neither such actual conflict between the two schemes of regulation that both cannot stand in the same area nor evidence of a congressional design to preempt the field.

We begin by putting aside two suggestions of the appellants which obscure more than aid in the solution of the problem. First, it is suggested that a

---

[24] Avocados not meeting this standard may not be sold in California. Substandard fruits are "declared to be a public nuisance," and they may be seized, condemned, and abated. Violators may be punished criminally, ($50 to $500 fine or imprisonment for not more than six months, or both), and by civil penalty action (market value of fruits).

federal license or certificate of compliance with minimum federal standards immunizes the licensed commerce from inconsistent or more demanding state regulations. While this suggestion draws some support from decisions which have invalidated direct state interference with the activities of interstate, carriers, even in that field of paramount federal concern, the suggestion has been significantly qualified. That no State may completely exclude federally licensed commerce is indisputable, but that principle has no application to this case.

Second, it is suggested that the coexistence of federal and state regulatory legislation should depend upon whether the purposes of the two laws are parallel or divergent. This Court has, on the one hand, sustained state statutes having objectives virtually identical to those of federal regulations, and has, on the other hand, struck down state statutes where the respective purposes were quite dissimilar. The test of whether both federal and state regulations may operate, or the state regulation must give way, is whether both regulations can be enforced without impairing the federal superintendence of the field, not whether they are aimed at similar or different objectives.

The principle to be derived from our decisions is that federal regulation of a field of commerce should not be deemed preemptive of state regulatory power in the absence of persuasive reasons—either that the nature of the regulated subject matter permits no other conclusion or that the Congress has unmistakably so ordained.

A

A holding of federal exclusion of state law is inescapable and requires no inquiry into congressional design where compliance with both federal and state regulations is a physical impossibility for one engaged in interstate commerce. That would be the situation here if, for example, the federal orders forbade the picking and marketing of any avocado testing more than 7 percent oil, which the California test excluded from the State any avocado measuring less than 8 percent oil content. No such impossibility of dual compliance is presented on this record, however. As to those Florida avocados of the hybrid and Guatemalan varieties which were actually rejected by the California test, the District Court indicated that the Florida growers might have avoided such rejections by leaving the fruit on the trees beyond the earliest picking date permitted by the federal regulations, and nothing in the record contradicts that suggestion. Nor is there a lack of evidentiary support for the District Court's finding that the Florida varieties marketed in California "attain or exceed 8 percent oil content while in a prime commercial marketing condition," even though they may be "mature enough to be acceptable prior to the time that they reach that content. ..." Thus the present record demonstrates no inevitable collision between the two schemes of regulation, despite the dissimilarity of the standards.

B

The issue under the head of the Supremacy Clause is narrowed then to this: Does either the nature of the subject matter, namely the maturity of avocados, or any explicit declaration of congressional design to displace state regulation, require § 792 to yield to the federal marketing orders? The maturity of avocados seems to be an inherently unlikely candidate for exclusive federal regulation. Certainly it is not a subject by its very nature admitting only of national supervision. Nor is it a subject demanding exclusive federal regulation in order to achieve uniformity vital to national interests.

On the contrary, the maturity of avocados is a subject matter of the kind this Court has traditionally regarded as properly within the scope of State superintendence. Specifically, the supervision of the readying of foodstuffs for market has always been deemed a matter of peculiarly local concern. Many decades ago, for example, this Court sustained a State's prohibition against the importation of artificially colored oleomargarine (which posed no health problem), over claims of federal preemption and burden on commerce. In the course of the opinion, the Court recognized that the States have always possessed a legitimate interest in "the protection of (their) people against fraud and deception in the sale of food products" at retail markets within their borders.

It is true that more recently we sustained a federal statute broadly regulating the production of renovated butter. But we were scrupulous in pointing out that a State might nevertheless—at least in the absence of an express contrary command of Congress—confiscate or exclude from market the processed butter which had complied with all the federal processing standards, "because of a higher standard demanded by a state for its consumers." A State regulation so purposed was, we affirmed, "permissible under all the authorities." *Cloverleaf Butter Co. v. Patterson*, 315 U.S. 148, 162. That distinction is a fundamental one, which illumines and delineates the problem of the present case. Federal regulation by means of minimum standards of the picking, processing, and transportation of agricultural commodities, however comprehensive for those purposes that regulation may be, does not of itself import displacement of state control over the distribution and retail sale of those commodities in the interests of the consumers of the commodities within the State. Thus, while Florida may perhaps not prevent the exportation of federally certified fruit by superimposing a higher maturity standard, nothing in *Cloverleaf* forbids California to regulate their marketing. Congressional regulation of one end of the stream of commerce does not, ipso facto, oust all State regulation at the other end. Such a displacement may not be inferred automatically from the fact that Congress has regulated production and packing of commodities for the interstate market. We do not mean to suggest that certain local regulations may not unreasonably or arbitrarily burden interstate commerce; we consider that question separately. Here we are concerned only whether partial congressional superintendence of the field (maturity for the purpose of introduction of Florida fruit into the stream of interstate commerce) automatically forecloses regulation of maturity by another State in the interests of that State's consumers of the fruit. . . .

C

Since no irreconcilable conflict with the federal regulation requires a conclusion that § 792 was displaced, we turn to the question whether Congress has nevertheless ordained that the state regulation shall yield. The settled mandate governing this inquiry, in deference to the fact that a state regulation of this kind is an exercise of the "historic police powers of the States," is not to decree such a federal displacement "unless that was the clear and manifest purpose of Congress." In other words, we are not to conclude that Congress legislated the ouster of this California statute by the marketing orders in the absence of an unambiguous congressional mandate to that effect. We search in vain for such a mandate. . . .

Nothing in the language of the Agricultural Adjustment Act—passed by the same Congress the very next day—discloses a similarly comprehensive congressional design. . . . By its very terms, in fact, the statute purports only to establish minimum standards. . . .

"The Act itself does not impose regulations over the marketing of any agricultural commodity. It merely provides the authority under which an industry can develop regulations to fit its own situation and solve its own marketing problems." . . .

A third factor which strongly suggests that Congress did not mandate uniformity for each marketing order arises from the legislative history. The provisions concerning the limited duration and local application of marketing agreements received much attention from both House and Senate Committees reporting on the bill. . . . The Committee Reports also discussed § 10(i), 7 U.S.C. § 610(i), which authorized federal-state cooperation in the administration of the program, and cautioned significantly:

"Notwithstanding the authorization of cooperation contained in this section, there is nothing in it to permit or require the Federal government to invade the field of the States, for the limitations of the act and the Constitution forbid federal regulation in that field, and this provision does not indicate the contrary. Nor is there anything in the provision to force States to cooperate. Each sovereignty operates in its own sphere but can exert its authority in conformity rather than in conflict with that of the other."

. . . In the absence of any such manifestations, it would be unreasonable to infer that Congress delegated to the growers in a particular region the authority to deprive the States of their traditional power to enforce otherwise valid regulations designed for the protection of consumers. . . .

This case requires no consideration of the scope of the constitutional power of Congress to oust all state regulation of maturity, and we intimate no view upon that question. It is enough to decide this aspect of the present case that we conclude that Congress has not attempted to oust or displace state powers to enact the regulation embodied in § 792. The most plausible inference from the legislative scheme is that the Congress contemplated that state power to enact such regulations should remain unimpaired. . . .

[Justice WHITE, with whom Mr. Justice BLACK, Mr. Justice DOUGLAS, and Mr. Justice CLARK join, dissented, concluding that the California statute was inconsistent with the federal law, and thus preempted.]

* * * * *

In the *Hillsborough Co. v. Automated Medical Labs, Inc.*[25] decision, the U.S. Supreme Court reviewed the issue of federal preemption of the FD&C Act. This decision clarified certain limitations on federal preemption. In particular, *Hillsborough* illustrates that preemption based on a comprehensive scheme may turn on whether there is FDA support for the claim.

In *Hillsborough*, an action was brought challenging constitutionality of local ordinances governing collection of blood plasma from paid donors. However, FDA disavowed any desire to preempt more restrictive local laws. The Supreme Court held that federal regulations governing collection of blood plasma from paid donors did not preempt the local ordinances.

* * * * *

Hillsborough Co. v. Automated Medical Labs, Inc.

471 U.S. 707 (1985)

Justice MARSHALL delivered the opinion of the Court.

The question presented is whether the federal regulations governing the collection of blood plasma from paid donors preempt certain local ordinances.

I

Appellee Automated Medical Laboratories, Inc., is a Florida corporation that operates, through subsidiaries, eight blood plasma centers in the United States. One of the centers, Tampa Plasma Corporation (TPC), is located in Hillsborough County, Florida. Appellee's plasma centers collect blood plasma from donors by employing a procedure called plasmapheresis. Under this procedure, whole blood removed from the donor is separated into plasma and other components, and "at least the red blood cells are returned to the donor." Appellee sells the plasma to pharmaceutical manufacturers.

[25] 471 U.S. 707 (1985).

Vendors of blood products, such as TPC, are subject to federal supervision. Under § 351(a) of the Public Health Service Act, such vendors must be licensed by the Secretary of Health and Human Services (HHS). Licenses are issued only on a showing that the vendor's establishment and blood products meet certain safety, purity, and potency standards established by the Secretary. HHS is authorized to inspect such establishments for compliance.

Pursuant to § 351 of the Act, the Food and Drug Administration (FDA), as the designee of the Secretary, has established standards for the collection of plasma. . . .

It is a familiar and well-established principle that the Supremacy Clause, U.S. Const., art. VI, cl. 2, invalidates state laws that "interfere with, or are contrary to," federal law. Under the Supremacy Clause, federal law may supersede state law in several different ways. First, when acting within constitutional limits, Congress is empowered to preempt state law by so stating in express terms. In the absence of express preemptive language, Congress' intent to preempt all state law in a particular area may be inferred where the scheme of federal regulation is sufficiently comprehensive to make reasonable the inference that Congress "left no room" for supplementary state regulation. Preemption of a whole field also will be inferred where the field is one in which "the federal interest is so dominant that the federal system will be assumed to preclude enforcement of state laws on the same subject."

Even where Congress has not completely displaced state regulation in a specific area, state law is nullified to the extent that it actually conflicts with federal law. Such a conflict arises when "compliance with both federal and state regulations is a physical impossibility," or when state law "stands as an obstacle to the accomplishment and execution of the full purposes and objectives of Congress."

We have held repeatedly that state laws can be preempted by federal regulations as well as by federal statutes. Also, for the purposes of the Supremacy Clause, the constitutionality of local ordinances is analyzed in the same way as that of statewide laws.

III

In arguing that the Hillsborough County ordinances and regulations are preempted, appellee faces an uphill battle. The first hurdle that appellee must overcome is the FDA's statement, when it promulgated the plasmapheresis regulations in 1973, that it did not intend its regulations to be exclusive. In response to comments expressing concern that the regulations governing the licensing of plasmapheresis facilities "would preempt State and local laws governing plasmapheresis," the FDA explained in a statement accompanying the regulations that "[t]hese regulations are not intended to usurp the powers of State or local authorities to regulate plasmapheresis procedures in their localities."

The question whether the regulation of an entire field has been reserved by the Federal government is, essentially, a question of ascertaining the intent underlying the federal scheme. In this case, appellee concedes that neither Congress nor the FDA expressly preempted state and local regulation of plasmapheresis. Thus, if the county ordinances challenged here are to fail they must do so either because Congress or the FDA implicitly preempted the whole field of plasmapheresis regulation, or because particular provisions in the local ordinances conflict with the federal scheme. According to appellee, two separate factors support the inference of a federal intent to preempt the whole field: the pervasiveness of the FDA's regulations and the dominance of the federal interest in this area. Appellee also argues that the challenged ordinances reduce the number of plasma donors, and that this effect conflicts with the congressional goal of ensuring an adequate supply of plasma.

The FDA's statement is dispositive on the question of implicit intent to preempt unless either the agency's position is inconsistent with clearly expressed congressional intent or subsequent developments reveal a change in that position. Given appellee's first argument for implicit preemption—that the comprehensiveness of the FDA's regulations evinces an intent to preempt—any preemptive effect must result from the change since 1973 in the comprehensiveness of the federal regulations. To prevail on its second argument for implicit preemption—the dominance of the federal interest in plasmapheresis regulation—appellee must show either that this interest became more compelling since 1973, or that, in 1973, the FDA seriously underestimated the federal interest in plasmapheresis regulation.

The second obstacle in appellee's path is the presumption that state or local regulation of matters related to health and safety is not invalidated under the Supremacy Clause. Through the challenged ordinances, Hillsborough County has attempted to protect the health of its plasma donors by preventing them from donating too frequently. It also has attempted to ensure the quality of the plasma collected so as to protect, in turn, the recipients of such plasma. "Where . . . the field that Congress is said to have preempted has been traditionally occupied by the States 'we start with the assumption that the historic police powers of the States were not to be superseded by the Federal Act unless that was the clear and manifest purpose of Congress.'" Of course, the same principles apply where, as here, the field is said to have been preempted by an agency, acting pursuant to congressional delegation. Appellee must thus present a showing of implicit preemption of the whole field, or of a conflict between a particular local provision and the federal scheme, that is strong enough to overcome the presumption that state and local regulation of health and safety matters can constitutionally coexist with federal regulation.

Given the clear indication of the FDA's intention not to preempt and the deference with which we must review the challenged ordinances, we conclude that these ordinances are not preempted by the federal scheme.

We reject the argument that an intent to preempt may be inferred from the comprehensiveness of the FDA's regulations at issue here. . . . As we have pointed out, given the FDA's 1973 statement, the relevant inquiry is whether a finding of preemption is justified by the increase, since 1973, in the comprehensiveness of the Federal regulations. Admittedly, these regulations have been broadened over the years . . . The FDA has not indicated that the new regulations affected its disavowal in 1973 of any intent to preempt State and local regulation, and the fact that the Federal scheme was expanded to reach other uses of plasma does not cast doubt on the continued validity of that disavowal. Indeed, even in the absence of the 1973 statement, the comprehensiveness of the FDA's regulations would not justify preemption. . . . merely because the Federal provisions were sufficiently comprehensive to meet the need identified by Congress did not mean that States and localities were barred from identifying additional needs or imposing further requirements in the field.

We are even more reluctant to infer preemption from the comprehensiveness of regulations than from the comprehensiveness of statutes. As a result of their specialized functions, agencies normally deal with problems in far more detail than does Congress. To infer preemption whenever an agency deals with a problem comprehensively is virtually tantamount to saying that whenever a Federal agency decides to step into a field, its regulations will be exclusive. Such a rule, of course, would be inconsistent with the Federal–State balance embodied in our Supremacy Clause jurisprudence.

Moreover, because agencies normally address problems in a detailed manner and can speak through a variety of means, including regulations, preambles, interpretive statements, and responses to comments, we can expect that they will make their intentions clear if they intend for their regulations to be exclusive. Thus, if an agency does not speak to the question of preemption, we will pause before saying that the mere volume and complexity of its regulations indicate that the agency did in fact intend to preempt. Given the presumption that state and local regulation related to matters of health and safety can normally coexist with Federal regulations, we will seldom infer, solely from the comprehensiveness of federal regulations, an intent to preempt in its entirety a field related to health and safety. . . .

Appellee's second argument for preemption of the whole field of plasmapheresis regulation is that an intent to preempt can be inferred from the dominant federal interest in this field. We are unpersuaded by the argument. Undoubtedly, every subject that merits congressional legislation is, by definition, a subject of national concern. That cannot mean, however, that every Federal statute ousts all related State law. Neither does the Supremacy Clause require us to rank congressional enactments in order of "importance" and hold that, for those at the top of the scale, Federal regulation must be exclusive.

Instead, we must look for special features warranting preemption. Our case law provides us with clear standards to guide our inquiry in this area. For example, in the seminal case of *Hines v. Davidowitz*, 312 U.S. 52 (1941), the

Court inferred an intent to preempt from the dominance of the federal interest in foreign affairs because "the supremacy of the national power in the general field of foreign affairs . . . is made clear by the Constitution," and the regulation of that field is "intimately blended and intertwined with responsibilities of the national government." Needless to say, those factors are absent here. Rather, as we have stated, the regulation of health and safety matters is primarily, and historically, a matter of local concern.

There is also no merit in appellee's reliance on the National Blood Policy as an indication of the dominance of the federal interest in this area. Nothing in that policy takes plasma regulation out of the health-and-safety category and converts it into an area of overriding national concern.

Appellee's final argument is that even if the regulations are not comprehensive enough and the federal interest is not dominant enough to preempt the entire field of plasmapheresis regulation, the Hillsborough County ordinances must be struck down because they conflict with the federal scheme. Appellee argues principally that the challenged ordinances impose on plasma centers and donors requirements more stringent than those imposed by the federal regulations, and therefore that they present a serious obstacle to the federal goal of ensuring an "adequate supply of plasma." We find this concern too speculative to support preemption. . . .

More importantly, even if the Hillsborough County ordinances had, in fact, reduced the supply of plasma in that county, it would not necessarily follow that they interfere with the Federal goal of maintaining an adequate supply of plasma. Undoubtedly, overly restrictive local legislation could threaten the national plasma supply. Neither Congress nor the FDA, however, has struck a particular balance between safety and quantity; as we have noted, the regulations, which contemplated additional State and local requirements, merely establish minimum safety standards. . . .

Finally, the FDA possesses the authority to promulgate regulations preempting local legislation that imperils the supply of plasma and can do so with relative ease. Moreover, the agency can be expected to monitor, on a continuing basis, the effects on the federal program of local requirements. Thus, since the agency has not suggested that the county ordinances interfere with federal goals, we are reluctant in the absence of strong evidence to find a threat to the federal goal of ensuring sufficient plasma.

Our analysis would be somewhat different had Congress not delegated to the FDA the administration of the federal program. Congress, unlike an agency, normally does not follow, years after the enactment of federal legislation, the effects of external factors on the goals that the federal legislation sought to promote. Moreover, it is more difficult for Congress to make its intentions known—for example, by amending a statute—than it is for an agency to amend its regulations or to otherwise indicate its position.

In summary, given the findings of the District Court, the lack of any evidence in the record of a threat to the "adequacy" of the plasma supply, and the significance that we attach to the lack of a statement by the FDA, we conclude that the Hillsborough County requirements do not imperil the federal goal of ensuring sufficient plasma.[26] . . .

We hold that Hillsborough County Ordinances 80-11 and 80-12, and their implementing regulations, are not preempted by the scheme for Federal regulation of plasmapheresis. . . .

14.4.3 Direct Conflict between Federal and State Law

A direct conflict between federal and state law occurs when it is physically impossible to comply with both laws. Indirect conflict does not invalidate a state a law without proof of a comprehensive federal scheme occupies the field or the state law is an obstacle to Congress's objectives in passing the federal law.

The courts are reluctant to invalidate state laws as long as there is no major burden on interstate commerce. The connection with interstate commerce often determines the outcome of close cases. For example, laws with stricter standards than the FD&C Act have been upheld for in-state products. However, the in-state plants would be allowed to make products for export meeting the federal standard. In addition, state law cannot prohibit the importation of legal out-of-state products.

For instance, a Michigan law could not prohibit the sale of yellow margarine produced in Ohio and shipped into Michigan because the federal law permits artificial coloring.[27] On the other hand, although a Michigan law might prohibit the sale of yellow margarine in Michigan from Michigan plants, the state of Michigan cannot prevent Michigan firms from manufacturing yellow margarine for export.[28]

14.4.4 State Law an Obstacle to the Purposes and Objectives of Congress

When a state law "stands as an obstacle to the accomplishment of the full purposes and objectives of Congress," the federal law is supreme and invalidates the state law.[29] This is the most controversial area of federal preemption. Generally, this type of preemption is unlikely to be successful absent FDA support for the claim.[30] The reticence to apply this doctrine is demonstrated in the *Hillsborough* case.

---

[26] Two of the amici argue that the county ordinances interfere with the federal interest in uniform plasma standards. There is no merit to that argument. The Federal interest at stake here is to ensure minimum standards, not uniform standards. Indeed, the FDA's 1973 statement makes clear that additional, nonconflicting requirements do not interfere with Federal goals, and we have found no reason to doubt the continued validity of that statement.

[27] *See* Borden Co. v. Liddy, 239 F. Supp. 289 (S.D. Iowa 1965).

[28] *See* People v. Breen, 326 Mich. 720; 40 N.W.2d 788 (1950).

[29] Hillsborough Co. v. Automated Medical Labs, Inc., 471 U.S. at 707 (1985).

[30] *Id.*

The "obstacle to the full purposes and objectives of Congress" form of preemption was laid out in *Jones v. Rath Packing Co.*[31] In *Rath*, a meat processor and flour millers brought suit to enjoin the enforcement of California laws on the labeling of packaged foods by weight. The U.S. Supreme Court held that the California statute and regulation, which made no allowance for loss of weight resulting from moisture loss during the course of good distribution practices, was preempted by the Wholesome Meat Act, as applied to the meat processor. The Court also held, that although the Fair Packaging and Labeling Act did not preempt the California law regarding the flour, enforcement of the California law against the millers would prevent the accomplishment and execution of the full purposes and objectives of Congress, and, therefore, the state law was required to yield to the federal.

* * * * *

Jones v. Rath Packing Co.

430 U.S. 519 (1977)

Mr. Justice MARSHALL delivered the opinion of the Court:

Petition Jones is Director of the Department of Weights and Measures in Riverside County, Cal. In that capacity he ordered removed from sale bacon packaged by respondent Rath Packing Co. and flour packaged by three millers, respondents General Mills, Inc., Pillsbury Co., and Seaboard Allied Milling Corp. (hereafter millers). Jones acted after determining, by means of procedures set forth in 4 Cal. Admin. Code c. 8, Art. 5, that the packages were contained in lots whose average net weight was less than the net weight stated on the packages. The removal orders were authorized by Cal. Bus. & Prof. Code § 12211 (West Supp. 1977). . . .

Rath and the millers responded by filing suits in the District Court for the Central District of California. They sought both declarations that § 12211 and Art. 5 are preempted by federal laws regulating net weight labeling and injunctions prohibiting Jones from enforcing those provisions. . . .

I

In its present posture, this litigation contains no claim that the Constitution alone denies California power to enact the challenged provisions.[32] We are required to decide only whether the federal laws which govern respondents' packing operations preclude California from enforcing § 12211, as implemented by Art. 5.

---

[31] 430 U.S. 519, 540–43 (1977).

[32] The Court of Appeals affirmed the District Court's holding, *see* n. 5, *supra*, that the California provisions violate neither the Commerce Clause nor the Fourteenth Amendment. 530 F.2d at 1322–1323. The millers do not challenge these holdings here.

Our prior decisions have clearly laid out the path we must follow to answer this question. The first inquiry is whether Congress, pursuant to its power to regulate commerce, U.S. Const., art. 1, § 8, has prohibited state regulation of the particular aspects of commerce involved in this case. Where, as here, the field which Congress is said to have preempted has been traditionally occupied by the states, "we start with the assumption that the historic police powers of the states were not to be superseded by the federal Act unless that was the clear and manifest purpose of Congress." This assumption provides assurance that "the federal–state balance," will not be disturbed unintentionally by Congress or unnecessarily by the courts. But when Congress has "unmistakably . . . ordained" that its enactments alone are to regulate a part of commerce, state laws regulating that aspect of commerce must fall. This result is compelled whether Congress' command is explicitly stated in the statute's language or implicitly contained in its structure and purpose.

Congressional enactments that do not exclude all state legislation in same field nevertheless override state laws with which they conflict. U.S. Const., art. VI. The criterion for determining whether state and federal laws are so inconsistent that the state law must give way is firmly established in our decisions. Our task is "to determine whether under the circumstances of this particular case, (the state's) law stands as an obstacle to the accomplishment and execution of the full purposes and objectives of Congress." This inquiry requires us to consider the relationship between state and federal laws as they are interpreted and applied, not merely as they are written.

II

Section 12211 . . . applies to both Rath's bacon and the millers' flour. The standard it establishes is straightforward: "(T)he average weight or measure of the packages or containers in a lot of any . . . commodity sampled shall not be less, at the time of sale or offer for sale, than the net weight or measure stated upon the package."[33] . . .

After it is packed, bacon loses moisture. Some of that moisture is absorbed by the insert on which the bacon is placed. A wax board insert will absorb approximately 5/16 of an ounce from the product, whereas a polyethylene insert will absorb approximately 1/16 of an ounce. In addition, moisture is lost to the atmosphere or, in a hermetically sealed package, by condensation onto the packing material.

California's inspectors include in the weight of the material any moisture or grease which the bacon has lost to it. Federal inspectors at the packing plant, by contrast, determine the tare by weighing the packing material dry. It is not feasible for field inspectors to use a dry tare method. . . .

---

[33] "Tare" is the weight of the packing material in which the product is contained. In order to determine the tare, the inspector weighs each package and then removes and weighs the contents of each package. By subtracting the net weight from the gross weight, he obtains the tare.

Enforcement action is taken against packages with unreasonably large minus errors.

III

A. Rath's bacon is produced at plants subject to federal inspection under the Federal Meat Inspection Act (FMIA or Act), as amended by the Wholesome Meat Act. Among the requirements imposed on federally inspected plants, and enforced by Department of Agriculture inspectors, are standards of accuracy in labeling. On the record before us, we may assume that Rath's bacon complies with these standards. . . .

The Secretary of Agriculture has used his discretionary authority to permit "reasonable variations" in the accuracy of the required statement of quantity:

> The statement of net quantity of contents) as it is shown on a label shall not be false or misleading and shall express an accurate statement of the quantity of contents of the container exclusive of wrappers and packing substances. Reasonable variations caused by loss or gain of moisture during the course of good distribution practices or by unavoidable deviations in good manufacturing practice will be recognized. Variations from stated quantity of contents shall not be unreasonably large.

Thus, the FMIA, as implemented by statutorily authorized regulations, requires the label of a meat product accurately to indicate the net weight of the contents unless the difference between stated and actual weights is reasonable and results from the specified causes.

B. Section 408 of the FMIA prohibits the imposition of "(m)arking, labeling, packaging, or ingredient requirements in addition to, or different than, those made under" the Act. This explicit preemption provision dictates the result in the controversy between Jones and Rath. California's use of a statistical sampling process to determine the average net weight of a lot implicitly allows for variations from stated weight caused by unavoidable deviations in the manufacturing process. But California makes no allowance for loss of weight resulting from moisture loss during the course of good distribution practice. Thus the state law's requirement that the label accurately state the net weight, with implicit allowance only for reasonable manufacturing variations is "different than" the federal requirement, which permits manufacturing deviations and variations caused by moisture loss during good distribution practice. . . .

We therefore conclude that with respect to Rath's packaged bacon, § 12211 and Art. 5 are preempted by federal law.

IV

A. The federal law governing net weight labeling of the millers' flour is contained in two statutes, the Federal Food, Drug, and Cosmetic Act (FD&C Act), and the Fair Packaging and Labeling Act (FPLA), 15 U.S.C. §§ 1451–1461. For the reasons stated below, we conclude that the federal weight-labeling standard for flour is the same as that for meat.

The FD&C Act prohibits the introduction or delivery for introduction into interstate commerce of any food that is misbranded. A food is misbranded under the FD&C Act,

[i]f in package form unless it bears a label containing . . . an accurate statement of the quantity of the contents in terms of weight, measure, or numerical count: Provided, That . . . reasonable variations shall be permitted, and exemptions as to small packages shall be established by regulations prescribed by the Secretary.

§ 343(e).

This provision is identical to the parallel provision in the FMIA, see *supra*, at 1311, except that the FD&C Act mandates rather than allows the promulgation of implementing regulations. The regulation issued in response to this statutory mandate is also substantially identical to its counterpart under the FMIA:

The declaration of net quantity of contents shall express an accurate statement of the quantity of contents of the package. Reasonable variations caused by loss or gain of moisture during the course of good distribution practice or by unavoidable deviations in good manufacturing practice will be recognized. Variations from stated quantity of contents shall not be unreasonably large.

21 C.F.R. § 1.8b(q) (1976).

Since flour is a food under the FD&C Act, its manufacture is also subject to the provisions of the FPLA. That statute states a congressional policy that "(p)ackages and their labels should enable consumers to obtain accurate information as to the quantity of the contents and should facilitate value comparisons." § 1451. To accomplish those goals, insofar as is relevant here, the FPLA bans the distribution in commerce of any packaged commodity unless it complies with regulations

"Which shall provide that

"(2) The net quantity of contents (in terms of weight, measure, or numerical count) shall be separately and accurately stated in a uniform location upon the principal display panel of (the required) label." § 1453(a).

The FPLA also contains a saving clause which specifies that nothing in the FPLA "shall be construed to repeal, invalidate, or supersede" the FD&C Act. § 1460. Nothing in the FPLA explicitly permits any variation between stated weight and actual weight.

The *amici* States contend that since the FPLA does not allow any variations from stated weight, there is no difference between federal law governing labeling of flour and California law. The Court of Appeals, however, held that because of the savings clause, compliance with the FD&C Act, which does allow reasonable variations, satisfies the requirements of the FPLA. . . .

It is clear that 21 C.F.R. § 1.8b(q) (1976), insofar as it is based on the FD&C Act, has the force of law and allows reasonable variations. Thus, whether the statutory standard is viewed as strict, with the regulation considered a restriction on the power to prosecute, or whether the standard is itself viewed as incorporating the flexibility of the proviso and its implementing regulation, the result is the same. Under the FD&C Act, reasonable variations from the stated net weight do not subject a miller to prosecution, whether civil or criminal, if the variations arise from the permitted causes. The question raised by the arguments of *amici* is whether by enacting the FPLA, Congress intended to eliminate the area of freedom from prosecution created by the FD&C Act and its implementing regulation.

Over 60 years ago, Congress concluded that variations must be allowed because of the nature of certain foods and the impossibility of developing completely accurate means of packing. Since 1914, regulations under the food and drug laws have permitted reasonable variations from stated net weight resulting from packing deviations or gain or loss of moisture occurring despite good commercial practice. If Congress had intended to overrule this long-standing administrative practice, founded on a legislative statement of necessity, we would expect it to have done so clearly. Instead, it explicitly preserved existing law, with "no changes." The legislative history of the FPLA contains some indication that the saving clause was understood to preserve the reasonable-variation regulation under the FD&C Act, and no evidence that Congress affirmatively intended to overrule that regulation. We can only conclude that under the FPLA, as under the FD&C Act, a manufacturer of food is not subject to enforcement action for violation of the net-weight labeling requirements if the label accurately states the net weight, with allowance for the specified reasonable variations.

B. The FD&C Act contains no preemptive language. The FPLA, on the other hand, declares that

> it is the express intent of Congress to supersede any and all laws of the States or political subdivisions thereof insofar as they may now or hereafter provide for the labeling of the net qua(nt)ity of contents of the package of any consumer commodity covered by this chapter which are less stringent than or require information different from the requirements of section 1453 of this title or regulations promulgated pursuant thereto.

15 U.S.C. § 1461. . . .

The basis for the Court of Appeals' holding is unclear. Its opinion may be read as based on the conclusion that the state law is inadequate because its enforcement relies on a statistical averaging procedure. We have rejected that conclusion. Alternatively, the Court of Appeals may have found California's approach less stringent because the State takes no enforcement action against lots whose average net weight exceeds the weight stated on the label, even if that excess is not a reasonable variation attributable to a federally allowed cause.

We have some doubt that by preempting less stringent state laws, Congress intended to compel the States to expend scarce enforcement resources to prevent the sale of packages which contain more than the stated net weight. We do not have to reach that question, however, because in this respect California law apparently differs not at all from federal law, as applied. . . . Since neither jurisdiction is concerned with overweighting in the administration of its weights and measures laws, we cannot say that California's statutory lack of concern for that "problem" makes its laws less stringent than the federal.

Respondents argue that California's law is preempted because it requires information different from that required by federal law. . . . The legislative history, however, suggests that the statute expressly preempts as requiring "different information" only state laws governing net quantity labeling which impose requirements inconsistent with those imposed by federal law. Since it would be possible to comply with the state law without triggering federal enforcement action we conclude that the state requirement is not inconsistent with federal law. We therefore hold that 15 U.S.C. § 1461 does not preempt California's § 12211 as implemented by Art. 5.

That holding does not, however, resolve this case, for we still must determine whether the state law "stands as an obstacle to the accomplishment and execution of the full purposes and objectives of Congress." As Congress clearly stated, a major purpose of the FPLA is to facilitate value comparisons among similar products. Obviously, this goal cannot be accomplished unless packages that bear the same indicated weight in fact contain the same quantity of the product for which the consumer is paying. The significance of this requirement for our purposes results from the physical attributes of flour. . . .

Flour is composed of flour solids and moisture. The average water content of wheat kernels used to make flour is 12.5 percent by weight, with a range from 10 percent to 14.5 percent. Efficient milling practice requires adding water to raise the moisture content to 15 percent to 16 percent; if the wheat is too wet or too dry, milling will be hindered. During milling, the moisture content is reduced to 13 percent to 14 percent. App. 28–29.

The moisture content of flour does not remain constant after milling is completed. If the relative humidity of the atmosphere in which it is stored is greater than 60 percent, flour will gain moisture, and if the humidity is less than 60 percent, it will lose moisture. The federal net-weight labeling standard permits variations from stated weight caused by this gain or loss of moisture.

Packages that meet the federal labeling requirements and that have the same stated quantity of contents can be expected to contain the same amount of flour solids. Manufacturers will produce flour with a moisture content fixed by the requirements of the milling process. Since manufacturers have reason not to pack significantly more than is required and federal law prohibits under-packing, they will pack the same amount of this similarly composed flour into packages of any given size. Despite any changes in weight resulting from changes in moisture content during distribution, the packages will contain the same amount of flour solids when they reach the consumer. This identity of contents facilitates consumer value comparisons.

The State's refusal to permit reasonable weight variations resulting from loss of moisture during distribution produces a different effect. In order to be certain of meeting the California standard, a miller must ensure that loss of moisture during distribution will not bring the weight of the contents below the stated weight. Local millers, which serve a limited area, could do so by adjusting their packing practices to the specific humidity conditions of their region. For example, a miller in an area where the humidity is typically higher than 60 percent would not need to overpack at all. By contrast, a miller with a national marketing area would not know the destination of its flour when it was packaged and would therefore have to assume that the flour would lose weight during distribution. The national manufacturer, therefore, would have to overpack.

Similarly, manufacturers who distributed only in States that followed the federal standard would not be concerned with compensating for possible moisture loss during distribution. National manufacturers who did not exclude the nonconforming States from their marketing area, on the other hand, would have to overpack. Thus, as a result of the application of the California standard, consumers throughout the country who attempted to compare the value of identically labeled packages of flour would not be comparing packages which contained identical amounts of flour solids. Value comparisons which did not account for this difference and there would be no way for the consumer to make the necessary calculations would be misleading.

We therefore conclude that with respect to the millers' flour, enforcement of § 12211, as implemented by Art. 5, would prevent "the accomplishment and execution of the full purposes and objectives of Congress" in passing the FPLA. Under the Constitution, that result is impermissible, and the state law must yield to the federal.

The judgments are affirmed.

It is so ordered.

Mr. Justice REHNQUIST, with whom Mr. Justice STEWART joins, concurring in part and dissenting in part.

I agree that with respect to Rath's packaged bacon, § 12211 of the Cal.Bus. & Prof. Code and Art. 5 of 4 Cal. Admin. Code, c. 8, are preempted by the express preemptive provision of the Federal Meat Inspection Act. I also agree that with respect to General Mills' flour, § 12211 and Art. 5 are not preempted by the express preemptive provision of the Fair Packaging and Labeling Act (FPLA). I am unable to agree, however, with the implicit preemption the Court finds with respect to the flour. This latter preemption is founded in unwarranted speculations that hardly rise to that clear demonstration of conflict that must exist before the mere existence of a federal law may be said to preempt state law operating in the same field.

With respect to labeling requirements for flour under the scheme contemplated by the FPLA in conjunction with the Federal Food, Drug, and Cosmetic Act, the Court determines that the state-law labeling requirements are neither "less stringent than" nor inconsistent with those federal requirements. This conclusion quite properly dictates the Court's holding that Congress has not expressly prohibited state regulation in this field. The remaining inquiry, then, is whether the two statutory schemes are in utter conflict.[34] As this Court noted in *Kelly v. Washington*, 302 U.S. 1, 10:

"The principle is thoroughly established that the exercise by the state of its police power, which would be valid if not superseded by federal action, is superseded only where the repugnance or conflict is so 'direct and positive' that the two acts cannot 'be reconciled or consistently stand together.'"

When we deal, as we do here, with congressional action "in a field which the States have traditionally occupied," the basic assumption from which preemption must be viewed is "that the historic police powers of the States were not to be superseded by the Federal Act unless that was the clear and manifest purpose of Congress." I am simply unable to find that this stringent standard has been met in this case.

The Court's opinion demonstrates that it is physically possible to comply with the state-law requirement "without triggering federal enforcement action." This leads the Court to conclude that the "state requirement is not inconsistent with federal law." It also must lead to the conclusion that this is not a case "where compliance with both federal and state regulations is a physical impossibility for one engaged in interstate commerce." Preemption, then, if it is to exist at all in this case, must exist because the operation of the state Act inexorably conflicts with the purposes underlying the federal Act. The Court relies on the fact that one of the purposes of the FPLA is to "facilitate value comparisons" among consumers. But merely identifying a purpose is not enough; it must also be shown that the state law inevitably frustrates that purpose. . . .

The Court's reliance on supposition and inference fails in two respects to demonstrate that respondents have carried their burden of demonstrating preemption. First, on the Court's own premises, there should be no finding of preemption. We are told that the relevant inquiry is "the relationship between state and federal laws as they are interpreted and applied, not merely as they are written," while we are further told, that there is, in fact, no "federal interest in preventing packages from being overfilled," since the federal government

---

[34] There is no contention that the subject of the regulation is in its "nature national, or admit(ting) only of one uniform system . . ." *Colley v. Board of Wardens*, 12 How. 299, 319, 113 L.Ed. 996 (1852). On the contrary, "the supervision of the readying of foodstuffs for market has always been deemed a matter of peculiarly local concern." *Florida Avocado Growers v. Paul*, 373 U.S. 132, 144 (1963).

is not "concerned with overweighting in the administration of its weights and measures laws. . . ." Under these premises, it is hard to accept the Court's conclusion that, because of the federal purpose to facilitate consumer value comparisons, the state law is preempted because some packages might contain more than the minimum weight stated and more than another company's similarly marked package. For, we have been told that, should a manufacturer deliberately overpack, for whatever reason, there will be no federal action taken against him even though value comparisons might then "be misleading." It is virtually impossible to say, as the Court does, that "neither the state nor the federal government is concerned with over weighting," and yet conclude that state-induced overweighting conflicts with a "value comparison" purpose, while, presumably, other overweighting does not. In viewing such a purpose to be sufficient to require preemption while the very purpose is ignored in practice by the administering federal agency reverses the normal presumption against finding preemption. The reasoning process which leads the Court to conclude that there is no express preemption leads me to conclude that there is no implied preemption. . . .

Similarly defective is the reasoning process by which the majority concludes that local millers could adjust their packaging practices to specific humidity conditions, while national millers could not, since the national millers "would not know the destination of (their) flour when it was packaged and would therefore have to assume that the flour would lose weight during distribution." This assumption, too, is unsupported by the record.[35] We simply have no basis for concluding that national distributors do not know, or could not know through the exertion of some modicum of effort, where their flour will end up. The possibility that a packer might have to incur some extra expense in meeting both systems simply does not mean that the "purposes of the act cannot otherwise be accomplished," nor does it demonstrates that "the two acts cannot be reconciled . . . ."[36]

---

[35] The Court's reliance on the possible differential effect of California's requirements on local and national millers is itself wholly speculative. To begin with, we do not know from the record that there are both "local" and "national" millers, however defined. Even if both exist, we simply do not know that local millers will ship flour only to areas with comparable humidity levels. Any miller might experience a variety of humidity conditions by shipping to two different areas, despite the fact that his operation may be considered local in that the two areas are relatively contiguous. Even in the same town, stores that are air-conditioned may have significantly different humidity conditions than exist elsewhere in the town. In such situations, the local millers would have to adjust their packing process to account for this differential, either by packing different quantities into different packages, and then tracing their distribution, or by overpacking all packages sufficiently to ensure that any possible humidity conditions could be met. The same would appear to be true for national millers. We simply, then, do not know that local millers and national millers would not be similarly affected. The Court's assertions to the contrary are nothing but speculations.

[36] For all that appears, packers could easily adjust their processes so as to insure compliance with the purposes of both Acts. Even if such adjustment should entail a minor economic inconvenience, it has nowhere been demonstrated that the imposition of a moderate economic burden conflicts

The assumptions in the Court's opinion not only are insufficient to compel a finding of implied preemption, they suggest an approach to the question of preemption wholly at odds with that enunciated in *Florida Lime & Avocado Growers, Inc. v. Paul*. There, this Court . . . rejected a test which looked to the similarity of purposes, and noted instead that a manufacturer could have complied with both statutes by modifying procedures somewhat, which demonstrated that there was "no inevitable collision between the two schemes of regulation, despite the dissimilarity of the standards." Nothing has been shown to demonstrate that this conclusion is not equally justified in the instant case.

The Court today demonstrates only that there could be not that there must be a conflict between state and federal laws. Because reliance on this test to find preemption, absent an explicit preemptive clause, seriously misapprehends the carefully delimited nature of the doctrine of preemption, I dissent from the holding that § 12211 and Art. 5 are preempted with respect to General Mills' flour. . . .

* * * * *

14.4.5 Unreasonable Burden on Interstate Commerce

Earlier, the preemption of the Michigan Comminuted Meat Law by the FMIA in 1967 was discussed as an example of express preemption.[37] This story, however, has more than one interesting twist. The preempted higher Michigan standards for comminuted meat, particularly ground beef and sausage, were a source of great local pride.

Consequently, after Michigan's law was preempted, the state legislature amended the law to require that retailers selling comminuted meat whose ingredients failed to comply with the Michigan standards must notify consumers with a placard or printing on menus. The required statement read, "Do not meet Michigan's high meat ingredient standards but do meet lower federal standards." Although the court first determined that the mandated placards were not preempted by the federal law, eventually it was held that Michigan law violated the Commerce Clause because it imposed an unreasonable burden on interstate commerce.[38]

Determining whether a law is an unreasonable burden on interstate commerce requires determination whether the law is evenhanded, whether the

---

with the purpose of the federal statutory scheme. California, in the exercise of its police powers, may be deemed to have believed that the benefits of its enactment outweigh these costs. Unless it can be shown that additional cost itself conflicts with a clear congressional purpose, the presumption is that our federal system of government tolerates such costs. And if added costs will vitiate the conflict, I do not see how it can be said that the statutory schemes necessarily conflict rather than just "may possibly" conflict. *Goldstein v. California*, 412 U.S. 546, 554 (1973).

[37] Armour & Co. v. Ball, 438 F.2d 76 (6th Cir. 1972).

[38] American Meat Institute v. Pridgeon, 724 F.2d 45 (6th Cir. 1984).

burden is not excessive in relation to the local benefit, and whether the law is rational. "Only if the burden on interstate commerce clearly outweighs the state's legitimate purposes does such a regulation violate the Commerce Clause."[39]

The lack of a single federal food agency creates additional twists in this story. FDA and USDA have joint jurisdiction over the transport of meat products and after processing. However, FDA has exclusive jurisdiction over retail establishments (in federal jurisdiction). USDA may still regulate USDA-labeled packages that are found in retail establishments, but USDA lacks authority over the retail establishments directly, such as retail meat cutting rooms.

These jurisdictional limits result in the Michigan law being preempted for USDA-regulated products only. Thus, ground beef and other comminuted meats produced by a retail grocery must meet the more stringent Michigan requirements. Therefore products meeting the different USDA and Michigan standards are sold side by side in Michigan grocery stores.

This comparison of USDA and FDA preemption also highlight important differences in the styles or approaches of the federal agencies toward states. One important limit on federal power over the states is that the federal government cannot directly order the states to enforce a federal food law requirement. USDA largely carries out their statutes directly with federal employees. On the other hand, FDA relies heavily on the cooperation of states. Because the FDA relies heavily on state authority to enforce food safety requirements, so the FDA has rarely spoke out in favor of preemption of state food regulations.

14.5   FEDERAL LAWS DELEGATING AUTHORITY TO THE STATES

Two federal food and drug laws delegate specific powers to the states.

1. Prescription Drug Marketing Act of 1987. The Prescription Drug Marketing Act provides for enhanced listing and registration requirements for drug distributors by states.
2. Nutritional Labeling and Education Act of 1990 (NLEA). NLEA empowered states to enforce the provisions of NLEA and the regulations promulgated by FDA to implement NLEA.

The NLEA provision does require the states to provide FDA with advance notice of any enforcement action. This provision of enforcement of the federal law by the states holds more symbolic than actual enforcement value and is rarely invoked.

---

[39] Minnesota v. Clover Leaf Creamery Co., 449 U.S. 456, 474 (1981).

CHAPTER 15

Private Actions

15.1 INTRODUCTION

Government regulation is the primary topic of this text. However, not all regulation of the food supply is conducted by the government. Self-regulation, for example, is an important facet of the restraints placed on the food industry. Private actions also serve an important role in the overall food regulatory system.

The ultimate goal of any food safety system must be inducing a general attitude of food safety, responsibility, and knowledge in the regulated industry. Government resources cannot be expected to provide the level of oversight that would be necessary otherwise. Fortunately, most food business owners possess commitment to the community value of providing safe and wholesome food. Many trade associations and organizations such as the Better Business Bureau provide voluntary self-regulatory programs and guidance.

Earlier, the text discussed briefly that firms within an industry provide a measure of self-regulation through civil actions against other firms. For example, the federal Lanham Act provides for private suits to stop false or misleading advertising by competitors. A firm might bring a Lanham action to stop competitors from advertising their tomato products as "fresh" when actually made from tomato paste, not fresh tomatoes. Most actions in this area fall in the category of unfair trade practices, which is a field of tort law that protects against inequitable conduct. Essentially these torts define and police commercial ethics in the marketplace.

Products liability is another source of tort law that serves a regulatory role. Products liability refers to the civil liability of a manufacturer for injuries caused by "defects" in their products. Products liability law varies from state to state, but similar law exists in every state. These lawsuits provide important economic feedback to firms to invest more in food safety. This area of tort law provides both a method of compensation for damages, but also recognizes and

Food Regulation: Law, Science, Policy, and Practice, by Neal D. Fortin
Copyright © 2009 Published by John Wiley & Sons, Inc.

protects certain interests, and provides a "'prophylactic' factor of preventing future harm."[1]

15.2   PRODUCTS LIABILITY

15.2.1   The Hot Coffee Case

Tort liability for injury or illness from FDA-regulated products has been the focus of much media, public, and legal community attention in recent years. Unfortunately, much of the public discussion and debate on tort is based on anecdote and exaggeration. Outrageous jury awards, of course, capture the attention. One of the most talked about torts is the McDonald's "hot coffee case." Some believe this case is more influential in defining public perception of tort law than any other single factor.[2]

It is important to understand that the U.S. legal system is designed to toss frivolous lawsuits long before they reach trial. In addition the system allows judges to reduce excessive verdicts. In the hot coffee case, for example, the judge reduced the $2.9 million verdict by two-thirds to $640,000, and the parties reportedly settled out of court for less than $600,000.[3]

Particularly in the area of foodborne illness, the vast majority of injuries never reach trial. Foodborne pathogen determination requires expensive investigation and laboratory testing. The chance of finding the causative agent (and responsible party) is slight. Fewer than one in ten thousand foodborne illness cases are litigated and even fewer are paid compensation. For every million acute foodborne illnesses, approximately 10 to 45 torts ensue.[4] Put another way, out of 76 million serious foodborne illness cases annually in the United States, roughly 75,996,000 victims lack recourse to tort remedies.

15.2.2   Products Liability

Products liability is the area of law involving the civil liability of a manufacturer of a product sold to consumers for injuries caused by a defect in that product. "Defect," however, has a more specific and detailed meaning in products liability law than in everyday usage. In the simplest sense, a product defect exists when a problem, weakness, omission, or error exists that causes a safety issue with that product.[5]

---

[1] PAGE, KEETON, et al., PROSSER AND KEETON ON THE LAW OF TORTS § 95, at 677 (5th ed. 1984).
[2] *See, e.g.*, WILLIAM HALTOM and MICHAEL MCCANN, DISTORTING THE LAW: POLITICS, MEDIA, AND THE LITIGATION CRISIS (2004).
[3] *See, e.g.*, William Glaberson, *The $2.9 Million Cup of Coffee; When the Verdict Is Just a Fantasy*, NEW YORK TIMES, June 6, 1999.
[4] JEAN C. BUZBY & PAUL D. FRENZEN, *Food Safety and Product Liability*, 24 FOOD POLICY 637–651 (1999). (This is a conservative estimate and does not include confidential settlements. The annual number of cases is undeterminable because there is no national system documenting them.)
[5] J. O'REILLY & N. CODY, THE PRODUCTS LIABILITY RESOURCE MANUAL 3 (1993).

Products liability theories provide three main causes of action in which liability might apply to a manufacturer or distributor of an FDA-regulated product plus a fourth related theory of liability: (1) strict liability, (2) breach of implied warranty, (3) negligence, and (4) Misrepresentation or nondisclosure. Misrepresentation or nondisclosure is not always grouped under product liability directly, but the theory of liability runs through all products liability law as a means of committing the tortuous behavior.[6] Misrepresentation or nondisclosure causes of actions may arise with FDA-regulated products either from the premise that the manufacturer concealed material information from the FDA during the agency's review, or from allegation of misrepresentation or nondisclosure to the consumer. Of these causes of actions, strict liability and breach of implied warranty are the primary theories of recovery.[7]

Strict Liability   A person injured or made ill by a food containing a dangerous object or deleterious substance may recover damages from manufacturers or sellers of the product in action for strict liability in tort. As formulated in section 402A of the *Restatement Second of Torts*, a prima facie case in a strict liability in tort action for physical harm caused by the consumption of a dangerous or contaminated food or beverage product requires proof that:[8]

1. the defendant was engaged in the business of selling that food or beverage product;
2. the product consumed by the plaintiff was defective and unreasonably dangerous;
3. the product was defective when it left the defendant's control, and was in a substantially unchanged condition when it reached the plaintiff; and
4. consumption of the product caused the plaintiff to suffer physical harm.

The main defenses to a strict liability action are refutation of any of the four elements of proof of a claim.[9] In addition, assumption of the risk may be a defense, but contributory negligence is not a defense.

Breach of Implied Warrant   A persons injured or made ill by a food containing a dangerous object or deleterious substance may recover damages from manufacturers or sellers of the product in action for breach of the implied warranty of merchantability under the *Uniform Commercial Code*

---

[6] PAGE KEETON et al., *supra* note 1, at § 105, at 725.
[7] Westlaw, *Cause of Action for Physical Harm Caused by Eating or Drinking Dangerous or Contaminated Food or Beverage*, 4 CAUSES OF ACTION 787 (February 2004).
[8] *Id.*
[9] *Id.*

(U.C.C.) § 2-314.[10] The U.C.C. implied warranty of merchantability in the sale of goods can be viewed as a form of strict liability.[11]

A prima facie case in a breach of implied warranty action requires proof that:[12]

1. the defendant was a merchant with respect to the food or beverage product purchased by the plaintiff;
2. the product was unwholesome or unfit for human consumption and, therefore, breached the warranty; and
3. the failure of the product to conform to the warranty was the proximate cause of the physical harm suffered by the plaintiff.

In addition, the plaintiff may have to prove that the condition of the product was unchanged between the time it left the defendant's control and the time it was consumed by the plaintiff, and the defendant was given notice of the breach of warranty.

The main defenses to a breach of implied warranty action are (1) refutation of any of the three elements of proof of a claim, (2) that the injury-causing object is "natural" to the food product or was reasonably to be expected, and (3) that the product was not unwholesome or unfit for human consumption when it left defendant's control. Note, however, that absence of privity of contract between the plaintiff and the defendant generally is no longer a defense.

Under strict liability in tort, any person suffering physical harm from eating or drinking the product may recover, whether they actually purchased the product or not.[13] However, in a breach of implied warranty action, only

---

[10] U.C.C. § 2-314 (2001) provides as follows:

(1) Unless excluded or modified (Section 2-316), a warranty that the goods shall be merchantable is implied in a contract for their sale if the seller is a merchant with respect to goods of that kind. Under this section, the serving for value of food or drink to be consumed either on the premises or elsewhere is a sale.

(2) Goods to be merchantable must be at least such as
  (a) pass without objection in the trade under the contract description; and
  (b) in the case of fungible goods, are of fair average quality within the description; and
  (c) are fit for the ordinary purposes for which such goods are used; and
  (d) run, within the variations permitted by the agreement, of even kind, quality, and quantity within each unit and among all units involved; and
  (e) are adequately contained, packaged, and labeled as the agreement may require; and
  (f) conform to the promise or affirmations of fact made on the container or label if any.

(3) Unless excluded or modified (Section 2-316), other implied warranties may arise from course of dealing or usage of trade.

[11] PAGE, KEETON et al., *supra* note 1, at § 97.

[12] 4 CAUSES OF ACTION 787 *supra* note 7.

[13] *Id.*

persons who are reasonably expected to eat or drink the product generally may recover.[14]

Breach of Express Warranty   In some cases, a cause of action under breach of express warranty may arise. The U.C.C. describes express warranty as any affirmation of fact or promise made by the seller to the buyer, which relates to the goods.[15] For example, a label statement that a product is "ready to eat" creates an express warranty that the food conforms to the expectations of a ready-to-eat food. It is not necessary that the word warrantee or guarantee be used.

In most jurisdictions the burden of proof is greater for express warranty than strict liability. Therefore, when both claims arise, often the express warranty claim is not pursued.

Negligence   To determine whether a defendant is negligent, the case usually requires proof of five separate elements: (1) duty of care for the defendant manufacturer, producer, packer, or seller; (2) breach of that duty; (3) cause in fact; (4) scope of liability or scope of protection; and (5) damages.

The focus in negligence is on reasonable care (the duty of care). Reasonable care is generally considered the care exercised by a reasonable a manufacturer or seller under similar circumstances. Factors that may be considered in determining if reasonable care was exercised in particulars circumstances include current industry standards, current state of the technology and knowledge, HACCP implementation, and compliance with government regulations.

Negligence is a less used cause of action because of the higher burden of proof. However, in some situations there may be advantages of pleading negligence in addition to strict liability and warranty; for example, an advantageous statute of limitations.

15.2.3   Foreign-Natural Test versus Reasonable Expectation

Under regulatory standards for adulteration, the natural/nonnatural distinction is important in determining whether a food is adulterated under the FD&C Act[16] American society tends to be more accepting of naturally occurring risks and defects than those created by food manufacturers. This predilection has important bearing on imposition of a duty of care, but it is not the sole factor in determining whether products liability exists for a particular defect and injury.

---

[14] U.C.C. § 2-318 removes any privity requirement by providing that both express and implied warranties extend to "any person who may reasonably be expected to use, consume, or be affected by the goods and who is injured in person by breach of the warranty."

[15] U.C.C. § 2-313.

[16] *See* Chapter 8 for a discussion of this phenomenon.

Under tort law the test is generally not natural versus nonnatural (or foreign) but the consumers' reasonable expectation.[17] A harmful ingredient of the food product constitutes a defect if a reasonable consumer would not expect the food product to contain that ingredient.[18] For instance, liability may be found for a can of peas containing a pebble because it is considered a manufacturing defect.[19] Spoilage of a can or jar of food may also be a manufacturing defect.[20]

On the other hand, under the consumer expectations test for strict liability or breach of warranty, a reasonable consumer cannot expect a fish filet to be free of a half-inch fish bone,[21] or a chicken enchilada to be chicken bone free.[22] However, a jury may find a breach of implied warranty when a hamburger contains a piece of bone[23] or a cocktail olive that has a hole and seems to have been pitted but contains a pit.[24] Keep in mind that negligence still applies when a natural substance food causes injury, if the presence of the natural substance was due to failure to exercise reasonable care in food preparation.[25]

The next two cases provide examples of the application of the foreign-natural test and reasonable expectation test. Although nearly all states now apply the reasonable expectation test, Louisiana still holds to the older foreign-natural test.

In *Porteous v. St. Ann's Café & Deli*, a customer brought an action against the restaurant for injuries to customer's tooth when he bit into pearl contained in oyster.[26] The court ruled that restaurant was negligent. On appeal the Louisiana Supreme Court held that a food provider has a duty to act as would a reasonably prudent man skilled in the culinary art in the selection and preparation of food, but the restaurant did not breach its duty to diner.

* * * * *

Porteous v. St. Ann's Café & Deli

713 So. 2d 454 (La. 1998)

CALOGERO, Chief Justice.

On January 22, 1995, Donald C. Porteous, Jr. was dining at St. Ann's Cafe & Deli. While eating the second half of an oyster po-boy, he bit down onto a

---

[17] RESTATEMENT THIRD, TORTS: PRODUCTS LIABILITY § 7 expressly adopts this test.
[18] *Id.* § 7, Comment a.
[19] *Id.* § 7.
[20] *Id.* § 7, Comment b.
[21] Ex parte Morrison's Cafeteria of Montgomery, Inc., 431 So. 2d 975 (Ala. 1983).
[22] Mexicali Rose v. Superior Court, 822 P.2d 1292 (1992).
[23] Goodman v. Wenco Foods, Inc., 423 S.E.2d 444 (1992).
[24] Hochberg v. O'Donnell's Restaurant, Inc., 272 A.2d 846 (D.C. 1971).
[25] *Supra* note 22.
[26] 713 So. 2d 454 (La. 1998).

small, grey, and roughly round substance, which apparently was a pearl. When plaintiff bit onto the pearl, he broke a tooth and cracked it all the way down the shaft. The plaintiff reported the incident to a waiter. The waiter wrote an incident report and took possession of the remainder of the sandwich and the pearl. Two days later, plaintiff went to his dentist and thereafter underwent dental treatment, which included a root canal and placement of a crown atop the broken tooth. The plaintiff then sued St. Ann's Cafe & Deli and Lafayette Insurance Company, alleging that the defendant was negligent because of the lack of adequate food inspection procedures, which resulted in the presence of an injurious substance and his sustaining injury to his tooth.

In determining whether the defendant was liable for the plaintiff's injuries, the trial court applied the "foreign-natural" test. That test was adopted from the common law. Louisiana courts of appeal have used this common law test to determine the liability of a restaurant when a customer is injured by a harmful substance in the restaurant's food. Under the foreign-natural test, if the injurious substance is foreign to the food, then the restaurant is strictly liable. If the injurious substance is natural to the food, there is no strict liability. Rather, liability is imposed only if the restaurant was negligent in failing to discover and remove the harmful natural substance from the food.

After applying the foreign-natural test, the trial court held that although the injurious pearl was natural to the oyster, the restaurant was negligent, and therefore liable, because of the lack of adequate procedures to ensure that injurious substances, such as a pearl in the oyster, were not served on the po-boys. The plaintiff was then awarded damages plus costs and interest.

The court of appeal recited the facts found by the trial court and declared that the "trial court's determination of credibility and findings of fact will not be disturbed on appeal so long as they are reasonable in light of the record as a whole." The court of appeal concluded that the trial court's finding that the restaurant negligently failed to institute procedures to intercept harmful objects in the oysters was a reasonable finding and would not be disturbed on appeal. Thus, the trial court was affirmed in the court of appeal.

We granted certiorari to determine if the law and the facts were properly applied in this restaurant–harmful food product case, a precise matter which has not been addressed in recent decades by this Court. For the reasons that follow, we find that the lower courts erred in applying the common law foreign-natural test. Rather, the proper analysis to determine the defendant's liability is to be found in Louisiana's substantive law as found in the Louisiana Civil Code in the articles relating to liability and damages for offenses and quasi offenses—the traditional duty risk tort analysis. With the entire record now in hand, we hold that, under the traditional duty risk tort analysis, the plaintiff has failed to prove that the defendant breached its duty to act as would a reasonably prudent restauranteur in selecting, preparing, and cooking food, including removal of injurious substances. We therefore reverse the judgments of the lower courts.

Discussion

In the recent decades, this Court has not spoken on this issue and the Louisiana Courts of Appeal have borrowed the foreign-natural test from the common law. We decline to adopt that test.[27]

Under the foreign-natural test, the outset determination is whether the injurious substance is "foreign" or "natural" to the food. As this test evolved nationally, the cases held that if an injurious substance is natural to the food, the plaintiff is denied recovery in all events. But if the injurious substance is foreign, the restaurant is strictly liable. Louisiana Courts of Appeal chose to follow the foreign-natural test to determine the liability of restaurants, but embellished a bit on the strict common law foreign-natural test, in permitting the plaintiff to recover notwithstanding the fact that the injurious substance is natural to the food if the restaurant is negligent in its failing to discover and remove the injurious natural substance.

In time, the foreign-natural test was widely criticized and rejected by many states in favor of the reasonable expectation test. Under the reasonable expectation test, the query to determine liability is whether a reasonable consumer would anticipate, guard against, or expect to find the injurious substance in the type of food dish served. Whether the injurious substance is natural or foreign is irrelevant. Rather, liability will be imposed on the restaurant if the customer had a reasonable expectation that the injurious substance would not be found in the food product. On the other hand, if it can be shown that the customer should reasonably have expected the injurious substance in his food, that customer is barred from recovery.

The Civil Code is the chief repository of the substantive law of Louisiana, and as previously indicated, the theory of recovery available to an injured plaintiff to determine the liability of a restaurant in a case of this sort is the determination of negligence with the traditional duty risk tort analysis.

Tort Claim

Articles 2315 and 2316 are the codal bases for a claim in tort. Article 2315 states that "[e]very act whatever of man that causes damage to another obliges him by whose fault it happened to repair it." Article 2316 provides that "[e]very person is responsible for the damage he occasions not merely by his act, but

---

[27] The following is a more detailed explanation of the two common law tests to which reference was made earlier in this opinion, the foreign-natural test and the reasonable expectation test, which have been utilized by state courts to determine the restaurant's liability, when a plaintiff sustains injuries because of an injurious substance in food he is served in a restaurant. Under either test, courts have no difficulty holding a defendant strictly liable for injuries sustained because of "foreign" injurious substances (such as, glass or insects). *See LeBlanc v. Louisiana Coca Cola Bottling Co.*, 221 La. 919, 60 So.2d 873 (1952). But, if the injurious substance is "natural" to the food product, such as bones or shells, courts, depending on whether they follow the foreign-natural test or the reasonable expectation test, are divided as to whether liability should be imposed.

by his negligence, his imprudence, or his want of skill." To determine whether a defendant is negligent, the case usually requires proof of five separate elements: (1) duty, (2) breach of duty, (3) cause-in-fact, (4) scope of liability or scope of protection, and (5) damages. Relative to these five elements, the case at hand turns on two of the elements—the issue of the defendant's duty and the defendant's breach of duty—discussion regarding which follows.

Duty of the Defendant

A defendant's duty to conform his conduct to a specific standard may be express or implied, either statutorily or jurisprudentially. In Louisiana there is no statute which expressly addresses a commercial restaurant's duty to serve food free of injurious substances. There is, nonetheless, no doubt that there is and should be such a duty. We determine that the duty is the following: A food provider, in selecting, preparing, and cooking food, including the removal of injurious substances, has a duty to act as would a reasonably prudent man skilled in the culinary art in the selection and preparation of food.

Breach of Duty

Plaintiff alleges that the defendant restaurant breached its duty by acting unreasonably in the selection, preparation, and cooking of the food because the restaurant lacked adequate inspection procedures to detect and remove injurious substances from the food served to its customers. The defendant, on the other hand, asserts that it did not breach its duty because it acted reasonably in the selection and preparation of the food product at issue.

In determining whether a restaurant breached its duty by failing to act reasonably in the selection, preparation, and cooking of the food that contained a substance which caused injury, the court should consider, among other things, whether the injurious substance was natural to the food served and whether the customer would reasonably expect to find such a substance in the particular type of food served.

These are the determinative factors in the foreign-natural test and the reasonable expectation test, but are only factors to be considered by the court when using the duty risk analysis in negligence law.

In the present case, the plaintiff was injured when he bit onto a pearl while eating an oyster po-boy in the defendant restaurant. A pearl in an oyster is not entirely rare, but is, indeed, a naturally occurring phenomenon. So long as oysters are harvested and eaten, there will occasionally, though perhaps infrequently, be pearls found in oysters. Furthermore, when eating oysters, a customer should be aware of—and alert to the possibility—that a pearl may be found within the oyster.

Additionally, at trial, the restaurant manager, Ms. Marvez, testified that an accident like this had never occurred before in her restaurant. Ms. Marvez further stated that the restaurant buys its oysters pre-shucked, pre-washed, and pre-packed from a reputable seafood company. When Ms. Marvez was asked about the restaurant's procedures to ensure that there were no foreign objects in the oysters, she replied that "the cook has to physically hold the oyster and bread it and at that time they could feel if there's anything in there. If it's something large or if it's something—now if it's embedded in the oyster, no, we don't dissect the oyster. . . ." She also stated that the cooks have to grab the oysters to bread them, and if they were to find an object, they would remove it. She did not recall any time when she was told that a cook found an object in an oyster. Moreover, although the cooks do not wash the oysters before they are battered, the cooks do visually inspect the oysters and touch them before applying the batter.

In light of the above-described testimony, we determine that the defendant did not act unreasonably in selecting, preparing, and cooking the food. There was nothing more the defendant restaurant could reasonably have done to eliminate the small possibility that a customer might find a pearl in an oyster and be injured thereby. The law should not impose upon restaurants the responsibility of dissecting every oyster in order to determine whether there is a pearl formed or forming inside each one. We determine, therefore, under the traditional duty risk tort analysis, that the defendant restaurant did not breach its duty to this plaintiff and, thus, is not liable for the plaintiff's injury.

Decree

For the foregoing reasons, the judgments of the district court and the court of appeal in favor of plaintiff are reversed. Judgment is rendered in favor of the defendants, dismissing plaintiff's suit with prejudice and at his cost.

LEMMON, Justice, dissents and assigns reasons.

The critical issue is whether the pearl in the oyster sandwich caused the food to be unreasonably unsafe. The ordinary customer would not reasonably expect to encounter a pearl in an oyster sandwich. Therefore, an oyster sandwich containing a pearl i[s] not reasonably safe. The plaintiff, having proved that the food served to him was not reasonably safe, should recover, irrespective of proof of negligence.

As in cases involving unreasonably dangerous products, the innocent consumer in a food product case has no method to protect himself or herself. The risk of injury from unreasonably dangerous food should not fall on the back of the innocent consumer, but on the purveyor of the food product who can spread that risk (as was done in this case) with liability insurance.

* * * * *

The case of *Jackson v. Nestle-Beich, Inc.,*[28] provides an explanation of two tests: the foreign-natural test and the consumers' reasonable expectation test.

---

[28] 589 N.E.2d 547 (Ill. 1992).

The *Nestle-Beich* plaintiff allegedly broke a tooth on a pecan shell while biting into chocolate-covered, pecan-caramel candy manufactured by the defendant. The *Nestle-Beich* court held that the reasonable expectation test, rather than the foreign-natural doctrine, would be used to determine liability of vendor of food product for injuries caused by ingredient in the food.

* * * * *

Jackson v. Nestle-Beich, Inc.

589 N.E.2d 547 (Ill. 1992)

Justice FREEMAN, delivered the opinion of the court:
Appellant, Nestle-Beich, Inc. (Nestle), appeals the decision of the appellate court reversing the grant of summary judgment in its favor in a personal injury action brought by appellee, Elsie M. Jackson (Jackson). We affirm.

Factual Background

In May 1988, Jackson purchased a sealed can of Katydids, chocolate-covered, pecan and caramel candies manufactured by Nestle. Shortly thereafter, Jackson bit into one of the candies and allegedly broke a tooth on a pecan shell embedded in the candy. As a result, Jackson filed a complaint asserting breach of implied warranty (count I) and strict products liability (count II) against Nestle.

Nestle moved for summary judgment on the basis of the foreign-natural doctrine. That doctrine provides that, if a substance in a manufactured food product is natural to any of the ingredients of the product, there is no liability for injuries caused thereby; whereas, if the substance is foreign to any of the ingredients, the manufacturer will be liable for any injury caused thereby.

In granting Nestlé's motion, the trial court concluded that Illinois law is "that a food product is not rendered unwholesome by reason of inclusion therein of a substance natural to an ingredient" of the product.

In reversing, the appellate court thoroughly reviewed the rationales underlying the foreign-natural doctrine and the reasonable expectation test, which is applied in certain jurisdictions. The reasonable expectation test provides that, regardless whether a substance in a food product is natural to an ingredient thereof, liability will lie for injuries caused by the substance where the consumer of the product would not reasonably have expected to find the substance in the product.

The appellate court concluded "that the foreign-natural doctrine originally set forth in Mix and adopted . . . in Goodwin should not be followed." The court determined that the doctrine was based on the faulty assumption that consumers know that prepared food products will or might contain whatever any of their ingredients, in a natural state, contain. Ultimately, the court held that the naturalness of the harmful ingredient of a food product does not absolutely bar recovery but is only one factor to be considered in determining whether the presence of the ingredient breached a warranty or rendered the product unreasonably dangerous. We agree with the appellate court's conclusion that the foreign-natural doctrine is unsound and should be abandoned.

Nestle's Arguments

In appealing the appellate court's decision, Nestle first asserts that the decision "has, in practice, created a strict liability situation [ ]" because the court "failed to change the general test" for determining the existence of a breach of warranty with respect to food products, viz., the presence of foreign matter in the food or its diseased, decayed, or otherwise spoiled and poisonous condition. Nestle reasons that if, as the appellate court held, the naturalness of the harmful ingredient does not bar recovery, its mere presence will henceforth breach the warranty.

We are somewhat perplexed by Nestlé's assertion that the appellate court failed to change the test of breach of warranty, with respect to food products, in light of Nestlé's conclusion that, as a result of the court's decision, naturalness, the linchpin of that test, will no longer bar recovery. We would ask Nestle how that can be unless the appellate court's decision effectively changed the test. In addition, we find that test, as stated in Illinois Law & Practice, simply of no assistance to Nestle. That work cites Goodwin and it is the continuing validity of Goodwin which is at issue here.

Although not explicitly, the appellate court's decision does, effectively, establish the same test for both breach of warranty and strict products liability claims in food cases, as Nestle appears to argue. That test is the reasonable expectation of the consumer with respect to the ingredients of the food product involved. However, Nestle offers no sound argument for holding that the appellate court could not do so. Therefore, we find this line of argument completely unavailing to Nestle.

Nestle further asserts that the appellate court's decision "fails to acknowledge that the unique situation of natural food hazards is worthy of treatment different than other products ... by applying the foreign-natural doctrine, reasonable expectation test or some hybrid of the two" because "perfection in removing naturally occurring substances is impossible on each and every occasion."

We do not find this argument to be a valid criticism of the appellate court's opinion. In so arguing, Nestle itself fails to acknowledge that the appellate court effectively adopted the reasonable expectation test as the measure of the viability of both breach of warranty and strict products liability claims in food cases. That is, the appellate court's decision does treat manufacturers such as Nestle differently than manufacturers of other products.

The crux of Nestle's arguments on appeal is that we should adopt the Louisiana version of the foreign-natural doctrine. In Louisiana, if injury is caused by a foreign substance in a food product, the manufacturer is subject to being held strictly liable. In contrast, if the substance causing injury is natural to the product or its ingredients, the manufacturer may be held liable only if the presence of the substance resulted from its negligence in the manufacture of the product.

We decline Nestle's invitation to adopt the Louisiana version of the foreign-natural doctrine in place of the reasonable expectation test. We agree with Jackson that the Louisiana approach comes too close to the outdated and discredited doctrine of caveat emptor.

Moreover, contrary to Nestle's implication in arguing for the adoption of Louisiana's approach, the appellate court's decision in the instant case is not the first to extend the modern-day doctrine of strict liability to food products in Illinois. In *Warren v. Coca-Cola Bottling Co.*,[29] the court recognized that causes of action for breach of implied warranty, strict products liability, and negligence properly lay against the manufacturer of an article of food or drink intended for human consumption and sold in a sealed container. Accordingly, the decision in the instant case does not impose burdens upon manufacturers of food products sold or used in Illinois which they have not previously borne.

Additionally, we must reject the underlying theme of Nestle's appeal. Nestle asserts that manufacturers of food products whose manufacture involves a risk that harmful matter which is natural to any of their ingredients will not be eliminated from the finished product should be exempted from strict liability due to the difficulty of eliminating such matter.

Preliminarily, we would note that this argument is essentially an argument for recognizing a state of the art defense in food product cases. However, in Illinois the state of the art has never been a defense to strict products liability.

Moreover, we do not find that manufacturers of products such as Nestle describes serve so important a public service that they merit treatment substantially different from that of manufacturers of other products. In this regard, we believe the consumer's reasonable expectation as to the contents of food products, as the gauge of strict liability, adequately balances consumers' interest in defect-free products and such manufacturers' interest in reasonable costs of doing business.

With an awareness of that test, consumers and their attorneys need ask themselves only one question before deciding to bring an action of this type: Would a reasonable consumer expect that a given product might contain the substance or matter causing a particular injury? If the answer is in the affirmative, we would expect that consumers and their attorneys would think twice about suing the manufacturer. Similarly, with an awareness of that test, manufacturers can act accordingly with respect to their means of production. Additionally, if the answer to the foregoing question is in the negative, we would expect that manufacturers and their attorneys would think twice about declining to offer a settlement of this type of action. The test thus provides a reasonable and concrete standard to govern actions of this sort.

[29] 166 Ill.App.3d 566, 117 Ill. Dec. 30, 519 N.E.2d 1197 (1988).

Moreover, we believe that the fact that the reasonable expectation test comports with the rationale underlying strict products liability strongly recommends the test as the dispositive inquiry in this type of case. That rationale provides that an allegedly defective product must be dangerous to an extent beyond that which would be contemplated by the ordinary consumer who purchases it, with the ordinary knowledge common to the community as to its characteristics. (Restatement (Second) of Torts § 402A, Comment i, at 352 (1965).) The similarity between the language of the reasonable expectation test and the Restatement is striking and strongly recommends the former as the rule of decision in this type of case.

Nestle further argues that a fundamental basis of strict products liability, viz., that manufacturers who create risks posed by a defective product and who reap a profit by placing it in the stream of commerce should bear the losses caused thereby, does not apply in this case. That principle does not apply, Nestle reasons, because the manufacturer of natural food products does not create the risk. Rather, the risk is created by nature and all a manufacturer can do is its best to minimize the risk. We find this reasoning unpersuasive.

In so arguing, Nestle ignores that strict liability, in theory, is intended to apply to all products placed in the stream of commerce regardless whether they have undergone some processing or not. Accordingly, a supplier of poisonous mushrooms which are neither cooked, canned, packaged, nor otherwise treated is subject to strict liability. (Restatement (Second) of Torts § 402A, Comment e, at 350 (1965).) If such a supplier would be strictly liable for injuries caused by its product, we see no reason why Nestle should not be. Both would have sold a product which injured a consumer as a result of an aspect of the product's or an ingredient's natural state. In the mushroom supplier's case, that aspect would be their poisonousness. In Nestle's case, that aspect would be that the meat of pecans, an ingredient of Katydids, is found inside a hard shell. That Nestle actually processes the ingredients in its product before placing it in the stream of commerce and thereby has some opportunity to discover and eliminate the risk of injury posed by its ingredients, unlike the seller of poisonous mushrooms, actually militates in favor of, rather than against, imposing strict liability against Nestle.

Furthermore, we do not believe that Nestlé's product merits inclusion in that group of products "which, in the present state of human knowledge," are "incapable of being made safe for their intended and ordinary use," i.e., unavoidably unsafe products, and which are not subject to strict liability. Simply put, Nestlé's product lacks the social utility of those products which justifies their exemption from such liability. Restatement (Second) of Torts § 402A, Comment k, at 353 (1965); cf. Ill.Rev.Stat.1989, ch. 111 1/2, par. 5101 et seq. (exempting from strict liability those engaged in, inter alia, procuring, furnishing, processing, or distributing human whole blood, plasma, blood products, blood derivatives and products, corneas, bones, organs or other human tissue for the purpose of injection, transfusion, transplantation in the human body).

Nestle also argues that two other rationales for strict products liability, viz., that consumers cannot adequately protect themselves from defective products and that manufacturers such as Nestle are in a better position to identify and guard against potential defects, are inapplicable in this case. We disagree.

Rather than imposing an obligation upon consumers of Nestle's product to protect themselves by "think[ing] and chew[ing] carefully," we believe that the obligation of protection is better placed on Nestle and like manufacturers. In this regard, we agree with the following observation:

> With the prevalence of processed foods on the market today and the development of technology in the food industry, consumers increasingly rely upon food processors to inspect and purify the foods they consume. Many products today are even packaged in such a manner that inspection by the consumer is difficult if not impossible. One might imagine a consumer in a jurisdiction that applies the foreign-natural test tearing away the crust from a beef pot pie to search for tiny bones, or picking apart a cherry-nut ice cream cone to remove stray shells or pits.
>
> In an era of consumerism, the foreign-natural standard is an anachronism. It flatly and unjustifiably protects food processors and sellers from liability even when the technology may be readily available to remove injurious natural objects from foods. The consumer expectation test, on the other hand, imposes no greater burden upon processors or sellers than to guarantee that their food products meet the standards of safety that consumers customarily and reasonably have come to expect from the food industry.

With respect to Nestle's argument that it is in no better position than consumers of its Katydids to identify the risks associated therewith, we disagree that the common knowledge that pecans are hard-shelled nuts makes it common knowledge that processed foods containing pecan meats may also contain pecan shell. Moreover, contrary to Nestle's assertion, we do not believe that "the practical difficulties of food separation are common knowledge" to all consumers. As a result, we do not believe that consumers of its Katydids must be required to "think and chew carefully" when consuming them.

Nestle further argues that another of the rationales for strict liability, the difficulty of proving negligence, can be satisfied hereby requiring the manufacturer to prove its freedom from negligence, as is done in Louisiana. We disagree.

Even if we had no doubt that that is the approach taken in Louisiana, we do not believe its adoption in Illinois would be salutary to our jurisprudence. We do not believe it wise to begin carving out exceptions to strict products liability by placing upon manufacturers a burden of proving freedom from negligence. If we were to find Nestle entitled to such treatment, we would be hardpressed to reject the arguments of any manufacturer for similar treatment. Eventually, the exceptions would swallow the rule of strict liability. We cannot countenance such a possibility.

Lastly, Nestle argues that another of the rationales for strict liability, viz., that it is more effective than negligence liability in inducing the manufacture of safer products, also does not apply here given the nature of the product involved. We disagree.

Even under the reasonable expectation test of strict liability we approve in this case, manufacturers, such as Nestle, of products posing the risk involved in this case can take one simple and relatively inexpensive step to make their products safer and to avoid liability for injuries caused thereby. Specifically, they can place an adequate warning to the consumer on their product's container of the possibility or risk of injury posed thereby. (Restatement (Second) of Torts § 402A, Comment j, at 353 (1965).) The relative ease with which such a measure may be taken also militates in favor of holding manufacturers of such products subject to strict liability in the absence thereof.

In this regard, we note that, even if we agreed with Nestle that its Katydids merit classification as an unavoidably unsafe product, we would nonetheless find it subject to strict liability due to the absence of a warning of the unavoidable risk of injury it posed. *See* Restatement (Second) of Torts § 402A, Comment k, at 353 (1965) (an unavoidably unsafe product is not defective or unreasonably dangerous when properly prepared and accompanied by proper directions and warning).

For all of the reasons stated herein, we affirm the judgment of the appellate court.

Appellate court affirmed.

Justice HEIPLE, dissenting:

The majority decision overturns the long-standing doctrine in Illinois regarding the sale, purchase and consumption of food products. Since 1944, Illinois has followed the so-called foreign-natural doctrine. Stated in its simplest terms, the foreign-natural doctrine provides that the vendor of food is not liable for injuries due to unremoved but naturally occurring ingredients such as nut shells, fruit pits, fish bones, and so forth, but is liable for foreign objects in the food such as glass shards or pieces of metal. The majority opinion in the instant case discards the foreign-natural doctrine and substitutes the reasonable expectation test. In its essence that test provides that the vendor of a food product is liable for injuries caused by an ingredient in the food whether natural or foreign whenever the consumer of the product would not reasonably have expected to find the substance in the product.

In truth, the reasonable expectation test is what gave rise to the foreign-natural doctrine. That is to say, since it would be reasonable to expect to find a nut shell in a product containing nuts, there would be no liability. Rather than approach each broken tooth or other injury on a case-by-case basis, it was deemed more expeditious and efficient to crystallize the matter into the foreign-natural doctrine. That doctrine both did justice and promoted judicial economy.

A reversion to the reasonable expectation test simply means that each food-related injury in this State will be subject to a lawsuit to determine whether the consumer's reasonable expectation was violated. The costs will be significant, first to the manufacturers and second to the consuming public. It is axiomatic that all production costs eventually end up in the price of the product. Additionally, if the costs exceed profitability, the product leaves the market place altogether and the consumers lose choice, selection, and availability of products.

The effects of this decision will go far beyond the defendant Nestle-Beich Company, whose candy caused a broken tooth. It extends to all manufacturers and purveyors of food products including the neighborhood baker, the hot dog vendor and the popcorn man. Watch out Orville Redenbacher!

The continued march towards strict and absolute liability for others (others meaning anyone not injured who has assets) and the absence of any responsibility by the injured for their own welfare takes yet another step with this majority ruling. Accordingly, I dissent.

* * * * *

NOTES

15.1. Another case on the reasonable expectation. See *Mexicali Rose v. Superior Court*, 822 P.2d 1292 (1992), where the consumer brought an action against a restaurant for damages sustained when he swallowed a chicken bone in a chicken enchilada.

15.2.4 Negligence Per se

Under the doctrine of negligence per se, a violation of a statute provides the standard of care in a common law negligence action. States are divided on the application of negligence per se for the violation of a regulation.[30] Even in jurisdictions where negligence per se is applied, the doctrine is not applied strictly to every violation. The guiding principle in determining the applicability of the doctrine of negligence per se is whether its application is necessary to effectuate the legislative purpose.[31]

The next case, involving a medical device, demonstrates how proof of violation of the FD&C Act may constitute negligence per se.

* * * * *

Orthopedic Equipment Co. v. Eutsler

276 F.2d 455 (1960)

SOBELOFF, Chief Judge:

A manufacturer of surgical instruments and equipment . . . is accused in this diversity action of misbranding a surgical nail which became stuck in

[30] 57A AM. JUR. 2d NEGLIGENCE § 754 (Feb. 2008).
[31] *Id.*

the plaintiff's leg in the course of an operation. . . . On March 30, 1956, the twenty-one year old plaintiff was helping his father take down a tree on a farm near Orange, Virginia. He was injured by the tree falling upon him. At the University of Virginia Hospital, it was found that he had sustained a fracture of the leg and other injuries. In the judgment of the surgeons, the treatment indicated for the fracture was an operation known as intramedullary nailing by use of a Kuntscher Cloverleaf Intramedullary Nail. This involves the insertion of a long metal rod or nail into the medullary canal (containing the narrow) of the femur (or thigh bone), in order to stabilize the broken fragments. The advantage sought by this method is an early union and weight-bearing without the necessity of a plaster cast.

A team of orthopedists, experienced in this technique, operated on April 3, 1956. Having prepared the canal by use of a 9 mm. medullary reamer or drill, the surgeons began to insert into the medullary canal a Kuntscher Cloverleaf intramedullary nail manufactured by the defendant. These Kuntscher nails usually have imprinted upon them two figures signifying their dimensions, e. g., 9×40, 10×42, but the imprint or label does not explain the meaning of these figures. It is agreed by the parties that the larger figure is understood to represent the length of the nail in centimeters. According to the plaintiff's expert witnesses, the interpretation placed upon the smaller figure by orthopedists is that the nail will fit into a hole having a width or diameter corresponding in millimeters to the figure on the nail. This follows from the necessity that the nail shall fit tightly into the canal previously prepared by a reamer of corresponding diameter. These witnesses also testified that after the canal is reamed, the nail is selected on the basis of the measurement on its "label," or imprint, conforming to the measurement of the reamer used. Furthermore, plaintiff's experts testified, orthopedic surgeons invariably rely upon the figures imprinted on the nail, when there are figures imprinted, without making independent measurements. Thus, according to the surgeons, they relied in this instance too on the accuracy of the marking, "OEC 9×40," in selecting the nail.

As the nail was driven down the canal of the upper fragment of the thigh bone, the surgeons at first met normal resistance. When it penetrated further, however, greater resistance was encountered. Nevertheless, the doctors did not regard this as unusual, since they knew that they had used a 9 mm. reamer and the nail was marked to indicate 9 mm.; they concluded that it must merely have met some slight obstruction which, as in past operations, would be passed or overcome without difficulty. Accordingly, as was customary in such cases, two or three slightly heavier blows were then struck.

Because the nail would progress no further even after these heavier blows, the surgeons decided to remove it. However, when persistent efforts to dislodge the nail proved unavailing, the portion of the nail protruding below the canal of the upper fragment was cut off, the wound closed, and a plaster cast applied in the hope that in a few weeks the bone would atrophy sufficiently to loosen the nail and permit its withdrawal.

About a month later, on May 4, the surgeons again tried to extract the nail, but were unsuccessful. Thereupon, one of the doctors designed a new instrument, and by its use removal of the nail was finally accomplished in a third operation on June 5. Measurements of cross sections of the nail, as testified to by a machinist, varied from a minimum of 9.27 mm. to a maximum of 10.12 mm.

Due to the nail's impaction, incurable osteomyelitis or bone infection resulted. The plaintiff has permanently lost the use of his leg, and its ultimate amputation is expected.

This action was brought against defendant for alleged "negligent manufacture, labeling, and launching on the market of said nail . . . ," plaintiff presumably at first intending to charge ordinary common law negligence only. Later, however, it was stipulated by counsel that, without formal amendment, the complaint should also be regarded as alleging a violation of the Federal Food, Drug, and Cosmetic Act. In this appeal, the defendant assigns five grounds of error, which we shall now discuss.

The Federal Food, Drug, and Cosmetic Act

Defendant contends, for two reasons, that the District Judge erred in basing his charge to the jury on the Federal Food, Drug, and Cosmetic Act, 21 U.S.C.A. 301–392. It asserts first, that the evidence of misbranding is insufficient, and second, that in any event, the Act does not apply to surgical instruments. . . .

We think that the evidence was sufficient to raise a jury question of misbranding. Notwithstanding the defendant's expert testimony to the contrary, the testimony of plaintiff's experts as to the understanding of the medical profession of the number 9 on the nail certainly presented an issue for the jury as to the "true" meaning of the number. . . .

Defendant insists that, in any event, the Federal Food, Drug, and Cosmetic Act does not apply to surgical nails marketed for use only by skilled surgeons. With this we are compelled to disagree. The definition of "devices" embraced by the Act is clearly of sufficient breadth and scope to include a surgical nail, which frequently remains in the patient for many months and is designed to and does affect both the "structure and function of the body."

It is urged by defendant that the regulations issued pursuant to section 352(f) of the Act nevertheless exempt manufacturers from the obligation to give directions for the use of surgical instruments since such devices are designed for use by a skilled profession. This specific exemption of surgical instruments from section 352(f), however, does not relieve the defendant from compliance with other provisions of the Act, including the remainder of section 352, and seems to us rather to indicate a contrary intention. In short, while the Act imposed no obligation upon defendant to label its nail, once it undertook to do so the Act required it to avoid misbranding.

The dicta quoted in defendant's brief to the effect that the Act was designed primarily to protect the public, especially consumers, do not support the inference that defendant seeks to draw, namely that surgical instruments are not meant to be covered by the Act since they are not ordinarily sold to members of the public. On the contrary, these expressions are more consistent with the inclusion of such instruments within the scope of the Act, for the patient as a member of the public is the ultimate consumer. As the District Judge said, in overruling defendant's motion for summary judgment,

". . . I think it is immaterial that the nail was not sold to, or purchased by the plaintiff, and I think it is incorrect to say that the nail was not manufactured for sale to, or use by, the general public. Actually it was, although it was to reach the general public through expert hands."

Having determined that the Federal Food, Drug, and Cosmetic Act applies in the instant case, and that there was sufficient evidence of misbranding, we turn now to the effect of a violation of the Act. The Federal Food, Drug, and Cosmetic Act does not expressly provide a civil remedy for injured consumers. However, the statute imposes an absolute duty on manufacturers not to misbrand their products, and the breach of this duty may give rise to civil liability.

The basic question is whether a violation of the strict duty created by the Act shall be deemed negligence per se under Virginia law, assuming as we must from the submission made to the jury and from its verdict, that the violation was the proximate cause of the plaintiff's injury. The majority of American courts which have passed on this question, in cases arising under state laws resembling the Federal Act, have held violations to be negligence per se. Apparently the Supreme Court of Appeals of Virginia has not had occasion to decide whether a violation of the Virginia Food Act, or the state statutory provisions dealing with misbranding and adulteration of drugs and cosmetics, constitutes negligence per se. The Virginia Court, however, has stated, in a case involving a motor vehicle statute, that:

"The violation of a statute, although negligence per se, will not support a recovery for damages unless such violation proximately causes or contributes to the injury complained of."

Since Virginia law seems to regard violation of motor vehicle statutes as negligence per se, again assuming from the jury's verdict here that the violation was found to be a proximate cause of the injury, and in light of the decisions in other states passing on this question, we think that a violation of the Federal Food, Drug, and Cosmetic Act is negligence per se in Virginia, and that the District Judge correctly based his charge on that premise. . . .

The defendant argued parenthetically that other manufacturers of surgical nails habitually label them imprecisely. It is enough to say that even if this were a case of ordinary common law negligence, defendant could not justify its mislabeling on this ground. Customary practice does not prescribe the duty of care. As stated by Mr. Justice *Holmes in Texas & Pacific Ry. Co. v. Behymer*:

"What usually is done may be evidence of what ought to be done, but what ought to be done is fixed by a standard of reasonable prudence, whether it usually is complied with or not." . . .

The judgment is
Affirmed.

* * * * *

In some states, violation of a statute is merely a rebuttable presumption of negligence, and violation of a regulation is merely evidence of negligence.[32] When violation of a statute designed for the protection of human life or property does not constitute negligence per se but is only prima facie evidence of negligence, the presumption may be rebutted by proof that the defendant acted reasonably under the circumstances, despite the violation.

Blommer Chocolate Co. v. Bongards Creameries, Inc.

635 F. Supp. 911, 917 (N.D. Ill. 1985)

MORAN, District Judge.

Introduction

This case stems from the *Salmonella* contamination of several hundred thousand pounds of plaintiff's chocolate products in early 1982. When ingested by humans, *Salmonella* bacteria can cause salmonellosis, commonly known as food poisoning. Salmonellosis on occasion can be fatal.

Plaintiff, Blommer Chocolate Company, as its name suggests, is a manufacturer of chocolate products. Its suit is directed against Bongards Creameries, Inc., which manufactured the dry whey powder, a milk product that allegedly was the source of the contamination, and J.M. Swank Company, Inc., from which Blommer ordered the whey. Counts I–III of the complaint allege that defendants breached express warranties, implied warranties of merchantability, and implied warranties of fitness for a particular purpose, respectively. In Counts IV and V Blommer alleges that Bongards negligently misrepresented the quality of the whey powder and was guilty of common law negligence. Count VI alleges that the defendants were negligent per se because in selling contaminated whey powder they violated Illinois pure food laws. Count VII alleges that defendants are strictly liable in tort. . . .

Facts

Linda Wolin, Blommer's purchasing agent, was responsible for purchasing the dry whey powder used by Blommer as an ingredient in its chocolate coatings. During 1980 and 1981, Blommer regularly purchased whey from Swank, a food broker. According to Wolin, in ordering from Swank she

---

[32] *See e.g.*, Kennedy v. Great Atlantic & Pacific Tea Co., 274 Mich. App. 710 (2007). (The violation of a statute creates a rebuttable presumption of negligence, and the violation of an administrative regulation constitutes evidence of negligence.)

stressed that the whey powder supplied Blommer had to be free of *Salmonella*. There is apparently no safe level of *Salmonella* contamination in products destined for human consumption.

In the summer of 1981, Wolin contacted Randy Hill, a Swank employee with whom she had had regular dealings, to arrange a large order of dry whey powder. According to Wolin, Hill recommended Bongards as a supplier of whey powder and assured Wolin that the Bongards whey would meet Blommer's quality standards. The August 11, 1982, purchase order associated with their discussion specified that the whey would be "extra grade," "guaranteed *Salmonella* negative," and "tested *Salmonella* negative before shipment to Blommer." The order also stated that "*Salmonella* statements are to accompany invoices or be written on the invoices."

It is clear from the record that "extra grade" whey denotes whey of the highest quality that is *Salmonella*-free and fit for human consumption. . . . Neither defendant has submitted any evidence that "extra grade" whey was not understood in the trade as being free of *Salmonella*. . . .

Three of the four invoices covering Bongards' sale of whey to Pacemaker describes the whey as being of "extra grade." . . . The deliveries were promptly followed by invoices from Swank. These invoices stated that the whey delivered was "guaranteed *Salmonella* Penicillin free and conforms to all other USDA and FDA specifications where it [sic] applies." The whey was packaged in heavy-duty plastic-lined bags labeled "extra grade," that had been sealed at Bongards and remained sealed until opened at Blommer.

According to Damien Gabis, executive vice-president of Silliker Laboratories, Inc., which acts as a consultant to Blommer on matters pertaining to the detection and control of *Salmonella*, there are two sources of *Salmonella* contamination at a food manufacturing plant like Blommer's. First, *Salmonella* may occur in the raw ingredients; second, *Salmonella* may exist in the processing environment and infiltrate the food products during their preparation.

Between 1967 and early 1982, Silliker had performed several thousand tests of Blommer's raw ingredients, processing environment and finished products. Prior to the contamination at issue here, Silliker never found *Salmonella* in either a finished product or in the processing environment. Well before 1982 Silliker did find *Salmonella* in one dry milk sample. *Salmonella* also was occasionally found in the dust of raw cocoa beans. *Salmonella* in cocoa beans is not unexpected and Blommer isolated the beans before roasting them at a temperature high enough to kill the *Salmonella* bacteria.

Silliker regularly tested the whey received by Blommer. Its test of the whey received by Blommer on January 13, 1982, revealed no *Salmonella* contamination. Blommer used all of the January 13, 1982, shipment to manufacture chocolate coatings between February 5, 1982, and February 15, 1982. Silliker also tested the February 2, 1982, shipment soon after it was received and found no *Salmonella*. Blommer used a portion of this shipment before mid-February.

On or about February 13, 1982, Silliker found *Salmonella* in several finished product samples from Blommer. Almost simultaneously one of Blommer's customers found *Salmonella* in a recently delivered shipment of Blommer's chocolate compound. Further tests of Blommer's finished products determined that only those products which contained Bongards' whey were contaminated. As a result of the contamination, Blommer was forced to recall its chocolate coatings, decontaminate its processing facilities, and assist several of its customers who were forced to decontaminate their facilities.

There are approximately 1,500 strains of *Salmonella* bacteria. The distinctiveness of the various strains assists in tracking down the source of the *Salmonella* contamination. The strain found in Blommer's chocolate compound is known as cubana.

Suspecting that the Bongards whey was the source of the contamination, Silliker tested what remained of the whey in the February 2, 1982, shipment. This re-test found that the shipment was contaminated with *Salmonella* cubana. All of the January 13, 1982, shipment had by this point been used up, so no test on this whey was possible. Silliker then notified the Chicago office of the Food and Drug Administration [FDA] of this data. The FDA dispatched investigators to Bongards. These investigators found *Salmonella* cubana in tailings taken from the dry whey sifter collection barrel. The FDA also dispatched investigators to Blommer. After examining the Blommer processing facilities they permitted production to continue, presumably convinced that the source of contamination was neither the processing environment nor other raw materials.

Bongards was not unacquainted with the problem of *Salmonella* contamination. During an inspection of Bongards from January 11, 1982, through January 14, 1982, inspectors from the U.S. Department of Agriculture [USDA] found *Salmonella* in dry whey powder that had been produced on January 10, 1982. Tests of whey powder produced on January 16, 1982, revealed *Salmonella*. Bongards' testing records also revealed occasional instances of *Salmonella* contamination. None of these tests included the serological typing, which would have revealed whether the *Salmonella* was of the cubana strain. Other USDA tests performed before and after January and February 1982 uncovered *Salmonella* contamination at Bongards.

Discussion

1. Blommer's Motion for Partial Summary Judgment against Swank

. . . Swank does not contest the applicability of contract law to the dispute. Blommer's purchase order, after all, did call for extra grade whey that was free from salmonella. Invoices that Swank issued to Blommer stated that it had supplied extra grade whey that was guaranteed *Salmonella*-free. Swank's president testified that extra grade whey was by definition fit for human consumption and free of *Salmonella*. The real dispute is whether Blommer has shown with a sufficient degree of certainty that the Bongards whey supplied by Swank was the source of the contamination.

As summarized above, the record shows that only those Blommer products that were made with Bongards' whey were contaminated. Tests show that the February 2, 1982, shipment of Bongards' whey was contaminated. In late February 1982, the FDA found that Bongards' whey-processing facilities were contaminated. The USDA found *Salmonella* contamination at Bongards twice in mid-January 1982 and at other times both before and after the Blommer incident. Bongards' own records show occasional contamination of the whey it produced during early 1982. In contrast, while the USDA has noted some sanitary deficiencies at Blommer, records reveal no instances of *Salmonella* contamination of Blommer's processing environment or finished product, and only one instance of raw material contamination, outside of cocoa beans, prior to this occurrence.

These facts alone point strongly toward summary judgment. Because the samples taken from the Blommer chocolate, Bongards' February 2, 1982, shipment of whey and Bongards' whey-processing facility, were all subjected to serological testing, the evidence becomes compelling. These samples all revealed that the *Salmonella* was of the cubana strain. Given this commonality, the existence of 1,500 strains of *Salmonella*, the well-accepted use of serological testing to track down the source of contamination, the likelihood that Bongards' whey contaminated Blommer's food products is extremely strong.

After extensive discovery Swank has still failed to do more than make minor dents in the armor of Blommer's case. It has not explained why only the products using Bongards' whey became contaminated. It does not dispute that *Salmonella* cubana was found in the February 2, 1982, shipment and in Bongards' processing facility. It has not rebutted the presumption of the culpability that springs quite naturally from finding the same strain of *Salmonella* in the Bongards' whey, in only those products using Bongards' whey, and in the Bongards processing facility.

Swank makes a valiant effort to avoid summary judgment, based primarily on the fact that the tests performed on the whey before its use by Blommer did not reveal salmonella contamination. However, even if Blommer's failure to discover before use that the whey was contaminated reveals the inadequacy of test procedures, this would not help Swank. First, Swank does not contest the accuracy of the later positive findings of *Salmonella* cubana contamination in the Bongards' whey, in only the products using Bongards' whey, and in Blommer's plant. Nor has Swank undercut the validity of the USDA inspections that found contamination of Bongards' processing facilities and of Bongards' own records that showed occasional product contamination.

Second, Swank's suggestion that the inadequacy of test procedures may mean that *Salmonella* contamination of Blommer's raw materials or its processing environment was never uncovered and that this contamination caused the contamination of the chocolate products, is the sort of speculation that cannot defeat a well-supported motion for summary judgment. Swank has yet to advance any hard evidence that Bongards' whey was not the source of the contamination. Swank has advanced nothing to suggest that the inconsistent results of the early Blommer tests stem from poor testing methods rather than from the nature of the contamination or some other factor.

While summary judgment is to be granted with caution, it is appropriate here. Swank, of course, does not bear the burden of proving that Bongards' whey was not the source of the contamination. Faced with the well-supported theory that Bongards was the source of the contamination, however, Swank has failed to even hint at the outlines of an exculpatory theory for which there is some evidentiary support. Although the conclusion that the *Salmonella* originated at Bongards rests upon inferences from undisputed facts it is no less compelling.

2. Bongards' Motion for Summary Judgment on all Counts of Blommer's Complaint

Bongards mounts a two-pronged attack on Blommer's complaint. It argues that the breach of warranty claim, Counts I–III, fail for lack of privity and the tort claims, Counts IV–VII, fail because Blommer's losses were only economic and not recoverable in tort. Because Blommer has not moved for summary judgment against Bongards, the issue in effect is whether under the existing record it is impossible as a matter of law for Blommer to recover either tort or contract damages against Bongards.

. . .

In fact, in *Vaughn v. General Motors Corp.*, 102 Ill.2d 431 (1984), the Illinois Supreme Court embraced an expansive view of what damages are compensable in a strict liability action. . . .

Count VI, however, must be dismissed for a different reason. That count alleges that Bongards was negligent per se, having violated state pure food laws. As this court noted in its earlier opinion, violation of a statute designed for the protection of human life or property does not constitute negligence per se but is only prima facie evidence of negligence, which may be rebutted by proof that the party acted reasonably under the circumstances, despite the violation.

Bongards argues that it should be granted summary judgment on Count I because it made no express warranties to Blommer as to the condition of the whey. It has introduced evidence that the so-called *Salmonella* statements that Blommer has used as a basis for the express warranty claim were in fact prepared after the contamination was discovered. Bongards, however, did expressly warrant, in invoices issued to Pacemaker, that its whey was "extra grade." Because it appears that "extra grade" whey was understood to be free of salmonella, it cannot be concluded that Bongards made no express warranties as to the condition of the whey.

The breach of warranty action is the appropriate way for Blommer to recover purely economic losses. Bongards argues that the breach of warranty claims fail because there was no privity of contract between Blommer and Bongards. Generally, a plaintiff must have been in privity of contract with the defendant in order to bring a breach of warranty action. Historically, as the economic relationships became more complex the privity requirement increasingly permitted manufacturers to escape liability to individuals harmed by their defective products. The tort theory of strict liability developed in part in response to limitations imposed by the privity requirement in breach of warranty actions.

Even before the Illinois Supreme Court embraced the doctrine of strict tort liability in *Suvada v. White Motor Co.*, 32 Ill.2d 612, (1965), Illinois courts had done away with the privity requirement in breach of warranty actions involving victuals.

Bongards argues that the privity requirement is waived with respect only to ultimate consumers of the product. This approach, however, does not appear to be followed in Illinois. In *Southland*, for example, the vegetable fat supplied by defendant injured plaintiff's chickens and not consumers of the chickens, and the court nevertheless permitted recovery. Here, the allegedly contaminated whey made chocolate destined for human consumption unfit.

Consequently, Blommer is not barred as a matter of law from recovering its damages from Bongards in tort, or its purely economic losses under a breach of warranty theory. At this point, the court need not consider objections to specific warranty and tort claims, especially because Bongards has advanced none.

. . . .

Conclusion

Blommer's motion for partial summary judgment against Swank is granted. Bongards' motion for summary judgment is granted as to Count VI of the complaint and denied otherwise. Bongards' motion for judgment on all counts of Swank's cross-claim is denied. Pacemaker's motion to dismiss Swank's third party complaint is denied.

* * * * *

15.3   PREEMPTION OF TORT CLAIMS

15.3.1   Generally

Most of the jurisprudence on FD&C Act preemption of tort claims applies to medical devices and drugs. These products face more extensive federal regulatory scheme than foods, including strict review of safety and efficacy, so different preemption questions arise.

The FD&C Act is silent on preemption of tort claims for foods. As a general matter, the FD&C Act will not preempt tort claims. There is a presumption against preemption. "Historically, common law liability has formed the bedrock of state regulation, and common law tort claims have been described as 'a critical component of the States' traditional ability to protect the health and safety of their citizens.'"[33]

Nonetheless, compliance with FDA regulations may be offered as evidence for the fulfillment of the tort standard for reasonable care. In addition the majority of courts have found that state tort law is preempted when the tort issue coincides with a matter approved by FDA, such as a product approval or label approval. For instance, once FDA has approved the safety of a food additive, this would generally preclude a tort claim that the additive was unsafe under the conditions of use approved by FDA, unless one could show that the FDA was arbitrary or capricious in its approval. This type of issue, however, does not often arise regarding tort claim.

15.3.2 State Law

Some state statutes provide a product liability rebuttable presumption of nonliability for compliance with relevant federal or state regulatory standards. The Michigan statute provided as an example below also grants a product liability shield to drug manufacturers and sellers for drugs approved by the FDA.

* * * * *

Revised Judicature Act of 1961 (Excerpt), Act 236 of 1961

MCL § 600.2946 Product liability action; admissible evidence

. . . .

(4) In a product liability action brought against a manufacturer or seller for harm allegedly caused by a product, there is a rebuttable presumption that the manufacturer or seller is not liable if, at the time the specific unit of the product was sold or delivered to the initial purchaser or user, the aspect of the product that allegedly caused the harm was in compliance with standards relevant to the event causing the death or injury set forth in a federal or state statute or was approved by, or was in compliance with regulations or standards relevant to the event causing the death or injury promulgated by, a federal or state agency responsible for reviewing the safety of the product. Noncompliance with a standard relevant to the event causing the death or injury set forth in a federal or state statute or lack of approval by, or noncompliance with regulations or standards relevant to the event causing the death or injury promulgated by, a federal or state agency does not raise a presumption of negligence on the part

---

[33] Desiano v. Warner-Lambert & Co., 467 F.3d 85 (2d. Cir. 2006) (citing Ciprollone v. Liggett Group, Inc., 505 U.S. 504 (1992).

of a manufacturer or seller. Evidence of compliance or noncompliance with a regulation or standard not relevant to the event causing the death or injury is not admissible.

(5) In a product liability action against a manufacturer or seller, a product that is a drug is not defective or unreasonably dangerous, and the manufacturer or seller is not liable, if the drug was approved for safety and efficacy by the United States Food and Drug Administration, and the drug and its labeling were in compliance with the United States Food and Drug Administration's approval at the time the drug left the control of the manufacturer or seller. However, this subsection does not apply to a drug that is sold in the United States after the effective date of an order of the United States Food and Drug Administration to remove the drug from the market or to withdraw its approval. This subsection does not apply if the defendant at any time before the event that allegedly caused the injury does any of the following:

(a) Intentionally withholds from or misrepresents to the United States Food and Drug Administration information concerning the drug that is required to be submitted under the Federal Food, Drug, and Cosmetic Act, ... and the drug would not have been approved, or the United States Food and Drug Administration would have withdrawn approval for the drug if the information were accurately submitted.

(b) Makes an illegal payment to an official or employee of the United States Food and Drug Administration for the purpose of securing or maintaining approval of the drug.

* * * * *

15.4 THE FALSE CLAIMS ACT

In 1863, President Lincoln signed the False Claims Act (FCA)[34] into law to stop war profiteering by military contractors. The FCA prohibits the knowing presentation to the United States of false or fraudulent claims for payment. The FCA was amended in 1986 to strengthen its qui tam provisions and enhance incentives to expose and rectify fraud against the government.

Qui tam is sometimes referred to a privatized attorney general action. The qui tam provisions permit private persons to sue on behalf of the United States to recover improper payments.[35] The action is called a *qui tam* action, from the Latin phrase *qui tam pro domino rege quam pro se ipso in hac parte sequitur*, which means, who as well for the king as for himself sues in this matter. A FCA qui tam plaintiff is called a *relator*. The qui tam case caption typically names the government and the relator as plaintiffs, for example, *United States of America ex rel. Neal D. Fortin, plaintiffs v. ABC Pharmaceuticals, Inc., defendant.*

[34] 31 U.S.C. § 3729 *et seq.*
[35] 31 U.S.C. § 3730(b).

In recent years, a number of prominent FCA cases brought by qui tam relators have recovered several billion dollars from defendants engaged in health care and defense industry fraud. The biggest impact has been in the health care industry, regarding pricing and marketing of pharmaceuticals.

The FCA creates liability for a person who knowingly presents or causes to be presented to an officer or employee of the United States government or a member of the Armed Forces of the United States a false or fraudulent claim for payment or approval.[36] The FCA also create liability for a person who knowingly makes, uses, or causes to be made or used, a false record or statement to get a false or fraudulent claim paid or approved by the government,[37] or who conspires to defraud the government by getting a false or fraudulent claim allowed or paid.[38]

15.5 NO PRIVATE CAUSE OF ACTION UNDER THE FD&C ACT

In addition to government action, numerous federal statutes provide for citizen suits as an alternative means enforce the law. In particular, virtually all federal environmental laws contain citizen suit provisions. The Clean Water Act,[39] Federal Water Pollution Control Act,[40] and Endangered Species Act,[41] all contain private cause of actions. The Federal Insecticide, Fungicide, and Rodenticide Act (FIFRA) is the sole exception.

Citizen suits take two forms. The citizen may sue an alleged violator of the law. Sometimes this is referred to as a private attorney general action. The second form of citizen action is a power to sue relevant government officials for failure to carry out nondiscretionary obligations. Such action-forcing litigation played a significant role with the Environmental Protection Agency (EPA) enforcement.

As the following case indicates, the FC&C Act does not create a private cause of action.

* * * * *

Florida ex rel. Broward Co. v. Eli Lilly & Co.

329 F. Supp. 364 (1971)

Atkins, District Judge.

[T]he State of Florida ... brought suit on its own behalf and on behalf of class of consumers and purchasers against the defendants to recover damages allegedly sustained in connection with the purchase, administration,

---

[36] 31 U.S.C. § 3729(a)(1).
[37] 31 U.S.C. § 3729(a)(2).
[38] 31 U.S.C. § 3729(a)(3).
[39] 42 U.S.C. § 7604 (1994).
[40] 33 U.S.C. § 1365 (1994).
[41] 16 U.S.C. § 1540(g) (1994).

and use of certain fixed-ratio combination drugs claimed to have been manufactured and sold by the defendants. In essence, the Florida complaint charged that the defendants fraudulently induced the plaintiff to purchase drugs by falsely representing their effectiveness and side effects and by failing to provide adequate directions for and warnings against their use. Such conduct was claimed to be actionable under provisions of the Federal Food, Drug, and Cosmetic Act, 21 U.S.C. § 301 et seq. The complaint also charged the defendants with common law fraud, negligence, and breach of warranty. . . .

The defendants moved to dismiss the First Amended Complaint. Specifically, the defendants argued (1) that the claims of the plaintiff could not be brought under the Federal Food, Drug, and Cosmetic Act and were not, therefore, within the Court's federal question jurisdiction under sections 1331 and 1337; and (2) that the State was not a "citizen" under section 1332 and could not therefore invoke this Court's diversity jurisdiction. . . .

The Federal Food, Drug, and Cosmetic Act does not create a private right of action and the claims pleaded in the First Amended Complaint do not, therefore, arise under federal law. Section 307 of the Act, 21 U.S.C. § 337, provides that "all" proceedings for the enforcement or to restrain violations of the Act shall be brought by the United States. Section 302(a), 21 U.S.C. § 332(a), limits the jurisdiction of district courts under the Act to injunctive proceedings involving purely prospective relief. The legislative history of the Act indicates that an express provision for a private right of action for damages was included in an early version of the bill but was omitted from all later versions after being attacked on the ground that it would create an unnecessary federal action duplicative of state remedies. Thus the terms and legislative history of the statute compel the conclusion that Congress did not intend to allow private rights of action for damages under the statute.

This conclusion is reinforced by the decisions of the only two other courts that have squarely faced this issue, and by the several other federal decisions which, in viewing the relationship between the Federal Food, Drug, and Cosmetic Act and applicable state remedies, have clearly indicated that violations of the Act do not constitute an independent basis for federal question jurisdiction.

Plaintiff's reliance upon cases arising under other federal regulatory statutes is misplaced. First, the federal statutes involved in those cases had neither provisions requiring all actions to be brought by the United States nor ones restricting federal district court jurisdiction to injunctive actions. Second, those decisions did not deal with legislative history like that of the Food, Drug and Cosmetic Act, showing an explicit rejection by Congress of a provision for private actions. Finally, such decisions typically involve claims for which no corresponding civil remedies are available in state courts. Since there is no private right of action under the Food, Drug, and Cosmetic Act, Florida's First Amended Complaint furnished no basis for the exercise of this Court's federal question jurisdiction under either section 1331 or 1337. . . .

* * * * *

If there were any doubts about the lack of a private cause of action in the FD&C Act, they were dispelled by a series of subsequent cases.

* * * * *

Merrell Dow Pharmaceuticals, Inc. v. Thompson

478 U.S. 804 (1986)

Justice STEVENS delivered the opinion of the Court.

The question presented is whether the incorporation of a federal standard in a state-law private action, when Congress has intended that there not be a federal private action for violations of that federal standard, makes the action one "arising under the Constitution, laws, or treaties of the United States," 28 U.S.C. § 1331. . . .

This case does not pose a federal question of the first kind; respondents do not allege that federal law creates any of the causes of action that they have asserted. This case thus poses what Justice Frankfurter called the "litigation-provoking problem"—the presence of a federal issue in a state-created cause of action. . . .

In this case, both parties agree with the Court of Appeals' conclusion that there is no federal cause of action for FDCA violations. For purposes of our decision, we assume that this is a correct interpretation of the FDCA. . . .

The significance of the necessary assumption that there is no federal private cause of action thus cannot be overstated. For the ultimate import of such a conclusion, as we have repeatedly emphasized, is that it would flout congressional intent to provide a private federal remedy for the violation of the federal statute. . . . .

* * * * *

NOTES

15.2. Other statutes without a private cause of action. In *Pacific Trading Co. v. Wilson & Co., Inc.*, 547 F.2d 367 (7th Cir. 1976), the court held that no private cause of action may be implied under the Packers and Stockyards Act, the United States Warehouse Act, the FD&C Act, or the Federal Meat Inspection Act. Among other cases, *Shoultz v. Monfort of Colorado*, Inc., 754 F.2d 318 (10th Cir. 1985) also rejected a private cause of action under the Federal Meat Inspection Act.

PART VI

GENERAL CHAPTERS

CHAPTER 16

International Food Law

> Through commerce, through globalization, through the spread of democratic
> institutions, through immigration to America, it's becoming more and more one
> world of many different kinds of people. And how they're going to live together
> across the world will be the challenge. . . .
> —U.S. Supreme Court Justice Breyer on ABC's "This Week," July 6, 2003

16.1 INTRODUCTION

With increasing international trade in food, it is essential to have at least a
general understanding of international food regulation. This chapter covers
the efforts to coordinate and harmonize regulatory efforts, the international
food standard setting bodies, and other issues in international food trade.

International initiatives to coordinate international food regulation and
facilitate trade can be divided into three categories: cooperation, mutual recognition, and harmonization. Informal cooperation has existed for many years
in a variety of forms. For example, various organizations from the Association
of Food and Drug Officials (AFDO) to the World Health Organization (WHO)
provide opportunities for government officials to exchange information. FDA
and USDA officials periodically meet with counterparts in other countries and
regions.

More formal cooperative arrangements are typically put into memoranda
of understanding (MOUs). These are much like MOUs between U.S. agencies,
but they must be approved by the U.S. Secretary of State. A listing of FDA
international agreements is found at http://www.fda.gov/oia/.

Mutual recognition is perhaps the most desirable from a regulated perspective, because this eliminates duplicative approval requirements. As a precondition of import, the law requires USDA equivalency recognition for foreign
meat inspection programs.[1] However, FDA lacks the statutory authority to

---

[1] *See* Chapter 11, Importation and Exportation.

Food Regulation: Law, Science, Policy, and Practice, by Neal D. Fortin
Copyright © 2009 Published by John Wiley & Sons, Inc.

require equivalency as a precondition for import of FDA-regulated products into the United States.

The Food and Drug Administration Modernization Act of 1997[2] required that FDA begin the process of acceptance of mutual recognition agreements to reduce the burden of regulation and to harmonize regulatory requirements. The practical significance of this requirement remains to be seen. In 1999 the United States and the European Community signed the "Agreement between the United States of America and the European Community on Sanitary Measures to Protect Public and Animal Health in Trade in Live Animals and Animal Products." This agreement covers a wide range of foods (all of animal origin), such as milk and dairy products, seafood, honey, wild game, snails, and frog legs.

Harmonization of food regulatory standards has played the most prominent role in recent efforts to facilitate trade. International standards for foods have been important since the 1960s, but newer trade agreements have enhanced the significance of these standards.

16.2   INTERNATIONAL FOOD STANDARDS

International standards play a role of growing importance, not only for export from the United States but within the United States. The trend toward globalization includes a growing international trade in food and agricultural products. With growing importance of international trade comes a greater need for a uniform reference point in standards. Differing food laws and standards by various countries impose barriers to trade and raise transaction costs. Increasing trade magnifies the burden of these trade barriers.

Initially, international food trade associations formed to deal with these barriers. The International Dairy Federation, for example, was founded in 1903 to work on harmonizing standards for milk and milk products. The lobbying activities of these trade associations increased the awareness of governments for the need of harmonized international standards. The growing importance of food trade to the economies of nations added to this awareness.

At the same time, there was a growing recognition of the need to ensure the safety of food. Minimum international standards were needed to protect consumers all over the world. The globalization of ingredient supply chains means that inferior standards in one country may end up adversely impacting the multi-ingredient foods sold in another country.

In the midst of these growing needs, Codex Alimentarius emerged and has taken a lead role in the international standards for trade in food.

16.2.1   Codex Alimentarius

The Commission was created in 1963 by the Food and Agriculture Organization (FAO) and the World Health Organization (WHO). The Commission was

---

[2] Public Law 105-115, section 410, codified at section 803 FD&C Act.

given two primary objectives: protecting the health of consumers and ensuring fair practices in food trade. The Commission accomplishes these objectives through the development and publication of international food standards and guidelines. These published standards are collectively referred to as Codex Alimentarius, or simply Codex. "Codex Alimentarius" is Latin for the "Food Book" or "Food Code."

Codex Membership In 2008 Codex Alimentarius Commission membership included 174 member countries and one member organization (the European Community). This represented 99 percent of the world's population. Membership is open to all member nations and associate members of FAO and WHO. Membership on the Commission confers no duties on a nation but allows a nation to fully contribute to the development of the standards. Participation is important to nations to ensure that the interests of their consumers, producers, processors, exporters, and governments are taken into account when Codex standards are developed. Observers, as a practice, are also allowed to voice their points of views, but only members may vote. National delegations may also include representatives from industry, consumers' groups, and academia.

Organizational Structure The full Codex Alimentarius Commission meets every two years, alternating between Geneva, Switzerland, and Rome, Italy. Between Commission sessions, the Codex Executive Committee meets to carry out the business of the Commission.

The Codex Alimentarius Commission decides on the adoption of new and amended standards, but most of the work of Codex is accomplished by the various Codex subsidiary bodies. Various member nations host the Codex committees. All Codex member nations are invited to participate, but committee attendance is not restricted to members. The subsidiary bodies are divided into three main types: general subject committees, commodity committees, and regional coordinating committees.

The work of the General Subject Committees relates to concepts and principles that apply to all foods or are all-embracing in scope. There are ten of these committees: General Principles, Food Additives, Contaminants in Foods, Food Hygiene, Food Labeling, Methods of Analysis and Sampling, Pesticide Residues, Residues of Veterinary Drugs in Foods, Food Import and Export Inspections and Certificate Systems, and Nutrition and Foods for Special Dietary Uses.

The Codex commodity committees work and have responsibility for matters within terms of reference revolving around a food commodity. Commodities committees are sometimes referred to as vertical committees because they have narrower areas of responsibility. Although the terms of reference define limits for each committee, these committees inevitably overlap with other committees.

Regional coordinating committees exist to discuss regional concerns and implementation. Coordinating committees play a role in encouraging countries to participate more actively and effectively in the work of Codex. The committees also help ensure that Codex is responsive to the regional concerns and issues.

Codex and National Standards Some countries adopt Codex standards legislatively. Others use the Codex as a model in developing their own standards. Predominantly, however, these have been developing countries, perhaps because these countries were initiating food laws and found Codex a good starting point.[3] Countries with established food laws have generally been unwilling to amend their laws to match Codex.

Initially, Codex served only as important model standards and guidelines. Member governments of Codex had no obligation to use Codex standards. However, after 1994, with adoption of the World Trade Organization (WTO) Sanitary and Phytosanitary (SPS) Agreement and other trade agreements, Codex gained heightened legal status. International trade agreements needed a reference point for food standards, and Codex provided the solution.

Consequently, in food trade disputes between WTO nations, Codex serves as the presumptive standard. Members are not required to adopt Codex standards, but members must be able to justify nonadoption according to defined criteria. In short, Codex standards are accepted as providing necessary protection. Higher standards may be adopted by a nation, but they must be justified on the basis of sound science and the use of appropriate risk assessment.

The Codex Alimentarius Commission recognized that the enhanced status of Codex must be met by greater use of science and risk assessment. However, there remain concerns whether Codex can be based solely on science and public health when countries may vote for other factors. For instance, a country that sells unpasteurized cheese may vote against any standard that would require pasteurization.[4] Additionally, the fact that Codex has two goals, ensuring fair international trade and protecting public health, raises the concern that trade may override health concerns.[5] Generally, however, Codex seems to move forward where there is agreement on the science and only falters where the issue is not a matter of science or there is uncertainty of risk.

NOTES AND QUESTIONS

16.1. How important do you believe the Codex Alimentarius is to firms that do not export from the United States?

16.2. How can the food industry, consumers, and related stakeholders be proactive in the Codex system?

---

[3] David Jukes, *The Codex Alimentarius Commission—Current Status*, FOOD SCIENCE AND TECHNOLOGY TODAY (Dec. 1998) *available at*: http://www.reading.ac.uk/foodlaw/codex-1.htm (last accessed Sept. 16, 2008).
[4] Lucinda Sikes, *FDA's Consideration of Codex Alimentarius Standards in Light of International Trade Agreements*, 53 FOOD AND DRUG LAW JOURNAL 327 (1998).
[5] *Id.*

16.3. Will the Codex Commission and committees be subject to political pressure?

16.4. Further information on Codex Alimentarius is available at http://www. codexalimentarius.net. See, in particular, *Understanding the Codex Alimentarius*, 3rd Ed. (2006), *available at*: ftp://ftp.fao.org/codex/ Publications/understanding/Understanding_EN.pdf (last accessed Nov. 26, 2007).

16.2.2   The WTO and International Trade Agreements

The Uruguay Round of Multilateral Trade Negotiations in Marrakech led to the establishment of the World Trade Organization (WTO) on January 1, 1995. WTO is the successor to the General Agreement on Tariffs and Trade (GATT). Although GATT was established on a provisional basis after World War II, GATT remained the only multilateral instrument governing international trade from 1948 until the establishment of the WTO in 1995.[6]

The WTO is the international body dealing with the rules of trade between nations. It is the forum for nations to negotiate trade agreements. At the heart of WTO are the WTO trade agreements, which are signed by most of the world's trading nations. These agreements are essentially contracts binding nations to ground rules for international trade. The third important side of WTO is it provides a procedure for dispute resolution.[7]

The WTO had 151 members in 2008. Decisions are made by the entire membership. The highest authority is the Ministerial Conference, composed of representatives of all WTO members. The day-to-day work falls to the General Council and a number of other subsidiary bodies.

As tariff barriers to trade were eliminated, added attention fell on the non-tariff barriers to trade. Among these were national standards for foods and agricultural products—ostensibly put in place to protect the health of consumers, animals, and plants—that could be disguised barriers to trade. To address these issues, a separate WTO agreement, the Agreement on the Application of Sanitary and Phytosanitary Measures (SPS Agreement) sets out the basic rules on food safety and animal and plant health standards

The SPS Agreement   The SPS Agreement allows countries to set their own standards, but regulations must be based on science, and they should applied only to the extent necessary to protect human, animal or plant life or health. Additionally, these regulations should not arbitrarily or unjustifiably discriminate between countries where identical or similar conditions prevail. Nations

---

[6] WTO, Understanding the WTO: The GATT Years: From Havana to Marrakesh, *available at:* http://www.wto.org/english/thewto_e/whatis_e/tif_e/fact4_e.htm (last accessed Mar. 29, 2008).
[7] WTO, Understanding the WTO: What Is the World Trade Organization? *available at*: http://www.wto.org/english/thewto_e/whatis_e/tif_e/fact1_e.htm (last accessed Mar. 29, 2008).

may determine what the appropriate level of protection is, but these standards must be based on scientific risk assessment.

SPS countries are encouraged to use international standards, guidelines, and recommendations. However, countries may enact higher standards if there is scientific justification based on appropriate risk assessment so long as the approach is consistent, not arbitrary. In dealing with scientific uncertainty, countries are allowed to apply precautionary measure on a temporary basis until scientific risk assessment is completed (article 5.7 of the SPS Agreement).

The SPS Agreement stipulates that national SPS measures should be based on international standards, guidelines, and recommendations (SPS article 3). Codex Alimentarius standards are considered to reflect international consensus regarding the scientific requirements for protecting human health. Thus, if a nation's food regulations are based on Codex standards, they are considered justified and consistent with the requirements of the SPS Agreement. Conversely, nations that do not consider Codex standards when framing national legislation and regulations run the risk of potential challenges and trade disputes.

The TBT Agreement   In addition to sanitary and phytosanitary regulations, countries create a variety of technical regulations related to packaging, marking, labeling requirements, and testing and certification procedures. If applied in an arbitrary manner, these technical matters could be used in a protectionist manner to create nontariff barriers to trade. The WTO Agreement on Technical Barriers to Trade (TBT Agreement) tries to ensure that regulations, standards, and testing and certification procedures do not create unnecessary obstacles.

The TBT Agreement recognizes the rights of nation to adopt such measures, to the extent they consider appropriate, but that is counterbalanced with disciplines. Governments are encouraged to apply international standards, such as those of the Codex. Additionally, whatever regulations a nations applies should not discriminate. The TBT Agreement also establishes a "code of good practice" for the preparation, adoption, and application of standards at the national and local levels.

DISCUSSION QUESTION

16.5. (How) Should factors other than assessment of risks and scientific evidence be considered when framing international regulations?

16.6. In addition to the FAO/WHO Codex Alimentarius Commission for food standards, the International Animal Health Organization (OIE for Office International des Epizooties) is the international standard setting body for animal health www.oie.int, and the FAO's Secretariat of the International Plant Protection Convention (IPPC) is the international standard setting body for plant health www.ippc.int.

16.7. Information on the U.S. Codex Office is available at: http://www. fsis.usda.gov/regulations & policies/Codex Alimentarius/index.asp (last accessed Mar. 30, 2008).

16.2.3 Jurisprudence

Understanding international law is no longer just a legal specialty . . . It is becoming a duty.

—Justice Sandra Day O'Conner[8]

Under article VI of the U.S. Constitution, the courts are required to accept treaties as part of the "supreme Law of the Land." Thus international treaties may trump U.S. laws. Only the Constitution cannot be overridden. In *Reid v. Covert*, 354 U.S. 1, 16 (1957), the Court held that "[N]o agreement with a foreign nation can confer power on the Congress, or any other branch of government, which is free from the restraints of the Constitution."

Moreover, international law has long held a fundamental place in American jurisprudence. An often-cited precedent for interpreting domestic statutes in accord with international law is found in *Murray v. Charming Betsy*, 6 U.S. 64, 118 (1804), where Justice Marshall and the Court decreed that "an act of congress ought never to be construed to violate the law of nations if any other possible construction remains." This principle of harmonization remains today.

NOTE

16.8. A recent Supreme Court decision, *Medellin v. Texas*, 552 U. S. ___(2008), in a majority opinion by Chief Justice Roberts, held that treaties are not "self-executing," that is, are not enforceable in US courts without additional congressional action beyond the ratification of the treaty itself, unless the text of the treaty clearly indicates the treaty is self-executing. Whether a treaty is self-executing will have to be decided treaty by treaty. The opinion is available at: http://www.supremecourtus.gov/opinions/07pdf/06-984.pdf (last accessed Mar. 30, 2008).

16.3 FOREIGN REGULATORY SYSTEMS

Most industrialized nations have regulatory systems comparable to the United States, but many are organized differently. Some countries have borrowed

---

[8] Dan Gilgoff, *Law Schools Go International*, U.S. News & World Reports (Apr. 12, 2004) (quote from a 2002 meeting).

from the FDA model, but others have developed their own unique system of organization. The most common variation from the U.S. approach is that the broad scope of FDA's authority (over foods, drugs, medical devices, and electronics) is distributed to multiple agencies or ministries. Another common variation is a split between science and policy decisions and enforcement. For instance, food additives may be approved by one agency, while another agency or ministry oversees and enforces the appropriate use of food additives.

16.3.1   Background on the European Union

The European Union (EU) is a grouping of countries bound together by treaties. This integration began with the Treaty of Rome in 1957, which established the structure and operation of the European Community. Member countries are referred to as member states. Member states have agreed to delegate some of their sovereignty to common institutions so that decisions on specific matters of joint interest can be made at a European level. The five main EU institutions are as follows:

European Parliament (elected by the peoples of the member states)

Council of the European Union (representing the governments of the member states)

European Commission (executive body)

Court of Justice

Court of Auditors (management of the EU budget)

There are currently 25 member states. The most recent enlargement was in 2004, when 10 countries joined: Cyprus, the Czech Republic, Estonia, Hungary, Latvia, Lithuania, Malta, Poland, Slovakia, and Slovenia.

NOTES

16.9. EU Web resources: The EU at a glance, available at: http://europa.eu.int/abc/index_en.htm; EU History—A chronology from 1946 to 2004: http://europa.eu.int/abc/history/index_en.htm; Seven key days in the making of Europe, available at: http://europa.eu.int/comm/publications/booklets/move/16/txt_en.htm; Key facts and figures about the European Union (2004 edition), available at: http://europa.eu.int/comm/publications/booklets/eu_glance/44/index_en.htm

16.3.2   EU Food Issues with the United States

The European Union (EU) and the United States share the largest two-way trade and investment relationship in the world. Thus harmonization of standards plays a significant role in trade relations between the United States and the European Union. Because the stakes are so high, differences in standards take on added importance. Examples of some food issues include the following:

- Implementation of EU import quotas for U.S. rice
- Restrictions affecting U.S. wine exports to the EU
- Approval process and labeling requirements for agricultural biotechnology patents
- Ban on growth promoting hormones in meat production
- Packaging labeling requirements
- Poultry regulations (French ban of U.S. poultry)

16.3.3 The GE Food Fight

In May 2003 the United States filed a complaint with the World Trade Organization against the 1998 EU moratorium on the farming and import of new genetically engineered (GE) crops. U.S. Trade Representative Robert Zoellick characterized the European position on GE foods as "Luddite."

GE advocates in the United States often characterize this opposition as an unfair trade practice by the EU to discriminate against U.S. products. While exploitation of cultural fears by EU trade interests may play a role, these fears appear more deeply rooted. Europe has experienced a number of food safety calamities in recent years. For instance, mad cow disease (BSE) has killed more than 135 Europeans since 1995. Many officials initially deemed BSE to be a minimal risk to people. This crisis not only fanned fears of hidden dangers lurking in the food supply, but additionally decreased confidence in food industry and government experts. Of course, this is just one of many factors that resulted in the EU consumers generally being less accepting of GE food than U.S. consumers.

The European Union requires labeling of nearly all GE food.[9] The label must indicate "[t]his product contains genetically modified organisms" or "produced from genetically modified [name of organism]."

This creates an important trade issue for the United States because a large amount of U.S. crops are GE. U.S. firms have opposed such regulations as increasing production costs and disrupting trade. Particularly with grain and soybeans, the crops are stored, handled, and processed as an interchangeable commodity without separation by source—separation would increase documentation and handling costs. In addition, U.S. producers argue that labels identifying foods as derived from biotechnology are construed by consumers as a warning label—implying falsely that the food is less safe.

---

[9] Council Regulation 1830/2003, 2003 O.J. (L 268) 24 (EC).

16.3.4   The Precautionary Principle

Many people have argued that the law should establish a duty to prevent not only known environmental harms and health risks but also to prevent conduct that *may* be harmful although conclusive scientific evidence is not available indicating actual harm. This concept is sometimes referred to as the "Precautionary Principle."

Unfortunately, the term Precautionary Principle has created confusion. First, there is no standard meaning, and different versions vary wildly. In a strict usage, any uncertainty on safety prohibits a potentially risky activity until it is proven safe. In the milder meanings, Precautionary Principle becomes nothing more than commonsense precaution, "better safe than sorry," which everyone agrees on. For clarity, the strict definition for the Precautionary Principle is used here, and milder forms of precaution are simply called precautionary.

The notion that precaution should prevail when addressing new food safety issues and environmental issues is not new. While the term Precautionary Principle appears to have developed only in the last two decades, application of precaution is not uncommon in international or U.S. law. The Delaney Clause of the FD&C Act, for example, provides an example of the most protective ingredient safety provision in any food law in the world. The FD&C Act's approach to approval of new drugs and new food additives is precautionary. Rather than placing new pharmaceuticals or new food additives with uncertain health risks on the market, the FD&C Act requires that these products be subjected to numerous tests and evaluations to ensure a certain level of safety.

New drug approvals highlight a dilemma with application of the Precautionary Principle. Delay in the new drug approval process can contribute to the death of people who are desperately in need of life-saving drugs. Precaution has a cost as well a benefit. How these costs and benefits are balanced can be a complex matter. Determining what is "safe," deciding what level of precaution is required, and answering other thorny questions about what level of risk is appropriate are matters that can tie regulatory decisions in knots.

Complex concerns highlight the Precautionary Principle's major shortcoming. The Precautionary Principle offers false guidance for situations without absolute safety because it falsely assumes that inaction has zero risk. In complex concerns, new development must measured by risk and benefit versus alternate risks and benefits.

An extreme example of the alternate risks created by Precautionary Principle inaction occurred in 2002. The US donated thousands of tons of corn to Zambia, but the Zambian government refused the corn because some likely was GE corn. According to the Food and Agriculture Organization (FAO), the refusal put up to 2.9 million people at risk of starvation. The World Health Organization (WHO) conservative scenario predicted at least 35,000 Zambians would die of starvation if more food was not be donated. The Zambian government was applying the Precautionary Principle, but was their decision truly precautionary?

DISCUSSION QUESTIONS

16.10. Delaney comparison. Considerable debate has occurred recently over the use of the Precautionary Principle by the European Union. Is the United States less precautionary than the European Union? How does the Delaney Clause compare to the EU use of the Precautionary Principle?

16.11. A hypothetical exercise. A new barley variety (SuperB) is developed using conventional breeding in the country of Xanadu. The laws of Xanadu provide no regulatory review over new varieties produced through conventional breeding techniques. SuperB contains protein sequences that trigger wheat allergies in some sensitive individuals. Xanadu prohibits GMOs, but because SuperB is "natural," it is assumed to be safe. SuperB is sold to the public without any special labeling or advisory. Which country has lower risk? Xanadu, which prohibits GMOs. Or Canada, which allows GMOs, but requires all novel crops to be tested for safety? Has Xanadu created a false sense of security?

16.3.5 EU Requirements on GE Food

EU Regulation 1829/2003 on genetically engineered food comprises the most recent EU rules concerning food and feed containing or from genetically modified organisms (GMOs). The Regulation stipulates that GE food/feed must not: have adverse effects on human health, animal health, or the environment; mislead the consumer; differ from the food/feed it is intended to replace to such an extent that its normal consumption would be nutritionally disadvantageous for the consumer/animals.

Applications for approval of a GE food or feed must include a monitoring plan, a labeling proposal, and a detection method for the new GE food or feed. The European Food Safety Authority (EFSA) is responsible for the scientific risk assessment covering both the environmental risk and human and animal health safety assessment. Its opinion is made available to the public, and the public has the opportunity to make comments.

Within three months of receiving the opinion of EFSA and based on that opinion, the EU Commission drafts a proposal for granting or refusing authorization. The proposal is reviewed by the Standing Committee on the Food chain and Animal Health. If the Committee gives a favorable opinion, the EU Commission adopts the Decision. If not, the draft Decision is submitted to the Council of Ministers for adoption or rejection by a qualified majority.

To date, products from seventeen GMOs have been approved for marketing in the European Union.

16.3.6   EU Labeling

Since 1997, EU law has made labeling of GE food mandatory for all products that consist of a GMO or contain a GMO and for products derived from a GMO if there is still DNA or protein from the genetic engineering present. The EU Regulation 1830/2003 on labeling and traceability provides for labeling all food and feed containing, consisting of, or produced from a GMO. The label must indicate that "[t]his product contains genetically modified organisms" or "produced from genetically modified [name of organism]."

The EU stated purpose is to inform consumers and farmers about the exact nature and characteristics of the food or feed so that they can make informed choices. The same rules apply to animal feed as human food. This is intended to provide livestock farmers with accurate information on the composition and properties of feed.

The Regulation recognizes that adventitious or unintended presence of GE material in products placed on the market in the European Union is largely unavoidable. Minute traces of GMOs in conventional food and feed could arise during cultivation, harvest, transport, and processing. Therefore, the presence of GE material in conventional food does not have to be labeled if it is below 0.9 percent and if it can be shown to be adventitious and technically unavoidable.

The Regulation does not require labeling of products such as meat, milk, or eggs obtained from animals fed with genetically modified feed or treated with genetically modified medicinal products.

16.3.7   The U.S. Response to the European Union

The United States had long warned that it would launch a World Trade Organization (WTO) challenge if the European Union did not lift its six-year-old de facto moratorium on approving GE corn and other crops already deemed safe by EU top scientists. However, the United States was reluctant to initiate a trade struggle because it might only further alienate wary EU consumers and environmental groups that already deemed these products as "Frankenstein foods."

Nevertheless, many are angry at the estimated loss of exports of corn to Europe worth around $300 million. Therefore, the United States finally did initiate a WTO challenge. Experts generally agreed that the United States would win their WTO lawsuit over GE foods, but the European Union broke its moratorium on approval of new GE crops.

Nonetheless, to a certain degree, the EU market remains closed to GE foods because of consumer rejection of them, rather than the EU moratorium on approvals. For this reason the United States has somewhat accepted the

European Union as being largely closed to the GE foods and focused more on the world markets. China and several other Asian countries appear to be softening their opposition to GE foods.

The US trade officials' major reasons why the EU approach should change are discussed below.

Precautionary Principle Not Defined   The United States has generally opposed use of the Precautionary Principle because it regards the application as unscientific and arbitrary. No quantitative or objective standard or definition for the Precautionary Principle exists. Lack of a definition lends itself to wide variations in interpretation and levels of precaution. Lack of definition also allows exploitation by those who would impose trade barriers disguised as precautions. In particular, the United States claims the European Union uses the Precautionary Principle to ban U.S. products but approves other genetically modified products for sale in its member states.

Public Opinion or Public Health?   EU Directive 178/2002 requires that precautionary restrictions should not be more trade restrictive than required to meet the desired level of health protection. The EU ban on U.S.-approved GE crops is largely based on public opinion and fear rather than protection of a concrete public health risk. U.S. trade representative Robert Zoellick pointed out that the EU GMO ban in 1993 was "unsupported even by the EU's own scientific studies."[10]

Incomplete Risk Assessment   Under the WTO Agreement on Sanitary and Phytosanitary Standards (SPS), when a member state adopts health-related measures based on incomplete available information, the state must undertake a risk assessment within a reasonable period of time. *Reasonable* usually means within fifteen months. While restrictions may be created to respond to a country's fears and public perceptions, generalized or vague fears are insufficient. Food safety measures must be based on scientific risk assessments.

Prudential or Precautionary?

> I want a one-armed scientist . . . who would not qualify his advice with "on the other hand."
> —Senator Edmund Muskie (paraphrasing President John F. Kennedy)

Proving 100 percent safety is impossible to achieve, even with conventional foods. Conventional foods can contain toxins and anti-nutrients, and some people suffer allergic reactions to them. Proving absolute safety of GE foods is unachievable, which is why comparable safety is the standard applied by

---

[10] Fred Pearce, NEW SCIENTIST ONLINE NEWS (Sept. 10, 2003).

most experts who examine the safety of GE foods. The prudential approach balances risks against benefits.

NOTES

16.12. The Organization for Economic Cooperation and Development (OECD) provides information on the food laws of the member countries of OECD: http://www.oecd.org.

16.13. The Foreign Agricultural Service (FAS) of the USDA provides information on the laws and regulations of many countries as well as other trade information at: http://www.fas.usda.gov/.

CHAPTER 17

Ethics

17.1 PROFESSIONALISM AND ETHICS

> If in your own judgment you cannot be an honest lawyer, resolve to be honest
> without being a lawyer.
>
> —Abraham Lincoln[1]

* * * * *

Abraham Lincoln's Notes for a Law Lecture[2]

I am not an accomplished lawyer. I find quite as much material for a lecture
in those points wherein I have failed, as in those wherein I have been mod-
erately successful. The leading rule for the lawyer, as for the man of every
other calling, is diligence. Leave nothing for to-morrow which can be done
to-day. Never let your correspondence fall behind. Whatever piece of business
you have in hand, before stopping, do all the labor pertaining to it which can
then be done. When you bring a common-law suit, if you have the facts for
doing so, write the declaration at once. If a law point be involved, examine
the books, and note the authority you rely on upon the declaration itself,
where you are sure to find it when wanted. The same of defenses and pleas.
In business not likely to be litigated,—ordinary collection cases, foreclosures,
partitions, and the like,—make all examinations of titles, and note them, and
even draft orders and decrees in advance. This course has a triple advantage;
it avoids omissions and neglect, saves your labor when once done, performs
the labor out of court when you have leisure, rather than in court when you
have not. Extemporaneous speaking should be practiced and cultivated. It is
the lawyer's avenue to the public. However able and faithful he may be in
other respects, people are slow to bring him business if he cannot make a

---

[1] *Lincoln's Notes for a Law Lecture*, THE COLLECTED WORKS OF ABRAHAM LINCOLN, edited by Roy
P. Basler.

[2] This document fragment was dated July 1, 1850 by Lincoln's White House secretaries, John
Nicolay and John Hay, who collected many of his manuscripts after his death.

---

speech. And yet there is not a more fatal error to young lawyers than relying too much on speech-making. If any one, upon his rare powers of speaking, shall claim an exemption from the drudgery of the law, his case is a failure in advance.

Discourage litigation. Persuade your neighbors to compromise whenever you can. Point out to them how the nominal winner is often a real loser—in fees, expenses, and waste of time. As a peacemaker the lawyer has a superior opportunity of being a good man. There will still be business enough.

Never stir up litigation. A worse man can scarcely be found than one who does this. Who can be more nearly a fiend than he who habitually overhauls the register of deeds in search of defects in titles, whereon to stir up strife, and put money in his pocket? A moral tone ought to be infused into the profession which should drive such men out of it.

The matter of fees is important, far beyond the mere question of bread and butter involved. Properly attended to, fuller justice is done to both lawyer and client. An exorbitant fee should never be claimed. As a general rule never take your whole fee in advance, nor any more than a small retainer. When fully paid beforehand, you are more than a common mortal if you can feel the same interest in the case, as if something was still in prospect for you, as well as for your client. And when you lack interest in the case the job will very likely lack skill and diligence in the performance. Settle the amount of fee and take a note in advance. Then you will feel that you are working for something, and you are sure to do your work faithfully and well. Never sell a fee note—at least not before the consideration service is performed. It leads to negligence and dishonesty—negligence by losing interest in the case, and dishonesty in refusing to refund when you have allowed the consideration to fail.

There is a vague popular belief that lawyers are necessarily dishonest. I say vague, because when we consider to what extent confidence and honors are reposed in and conferred upon lawyers by the people, it appears improbable that their impression of dishonesty is very distinct and vivid. Yet the impression is common, almost universal. Let no young man choosing the law for a calling for a moment yield to the popular belief—resolve to be honest at all events; and if in your own judgment you cannot be an honest lawyer, resolve to be honest without being a lawyer. Choose some other occupation, rather than one in the choosing of which you do, in advance, consent to be a knave.

* * * * *

Memorandum for the Heads of Executive Departments and Agencies

THE WHITE HOUSE, WASHINGTON, January 20, 2001

SUBJECT: Standards of Official Conduct

Everyone who enters into public service for the United States has a duty to the American people to maintain the highest standards of integrity in Government. I ask you to ensure that all personnel within your departments and agencies are familiar with, and faithfully observe, applicable ethics laws and regulations, including the following general principles from the Standards of Ethical Conduct for Employees of the Executive Branch:

Public service is a public trust, requiring employees to place loyalty to the Constitution, the laws, and ethical principles above private gain.

Employees shall not hold financial interests that conflict with the conscientious performance of duty.

Employees shall not engage in financial transactions using nonpublic Government information or allow the improper use of such information to further any private interest.

An employee shall not, except as permitted by applicable law or regulation, solicit or accept any gift or other item of monetary value from any person or entity seeking official action from, doing business with, or conducting activities regulated by the employee's agency, or whose interests may be substantially affected by the performance or nonperformance of the employee's duties.

Employees shall put forth honest effort in the performance of their duties.

Employees shall not knowingly make unauthorized commitments or promises of any kind purporting to bind the Government.

Employees shall not use public office for private gain.

Employees shall act impartially and not give preferential treatment to any private organization or individual.

Employees shall protect and conserve Federal property and shall not use it for other than authorized activities.

Employees shall not engage in outside employment or activities, including seeking or negotiating for employment, that conflict with official Government duties and responsibilities.

Employees shall disclose waste, fraud, abuse, and corruption to appropriate authorities.

Employees shall satisfy in good faith their obligations as citizens, including all just financial obligations, especially those—such as Federal, State, or local taxes—that are imposed by law.

Employees shall adhere to all laws and regulations that provide equal opportunity for all Americans regardless of race, color, religion, sex, national origin, age, or handicap.

Employees shall endeavor to avoid any actions creating the appearance that they are violating applicable law or the ethical standards in applicable regulations.

Executive branch employees should also be fully aware that their post-employment activities with respect to lobbying and other forms of representation will be bound by the restrictions of 18 U.S.C. § 207.

Please thank the personnel of your departments and agencies for their commitment to maintain the highest standards of integrity in Government as we serve the American people.

* * * * *

17.2 ETHICAL PRACTICE POINTERS

Always do right. This will gratify some people and astonish the rest.
—Mark Twain, US humorist, novelist, short story author, and wit (1835–1910)

- Update your skills and knowledge continuously.
- Uphold and better the laws and regulations.
- Evaluate all possible options.
- Listen closely to all positions.
- Research all issues thoroughly—always Shepardize.
- Treat people with respect and civility.
- Don't forget the big picture.

17.3 ATTORNEY RULES OF ETHICS

* * * * *

Model Rules of Professional Conduct[3]

Rule 1.7 Conflict of Interest: General Rule

(a) Except as provided in paragraph (b), a lawyer shall not represent a client if the representation involves a concurrent conflict of interest. A concurrent conflict of interest exists if:

    (1) the representation of one client will be directly adverse to another client; or

    (2) there is a significant risk that the representation of one or more clients will be materially limited by the lawyer's responsibilities to another client, a former client or a third person or by a personal interest of the lawyer.

(b) Notwithstanding the existence of a concurrent conflict of interest under paragraph (a), a lawyer may represent a client if:

    (1) the lawyer reasonably believes that the lawyer will be able to provide competent and diligent representation to each affected client;

---

[3] American Bar Association, MODEL RULES OF PROFESSIONAL CONDUCT, *available at*: http://www. abanet.org/cpr/mrpc/mrpc_toc.html (last visited Nov. 2, 2005).

    (2) the representation is not prohibited by law;

    (3) the representation does not involve the assertion of a claim by one client against another client represented by the lawyer in the same litigation or other proceeding before a tribunal; and

    (4) each affected client gives informed consent, confirmed in writing.

* * * * *

Rule 1.9 Conflict of Interest: Former Client

Rule 1.9 Duties to Former Clients

(a) A lawyer who has formerly represented a client in a matter shall not thereafter represent another person in the same or a substantially related matter in which that person's interests are materially adverse to the interests of the former client unless the former client gives informed consent, confirmed in writing.

(b) A lawyer shall not knowingly represent a person in the same or a substantially related matter in which a firm with which the lawyer formerly was associated had previously represented a client

    (1) whose interests are materially adverse to that person; and

    (2) about whom the lawyer had acquired information protected by Rules 1.6 and 1.9(c) that is material to the matter; unless the former client gives informed consent, confirmed in writing.

(c) A lawyer who has formerly represented a client in a matter or whose present or former firm has formerly represented a client in a matter shall not thereafter:

    (1) use information relating to the representation to the disadvantage of the former client except as these Rules would permit or require with respect to a client, or when the information has become generally known; or

    (2) reveal information relating to the representation except as these Rules would permit or require with respect to a client.

* * * * *

Note, in particular, that a lawyer should not represent a person who interests are materially adverse to a former client on a substantially related matter represented for former client. Loyalty to client survives termination of the attorney-client relationship. However, one may ethically oppose a former client on matters unrelated to former representation. The ethical concern is more about confidentiality than loyalty. In addition, a lawyer should not knowingly represent a person on same matter that the firm previously represented a client, if confidential information was acquired that is material to the matter at hand. MR 1.9(b).

* * * * *

Rule 1.10 Imputed Disqualification: General Rule

Rule 1.10 Imputation of Conflicts of Interest: General Rule

(a) While lawyers are associated in a firm, none of them shall knowingly represent a client when any one of them practicing alone would be prohibited from doing so by Rules 1.7 or 1.9, unless the prohibition is based on a personal interest of the prohibited lawyer and does not present a significant risk of materially limiting the representation of the client by the remaining lawyers in the firm.

(b) When a lawyer has terminated an association with a firm, the firm is not prohibited from thereafter representing a person with interests materially adverse to those of a client represented by the formerly associated lawyer and not currently represented by the firm, unless:

   (1) the matter is the same or substantially related to that in which the formerly associated lawyer represented the client; and

   (2) any lawyer remaining in the firm has information protected by Rules 1.6 and 1.9(c) that is material to the matter.

(c) A disqualification prescribed by this rule may be waived by the affected client under the conditions stated in Rule 1.7.

(d) The disqualification of lawyers associated in a firm with former or current government lawyers is governed by Rule 1.11.

* * * * *

Rule 1.11 Government Lawyers and the Revolving Door

Rule 1.11 Special Conflicts of Interest for Former and Current Government Officers and Employees

(a) Except as law may otherwise expressly permit, a lawyer who has formerly served as a public officer or employee of the government:

   (1) is subject to Rule 1.9(c); and

   (2) shall not otherwise represent a client in connection with a matter in which the lawyer participated personally and substantially as a public officer or employee, unless the appropriate government agency gives its informed consent, confirmed in writing, to the representation.

(b) When a lawyer is disqualified from representation under paragraph (a), no lawyer in a firm with which that lawyer is associated may knowingly undertake or continue representation in such a matter unless:

(1) the disqualified lawyer is timely screened from any participation in the matter and is apportioned no part of the fee therefrom; and

(2) written notice is promptly given to the appropriate government agency to enable it to ascertain compliance with the provisions of this rule.

(c) Except as law may otherwise expressly permit, a lawyer having information that the lawyer knows is confidential government information about a person acquired when the lawyer was a public officer or employee, may not represent a private client whose interests are adverse to that person in a matter in which the information could be used to the material disadvantage of that person. As used in this Rule, the term "confidential government information" means information that has been obtained under governmental authority and which, at the time this Rule is applied, the government is prohibited by law from disclosing to the public or has a legal privilege not to disclose and which is not otherwise available to the public. A firm with which that lawyer is associated may undertake or continue representation in the matter only if the disqualified lawyer is timely screened from any participation in the matter and is apportioned no part of the fee therefrom.

(d) Except as law may otherwise expressly permit, a lawyer currently serving as a public officer or employee:

(1) is subject to Rules 1.7 and 1.9; and

(2) shall not:

(i) participate in a matter in which the lawyer participated personally and substantially while in private practice or nongovernmental employment, unless the appropriate government agency gives its informed consent, confirmed in writing; or

(ii) negotiate for private employment with any person who is involved as a party or as lawyer for a party in a matter in which the lawyer is participating personally and substantially, except that a lawyer serving as a law clerk to a judge, other adjudicative officer or arbitrator may negotiate for private employment as permitted by Rule 1.12(b) and subject to the conditions stated in Rule 1.12(b).

(e) As used in this Rule, the term "matter" includes:

(1) any judicial or other proceeding, application, request for a ruling or other determination, contract, claim, controversy, investigation, charge, accusation, arrest or other particular matter involving a specific party or parties, and

(2) any other matter covered by the conflict of interest rules of the appropriate government agency.

* * * * *

DISCUSSION QUESTION

17.1. Why don't we provide the same protection going from the private section when moving to the government?

17.4   CRIMINAL STATUTES RELATED TO ETHICS

Title 18 of the United States Code contains the criminal conflict of interest statutes applicable to employees in the executive branch of the government. Included in title 18 is the prohibition against solicitation or receipt of bribes, the prohibition against acting as an agent or attorney before the government, postemployment restrictions, prohibition against participating in matters affecting a personal financial interest, and the prohibition against receiving supplementation of salary as compensation for government service. Relevant sections are available at: http://www.usoge.gov/pages/laws_regs_fedreg_stats/statutes.html.

* * * * *

Digest of Criminal Statutes Related to Ethics FDA, Ethics Program (May 2002)[4]

Conflict of Interest Provision 18 U.S.C. § 208; 5 C.F.R. § 2635.401

Employees cannot participate personally and substantially in a government matter the resolution of which would affect their own or imputed financial interests.

If your child owns IBM stock worth $16,000, you cannot be involved in computer acquisitions for the office.

If you are on a leave of absence from a university, your re-employment rights are a financial interest. You must recuse from any grant or contract awards to your university. However, you may draft regulations affecting higher education in general, provided there is no special or distinct effect on your university. While seeking or negotiating for a nongovernment job, you cannot work on a matter at the Department that affects a prospective employer.

* * * * *

Prohibition on Salary Supplementation 18 U.S.C. § 209; 5 C.F.R. § 2635.503

Employees may not accept any contribution or supplementation of their government salary as compensation for services they perform officially.

While you may continue to participate in your former employer's bona fide pension and benefit plans, and receive appropriate severance payments paid

---

[4] *Available at*: http://www.fda.gov/opacom/ethics/criminaldig.html (last accessed Mar. 30, 2008).

by the company to all departing employees, you cannot receive a bonus paid to cushion the blow of a lower government salary.

* * * * *

Representation Restrictions 18 U.S.C. § 203, 205

Representation, whether compensated or not, of private individuals or organizations in matters before the federal government is prohibited.

You cannot appear for a friend who is being audited by the IRS, inquire about your aunt's social security check, or represent a nonprofit organization on a grant application before HHS or any other federal agency. However, you can help your friend fill out a tax form and assist your aunt in preparing an application for government benefits. You can always represent yourself, your parents, spouse, or children. If you are executor of a deceased uncle's estate, you are free to contact government agencies as a fiduciary.

* * * * *

Post-employment Restrictions 18 U.S.C. § 207; 5 C.F.R. § Part 2641

Former employees may not "switch sides" on matters they worked on or supervised. Former senior employees have additional "no contact" restrictions. Depending on prior pay status and the degree to which the post-employment activities touch upon former federal responsibilities, the restraints could last permanently, two years, or one year.

If you approved a grant application and thereafter go to work for the grantee, you can advise the grantee how to carry out the project properly and how to adhere to government procedures; you may not, however, sign any documents directed back to the agency requesting grant modifications or make calls asking for additional funding.

* * * * *

17.5 RESOURCES

* * * * *

FDA Ethics Program

FDA, http://www.fda.gov/opacom/ethics/

The Food and Drug Administration's ethics program helps ensure that decisions made by agency employees are not, nor appear to be, tainted by any question of conflict of interest. The ethics laws and regulations were established to promote and strengthen the public's confidence in the integrity of the federal government.

The Principles of Ethical Conduct were established by Executive Order 12674, modified by Executive Order 12731, as basic principles regarding the conduct of federal employees. Observance of these principles by Federal employees is important so as to promote confidence in the American public in the integrity of the federal government

* * * * *

- Office of Government Ethics: http://www.usoge.gov/home.html
- Office of Special Counsel: http://www.osc.gov/
- Department of Interior: http://www.doi.gov/ethics/
- Department of Justice: http://www.usdoj.gov/jmd/ethics/
- Department of Agriculture: http://www.usda-ethics.net/
- National Institutes of Health: http://ethics.od.nih.gov/
- Bar Associations

State bar associations typically offer a wide variety of resources on legal ethics. For an example, see the Michigan State Bar Association ethics Web page at: http://www.michbar.org/opinions/ethicsopinions.cfm. Members may contact the SBM Attorney Ethics Helpline at (877) 558-4760 or the Judges Ethics Helpline at (877) 558-4761 to receive an informal, advisory opinion from a staff attorney regarding an ethics issue pertaining to the inquirer's prospective conduct.

Table of Abbreviations

483	Form for Report of Inspectional Observations
AAFCO	Association of American Feed Control Officials
ADI	Acceptable Daily Intake
AFDO	Association of Food and Drug Officials
AMS	Agricultural Marketing Service (USDA)
ANPR	Advance Notice of Proposed Rulemaking
APA	Administrative Procedures Act
APHIS	Animal and Plant Health Inspection Service (USDA)
ARS	Agricultural Research Service (USDA)
ATF	Bureau of Alcohol, Tobacco, Firearms, and Explosives (U.S. Department of Justice)
BCP	Bureau of Consumer Protection (FTC)
BSE	Bovine spongiform encephalopathy
BST	Bovine somatotropin
Bt	*Bacillus thuringiensis*
C.F.R.	Code of Federal Regulations
CAC	Codex Alimentarius Commission
CBP	Bureau of Customs and Border Protection
CCFAC	Codex Committee on Food Additives and Contaminants
CCFICS	Codex Committee on Food Import and Export Certification and Inspection Systems
CCFL	Codex Committee on Food Labeling
CCP	Critical Control Point
CDC	Centers for Disease Control and Prevention
CEDI/ADI	Cumulative Estimated Daily Intakes (CEDIs) and Acceptable Daily Intakes (ADIs).

Food Regulation: Law, Science, Policy, and Practice, by Neal D. Fortin
Copyright © 2009 Published by John Wiley & Sons, Inc.

CHD	Coronary heart disease
CFCs	Chlorofluorocarbons
C.F.R.	Code of Federal Regulations
CFSAN	Center for Food Safety and Applied Nutrition (FDA)
CHD	Coronary heart disease
cGMPs	Current good manufacturing practices
Codex	Codex Alimentarius
COOL	Country-of-origin labeling
CSREES	Cooperative State Research, Education, and Extension Services (USDA)
CVM	Center for Veterinary Medicine (FDA)
DHHS	U.S. Department of Health and Human Services
DOJ	Department of Justice
DOT	Department of Transportation
DRI	Daily Reference Intake
DRV	Daily Reference Value
DSHEA	Dietary Supplements Health and Education Act of 1994
DV	Daily Value
EA	Environmental Assessment
EAFUS	Everything Added to Food in the United States database, an informational database maintained by CFSAN
EIS	Environmental Impact Statement
EC	European Community
EDI	Estimated Daily Intake
EEC	European Economic Community
EIR	Establishment Inspection Report
EPA	Environmental Protection Agency
EPIA	Egg Products Inspection Act
EU	European Union
FACA	Federal Advisory Committee Act
FALCPA	Food Allergen Labeling and Consumer Protection Act
FAO	Food and Agriculture Organization (United Nations)
FAP	Food Additive Petition
FCN	Food Contact Notification
FCS	Food Contact Substance, FD&C Act § 409
FD&C Act	Federal Food, Drug, and Cosmetic Act
FDA	Food and Drug Administration (DHHS)
FDAMA	Food and Drug Administration Modernization Act of 1997

FFDCA	Federal Food, Drug and Cosmetic Act, also FDCA and FD&C Act
FIFRA	Federal Insecticide, Fungicide, and Rodenticide Act
FMIA	Federal Meat Inspection Act
FOIA	Freedom of Information Act
FPLA	Fair Packaging and Labeling Act
FQPA	Food Quality Protection Act of 1996
FR	Federal Register
FSIS	Food Safety Inspection Service (USDA)
FTC	Federal Trade Commission
FTCA	Federal Trade Commission Act
GAO	General Accounting Office
GAP	Good agricultural practice
GATT	General Agreement of Trade and Tariffs
GE	Genetic Engineering, which refers to genetic modification through use of recombinant deoxyribonucleic acid (rDNA) techniques, or gene splicing, to give desired traits. Genetically engineered foods are also called biotech, bioengineered, and genetically modified.
GEMS	Global Environment Monitoring System (WHO)
GAO	General Accounting Office
GEO	Genetically Engineered Organisms (genetically engineered through use of rDNA techniques); a synonym or GMO
GIPSA	Grain Inspection, Packers, and Stockyards Administration (USDA)
GM	Genetically modified, a synonym for GE.
GMOs	Genetically modified organisms (genetically modified through use of rDNA techniques)
GMPs	Good Manufacturing Practices
GPO	Government Printing Office
GRAS	Generally recognized as safe; under §§ 201(s) and 409 of the FD&C Act, generally recognized, among qualified experts, as having been adequately shown to be safe under the conditions of its intended use
GRNs	Generally Recognized as Safe Notifications
HACCP	Hazard analysis critical control point
HHS	U.S. Department of Health and Human Services (sometime seen as DHHS), the parent department for FDA, CDC, and others agencies
IOM	Investigations Operations Manual

IP	Information panel
JECFA	Joint (FAO/WHO) Expert Committee on Food Additives
JECFA	Joint (FAO/WHO) Expert Committee on Food Additives
JIFSAN	Joint Institute for Food Safety and Nutrition
LACF	Low-acid canned foods and acidified foods
Lm	*Listeria monocytogenes*
LMOs	Living modified organisms
MAVs	Maximum allowable variations
MOU	Memorandum of understanding
MRLs	Maximum residue limits
NACMCF	National Advisory Committee on Microbiological Criteria for Foods
NAD	National Advertising Division (Council of Better Business Bureaus)
NAFTA	North American Free Trade Agreement
NARB	National Advertising Review Board (FTC)
NAS	National Academy of Sciences
NCWM	National Conference on Weights and Measures
NDA	New Drug Application
NEPA	National Environmental Policy Act
NGO	Nongovernment organization
NIH	National Institutes of Health
NIST	National Institute of Standards and Technology
NLEA	Nutrition Labeling and Education Act of 1990
NMFS	National Marine Fisheries Service (Commerce)
NOAEL	No observable adverse effect level
NOEL	No observed effect level
NOP	National Organic Program
NOSB	National Organic Standards Board
NPRM	Notice of proposed rulemaking
NRC	National Research Council
NRTE	Not ready-to-eat
NSIP	National Seafood Inspection Program
OCI	Office of Criminal Investigations (FDA)
ODS	Office of Dietary Supplements (NIH)
OFAS	Office of Food Additive Safety (CFSAN)
OFPA	Organic Foods Production Act of 1990
OIG	Office of the Inspector General (DHHS)

OMB	Office of Management and Budget
OSHA	Occupational Safety Hazard Administration
OTC	Over-the-counter
PAFA	Priority-based Assessment of Food Additive (PAFA) database that serves as CFSAN's institutional memory for the toxicological effects of food ingredients known to be used in the United States
PCBs	Polychlorinated biphenyls
PDP	Principal display panel
PHS	Public Health Service (DHHS)
PIPs	Plant-Incorporated Protectants.
PKU	Phenylketonuria
PMN	Premarket notification
PMO	Pasteurized Milk Ordinance
ppb	Parts per billion
PPIA	Poultry Products Inspection Act
ppm	Parts per million
PUFI	Packed under Federal Inspection
QCPs	Quality Control Programs
RACC	Reference amounts customarily consumed
RDI	Reference Daily Intake
rDNA	Recombinant deoxyribonucleic
RTE	Ready-to-eat
SCOGS	Select Committee on GRAS Substances
SDWA	Safe Drinking Water Act
SE	*Salmonella enteritidis*
SOPs	Standard operating procedures
SPS	Sanitary and Phytosanitary Agreement
SSOPs	Sanitation Standard Operating Procedures
STOP	Safe Tables Our Priority
TBT	Technical barriers to trade agreement
TRO	Temporary restraining order
TTB	Alcohol and Tobacco Tax and Trade Bureau (U.S. Department of Treasury)
U.S.C.	United States Code
USDA	United States Department of Agriculture
USP	United States Pharmacopeia
WHO	World Health Organization
WTO	World Trade Organization

Table of Cases

Food Regulation: Law, Science, Policy, and Practice, by Neal D. Fortin
Copyright © 2009 Published by John Wiley & Sons, Inc.

Food Regulation: Law, Science, Policy, and Practice, by Neal D. Fortin
Copyright © 2009 Published by John Wiley & Sons, Inc.

About the Author

Neal Fortin is professor and director of the Institute for Food Laws and Regulations at Michigan State University (www.iflr.msu.edu). The Institute provides a distance education program in international food law via the Internet. Mr. Fortin is also an adjunct professor of law at the Michigan State University College of Law and an adjunct professor at the University of Minnesota Public Health Institute. Previously, Mr. Fortin worked for the Michigan Department of Agriculture where he was the primary drafter of the Michigan Food Law of 2000 and manager of the food service regulatory program.

Mr. Fortin teaches United States Food Law, International Food Law, Codex Alimentarius, Food and Drug Law, and Nutrition Law and Policy. As an attorney, Mr. Fortin concentrated in food law, food safety, food labeling, ingredient evaluation, and advertising. He is a graduate from the Michigan State University College of Law and the University of Michigan.